Applied Technical Mathematics

Applied Technical Mathematics

Merwin J. Lyng *Mayville State College*

L. J. Meconi *University of Akron*

Earl J. Zwick *Indiana State University*

Waveland Press, Inc.
Prospect Heights, Illinois

For information about this book, write or call:
Waveland Press, Inc.
P.O. Box 400
Prospect Heights, Illinois 60070
(312) 634-0081

Copyright © 1978 by Houghton Mifflin Company. Adapted from *Career Mathematics: Industry and the Trades,* copyright © 1974 by Houghton Mifflin Company. Reprinted by arrangement with Houghton Mifflin Company, Boston.
1983 Waveland Press, Inc. printing

ISBN 0-88133-073-6

All rights reserved. No part of this book may be reproduced, stored in a retrieval system, or transmitted in any form or by any means without permission in writing from the publisher.

Printed in the United States of America.

Contents

PREFACE xv

Part 1 Technical and industrial computations 1

CHAPTER 1 READING AND USING MEASURING DEVICES 2

Reading meters 2

1-1 *Electric meters* 2
1-2 *Gas meters* 4
1-3 *Whole numbers* 5
1-4 *Tachometer* 8

Reading graphs 9

1-5 *Bar graphs* 9
1-6 *Broken-line graphs* 12
1-7 *Curved-line graphs* 14

Reading rules 15

1-8 *Rules and lengths* 15
1-9 *Equivalent fractions* 18
1-10 *Comparing lengths* 21
1-11 *Metric scales* 23

Reading technical drawings 26

1-12 *Lines* 26

1-13 Dimensions *28*
1-14 Curves and angles *30*

Taking inventory 32

Measuring your skills 32

CHAPTER 2   WORKING WITH INTEGERS   35

Reviewing whole numbers   35

2-1 Adding and subtracting whole numbers *35*
2-2 Multiplying and dividing whole numbers *38*

Introducing integers   39

2-3 Positive and negative integers *39*
2-4 Adding integers *41*
2-5 Subtracting integers *44*
2-6 Multiplying integers *46*
2-7 Dividing integers *48*

Taking inventory   49

Measuring your skills   50

CHAPTER 3   USING FRACTIONS   52

Adding and subtracting fractions   52

3-1 Fractions with the same denominator *52*
3-2 Fractions with different denominators *55*
3-3 The least common denominator *57*

Adding and subtracting mixed numbers   60

3-4 Whole numbers and mixed numbers *60*
3-5 Adding mixed numbers *63*
3-6 Subtracting mixed numbers *66*

Dimensions in feet and inches   68

3-7 Changing inches to feet and inches *68*
3-8 Conserving materials *71*

Multiplying fractions   73

3-9   Multiplying a fraction and a whole number *73*
3-10 Multiplying a mixed number and a whole number *74*
3-11 Multiplying two fractions *77*
3-12 Cancellation *79*

Dividing fractions   81

3-13 Reciprocals *81*

3–14 Dividing a fraction by a whole number 81
3–15 Dividing by a fraction 84

Combining operations with fractions 85

3–16 Estimating 85
3–17 Solving problems 88

Taking inventory 92

Measuring your skills 92

CHAPTER 4 DECIMALS, PERCENTS 95

Decimals and fractions 95

4–1 Reading and writing decimals 95
4–2 Equivalent decimals and fractions 97

Using decimals 101

4–3 Adding and subtracting decimals 101
4–4 Multiplying decimals 104
4–5 Micrometers 107
4–6 Dividing decimals 111

Percents 114

4–7 Meaning of percent 114
4–8 Discounts 117
4–9 Machine tolerance 119
4–10 Electrical tolerance 121
4–11 Efficiency 123

Taking inventory 124

Measuring your skills 125

CHAPTER 5 USING HAND-HELD CALCULATORS AND OTHER COMPUTING DEVICES 128

Hand-held calculators 128

5–1 Types of calculators 128
5–2 Making calculations 130
5–3 Solving problems 132

Advanced calculations 135

5–4 Long division 135
5–5 Working with fractions and decimals 136
5–6 Working with percents 139
5–7 Desk calculators and computers 141

Taking inventory 142

Measuring your skills 142

Measuring your progress 143

Part 2 Technical and industrial formulas 147

CHAPTER 6 EQUATIONS, FORMULAS 148

Components of equations 148

- 6-1 *Terms, expressions, and equations* 148
- 6-2 *Exponents* 152
- 6-3 *The order of operations* 156

Solving equations 158

- 6-4 *Solving equations by subtraction* 158
- 6-5 *Solving equations by addition* 161
- 6-6 *Solving equations by division* 163
- 6-7 *Solving equations by multiplication* 165
- 6-8 *Solving equations by several methods* 167

Working with formulas 171

- 6-9 *Writing formulas from rules* 171
- 6-10 *Solving problems with formulas* 174

Taking inventory 177

Measuring your skills 178

CHAPTER 7 LENGTH, AREA, VOLUME 180

Length 180

- 7-1 *Linear measures* 180
- 7-2 *Linear conversions* 184
- 7-3 *Perimeter of a rectangle or square* 186
- 7-4 *Circumference* 189

Area 191

- 7-5 *Area measurement* 191
- 7-6 *Area of parallelograms* 194
- 7-7 *Area of triangles and trapezoids* 196
- 7-8 *Areas of special regions* 198
- 7-9 *Areas of circles* 202
- 7-10 *Areas of cylinders, spheres, and cones* 204

Volume 207

7–11 Volume measurement 207
7–12 Volumes of rectangular solids 210
7–13 Volumes of cylinders 213
7–14 Volumes of cones, pyramids, and spheres 214

Weight, capacity, temperature 217

7–15 Weight measurement 217
7–16 Capacity measurement 220
7–17 Temperature formulas 222

Taking inventory 223

Measuring your skills 225

CHAPTER 8 POLYNOMIALS 227

Operations with polynomials 227

8–1 Adding and subtracting polynomials 227
8–2 Multiplying polynomials 231

Factoring 237

8–3 Factoring polynomials 237
8–4 Factoring the difference of two squares 244

Quadratic equations 247

8–5 Solving quadratic equations by factoring 247
8–6 The quadratic formula 251

Taking inventory 256

Measuring your skills 257

CHAPTER 9 GRAPHING 258

Linear equations 258

9–1 The coordinate plane 258
9–2 Graphing linear equations 263
9–3 Solving pairs of linear equations graphically 267
9–4 Solving pairs of linear equations algebraically 270

Nonlinear equations 275

9–5 Graphing nonlinear equations 275

Taking inventory 282

Measuring your skills 283

CHAPTER 10 GEOMETRY 285

Angles 285

10-1 Measuring angles 285
10-2 Central angles 287
10-3 Angle relationships 288

Constructions involving perpendiculars 291

10-4 Perpendicular lines 291
10-5 Circles 295

Basic angle constructions 298

10-6 Angles 298
10-7 Parallel lines 301

Triangles 303

10-8 Congruent and similar triangles 303
10-9 Right triangles 304

Taking inventory 307

Measuring your skills 308

CHAPTER 11 RATIO, PROPORTION, SCALE 309

Ratios 309

11-1 Meaning of ratio 309
11-2 Screw threads 312

Proportions 313

11-3 The meaning of proportion 313
11-4 Inverse proportions 316

Similar figures 320

11-5 Finding lengths with similar figures 320
11-6 Finding areas with similar figures 321

Scales 323

11-7 The architect's scale 323
11-8 Scale drawings 326

Taking inventory 330

Measuring your skills 330

CHAPTER 12 TRIGONOMETRY 332

Right triangles 332

12-1 The tangent ratio 332
12-2 The sine and cosine ratios 340
12-3 Special triangles 348

General triangles 354

12-4 The law of sines 354
12-5 The law of cosines 360

Taking inventory 365

Measuring your skills 365

Measuring your progress 367

Part 3 Technical and industrial applications 369

CHAPTER 13 POWER AND ENERGY 370

Power 370

13-1 Power and energy 370
13-2 Measuring horsepower 375

Factors that affect horsepower 378

13-3 Engine displacement 378
13-4 Compression ratios 381

Gears and pulleys 384

13-5 Gears 384
13-6 Pulleys 388

Fluid and electrical power 389

13-7 Hydraulic power 389
13-8 Pneumatic power 393
13-9 Electrical power 395

Taking inventory 398

Measuring your skills 398

CHAPTER 14 CONSTRUCTION 400

From drawing to foundations 400

14-1 Plans and estimating costs 400
14-2 Excavations 402
14-3 Footings 403
14-4 Foundations 405

Contents xi

Framing and exterior 408

14-5   Framing   408
14-6   Roofing   412
14-7   Windows and doors   414
14-8   Siding   415
14-9   Brickwork   417

Finishing the job   418

14-10   Electrical wiring   418
14-11   Plumbing   419
14-12   Interior finishing   420
14-13   Landscaping   423

Taking inventory   426

Measuring your skills   426

CHAPTER 15   MANUFACTURING   428

Forming and conditioning   428

15-1   Casting   428
15-2   Compressing and stretching   430
15-3   Conditioning   431

Separating materials   432

15-4   Shearing and chip removal   432
15-5   Separating by heat, chemicals, and electricity   433

Combining materials   434

15-6   Mixing and coating   434
15-7   Bonding   435
15-8   Mechanical fasteners   436

Testing and marketing the product   437

15-9   Quality control   437
15-10   Packaging   438
15-11   Marketing the product   439

Taking inventory   441

Measuring your skills   442

CHAPTER 16   GRAPHIC ARTS   443

Measuring type   443

16-1   The point system   443

16-2   The em   445
16-3   Column inch   446

Printing processes   447

16-4   Letterpress   447
16-5   Lithography   448
16-6   Gravure   449

Steps in printing   451

16-7   Preparing copy   451
16-8   Composition   452
16-9   Photoengraving   453
16-10   Paper   455
16-11   Finishing   457

Taking inventory   459

Measuring your skills   460

Measuring your progress   460

ANSWERS TO SELECTED EXERCISES   463

GLOSSARY   482

INDEX   489

CREDITS   495

Preface

The purpose of this text is to provide the basic mathematical skills necessary for technical careers. It is intended to be used in courses with a career orientation. The examples and exercises are designed to be as practical and work oriented as possible. While focusing on important industrial applications, the text reviews the fundamental computational skills that are essential to success in all fields of endeavor.

The text is divided logically into three parts. Part 1 (Chapters 1–5) is entitled "Technical and Industrial Computations." Its chapters thoroughly redevelop computational processes involving whole numbers, fractions, decimals, and percents. The operation of the hand-held calculator is introduced in Chapter 5 and will facilitate much of the work in the remaining chapters of the book. Part 2 (Chapters 6–12), entitled "Technical and Industrial Formulas," deals with methods for solving industrial and technical problems using algebraic, geometric, and trigonometric formulas. Part 3 (Chapters 13–16) is entitled "Technical and Industrial Applications." This section focuses on particular applications in four important industrial and technical fields: power and energy, construction, manufacturing, and the graphic arts. Each chapter of Part 3 can stand alone, and any combination of its chapters may be used to provide the flexibility that may be needed for diverse programs and student needs. Because of the flexibility of the text and the varying backgrounds of students, the pace at which the book may be used must be determined by each instructor.

Each chapter begins with a set of learning objectives. These objectives define the skills that are developed and practiced in each chapter. At the end of each chapter there is a review of the key concepts in the chapter ("Taking Inventory") and a set of exercises designed to reinforce the skills developed in the chapter ("Measuring Your Skills"). At the end of each part there is a set of exercises designed to reinforce the skills developed in that part ("Measuring Your Progress").

Answers to selected exercises, along with a glossary of key terms and expressions, are provided at the end of the text.

The mathematical concepts included in this book are developed by being related to practical applications in the trades and in technical and industrial areas. The text includes nearly 300 such examples. Also inserted throughout the text are "Tricks of the Trade" which introduce students to practical solutions of technical and industrial problems. In addition, there is a set of exercises at the end of every section, which combine to make a total of nearly 2600 exercises of varying difficulty. Dual U.S. system and metric system exercises are also included where appropriate.

The authors wish to acknowledge John Antonopoulis (Staten Island Community College), Chris Boldt (Eastfield College), Edward Doran (Community College of Denver), Theodore J. Hansink (Mira Costa College), and Elias Margolin (San Diego City College), who read the initial manuscript and offered valuable suggestions for the improvement of the text. We are also indebted to our families, colleagues, and friends for their words of encouragement while the text was being written. A special thanks must go to Myrna Lyng for her part in typing and assembling the finished manuscript.

M. J. L.
L. J. M.
E. J. Z.

1 Technical and industrial computations

In Part 1 we present basic mathematical ideas concerning whole numbers, fractions, decimals, percents, and integers. To emphasize the need to master these ideas, we discuss examples of their practical application in every chapter. Reading and interpreting meters, graphs, and technical drawings; taking precise measurements; minimizing waste; using micrometers; and determining discounts, tolerances, and efficiencies are just a few of the applications of basic mathematical ideas presented in Part 1.

We include complete redevelopment of the basic computational algorithms in each chapter, so that you will have a thorough foundation in mathematical skills. The inclusion of work with integers in Chapter 4 will pave the way for the more advanced calculations and applications you will encounter in Part 2.

In Chapter 5 we present a discussion of the hand-held calculator. Since this is a very valuable aid to computation, we provide introductory activities and give applications to problem solving. These activities will provide you with the necessary groundwork so that you can continue to use the hand-held calculator as an aid when you learn new and more advanced algorithms.

1 Reading and using measuring devices

After completing this chapter, you should be able to:
1. Read several kinds of industrial meters.
2. Read and interpret bar and line graphs.
3. Make precise measurements with inch and centimeter rules.
4. Read technical drawings.

Reading meters

1-1 ELECTRIC METERS

Many industries depend on electricity as a source of power. The amount of electricity that they use is measured in *kilowatt-hours (kWh)*. Figure 1-1 shows the dials on a common electric meter.

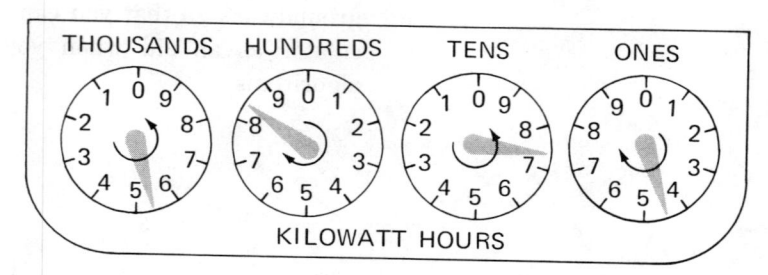

Figure 1-1

On this meter the last dial on the right records kilowatt-hours from 0 to 10. The other dials record the number of tens, hundreds, and thousands of kilowatt-hours used.

Here's how to read the meter. Note the position of the pointer on each dial. Starting at the far left, set the figures down in order. When the pointer is between two figures, use the smaller number. For example, the meter in Figure 1-1 reads 5874 kWh.

EXERCISES Read the electric meters.

1.

2.

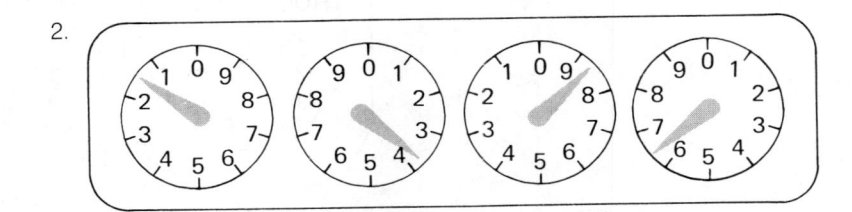

3.

4.

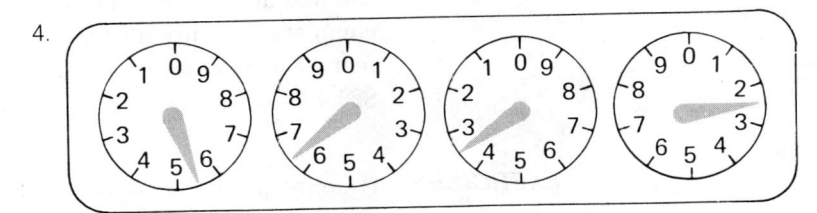

5.

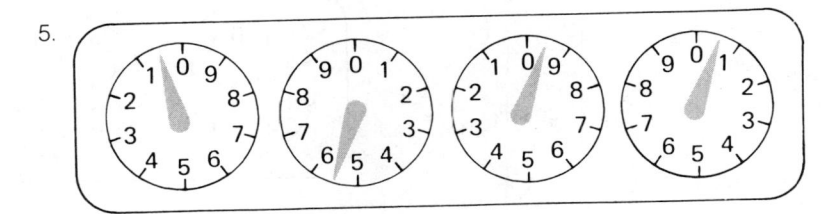

Reading and using measuring devices 3

6.

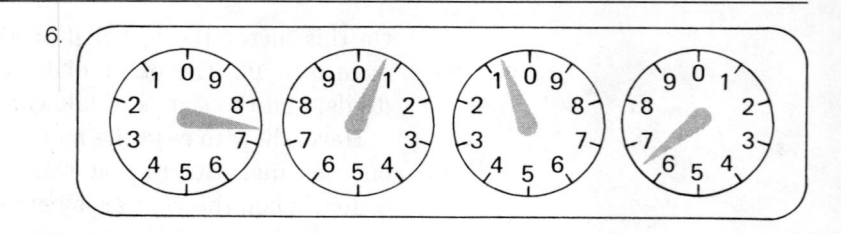

1-2 GAS METERS

Natural gas is another energy source that is commonly used in industry. Gas use is measured in units of 100 cubic feet, which are symbolized CCF. Figure 1-2 shows the dials on a common type of gas meter.

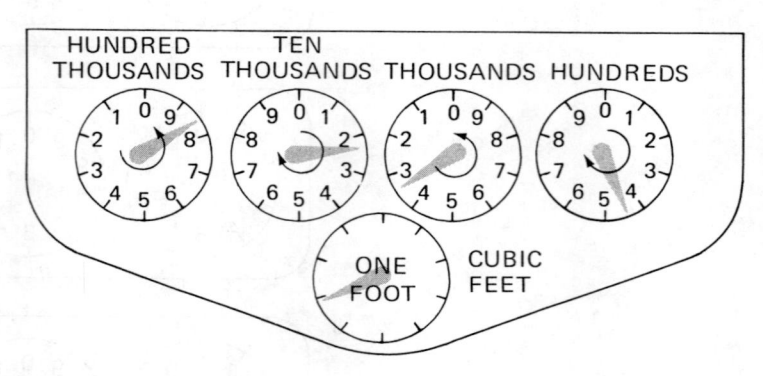

Figure 1-2

The upper four dials are used for reading the meter. The lower dial is used only to test the accuracy of the meter. Each mark on the dial at the far right represents 1 CCF, or 100 cubic feet. A complete revolution of this dial records 10 CCF, or 1000 cubic feet. The reading on the meter in Figure 1-2 is 8234 CCF, or 823,400 cubic feet.

EXERCISES Read the gas meters. (The test dial has been omitted.)

1.

4 Technical and industrial computations

2.

3.

4.

5.

6.

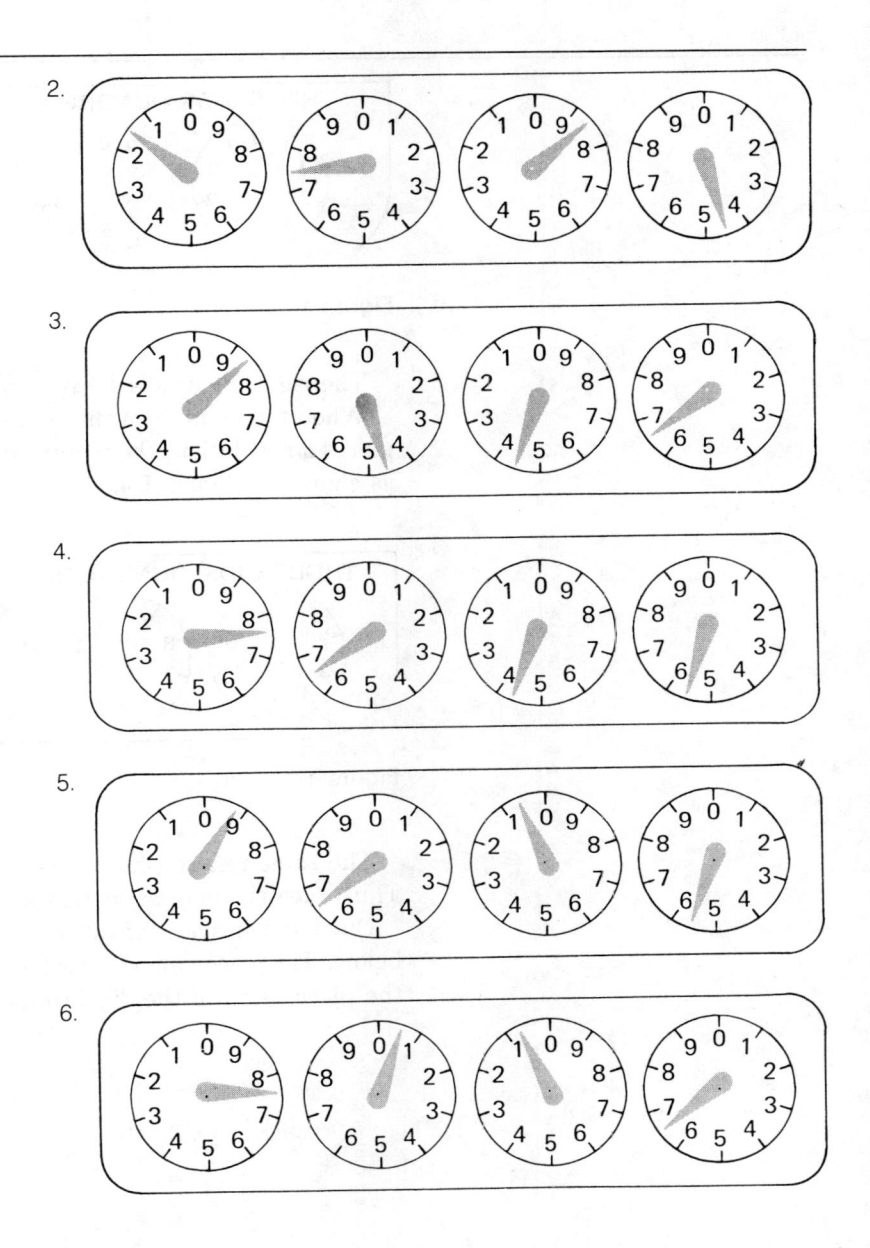

1-3 WHOLE NUMBERS

Do you notice any similarity between the way we read electric and gas meters and the way we express numbers? Recall that the unit shown on a dial is 10 times greater than the unit on the dial to the right. Look at the electric meter in Figure 1-3. What is the reading?

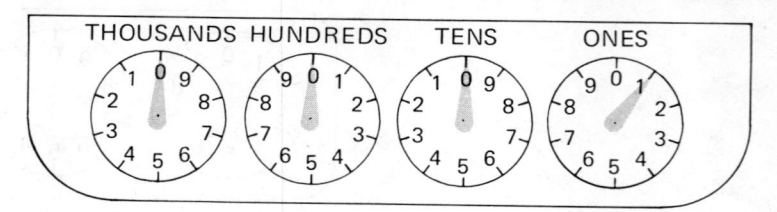

Figure 1-3

The right answer is 1 kWh.

When the pointer on the "ones" dial records 10 kWh and completes a revolution, the pointer on the "tens" dial moves 1 place, as shown in Figure 1-4.

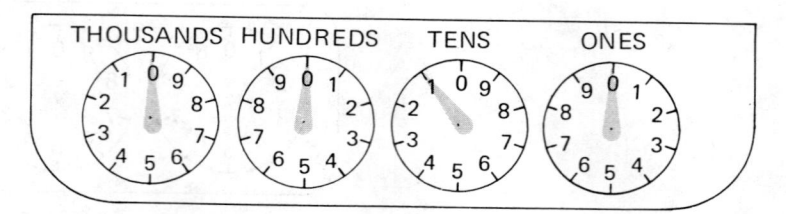

Figure 1-4

The same relationship holds true for the other adjacent dials. Thus, electric and gas meters are based on a system of tens.

Our number system is also based on tens. Look at the diagram below. The place value for each digit in the expression is 10 times the place value of the digit to its right.

billions	hundred millions	ten millions	millions	hundred thousands	ten thousands	thousands	hundreds	tens	ones
6,	3	2	7,	4	0	8,	2	5	6

We read this as "6 billion, 327 million, 408 thousand, 256." Note that we use a zero to show that there are no "ten thousands" in this expression. Large numbers like this are common in industry—in measurements and sales reports, for example.

EXERCISES   Read the following expressions aloud.

1. 2,324,617 automobiles
2. 379,340 pressure pumps
3. 627,229,367 rivets
4. 24,392 dado blades
5. 47,287,539 fasteners

Use numbers to write the following expressions.

6. 327 thousand, 216 ingots
7. 432 million, 627 thousand, 456 plastic bottles
8. 52 thousand, 293 brake shoes
9. three hundred twenty-nine thousand, six hundred fourteen employees
10. eighty-three thousand, twenty-nine cable reels

Frequently in our work we will need to round numbers. The flow chart in Figure 1-5 can help us.

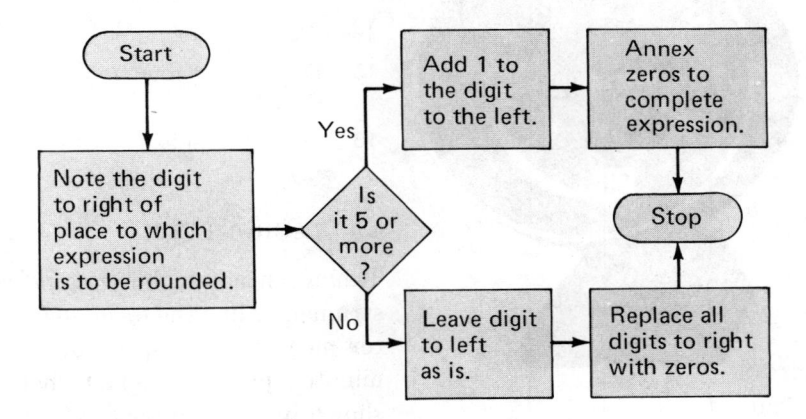

Figure 1-5

Example 1

Round 3276 to the nearest hundred.

Solution

The digit to the right of the hundreds' place is 7. Since 7 is greater than 5, we add 1 to the hundreds' place and complete the expression with zeros: 3276 → 3300.

Example 2

Round 3276 to the nearest thousand.

Solution

The digit to the right of the thousands' place is 2. Since 2 is less than 5, we leave the 3 in the thousands' place and complete the expression with zeros: 3276 → 3000.

EXERCISES   Round to the nearest hundred.

1. 482   2. 927   3. 6284   4. 5728

Round to the nearest thousand.

5. 3298   6. 5899   7. 29,507   8. 68,426

Round to the nearest 10 feet.

9. 32 ft   10. 527 ft   11. 3291 ft

Round to the nearest 100 miles.

12. 428 mi   13. 3261 mi   14. 4556 mi

15. Do you think rounding is used in reading electric and gas meters? Explain.

16. Give several examples to show how rounding is used in industry.

1-4 TACHOMETER

Tuning an automobile engine requires the use of precision instruments, like the *tachometer* shown in Figure 1-6. A tachometer measures the speed of a rotating object in revolutions per minute (rpm). To read a tachometer, take the reading on the dial shown by the pointer and multiply by 1000.

Figure 1-6

EXERCISES   Read the tachometers to the nearest hundred rpm.

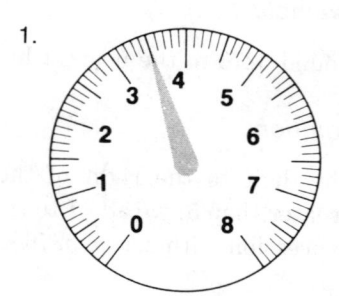

1.

2.

8   **Technical and industrial computations**

3.   4.

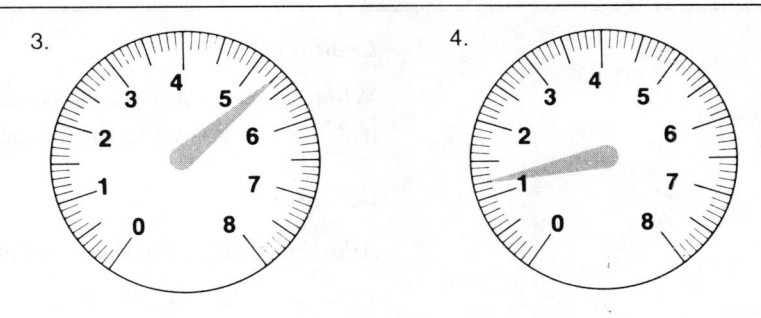

5-8. Round each of these readings to the nearest thousand rpm.

Reading graphs

1-5 BAR GRAPHS

If you saw an industrial *bar graph* like Figure 1-7, would you know how to read it?

1. What information is given along the bottom of the graph?
2. What is the scale along the side of the graph?
3. In which year were appliance sales the greatest?
4. What is the amount of sales, in billions of dollars, for 1976?

We can get the answers to these questions directly from the graph. To learn even more from the graph, we need to use addition, subtraction, multiplication, and division.

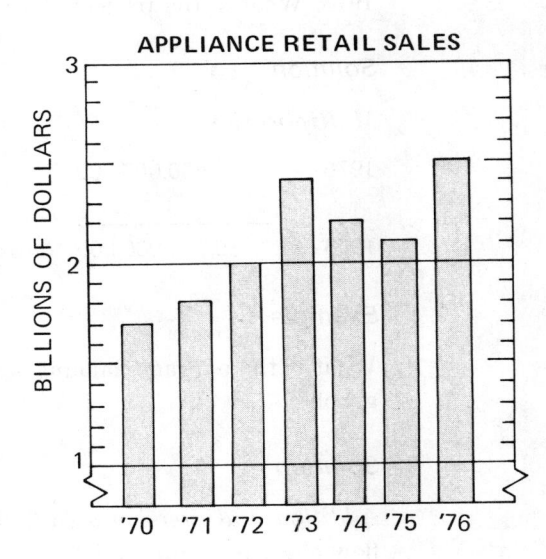

Figure 1-7

Example 1

What are the total appliance sales, in billions of dollars, for the last 3 years shown in the graph?

Solution

Addition   Sales for these years are:

1974 →	$2,200,000,000
1975 →	2,100,000,000
1976 →	+ 2,500,000,000
	$6,800,000,000 (total sales)

Example 2

By how much did the 1976 sales exceed the sales in 1970?

Solution

Subtraction   The sales figures for these years are:

1976 →	$2,500,000,000
1970 →	− 1,700,000,000
	$ 800,000,000 (difference in sales)

Example 3

The appliance industry expects to double its 1976 sales figure by 1978. What is the projected 1978 sales figure?

Solution

Multiplication

1976 →	$2,500,000,000
	× 2
1978 →	$5,000,000,000 (projected sales)

Example 4

What is the average amount of sales for the years shown on the graph?

Solution

Addition and division   To find an average, we can follow the flow chart in Figure 1-8.

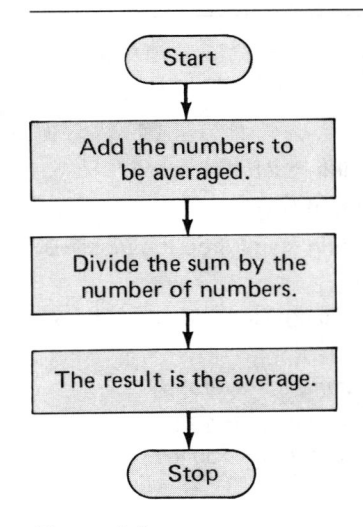

Figure 1-8

Step 1

$ 1,700,000,000
   1,800,000,000
   2,000,000,000
   2,400,000,000
   2,200,000,000
   2,100,000,000
+ 2,500,000,000
$14,700,000,000

Step 2

$ 2,100,000,000   (average sales)
7)$14,700,000,000

Bar graphs may be drawn either vertically or horizontally. Sometimes two sets of facts are shown on the same graph, as in Figure 1-9.

Figure 1-9

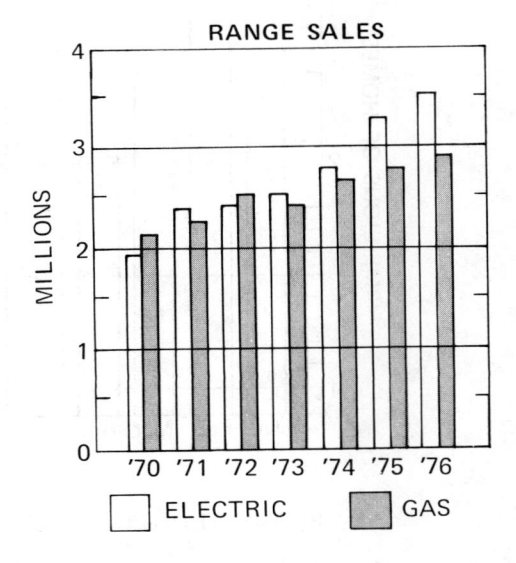

RANGE SALES

EXERCISES   Use Figure 1-9 to answer the questions.

1. How many electric ranges were sold in 1970? in 1975?
2. In which year were gas range sales the greatest? the least?
3. In which years did the sale of gas ranges exceed the sale of electric ranges? By how much in each case?
4. What are the total sales of electric and gas ranges in 1975?
5. If a gas range costs $250, about how much was spent for gas ranges in 1975?

6. Round the electric range sales for each of the years 1973 to 1976 to the nearest million dollars.

Exercises 7–12 refer to Figure 1-10.

7. During which month did housing starts reach their peak?
8. How many housing units were begun in May?
9. Between which two months did housing starts drop the most? By how much?
10. What is the average number of housing starts for the first six months?
11. Which month comes closest to matching the average?
12. Suppose the average asking price of a house is $28,990. What is the total value of the homes that were begun in June?

Figure 1-10

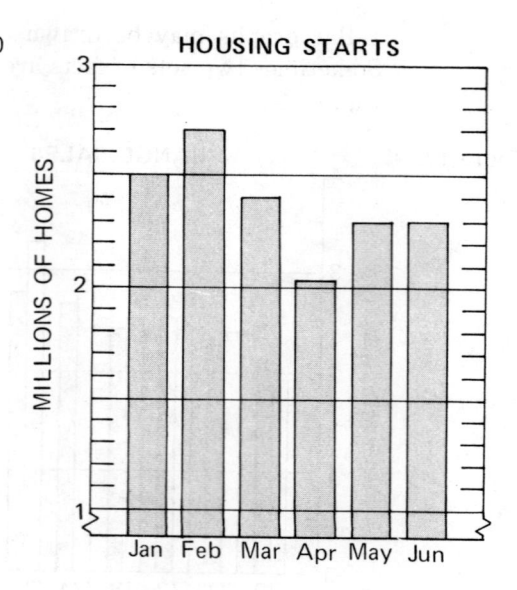

HOUSING STARTS

1-6 BROKEN-LINE GRAPHS

Broken-line graphs are used in nearly every industry to show changes in important trends. To make a broken-line graph, we join the tops of the bars of a vertical bar graph, as shown in Figures 1-11 and 1-12.

EXERCISES Use Figure 1-12 to answer the questions.

1. What were the sales for heavy equipment in 1976?
2. Between which two years did the sales increase the most? By how much?

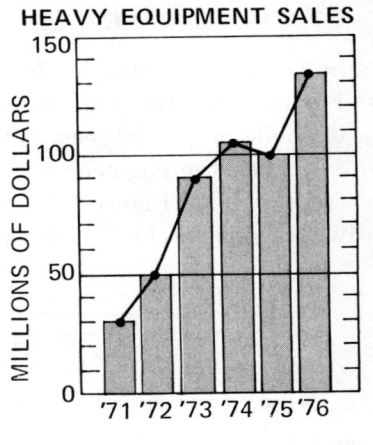

Figure 1-11 HEAVY EQUIPMENT SALES

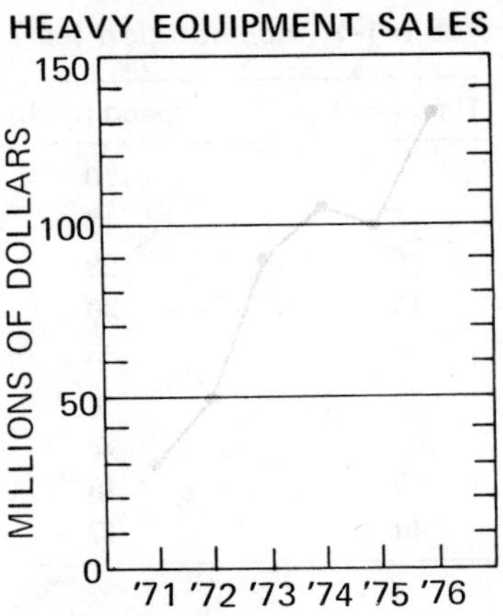

Figure 1-12 HEAVY EQUIPMENT SALES

3. How much of a decrease in sales took place between 1974 and 1975?
4. What will the sales be if the company doubles its 1976 figure?
5. What were the average sales in the period from 1974 to 1976?

Exercises 6–10 refer to Figure 1-13.

6. In which year was the supply of cement the greatest? the least?
7. In which year was the demand for cement the greatest? the least?
8. What is the difference between supply and demand in 1976?
9. What does the graph show concerning supply and demand? Explain.
10. Estimate when the demand will probably exceed the supply.

Figure 1-13 CEMENT SUPPLY AND DEMAND

SUPPLY ———

Reading and using measuring devices 13

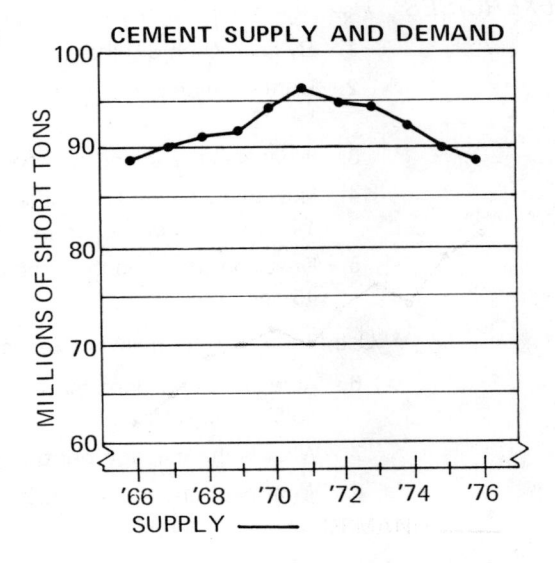

Table 1-1 *Acceleration test*

Time (sec)	Speed (mph)
0	0
5	12
10	23
15	30
20	35
25	39
30	42
35	45
40	47

1-7 CURVED-LINE GRAPHS

Some graphs cannot be drawn exactly using only broken lines. For example, when an engineer tests the acceleration of an automobile, he or she records the results first in a table like Table 1-1. Then the engineer transfers these results to a graph as a series of dots (Figure 1-14). Finally, he or she connects the dots with a smooth curve to form a *curved-line graph* (Figure 1-15).

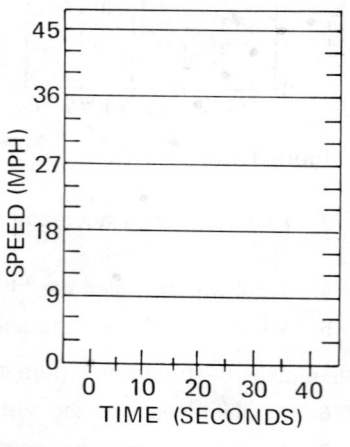

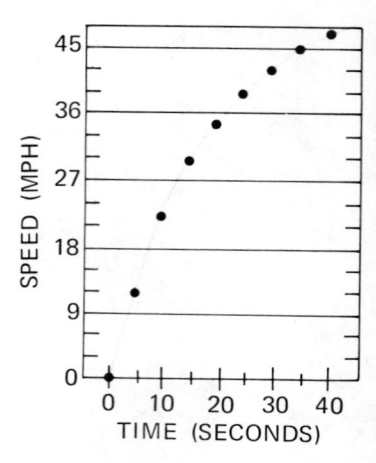

Figure 1-14

Figure 1-15

EXERCISES Use Figure 1-15 to answer the questions.

1. By how much did the speed increase during each 5-sec interval?
2. During which 5-sec interval was the acceleration the greatest? the least?
3. What was the speed at 8 sec? at 22 sec? at 37 sec?
4. How many seconds did it take the automobile to reach a speed of 14 mph? a speed of 26 mph? a speed of 35 mph?
5. Describe what you think the curved line will look like beyond the 45-sec line.

Use Figure 1-16 to answer Exercises 6-9.

6. At what speed does the automobile get the highest mileage per gallon (mpg)?
7. What is the mileage per gallon at 25 mph? at 45 mph?
8. What happens to the mileage per gallon as the speed increases beyond 30 mph?

Figure 1-16

GAS MILEAGE CHART

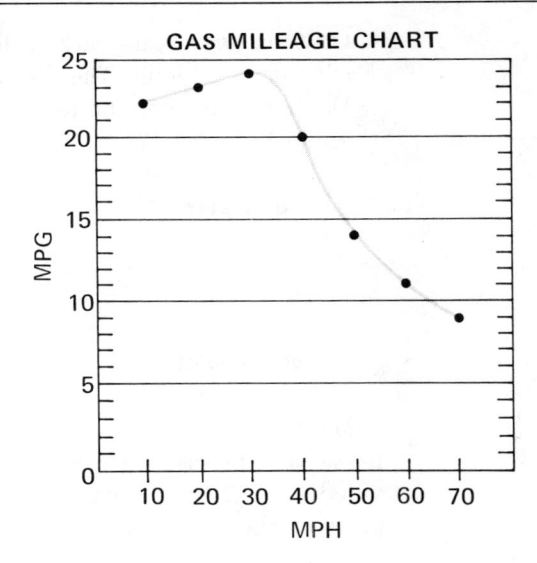

Table 1-2 *Fuel economy test*

Load (tons)	Miles per gallon (mpg)
2	18
4	17
6	15
8	12
10	8
12	7
14	6

9. Between which two speeds does the gas mileage per gallon drop the most?
10. Use the information in Table 1-2 to draw a curved-line graph.

Reading rules

1-8 RULES AND LENGTHS

To measure lengths precisely, we use a variety of rulers. Different rulers are made with different *scales*, depending on the intended use.

The ruler in Figure 1-17 has two scales. On the upper scale each inch is divided into 8 equal parts. Each part is $\frac{1}{8}$ inch long ($\frac{1}{8}''$). The distance AB is $\frac{5}{8}''$.

Figure 1-17

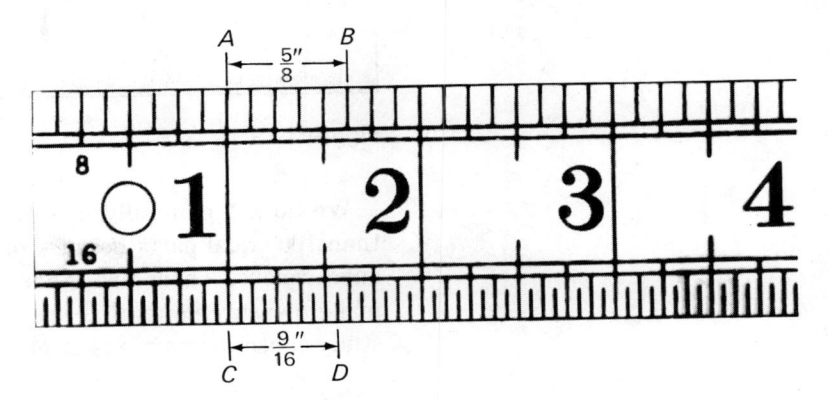

Reading and using measuring devices 15

On the lower scale each inch is divided into 16 equal parts. Each part is $\frac{1}{16}''$ long. The distance CD is $\frac{9}{16}''$.

Do you remember the terms we use to describe fractions like those in Figure 1-17?

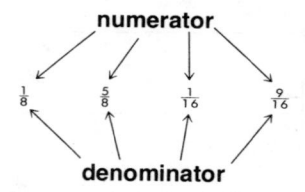

$$\text{numerator} \searrow \downarrow \downarrow \swarrow$$
$$\frac{1}{8} \quad \frac{5}{8} \quad \frac{1}{16} \quad \frac{9}{16}$$
$$\text{denominator} \nearrow \uparrow \uparrow \nwarrow$$

Many jobs require us to divide an inch into smaller fractions to make very precise measurements. In Figure 1-18 an inch is divided into 32 equal parts on the upper scale and into 64 equal parts on the lower scale. Each division on the upper scale is $\frac{1}{32}''$. Each division on the lower scale is $\frac{1}{64}''$.

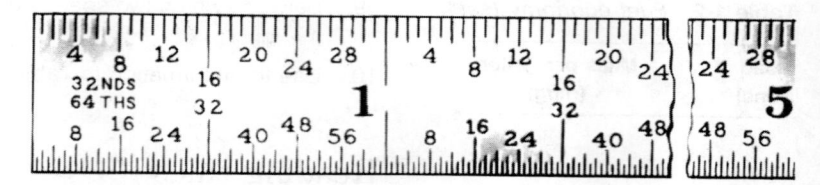

Figure 1-18

Some machinist's rules are divided into hundredths of an inch, as in Figure 1-19. The distance from the left edge of the scale to the number 10 is $\frac{10}{100}''$, or $\frac{1}{10}''$.

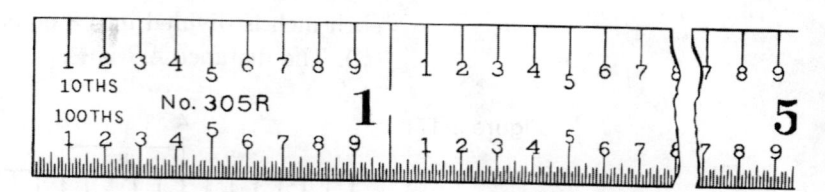

Figure 1-19

We do not generally use rules with scales divided into more than 100 equal parts because of the difficulty of seeing the small subdivisions. There are other instruments that can be used to make measurements that are accurate to $\frac{1}{1000}''$ or $\frac{1}{10,000}''$. We will study these instruments later.

Figure 1-20

EXERCISES Exercises 1-4 refer to Figure 1-20.

1. What is the smallest fraction of an inch
   a. on the upper scale?   b. on the lower scale?
2. On the upper scale, what distance is shown by
   a. 3 divisions?   b. 7 divisions?   c. 8 divisions?
3. On the lower scale, what distance is shown by
   a. 5 divisions?   b. 7 divisions?   c. 13 divisions?
4. Give the length of each line.
   a. AB   b. CD   c. EF
   d. GH   e. JK   f. LM

Figure 1-21

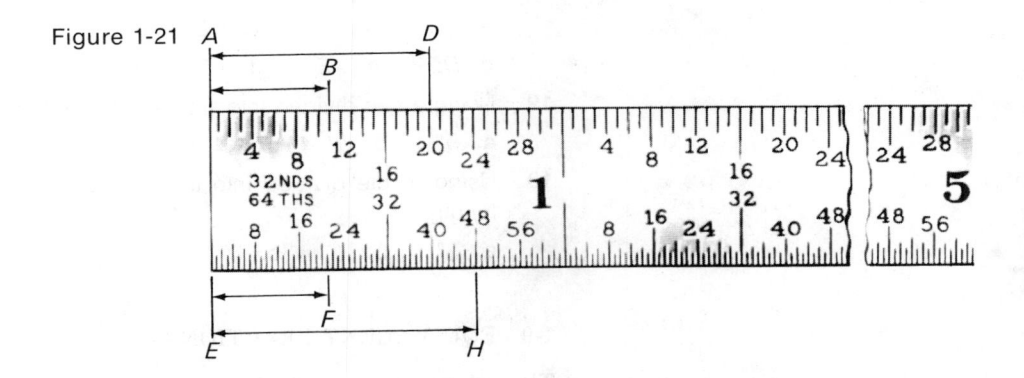

Exercises 5-8 refer to Figure 1-21.

5. What is the smallest division of an inch
   a. on the upper scale?   b. on the lower scale?
6. On the upper scale, what distance is shown by
   a. 5 divisions?   b. 15 divisions?   c. 23 divisions?

Reading and using measuring devices   17

7. On the lower scale, what distance is shown by
   a. 27 divisions?   b. 43 divisions?
8. Give the length of each line.
   a. *AB*   b. *AD*   c. *EF*   d. *EH*

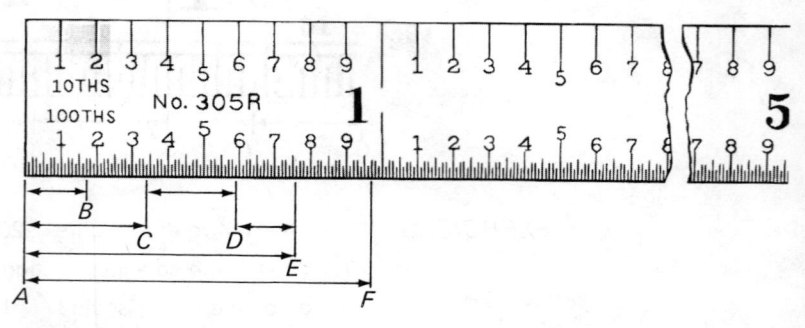

Figure 1-22

Exercises 9–11 refer to Figure 1-22.

9. What is the smallest division of an inch on the scale?
10. What distance is shown by
    a. 3 divisions?   b. 17 divisions?
    c. 29 divisions?   d. 57 divisions?
11. Give the length of each line.
    a. *AB*   b. *AC*   c. *CD*
    d. *DE*   e. *AE*   f. *AF*
12. Using a rule divided into sixteenths of an inch, draw each length.
    a. $\frac{3}{16}''$ b. $\frac{7}{16}''$ c. $\frac{11}{16}''$
13. Using a rule divided into thirty-seconds of an inch, draw each length.
    a. $\frac{7}{32}''$ b. $\frac{13}{32}''$ c. $\frac{25}{32}''$

1-9 EQUIVALENT FRACTIONS

The ruler in Figure 1-23 is divided into eighths and sixteenths of an inch. Distance *AB* is $\frac{3}{8}''$. But *AB* is also $\frac{6}{16}''$. So $\frac{3}{8}'' = \frac{6}{16}''$.

The fractions $\frac{3}{8}$ and $\frac{6}{16}$ are called *equivalent* because they have the same value. Do you think $\frac{5}{8}$ and $\frac{10}{16}$ are equivalent?

We can use an easy test to see if two fractions are equivalent. Cross-multiply the fractions. If the products are equal, the fractions are equivalent.

18 Technical and industrial computations

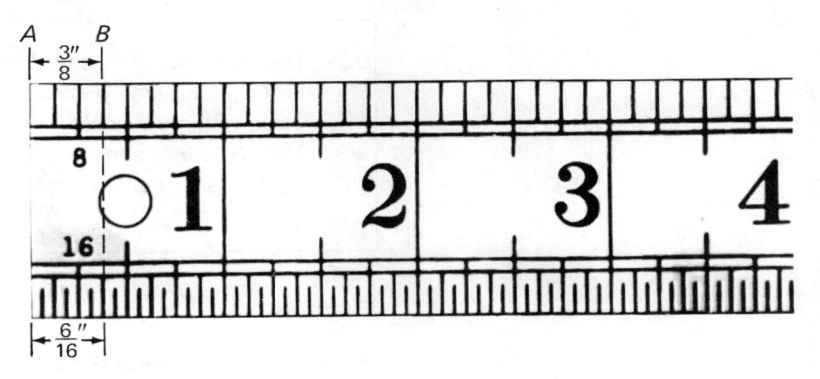

Figure 1-23

Example 1

Is $\frac{5}{8}$ equivalent to $\frac{10}{16}$?

Solution

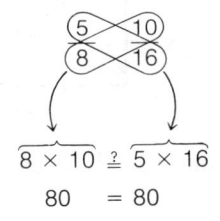

$$8 \times 10 \stackrel{?}{=} 5 \times 16$$
$$80 = 80$$

Yes, $\frac{5}{8}$ is equivalent to $\frac{10}{16}$.

Example 2

Is $\frac{9}{32}$ equivalent to $\frac{5}{16}$?

Solution

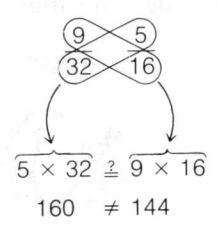

$$5 \times 32 \stackrel{?}{=} 9 \times 16$$
$$160 \neq 144$$

No, $\frac{9}{32}$ and $\frac{5}{16}$ are *not* equivalent.

We can use the following rules for finding a fraction that is equivalent to another.

Reading and using measuring devices 19

> 1. The numerator and denominator of any fraction may be multiplied by the same number (except zero) without changing the value of the fraction.
> 2. The numerator and denominator of any fraction may be divided by the same number (except zero) without changing the value of the fraction.

Example 3

A metal strip $\frac{11}{32}''$ wide is needed to fasten two parts of a blower. Can a strip $\frac{3}{8}''$ wide be ground down to the required width?

Solution

To change $\frac{3}{8}$ to thirty-seconds, multiply both the numerator and denominator by 4.

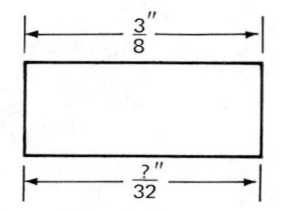

$$\frac{3 \times 4}{8 \times 4} = \frac{12}{32}$$

Since the $\frac{3}{8}''$ piece is $\frac{12}{32}''$, it can be ground down and used.

Example 4

The hole in a lock washer measures $\frac{8}{32}''$. Can this washer be used on a $\frac{1}{4}''$ bolt?

Solution

To change $\frac{8}{32}$ to an equivalent fraction, we may divide both the numerator and denominator by 8.

$$\frac{8 \div 8}{32 \div 8} = \frac{1}{4}$$

The washer diameter is $\frac{1}{4}''$. It will fit the $\frac{1}{4}''$ bolt.

When we can no longer change a fraction to an equivalent fraction by dividing its numerator and denominator by the same number, we say that the fraction is in *lowest terms*. Reducing fractions to lowest terms often saves time in calculating.

EXERCISES Determine whether the following pairs of fractions are equivalent.

1. $\frac{4}{8}, \frac{1}{2}$ 2. $\frac{9}{10}, \frac{90}{100}$ 3. $\frac{4}{16}, \frac{1}{4}$
4. $\frac{3}{7}, \frac{2}{14}$ 5. $\frac{1}{32}, \frac{2}{64}$ 6. $\frac{2}{3}, \frac{8}{12}$

Change each measurement to sixteenths of an inch.

7. $\frac{1}{8}''$ 8. $\frac{1}{4}''$ 9. $\frac{1}{2}''$
10. $\frac{3}{8}''$ 11. $\frac{3}{4}''$ 12. $\frac{5}{8}''$

Change each measurement to thirty-seconds of an inch.

13. $\frac{3}{16}''$ 14. $\frac{5}{8}''$ 15. $\frac{7}{16}''$
16. $\frac{1}{4}''$ 17. $\frac{1}{2}''$ 18. $\frac{9}{16}''$

Reduce each fraction to lowest terms.

19. $\frac{6}{8}$ 20. $\frac{4}{16}$ 21. $\frac{8}{32}$
22. $\frac{12}{16}$ 23. $\frac{20}{32}$ 24. $\frac{24}{64}$
25. $\frac{4}{10}$ 26. $\frac{20}{100}$ 27. $\frac{25}{100}$

Reduce each measurement to the term indicated.

28. $\frac{8}{16}''$ to eighths   29. $\frac{24}{32}''$ to fourths
30. $\frac{16}{32}''$ to halves   31. $\frac{4}{64}''$ to sixteenths
32. $\frac{48}{64}''$ to eighths   33. $\frac{49}{64}''$ to thirty-seconds
34. $\frac{4}{20}''$ to tenths   35. $\frac{8}{64}''$ to eighths

36. The square head of a machine bolt measures $\frac{24}{32}''$ across. Will a $\frac{3}{4}''$ open-end wrench fit this bolt head?

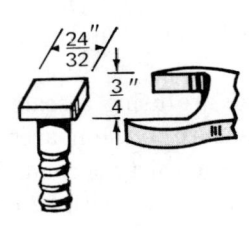

37. The owner's manual for a power lawn mower calls for an adjusting screw not less than $\frac{7}{8}''$ long. Will a screw $\frac{29}{32}''$ long fit?

1-10 COMPARING LENGTHS

Sometimes the end of the item you are measuring comes in between two marks on your rule. In such cases, you must estimate its actual length.

Reading and using measuring devices 21

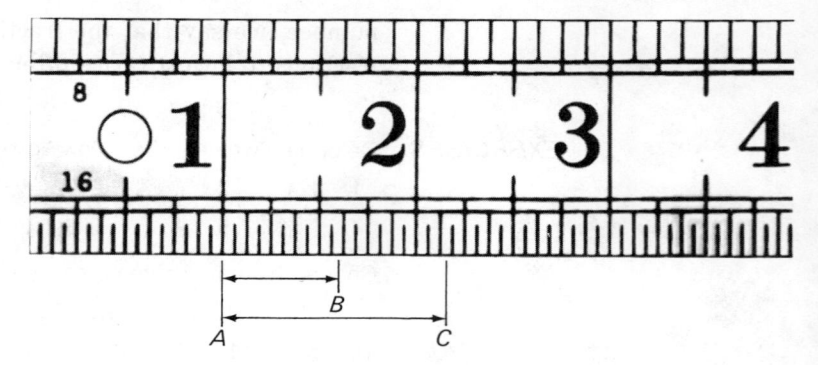

Figure 1-24

Look at the length AB in Figure 1-24. This length is a little more than $\frac{9}{16}''$, but less than $\frac{10}{16}''$. If you are measuring to the nearest sixteenth of an inch, you must decide which mark is nearer to B.

Now look at length AC. This length is more than $1\frac{1}{8}''$, but less than $1\frac{3}{16}''$. Point C falls between the $\frac{2}{16}''$ mark and the $\frac{3}{16}''$ mark. We estimate that it falls about halfway between and give the length in thirty-seconds of an inch. We can find equivalent fractions for $\frac{2}{16}$ and $\frac{3}{16}$.

$$\frac{2''}{16} = \frac{4''}{32}$$

$$\frac{3''}{16} = \frac{6''}{32}$$

Halfway between $\frac{4}{32}''$ and $\frac{6}{32}''$ is $\frac{5}{32}''$. The length AC, then, is about $1\frac{5}{32}''$.

When we are using a scale with 16 divisions to the inch, we usually measure to the nearest sixteenth of an inch, or estimate to the nearest thirty-second. For more precise measurements, you may use a scale with 32 or 64 subdivisions to the inch.

EXERCISES   Using a rule, measure each line to the nearest sixteenth of an inch.

1. ├─────────────────────────────────────┤
2. ├──────────────────┤
3. ├──────────────────────────────┤
4. ├───┤
5. ├──┤

22   Technical and industrial computations

6–10. Estimate the length of each line in Exercises 1–5 to the nearest thirty-second of an inch.

Draw the following lengths. Use a rule with a scale of 16 divisions to the inch.

11. $1\frac{5}{16}''$ **12.** $5\frac{1}{2}''$ **13.** $4\frac{1}{4}''$
14. $2\frac{3}{4}''$ **15.** $3\frac{1}{2}''$ **16.** $4\frac{5}{16}''$

17–22. Check the accuracy of your drawing by using a scale that has 32 divisions to the inch.

23. In working with sheet metal, you often need several pieces that are exactly alike. You make your pieces from a pattern, called a *template*. Measure each dimension in this template to the nearest thirty-second of an inch.

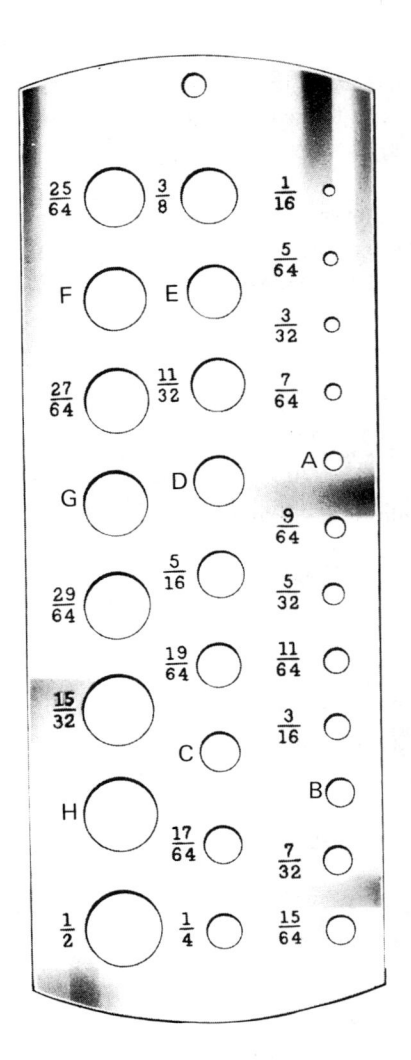

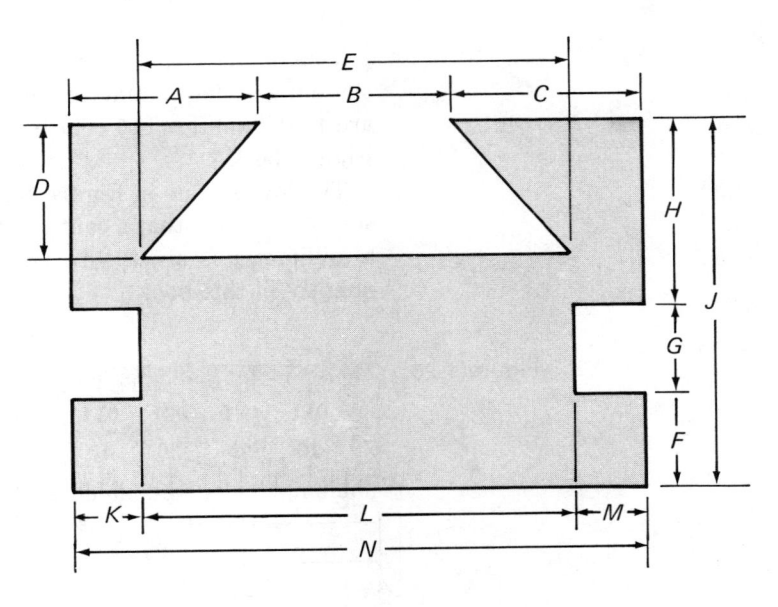

24. Machinists use a drill gauge to sort drill bits according to size. Each hole in the gauge is $\frac{1}{64}''$ larger than the next. Tell the size of the holes shown by letters.

1-11 METRIC SCALES

The metric system is the standard system of measurement throughout most of the world. In the United States the metric system is being used more and more in the major industries. Tomorrow's skilled technicians will need to understand this important system of measurement.

Figure 1-25

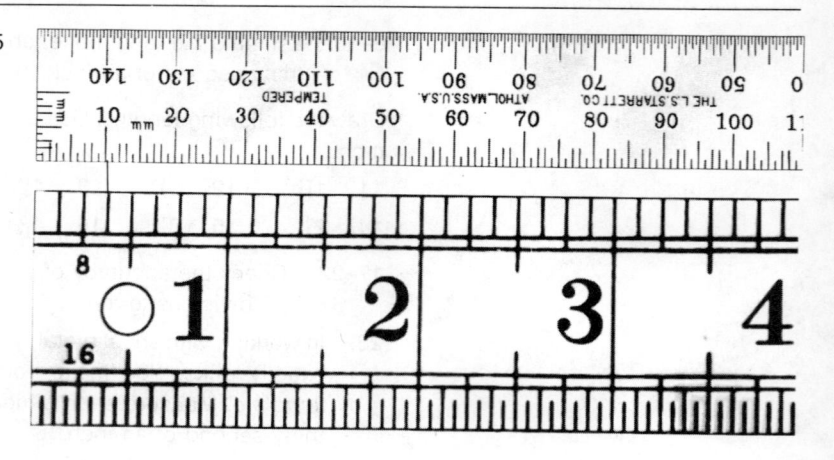

Two of the most commonly used units of length in the metric system are the millimeter (mm) and the centimeter (cm). There are 10 millimeters in 1 centimeter, as shown on the upper scale in Figure 1-25.

The lower scale in Figure 1-25 shows an inch rule for comparison. Do you see that a centimeter is about $\frac{3}{8}''$? Would you say that a millimeter is about $\frac{1}{32}''$? You will use these metric scales frequently in this book.

Figure 1-26

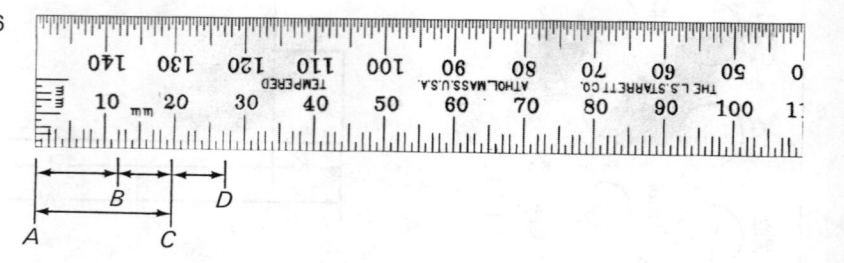

EXERCISES Exercises 1 and 2 refer to Figure 1-26.
1. On the metric scale give the distance between
   a. 3 divisions.   b. 10 divisions.
   c. 17 divisions.   d. 29 divisions.
2. Name the distance for
   a. *AB*.   b. *BC*.   c. *CD*.   d. *AC*.
3. Use a metric rule to measure the length of each line.
   a. ├─────────────┤
   b. ├──────────┤

24 Technical and industrial computations

c. ├────────────────────────────────┤
d. ├──────────────┤
e. ├──────────────────────────┤

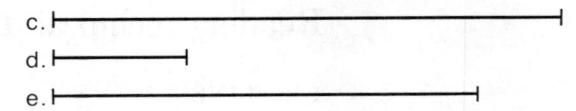

4. Use a metric rule to draw a line with a length of
   a. 18 mm.   b. 29 mm.   c. 4 cm.   d. 9 cm.

For Exercises 5–7, use a metric rule to find the lengths shown.

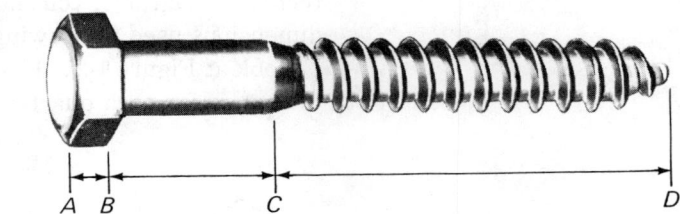

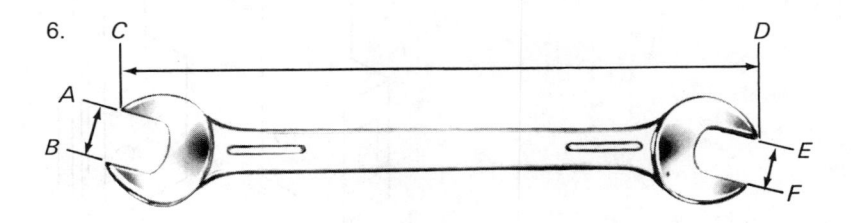

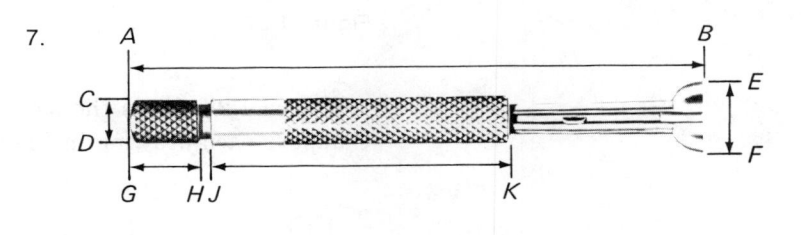

5.

A B   *C*   *D*

6.

7.

TRICKS OF THE TRADE

Here is a way to obtain two small square pieces of stock from a larger scrap. Just cut on the dashed lines and assemble the pieces, as shown.

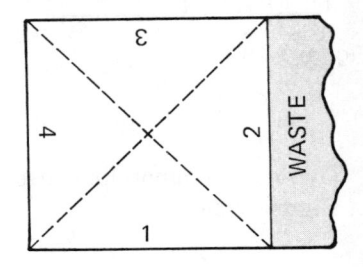

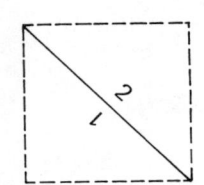

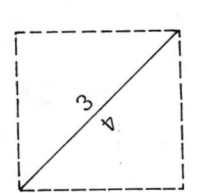

Reading and using measuring devices   25

Reading technical drawings

1-12 LINES

Work in industry must be precise. Measurements are made with sensitive instruments. Parts are machined with precision tools. And technical diagrams are drawn with care. To help you read technical diagrams, you should know about the kinds of lines and dimensions used in drawings.

Look at Figure 1-27. If we rotate the object, its shape appears to change for each quarter-turn.

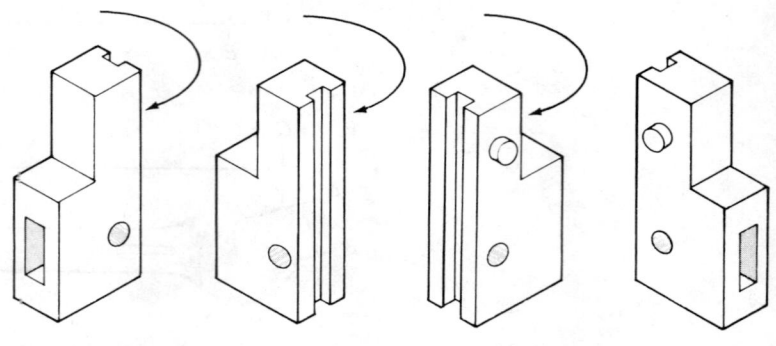

Figure 1-27

Figure 1-28 shows the same rotation in a technical drawing. Compare each drawing in Figure 1-28 with Figure 1-27.

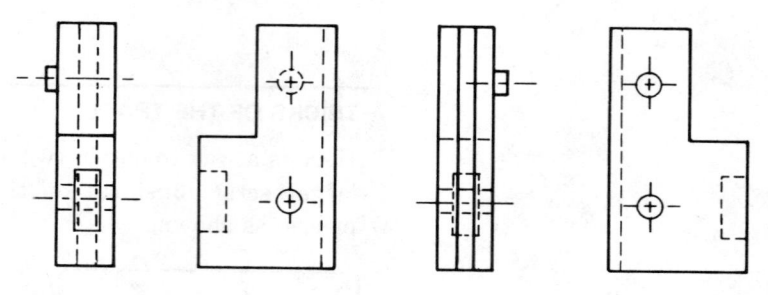

Figure 1-28

The most common lines used in technical drawings are shown in Figure 1-29.

26   Technical and industrial computations

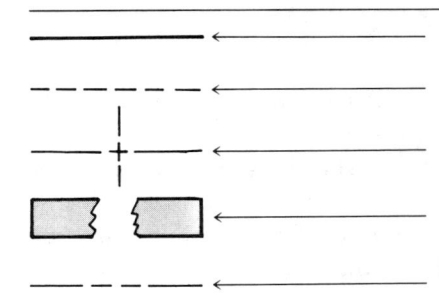

Figure 1-29

The *outline* and visible edges of a part are shown by solid lines.

Hidden lines are shown by evenly spread dashes. They indicate a part of the figure that cannot be seen in this view.

Center lines are shown by alternate short and long dashes. An object is rotated about the center line.

Jagged *break lines* are used when the object is too large to be shown accurately in one drawing.

A *phantom line* shows the alternate position of a moving part. Phantom lines are shown by a long dash and two short dashes.

EXERCISES Match the technical drawing with the figure.

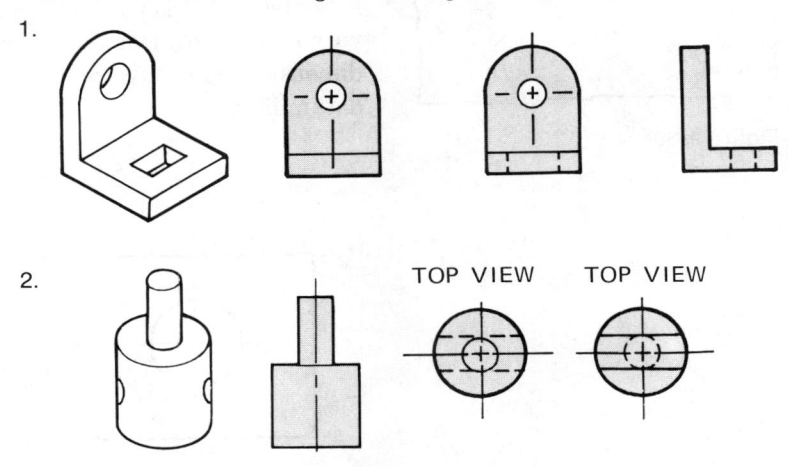

1.

2.

Sketch four views of each object. Show hidden lines, center lines, and so on.

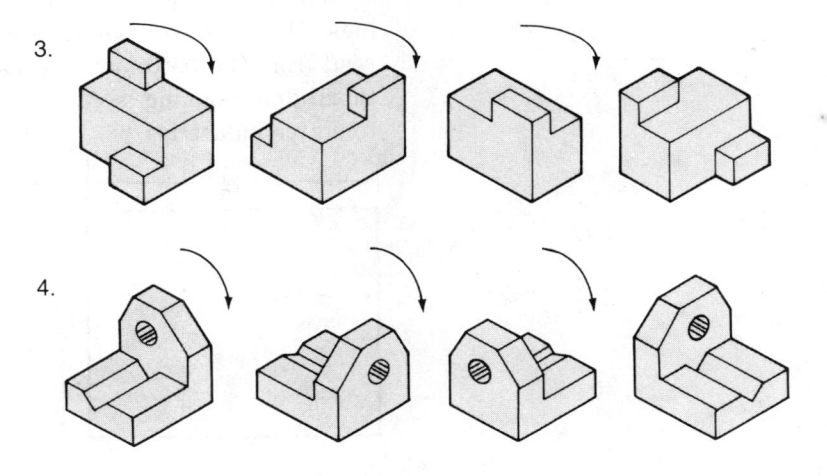

3.

4.

Reading and using measuring devices 27

Figure 1-30

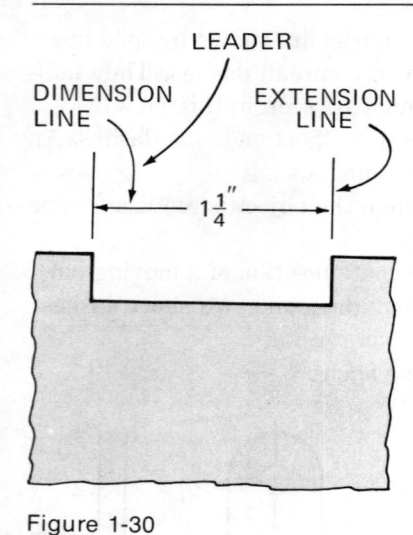

1-13 DIMENSIONS

Special lines are used to show dimensions on a drawing like Figure 1-30.

Dimension lines with the arrowheads show the specified length of a part of the object.

Extension lines are used so that direction lines are not crowded into the space around the object. This adds to the overall clarity of the drawing.

A *leader* is used to direct information and symbols to a place in the drawing.

Sometimes it is more convenient to give a series of dimensions with respect to one fixed line, called a *datum line*. In this case the dimension lines have only one arrowhead. Figure 1-31 shows a datum line.

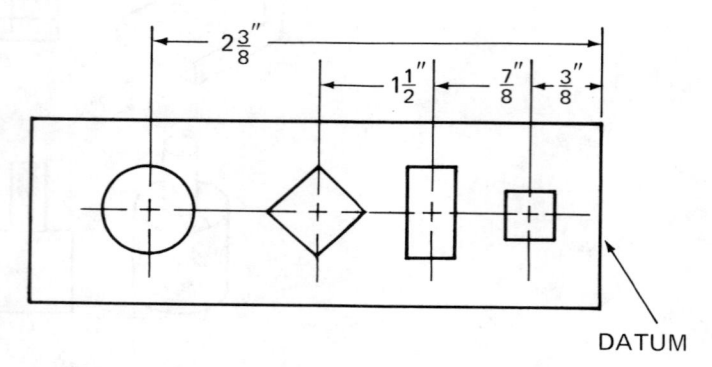

Figure 1-31

Dimensions are shown in one of two ways in technical drawings. With *unidirectional dimensioning* all dimensions can be read from the bottom of the page, as in Figure 1-32. Unidirectional dimensioning is commonly used in the aerospace and automotive industries.

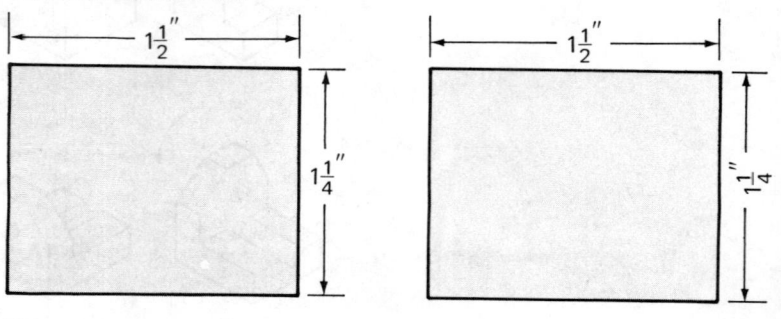

Figure 1-32   Figure 1-33

28 Technical and industrial computations

With *aligned dimensioning* the dimensions are readable from either the bottom or the right side of the drawing, as in Figure 1-33. This type of dimensioning is used in blueprints and drafting work.

You may use either system, but you should not use them both in the same drawing. In this book you will find examples of both systems of dimensioning.

EXERCISES Copy the figure. Then sketch in extension lines, dimension lines, and leaders to complete each drawing.

1.

2.

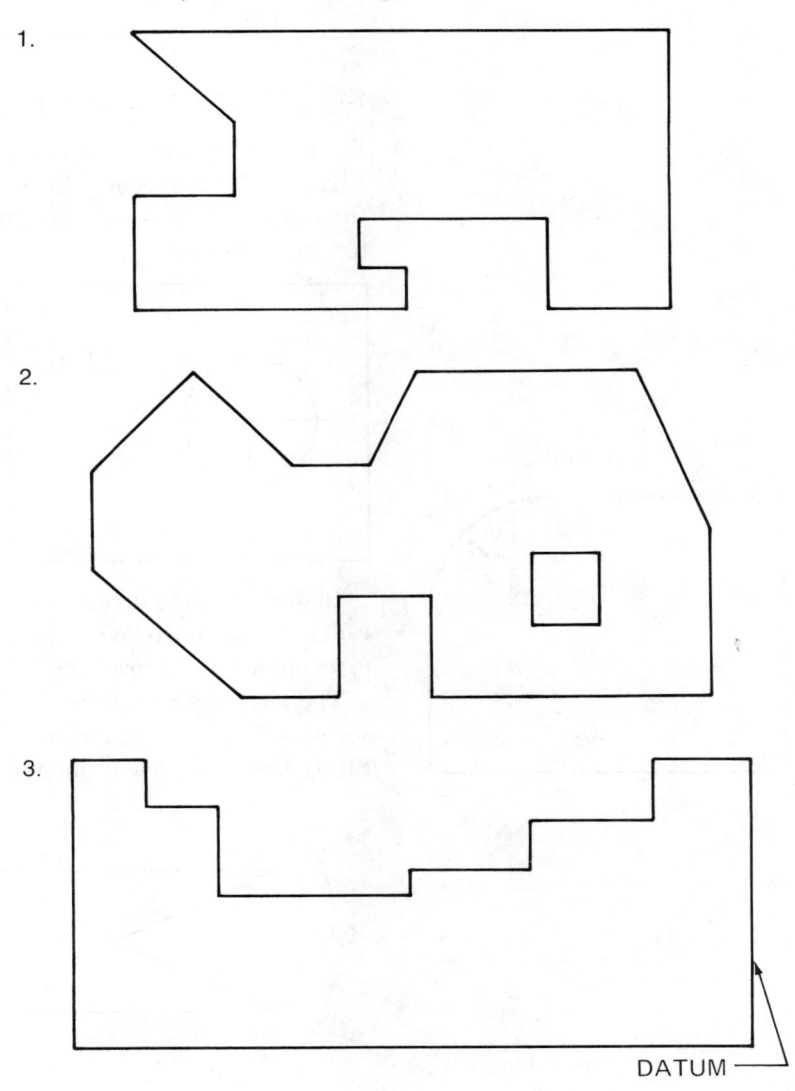

3.

DATUM

4. Redraw the figure, using aligned dimensions.

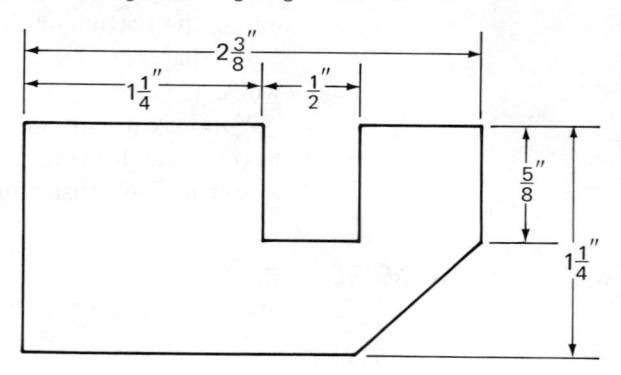

1-14 CURVES AND ANGLES

We give dimensions for circles in terms of the diameter, as in Figure 1-34. The center lines show the location of the center of the circle. The abbreviation DIA must always follow the diameter measurement.

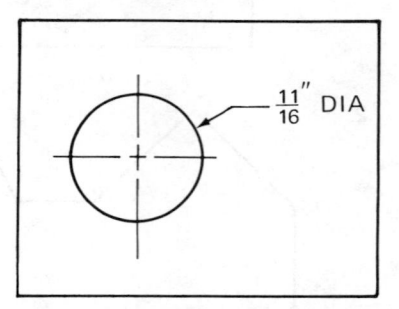

Figure 1-34

In dimensioning arcs, we show the center of curvature and the radius of the curve. We indicate the radius with the abbreviation r, as shown in Figure 1-35.

The size of an angle is given in degrees (°), minutes ('), and seconds ("), just as you would find them on a precision protractor. Study the examples of angle dimensions in Figure 1-36.

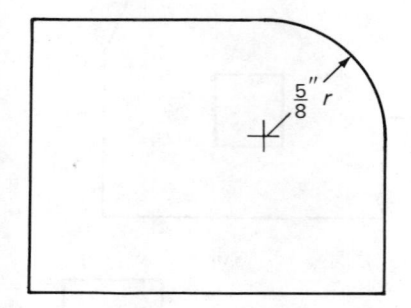

Figure 1-35

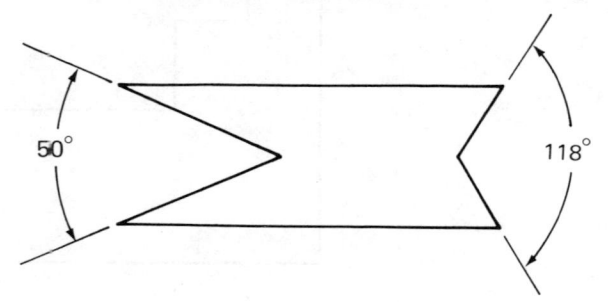

Figure 1-36

30 Technical and industrial computations

EXERCISES Copy each diagram. Use a protractor and ruler to find the distances shown. Add these dimensions to your drawing.

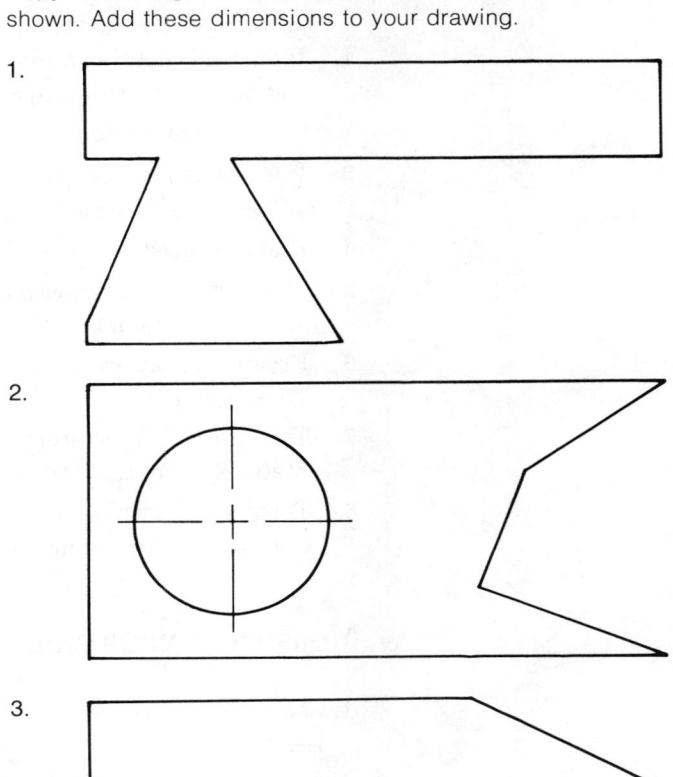

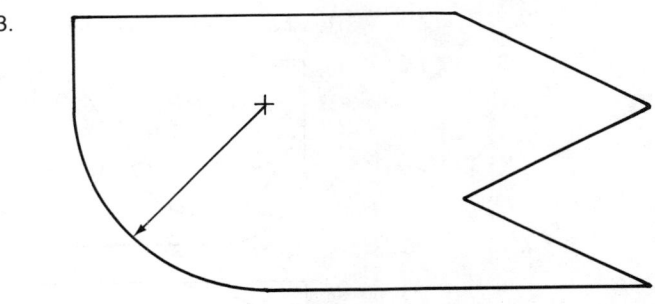

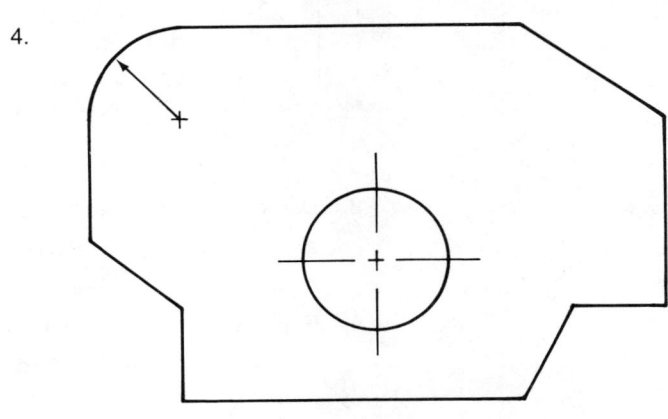

1.

2.

3.

4.

Taking inventory

1. Industrial meters can be used to measure fuel use, speeds, and other quantities. (pp. 2-8)
2. Our number system is based on tens. (p. 6)
3. Information can be shown on bar graphs, broken-line graphs, and curved-line graphs. (pp. 9-15)
4. Rulers consist of scales divided into equal parts. (pp. 16-17)
5. The terms of a fraction are the *numerator* and the *denominator*. (p. 16)
6. Fractions that are equal in value are called *equivalent* fractions. (pp. 18-21)
7. Two important systems of measurement are the United States system and the metric system. (pp. 23-24)
8. Technical drawings show the dimensions and other information that is used in manufacturing an object. (pp. 26-30)

Measuring your skills

Read the meters. (1-1, 1-2)

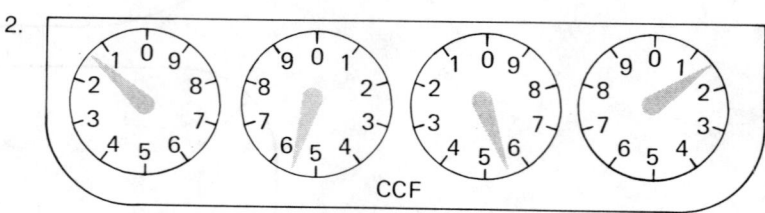

1. KWH

2. CCF

3. Use numbers to write the following expressions. (1-3)
    a. six hundred thousand, four hundred sixteen
    b. five million, five hundred thousand, five hundred
4. Round to the nearest thousand. (1-3)
    a. 5927   b. 64,274   c. 432,972

32 **Technical and industrial computations**

5. Read the tachometer. (1-4)

Exercises 6–10 refer to Figure 1-37. (1-5, 1-6, 1-7)

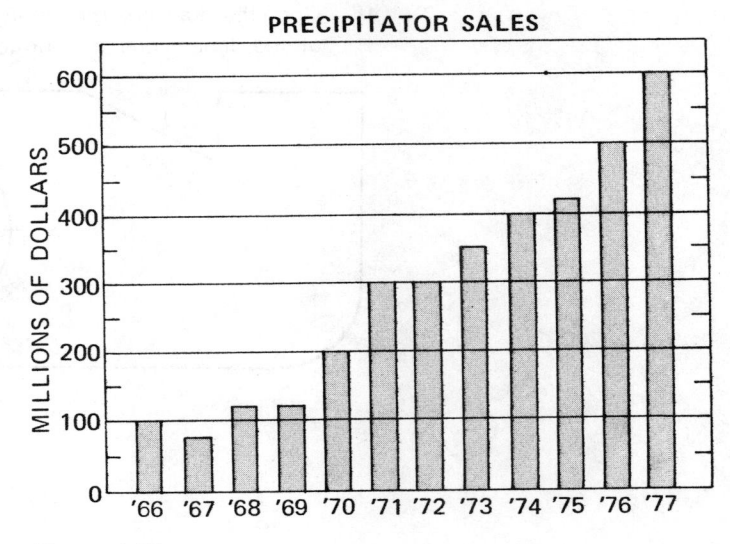

Figure 1-37

6. How many dollars were spent for industrial pollution-control devices (precipitators) in 1972?
7. How much more was spent in 1972 than in 1970?
8. How many times greater is the 1972 figure than the 1967 figure?
9. Estimate the amount that will be spent in 1978.
10. Draw the bar graph in Figure 1-37 as a broken-line graph. Show the sales for 1972–1977 as a curved line.
11. Using a rule, measure each line to the nearest thirty-second of an inch. (1-8)

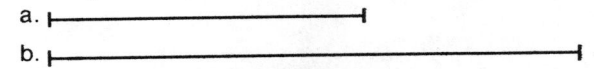

    a. ├────────────────┤
    b. ├──────────────────────────────────┤

12. Write a fraction equivalent to $\frac{5}{8}$. (1-9)
13. Change $\frac{7}{8}''$ to thirty-seconds of an inch. (1-9)
14. Change $\frac{56}{64}''$ to eighths of an inch. (1-9)
15. Reduce to lowest terms. (1-9)

    a. $\frac{80}{100}$ b. $\frac{16}{64}$ c. $\frac{26}{32}$

16. In a set of combination wrenches, which size is larger than $\frac{3}{8}''$ and less than $\frac{1}{2}''$? (1-10)

17. Give the length of each line in millimeters. (1-11)

   a. |————————————————————————————|

   b. |————————————|

18. Copy the drawing. Then measure the parts of the diagram. Show all the dimensions on your copy. (1-12, 1-13, 1-14)

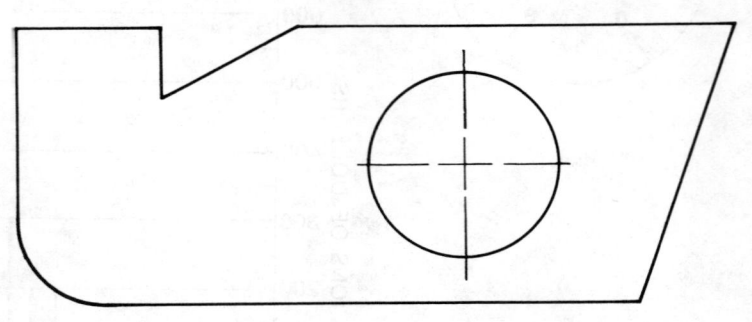

34   Technical and industrial computations

2 Working with integers

After completing this chapter, you should be able to:
1. Add and subtract whole numbers.
2. Multiply and divide whole numbers.
3. Recognize the need for integers.
4. Add and subtract integers.
5. Multiply and divide integers.

Reviewing whole numbers

2-1 ADDING AND SUBTRACTING WHOLE NUMBERS

Many of the measurements we made in Chapter 1 were integral measurements. That is to say, whole numbers were used to express the measurements. It is very important for us to be able to add, subtract, multiply, and divide whole numbers.

> To add whole numbers:
> 1. Arrange the numbers in columns starting at the right with the unit number, or ones; then
> 2. Add all the ones, tens, hundreds, and so on, in order.

Example 1

Add: 523 ft, 72 ft, 1070 ft, 326 ft.

Solution

	thousands hundreds tens ones	
Addends	5 2 3 ft	5 hundreds + 2 tens + 3 ones
	7 2 ft	7 tens + 2 ones
	1 0 7 0 ft	1 thousand + 0 hundreds + 7 tens + 0 ones
	+ 3 2 6 ft	+ 3 hundreds + 2 tens + 6 ones
Sum	1 9 9 1 ft	1 thousand + 8 hundreds + 18 tens + 11 ones
		+ 1 ten ← 10 ones
		1 thousand + 8 hundreds + 19 tens + 1 one
		1 hundred ← 10 tens
		1 thousand + 9 hundreds + 9 tens + 1 one

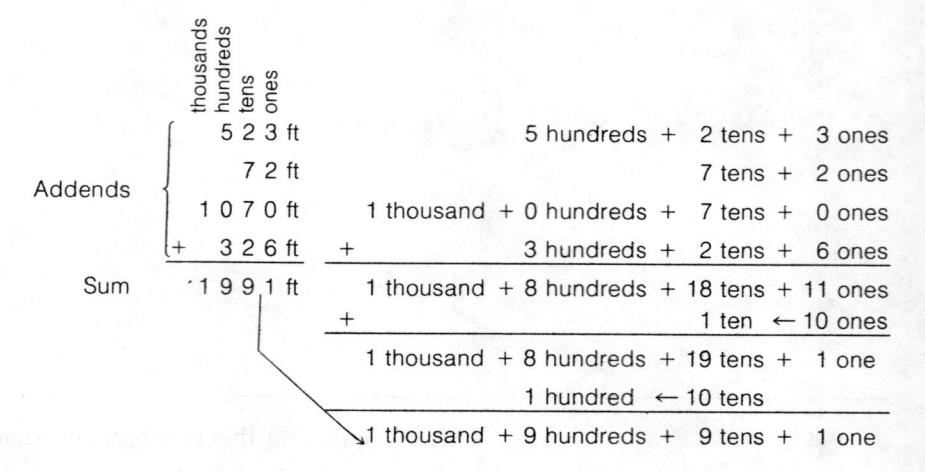

To subtract two whole numbers,
1. arrange the numbers in columns; then
2. subtract the ones, tens, hundreds, and so on, in order.

Example 2

Subtract: 763 mi − 341 mi.

Solution

	hundreds tens ones	
Minuend	7 6 3 mi	7 hundreds + 6 tens + 3 ones
Subtrahend	− 3 4 1 mi	− (3 hundreds + 4 tens + 1 one)
Difference	4 2 2 mi	4 hundreds + 2 tens + 2 ones

Example 3

Subtract: 572 km − 345 km.

36 Technical and industrial computations

Solution

```
      6
  5 7̸ 12 km        5 hundreds + 7 tens +  2 ones
 − 3 4  5 km                   1 ten  → 10 ones
   2 2  7 km       5 hundreds + 6 tens + 12 ones
                 − (3 hundreds + 4 tens + 5 ones)
                   2 hundreds + 2 tens +  7 ones
```

EXERCISES Add or subtract.

1. 56 in. + 8 in. + 43 in. + 29 in.
2. 2000 lb + 26 lb + 810 lb
3. 596 m − 321 m
4. 26 yd
   + 72 yd
5. 596 ft
   + 207 ft
6. 457 rpm
   − 126 rpm
7. 156 days + 27 days + 312 days + 1010 days + 196 days
8. 12 tons + 121 tons + 56 tons + 7 tons + 205 tons
9. 981 m − 546 m
10. 234 cc − 78 cc

Find the lengths shown by letters on the diagrams.

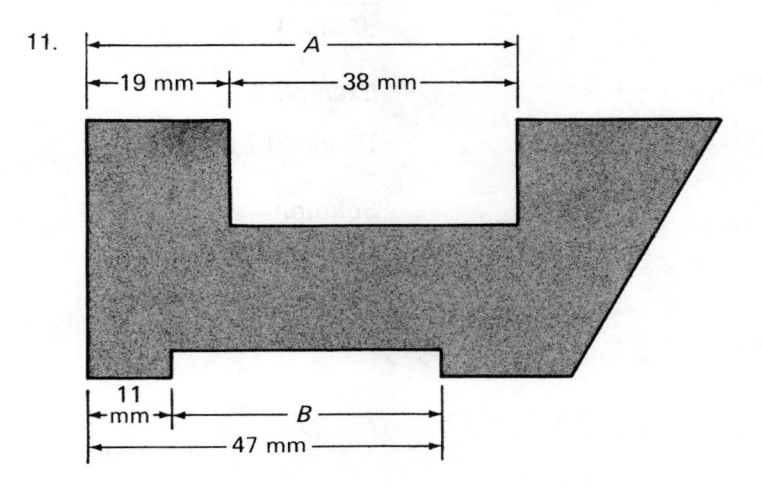

11.

12.

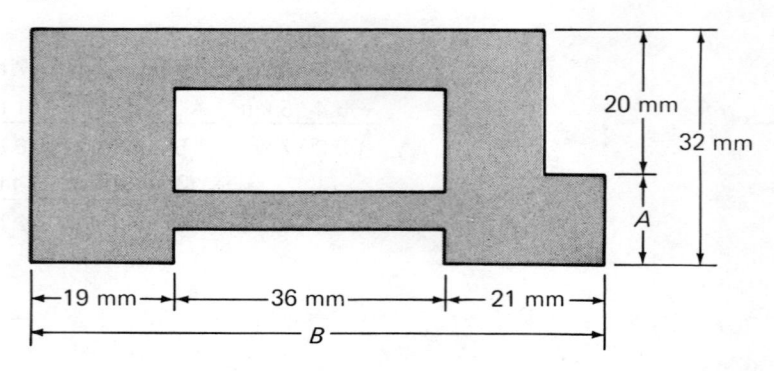

Figure showing a shaped piece with dimensions: 20 mm, 32 mm, A, 19 mm, 36 mm, 21 mm, B

2-2 MULTIPLYING AND DIVIDING WHOLE NUMBERS

In your technical work you will be computing distances, areas, volumes, forces, and many other quantities. These computations will require skill in multiplying and dividing whole numbers.

Example 1

Multiply 326 by 48.

Solution

```
    326   Factor
  × 48    Factor
   2608
   1304
  15648   Product
```

Example 2

Divide 91 by 7.

Solution

```
                13    Quotient
   Divisor   7)91     Dividend
                7
               21
               21
                0     Remainder
```

38 Technical and industrial computations

Example 3

784 ÷ 32 = ?

Solution

```
        24 r 16
    32)784
        64
       ---
        144
        128
       ---
         16
```

EXERCISES   Multiply or divide.

1. 23 × 6
2. 325 × 14
3. 156
   × 7
4. 257
   × 26
5. 84 ÷ 6
6. 144 ÷ 12
7. 8)328
8. 23)529
9. 956
   × 72
10. 4158
    × 137
11. 10,179 × 253
12. 15,876 ÷ 24
13. 121)7260
14. 109)72,594

Introducing integers

2-3 POSITIVE AND NEGATIVE INTEGERS

In technical work we need to use numbers in many different ways. For example, if we read 20° on a thermometer, we need to know if the reading is above zero or below zero. We indicate 20° above zero by $^+20°$, and 20° below zero by $^-20°$.

We use similar notation for gains and losses in voltage readings in electric circuits. For example, a gain of 10 volts is written $^+10$ and a loss of 15 volts is written $^-15$.

Whole numbers with a plus sign, $^+$, written in front of them are called *positive integers*. Whole numbers with a minus sign, $^-$, in front of them are called *negative integers*. The number lines in Figure 2-1 indicate positive and negative integers. Notice that 0 is neither positive nor negative.

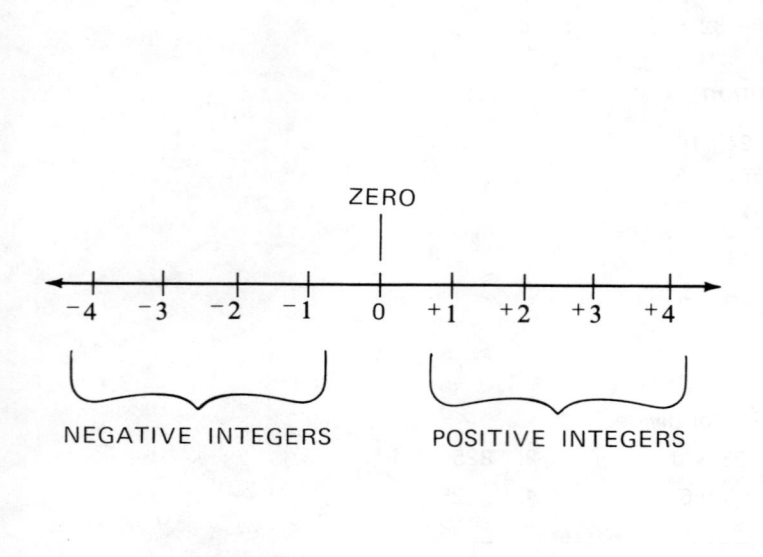

Figure 2-1

Example 1

If a three-person exploratory submarine is lowered 150 ft below sea level, what number is used to indicate its depth?

Solution

150 ft *below* sea level is indicated by $^{-}150$.

Example 2

If a bank deposit of $25 is represented by $^{+}25$, determine the integer that represents a bank withdrawal of $40.

Solution

$^{-}40$ represents a bank withdrawal of $40.

EXERCISES Draw a number line and locate the following integers on it.

1. $^{+}3$	2. $^{-}4$	3. $^{-}6$
4. $^{+}12$	5. $^{+}1$	6. $^{-}15$
7. 0	8. $^{-}1$	9. $^{+}4$

Technical and industrial computations

Which is greater:

10. $^+4$ or $^+3$? 11. 0 or $^+2$? 12. $^+2$ or $^-1$?
13. $^+2$ or $^-2$? 14. $^-3$ or $^+1$? 15. $^+4$ or $^-4$?
16. $^-3$ or $^-2$? 17. $^-1$ or $^-8$? 18. $^-7$ or $^-8$?

19. If the distance above ground is represented by a positive integer, what integer represents the depth of a mine shaft 1000 ft deep?

20. If a reference point is designated on a certain type of voltage meter, how would 125 volts above the reference point be designated? 35 volts below the reference point?

21. Below sea level is designated by negative integers; above sea level is designated by positive integers. Town A has an altitude reading of $^-150$ ft. Town B has an altitude reading of $^-200$ ft. Which town has the higher altitude?

2-4 ADDING INTEGERS

There are two cases involving addition of integers. One involves adding integers whose signs are alike, and the other involves adding integers whose signs are unlike.

> To add integers whose signs are alike, add the numbers and use the sign of the original integers.

Example 1

Add: $^+8 + {}^+4$.

Solution

$$\begin{array}{r} ^+8 \\ + {}^+4 \\ \hline ^+12 \end{array}$$

Example 2

Add: $^-12 + {}^-9$.

Solution

$$\begin{array}{r} ^-12 \\ + \ {}^-9 \\ \hline ^-21 \end{array}$$

Working with integers

Example 3

Find the sum: $^+2 + {}^+7 + {}^+1 + {}^+4$.

Solution

$^+2 + {}^+7 + {}^+1 + {}^+4$

$= {}^+9 + {}^+1 + {}^+4$

$= {}^+10 + {}^+4$

$= {}^+14$

Example 4

Find the sum: $^-3 + {}^-4 + {}^-8 + {}^-1$.

Solution

$^-3 + {}^-4 + {}^-8 + {}^-1$

$= {}^-7 + {}^-8 + {}^-1$

$= {}^-15 + {}^-1$

$= {}^-16$

If we disregard the sign of an integer, the resulting number is called the *absolute value* of the integer. For example, 8 is the absolute value of $^-8$, and 5 is the absolute value of $^+5$. Absolute values are helpful when we are adding integers whose signs are unlike.

> To add integers whose signs are unlike, subtract the smaller absolute value from the larger, and attach the sign of the larger absolute value to the answer.

Example 1

Add: $^-8 + {}^+3$.

Solution

$8 - 3 = 5$ (larger absolute value minus smaller)

therefore
$$\begin{array}{r} ^-8 \\ + {}^+3 \\ \hline ^-5 \end{array}$$

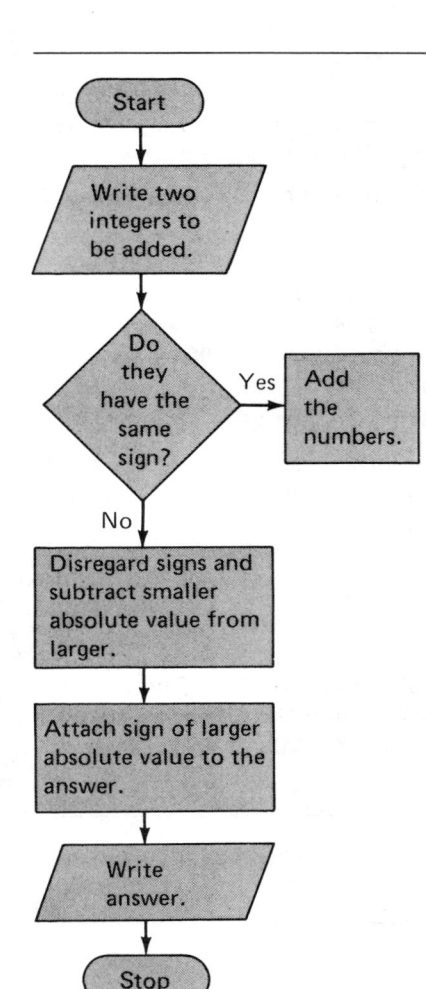

Figure 2-2

[Flow chart:]
- Start
- Write two integers to be added.
- Do they have the same sign? — Yes → Add the numbers.
- No ↓
- Disregard signs and subtract smaller absolute value from larger.
- Attach sign of larger absolute value to the answer.
- Write answer.
- Stop

Figure 2-2

Example 2

Add: $^+8 + {}^-3$.

Solution

$8 - 3 = 5$ (larger absolute value minus smaller)

therefore
$$\begin{array}{r} {}^+8 \\ + {}^-3 \\ \hline {}^+5 \end{array}$$

Example 3

Add: $^+12 + {}^-15$.

Solution

$15 - 12 = 3$ (larger absolute value minus smaller)

therefore
$$\begin{array}{r} {}^+12 \\ + {}^-15 \\ \hline {}^-3 \end{array}$$

Example 4

Find the sum: $^-3 + {}^-4 + {}^+11$.

Solution

$^-3 + {}^-4 + {}^+11$

$= {}^-7 + {}^+11$

$= {}^+4$

EXERCISES

Add. Follow the flow chart in Figure 2-2 if you need help.

1. $^+2 + {}^+8$
2. $^+14 + {}^+21$
3. $^-6 + {}^-7$
4. $^-23 + {}^-8$
5. $^-1 + {}^-100$
6. $^+46 + {}^+71$
7. $^+6 + {}^+5 + {}^+3$
8. $^-11 + {}^-2 + {}^-31$
9. $^+2 + {}^+3 + {}^-1 + {}^-6$
10. $^-5 + {}^+5$
11. $^+6 + {}^-7$
12. $^+18 + {}^-9$
13. $^-12 + {}^+8$
14. $^-12 + {}^+28$
15. $^+15 + {}^-15$
16. $^-8 + {}^+4 + {}^+10$
17. $^-5 + {}^+8 + {}^-6$
18. $^+9 + {}^-11 + {}^+8$

Working with integers

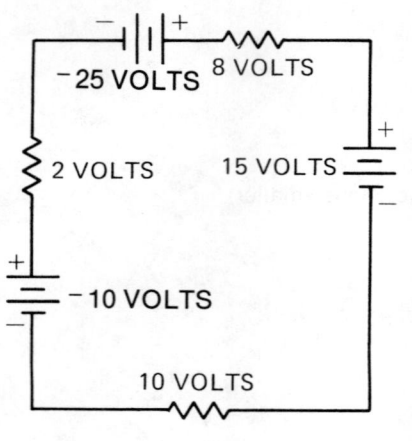

19. When resistors are connected in a series, the signs of the voltage through the resistors are determined by the direction of current flow. Find the sum of the voltages in the figure to the left.
20. If a temperature of $^-12°F$ is increased by 32°F, what is the resulting temperature?
21. A Brinell gauge reading of 220 is set for testing the hardness of steel cover plates. Readings above or below 220 are given by signed numbers. A reading of $^-5$ means the hardness test indicates the durability to be 220 + $^-5$ = 215. Find the durability for each of the following readings:

   a. $^-10$ b. $^-8$ c. $^+12$ d. $^+25$ e. $^-11$

22. A communications satellite is being carried by a rocket with launch time "T minus 8 minutes." Assuming no delays, how long has the rocket been in the air 20 minutes later?

2-5 SUBTRACTING INTEGERS

The temperature inside an assembly plant is 70°F. The temperature outside is $^-6°F$. If we wanted to find the difference between the two temperatures, we would subtract: $^+70 - {}^-6$.

To subtract one integer from another, change the sign of the second and proceed as in addition.

In the previous example, we have $^+70 - {}^-6 = {}^+70 + {}^+6 = {}^+76$. Therefore, the difference in temperatures is 76°.

Example 1

Subtract: $^-22 - {}^-13$.

Solution

$$^-22 - {}^-13 \qquad\qquad {}^-22 \qquad {}^-22$$
$$= {}^-22 + {}^+13 \quad \text{or} \quad -\,{}^-13 \quad +\,{}^+13$$
$$= {}^-9 \qquad\qquad\qquad\qquad\qquad {}^-9$$

Example 2

Subtract: $^-37 - {}^+16$.

Solution

$$^-37 - {}^+16$$
$$= {}^-37 + {}^-16 \quad \text{or} \quad \begin{array}{r} {}^-37 \\ - {}^+16 \\ \hline \end{array} \quad \begin{array}{r} {}^-37 \\ + {}^-16 \\ \hline {}^-53 \end{array}$$
$$= {}^-53$$

Example 3

Subtract: $^+15 - {}^-3$.

Solution

$$^+15 - {}^-3$$
$$= {}^+15 + {}^+3 \quad \text{or} \quad \begin{array}{r} {}^+15 \\ - {}^-3 \\ \hline \end{array} \quad \begin{array}{r} {}^+15 \\ + {}^+3 \\ \hline {}^+18 \end{array}$$
$$= {}^+18$$

Example 4

Subtract: $^+22 - {}^-11$.

Solution

$$^+22 - {}^-11$$
$$= {}^+22 + {}^+11 \quad \text{or} \quad \begin{array}{r} {}^+22 \\ - {}^-11 \\ \hline \end{array} \quad \begin{array}{r} {}^+22 \\ + {}^+11 \\ \hline {}^+33 \end{array}$$
$$= {}^+33$$

EXERCISES Subtract.

1. $^+10 - {}^-5$
2. $^+12 - {}^+7$
3. $^+17 - {}^-18$
4. $^+16 - {}^+37$
5. $^-2 - {}^-3$
6. $^-7 - {}^+13$
7. $^-27 - {}^-12$
8. $^-54 - {}^+21$
9. $^+15 - {}^-15$
10. $\begin{array}{r} {}^+127 \\ - {}^+74 \\ \hline \end{array}$
11. $\begin{array}{r} {}^-212 \\ - {}^-114 \\ \hline \end{array}$
12. $\begin{array}{r} {}^+5280 \\ - {}^-2640 \\ \hline \end{array}$
13. $^+2 - {}^+7 + {}^-3$
14. $^-7 - {}^-1 - {}^+6$
15. $^-9 - {}^-5 + {}^-8 - {}^+5$
16. $^+100 - {}^-80 + {}^-50 - {}^-35$

17. A gas meter read 2108 cu ft on December 4. On January 7 the meter read 2478 cu ft. How many cubic feet of gas were used during the month?

18. The temperature of a supercooled chemical was recorded as $^-112°$. Gradually it was heated to $^+21°$. What was the change in temperature of the chemical?

19. A man paid off a debt of $25. Now he has $82 left. Show by subtraction that he originally had $107.

Working with integers

20. A test car was driven over the same track for four days in order to obtain controlled gas mileage averages. The first three days the car traveled the following numbers of miles: 324, 306, 288. How many miles did the test car travel on the fourth day if the total mileage was 1260?

21. The velocity of an object was measured at $^-132$ ft/sec. Then 2 sec later the velocity was measured at $^-116$ ft/sec. How much did the velocity vary between readings?

2-6 MULTIPLYING INTEGERS

You can solve many technical problems by multiplying positive and negative integers. For example, if the temperature decreases a steady 2° per day over a 3-day period, you can find the total decrease by multiplying $3 \times {}^-2$. Over the 3-day period the temperature decreased 6°. This can be illustrated as $3 \times {}^-2 = {}^-6$.

There are three rules for multiplying integers.

---
1. If the signs of all the integers are positive, multiply the numbers and attach a positive sign to the answer.
2. If there is an odd number of negative signs, multiply the numbers and attach a negative sign to the answer.
3. If there is an even number of negative signs, multiply the numbers and attach a positive sign to the answer.

Example 1

Multiply: $^+3 \times {}^+6$.

Solution

$^+3 \times {}^+6 = {}^+18$ (all positive signs)

Example 2

Multiply: $^+3 \times {}^-8$.

Solution

$^+3 \times {}^-8 = {}^-24$ (odd number of negative signs)

Example 3

Multiply: $^-9 \times {}^-4$.

Solution

⁻9 × ⁻4 = ⁺36   (even number of negative signs)

Example 4

Multiply: ⁺2 × ⁻3 × ⁻4 × ⁺2 × ⁻1.

Solution

⁺2 × ⁻3 × ⁻4 × ⁺2 × ⁻1   (odd number of negative signs)

= ⁻48

Example 5

Multiply: ⁻23 × ⁺6.

Solution
$$\begin{array}{r} ^-23 \\ \times\ ^+6 \\ \hline ^-138 \end{array}$$

EXERCISES   Multiply.

1. ⁺7 × ⁻8
2. ⁺6 × ⁻5
3. ⁻9 × ⁻9
4. ⁻3 × ⁺3
5. ⁺8 × 0
6. ⁺11 × ⁻1
7. ⁺8 × ⁺7 × ⁻2
8. ⁺7 × ⁻5 × ⁻4
9. ⁻6 × ⁻2 × ⁻3
10. ⁺17 × ⁻11
11. ⁻2756 × ⁻23
12. ⁻504 × ⁺103
13. ⁺5 × ⁺5 × ⁺5
14. ⁻4 × ⁻4 × ⁻4
15. ⁺12 × ⁻3 × ⁻20
16. ⁺8 × ⁻8 × ⁺2 × ⁻2

17. A defective compressor in a refrigerator unit caused the temperature to increase 3° each day. Two weeks passed before the defect was discovered. What was the temperature in the unit when the defect was discovered if the original temperature was ⁻10°?

18. A narrow metal plate expands 2 mm for each 10° above 60°F. If the plate is 34 mm in length at 60°F, what will its length be when the temperature reaches 90°?

19. An electronic meter contains 32 transistors. The transistors are purchased from two suppliers. Supplier A charges $1.81 for each transistor, and supplier B charges $2.05 for each transistor. How much would be saved if all the transistors were purchased from supplier A instead of from supplier B?

Working with integers

20. During an experiment a weather balloon carrying instruments is dropped from an airplane. The balloon is to pass a 200-ft tower on its way down. After 6 sec, the distance it drops is found by evaluating 500 + [($^-$16) × 5 × 5]. **Is the baloon above or below the tower at this time?**

2-7 DIVIDING INTEGERS

The rules for dividing integers are similar to those for multiplying integers.

> When dividing two integers, the answer is positive if both numbers have the same sign.
> The answer is negative if the two numbers have opposite signs.

Example 1

Divide: $^-4 \div {^+2}$.

Solution

$^-4 \div {^+2} = {^-2}$ (signs different)

or $\dfrac{^-4}{^+2} = {^-2}$

Example 2

Divide: $^+8 \div {^-2}$.

Solution

$^+8 \div {^-2} = {^-4}$ (signs different)

or $\dfrac{^+8}{^-2} = {^-4}$

Example 3

Divide: $^-10 \div {^-5}$.

Solution

$^-10 \div {^-5} = {^+2}$ (signs alike)

or $\dfrac{^-10}{^-5} = {^+2}$

Example 4

Divide: $^-27\overline{)^+81}$.

Solution

$$\begin{array}{r} ^-3 \\ ^-27\overline{)^+81} \\ ^+81 \end{array} \quad \text{(signs different)}$$

Note: $81 \div 27 = 3$, since $81 = 27 \times 3$. In example 4 we have $^+81 \div {^-27} = {^-3}$, since $^+81 = {^-27} \times {^-3}$.

EXERCISES Divide.

1. $^+20 \div {^+4}$
2. $^-15 \div {^+5}$
3. $^+27 \div {^-9}$
4. $^-84 \div {^-3}$
5. $^+12 \div {^-12}$
6. $0 \div {^-6}$
7. $^-144 \div {^-12}$
8. $^+56 \div {^-7}$
9. $^-576 \div {^+4}$
10. $\dfrac{^-400}{^-25}$
11. $\dfrac{^+121}{^-11}$
12. $\dfrac{^-125}{^+5}$
13. $\dfrac{^+5280}{^-20}$
14. $^-12\overline{)97248}$
15. $27\overline{)^-945}$

16. If $^-30°$ indicates a 30-degree drop in temperature over a 10-day period, what is the average drop in temperature per day?

17. The current in amperes for a certain electric circuit is determined by finding the answer to the problem

$$\dfrac{^+10 - (^-5 \times {^+12})}{^+12 + {^+2}}$$

Find the current.

Taking inventory

1. To add (subtract) whole numbers, arrange the numbers in columns and add (subtract) the ones, tens, hundreds, and so on, in order. (p. 35)

2. In multiplication of two whole numbers, the two numbers are called *factors* and the answer is called the *product*. (p. 38)

3. In a division problem, the number being divided is called the *dividend*, the number being divided into the dividend is called the *divisor*, and the answer is called the *quotient*. (p. 38)

4. In the division problem $7 \div 3$, the quotient is 2 and there is a *remainder* of 1. (p. 38)

Working with integers

5. Whole numbers written with a plus sign, $^+$, are called *positive integers*, and whole numbers written with a minus sign, $^-$, are called *negative integers*. (**p. 39**)

6. To add integers whose signs are alike, add the numbers and use the sign of the original integers. (**p. 41**)

7. To add integers whose signs are unlike, subtract the smaller absolute value from the larger, and attach the sign of the integer with the larger absolute value to the answer. (**p. 42**)

8. To subtract one number from another, change the sign of the second number and proceed as in addition. (**p. 44**)

9. When you multiply integers, the answer is positive if all the signs are positive; the answer is negative if there is an odd number of negative signs; and the answer is positive if there is an even number of negative signs. (**p. 46**)

10. When you divide two integers, the answer is positive if both numbers have the same sign, and the answer is negative if the two numbers have opposite signs. (**p. 48**)

Measuring your skills

Perform the indicated operations. (2-1, 2-2)

1. 846 + 397
2. 543 − 137
3. 642 × 34
4. 56)1904

Add or subtract. (2-4, 2-5)

5. $^+24 + {}^-12$
6. $^+36 + {}^-50$
7. $^-42 + {}^+18$
8. $^-37 + {}^-3$
9. $^+72 - {}^+12$
10. $^+13 - {}^-15$
11. $^-5 - {}^-18$
12. $^-15 - {}^-15$
13. $^-10 - {}^-4 - {}^-7$

Multiply or divide. (2-6, 2-7)

14. $^-8 \times {}^-7$
15. $^+9 \times {}^-5$
16. $^-15 \times {}^+4$
17. $^-2 \times {}^-6 \times {}^+8$
18. $^-1 \times {}^+4 \times {}^+7$
19. $^-10 \times {}^+6 \times {}^-2 \times {}^-3$
20. $^+49 \div {}^-7$
21. $^-56 \div {}^-8$
22. $^+72 \div {}^-6$
23. $\dfrac{{}^-5 \times {}^-6}{{}^+3}$
24. $\dfrac{{}^+8 \times {}^-6}{{}^-4 \times {}^-3}$
25. $\dfrac{{}^-12 \times {}^-15 \times {}^+2}{{}^-3 \times {}^+4 \times {}^-2}$

26. If $^-10$ means 10 minutes before take-off, how would you express 12 minutes after take-off? (2-3)

27. If $^+50$ mi is used to mean 50 miles north of a certain point, what would $^-50$ mean? **(2-3)**

28. If the freezing temperatures of two substances are recorded as 30°F and $^-31$°F, how many degrees difference is there between the two freezing points? **(2-5)**

29. If a company has $1200 in a bank account and withdraws $300 each week for 5 weeks, what will its new balance be? **(2-5, 2-6)**

30. The amount of pressure, in pounds per square inch, needed to compress 300 cu in of a certain gas by 6 cu in is determined by calculating

$$\frac{^-25000 \times ^-6}{300}$$

Find this pressure. **(2-6, 2-7)**

3 Using fractions

After completing this chapter, you should be able to:
1. Add and subtract fractions.
2. Add and subtract mixed numbers.
3. Work with dimensions expressed in inches or in feet and inches.
4. Plan projects carefully in order to minimize waste.
5. Multiply fractions and mixed numbers.
6. Use cancellation when multiplying fractions.
7. Divide fractions and mixed numbers.
8. Make industrial calculations involving fractions and estimation.

Adding and subtracting fractions

3-1 FRACTIONS WITH THE SAME DENOMINATOR

Materials like aluminum and plastic can be shaped by forcing them through an open die, in a process called *extrusion*. To make the die, the machinist refers to a diagram that shows the required dimensions. By adding and subtracting fractions, the machinist finds the dimensions that are not shown.

Figure 3-1

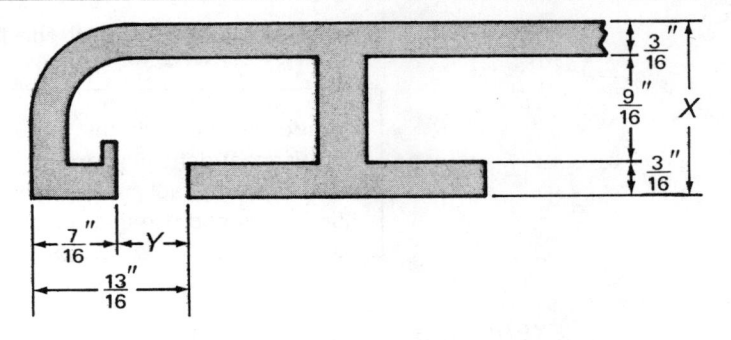

Example 1

Find length X.

Solution

$$\text{Length } X = \frac{3}{16} + \frac{9}{16} + \frac{3}{16}$$

$$= \frac{3 + 9 + 3}{16}$$

$$= \frac{15}{16}$$

Length X is $\frac{15''}{16}$.

Example 2

Find length Y.

Solution

$$\text{Length } Y = \frac{13}{16} - \frac{7}{16}$$

$$= \frac{13 - 7}{16}$$

$$= \frac{6}{16}$$

$$= \frac{3}{8}$$

Length Y is $\frac{3''}{8}$.

Using fractions

Examples 1 and 2 suggest the following rule.

> To add fractions with the same denominator, add the numerators and use that denominator.
> To subtract fractions with the same denominator, subtract the numerators and use that denominator.

EXERCISES Add or subtract. Write the answer in the simplest form.

1. $\frac{2}{8} + \frac{3}{8}$
2. $\frac{7}{16} - \frac{5}{16}$
3. $\frac{3}{32} + \frac{5}{32}$
4. $\frac{15}{64} - \frac{7}{64}$
5. $\frac{3}{10} + \frac{1}{10}$
6. $\frac{21}{32} - \frac{17}{32}$
7. $\frac{29}{100} + \frac{47}{100}$
8. $\frac{29}{64} - \frac{13}{64}$
9. $\frac{19}{32} - \frac{11}{32}$
10. $\frac{27}{64} - \frac{23}{64}$
11. $\frac{9}{64} + \frac{15}{64} + \frac{33}{64}$
12. $\frac{27}{100} + \frac{37}{100} + \frac{17}{100}$

Find the lengths. (Do the work inside the parentheses first.)

13. $\frac{29''}{32} - (\frac{3''}{32} + \frac{7''}{32})$
14. $\frac{87''}{100} - (\frac{33''}{100} + \frac{23''}{100})$
15. $\frac{33''}{64} - (\frac{5''}{64} + \frac{9''}{64} + \frac{5''}{64})$
16. $(\frac{49''}{100} + \frac{17''}{100}) - (\frac{27''}{100} + \frac{13''}{100})$

Find the lengths shown by letters on the diagrams.

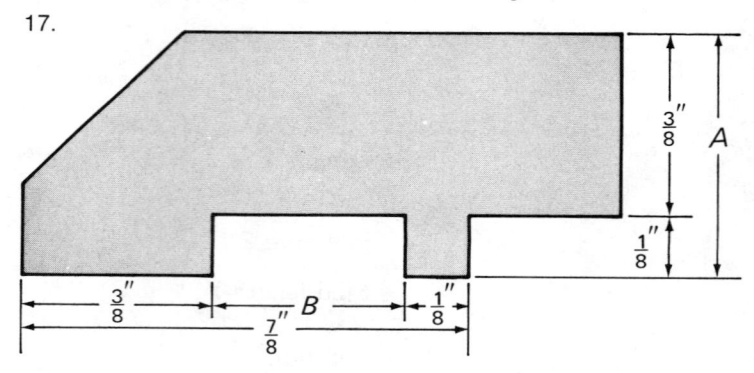

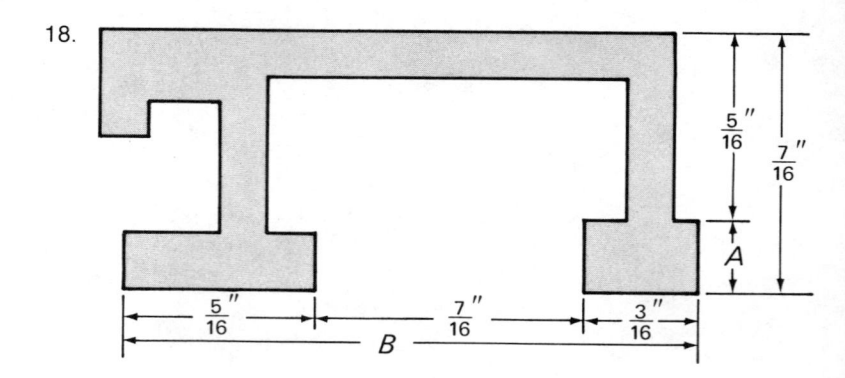

17.

18.

3-2 FRACTIONS WITH DIFFERENT DENOMINATORS

Figure 3-2 shows a carriage bolt, which is used widely in industry. In order to find the missing dimensions A and B, we have to add and subtract fractions with different denominators. The following rule can help us.

> To add or subtract fractions with different denominators:
> 1. change the fractions to equivalent fractions that have the same denominator; then
> 2. add or subtract the fractions, using the common denominator.

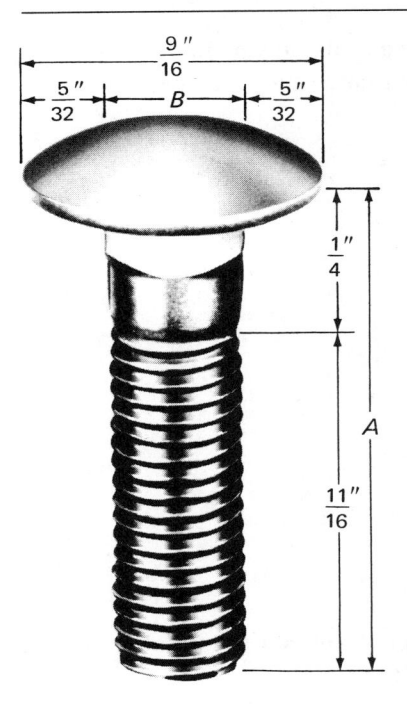

Figure 3-2

Example 1

Find length A in Figure 3-2.

Solution

$$\text{Length } A = \frac{1}{4} + \frac{11}{16}$$

$$= \frac{4}{16} + \frac{11}{16}$$

$$= \frac{4 + 11}{16} = \frac{15}{16}$$

Length A is $\frac{15}{16}''$.

Example 2

Find length B in Figure 3-2.

Solution

To find B, we must subtract the length of the flanges from the diameter of the bolt head. Thus,

$$\text{Length } B = \frac{9}{16} - \left(\frac{5}{32} + \frac{5}{32}\right)$$

$$= \frac{18}{32} - \frac{10}{32}$$

$$= \frac{8}{32} = \frac{1}{4}$$

Length B is $\frac{1}{4}''$.

Using fractions

Sometimes it is difficult to find a common denominator quickly. In such cases we may use a short cut. Study this example.

Example 3

$\frac{7}{10} + \frac{1}{8} = ?$

Solution

$$\frac{7}{10} + \frac{1}{8} = \frac{(7 \times 8) + (1 \times 10)}{10 \times 8}$$

$$= \frac{56 + 10}{80}$$

$$= \frac{66}{80}$$

$$= \frac{33}{40}$$

The arrows show how we multiply to obtain the numbers in the numerator and in the denominator.

EXERCISES Use the short cut to find a common denominator, then add or subtract. Write the answer in lowest terms.

1. $\frac{1}{4} + \frac{5}{8}$ 2. $\frac{3}{4} - \frac{3}{8}$ 3. $\frac{1}{16} + \frac{7}{8}$
4. $\frac{5}{8} - \frac{5}{16}$ 5. $\frac{7}{16} + \frac{1}{4}$ 6. $\frac{9}{16} - \frac{3}{8}$
7. $\frac{1}{2} + \frac{1}{8}$ 8. $\frac{1}{2} - \frac{3}{16}$ 9. $\frac{7}{10} + \frac{1}{4}$
10. $\frac{5}{8} - \frac{3}{10}$ 11. $\frac{35}{100} + \frac{3}{10}$ 12. $\frac{9}{32} - \frac{1}{4}$

Find the lengths. (Do the work in the parentheses first.)

13. $\frac{5}{8}'' + \frac{1}{16}'' + \frac{1}{8}''$ 14. $\frac{3}{4}'' - (\frac{1}{16}'' + \frac{1}{4}'')$
15. $\frac{3}{16}'' + \frac{5}{8}'' + \frac{1}{16}''$ 16. $(\frac{5}{32}'' + \frac{3}{16}'') - \frac{1}{8}''$
17. $(\frac{7}{16}'' - \frac{3}{8}'') + \frac{1}{4}''$ 18. $\frac{9}{32}'' - (\frac{1}{16}'' + \frac{1}{8}'')$

Find the missing dimensions in the following figures.

19.

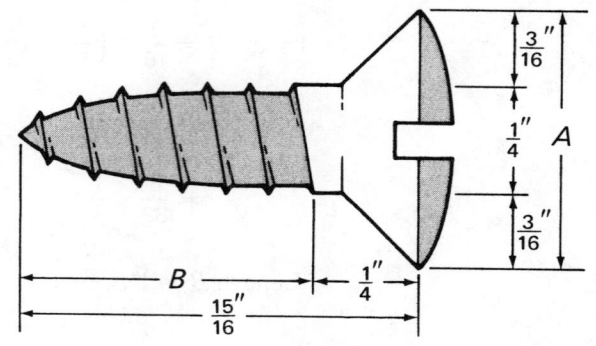

56   **Technical and industrial computations**

20.

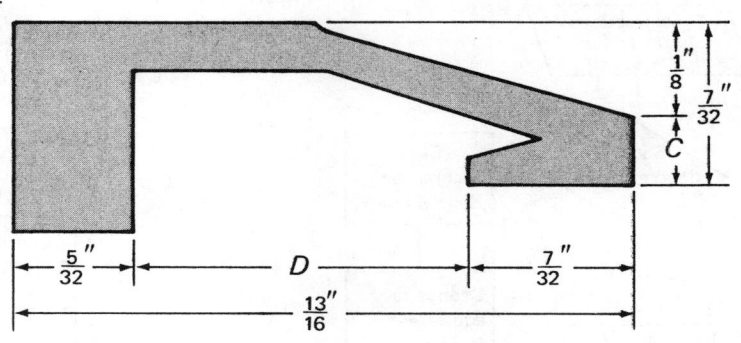

21.

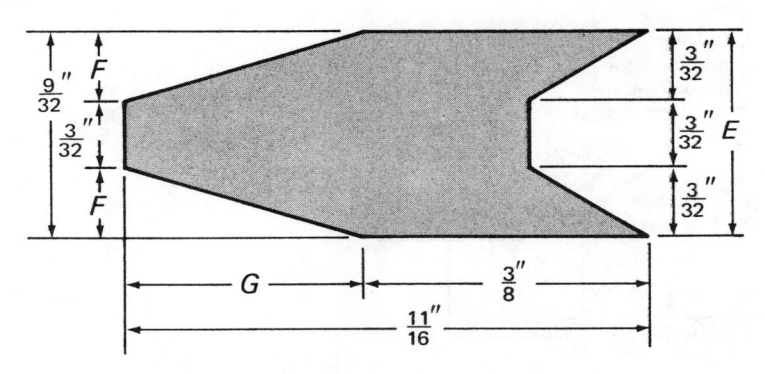

3-3 THE LEAST COMMON DENOMINATOR

To find the total length of the hydraulic valve lifter in Figure 3-3, we have to add three fractions that have different denominators.

Usually we can calculate more easily if we use the smallest number that can be divided evenly by the denominators of the fractions. This number is called the *least common denominator (LCD)*. For the hydraulic lifter in Figure 3-3 the least common denominator is 32. Do you agree that 32 is the smallest number that can be divided by 16, 8, and 32 evenly?

Example 1

Find the length of the hydraulic valve lifter in Figure 3-3.

Using fractions

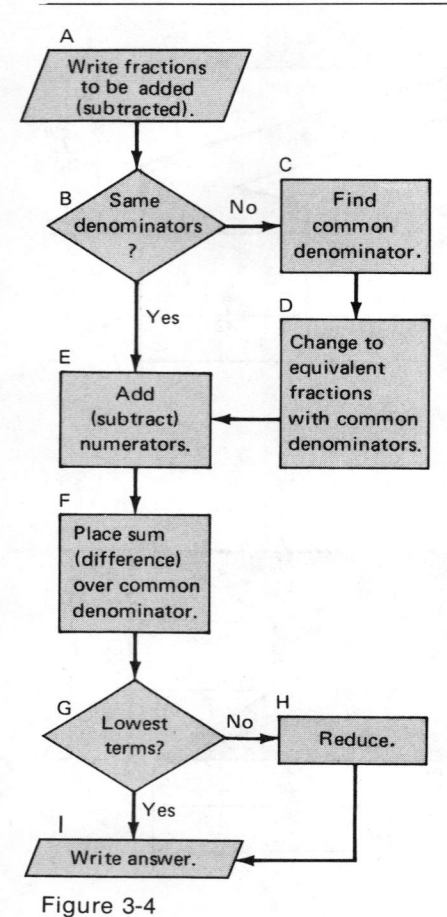

Figure 3-3

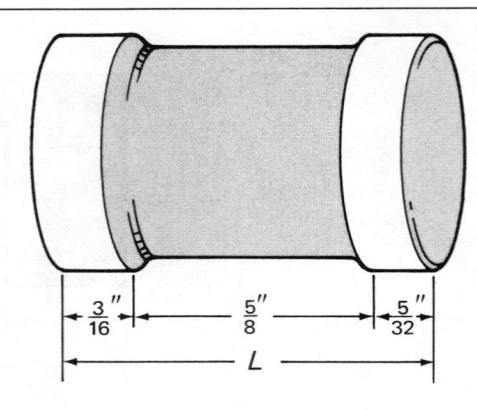

Solution

$$L = \frac{3}{16} + \frac{5}{8} + \frac{5}{32}$$

$$= \frac{6}{32} + \frac{20}{32} + \frac{5}{32} \text{ (32 is LCD)}$$

$$= \frac{6 + 20 + 5}{32}$$

$$= \frac{31}{32}$$

Length L is $\frac{31}{32}''$.

We also use the LCD when we want to subtract fractions that have different denominators.

To help us further in adding and subtracting fractions, we can use the flow chart in Figure 3-4. Follow the steps.

Figure 3-4

Example 2

Find length T on the template.

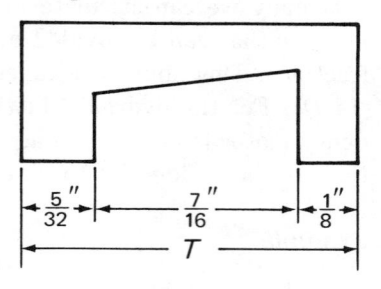

Solution

A. $\frac{5}{32}, \frac{7}{16}, \frac{1}{8}$
B. No
C. 32
D. $\frac{5}{32}, \frac{14}{32}, \frac{4}{32}$
E. $5 + 14 + 4 = 23$
F. $\frac{23}{32}$
G. Yes
H. —
I. $\frac{23}{32}''$

58 Technical and industrial computations

Example 3

Find length r on the template.

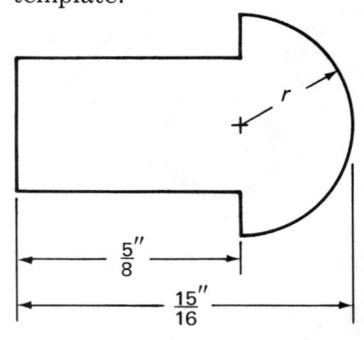

Solution
A. $\frac{15}{16}, \frac{5}{8}$
B. No
C. 16
D. $\frac{15}{16}, \frac{10}{16}$
E. $15 - 10 = 5$
F. $\frac{5}{16}$
G. Yes
H. —
I. $\frac{5}{16}''$

EXERCISES Find the least common denominator for the fractions, then add or subtract. Write the answers in lowest terms.

1. $\frac{3}{4} + \frac{1}{8}$
2. $\frac{9}{16} - \frac{3}{8}$
3. $\frac{23}{32} + \frac{1}{16}$
4. $\frac{45}{64} - \frac{5}{16}$
5. $\frac{1}{2} + \frac{3}{16}$
6. $\frac{13}{16} - \frac{3}{8}$
7. $\frac{43}{64} + \frac{1}{8}$
8. $\frac{9}{32} - \frac{3}{16}$
9. $\frac{3}{10} + \frac{1}{10} + \frac{53}{100}$
10. $\frac{1}{32} + \frac{1}{4}$
11. $\frac{29}{32} - (\frac{5}{64} + \frac{3}{16})$
12. $(\frac{3}{8} + \frac{3}{16}) - \frac{3}{64}$

Use the least common denominator to find the lengths.

13. $\frac{83}{100}'' + \frac{1}{10}'' = $ ___?___
14. $\frac{5}{8}'' + \frac{11}{32}'' = $ ___?___
15. $\frac{17}{100}'' + \frac{1}{10}'' = $ ___?___
16. $\frac{9}{16}'' - \frac{7}{32}'' = $ ___?___
17. $\frac{43}{64}'' - \frac{15}{32}'' = $ ___?___
18. $\frac{11}{16}'' - \frac{37}{64}'' = $ ___?___

TRICKS OF THE TRADE

Here is an easy way to find the center of a piece of round stock. Hold a carpenter's square against the stock. Mark the points 1 and 2 where it touches. Now rotate the stock so that only one mark touches the square. Mark the point 3, as shown. Rotate again, and mark the point 4. Connect points 1 and 3 and points 2 and 4. The center is where the two lines intersect.

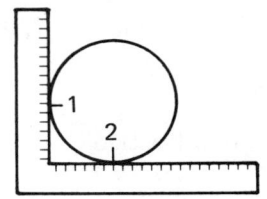

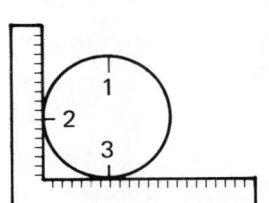

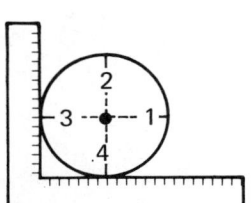

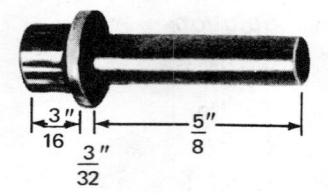

19. What is the total length of the rivet?
20. Find the lengths R and Q on the gasket.
21. Find the lengths, X, Y, and Z on the latch plate. 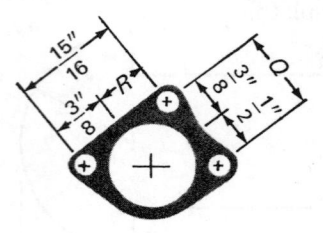
22. Draw a flow chart for finding the least common denominator of two fractions. 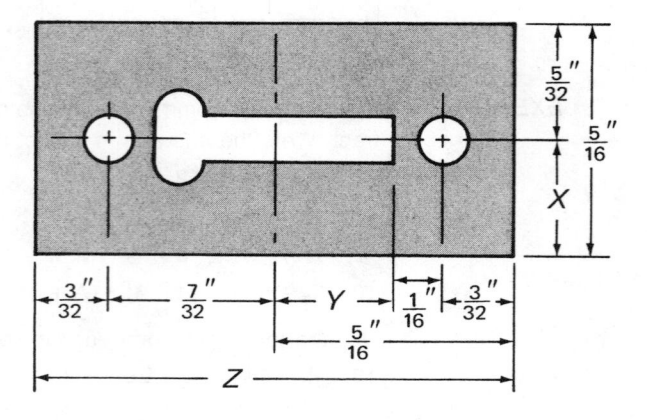

Adding and subtracting mixed numbers

3-4 WHOLE NUMBERS AND MIXED NUMBERS

Most industrial measurements involve not only fractions but also whole numbers and mixed numbers. For example, to manufacture the radio case in Figure 3-5, we must be able to work with numbers like $5\frac{7}{8}''$, $3\frac{3}{16}''$, $2''$, $\frac{3}{4}''$, and so on.

Example 1

Change 2 to a fraction.

60   Technical and industrial computations

Solution

We can write any whole number as a fraction by placing it over 1 as the denominator. Thus, $2 = \frac{2}{1}$.

Example 2

Change $5\frac{7}{8}$ to a fraction.

Solution

Method 1 Write the whole number as a fraction, find the common denominator, and then add.

$$5\frac{7}{8} = 5 + \frac{7}{8}$$

$$= \frac{5}{1} + \frac{7}{8}$$

$$= \frac{40}{8} + \frac{7}{8}$$

$$= \frac{40 + 7}{8}$$

$$= \frac{47}{8}$$

Figure 3-5

Method 2 Multiply the whole number by the denominator, then add the numerator.

$$5\frac{7}{8} = \frac{(5 \times 8) + 7}{8}$$

$$= \frac{40 + 7}{8}$$

$$= \frac{47}{8}$$

To simplify our work we often need to change fractions to whole numbers or mixed numbers.

Example 3

Simplify $\frac{32}{8}$.

Solution

Method 1 Write equivalent fractions using the same denominator.

$$\frac{32}{8} = \frac{8 + 8 + 8 + 8}{8}$$

$$= \frac{8}{8} + \frac{8}{8} + \frac{8}{8} + \frac{8}{8}$$

$$= 1 + 1 + 1 + 1 = 4$$

Method 2 Divide the numerator by the denominator. If there is a remainder, use it as the numerator.

$$\frac{32}{8} = 8\overline{)32} = 4$$
$$\phantom{\frac{32}{8} = 8)}\underline{-32}$$
$$\phantom{\frac{32}{8} = 8)\,\,}0$$

Example 4

Simplify $\frac{43}{16}$.

Solution

Method 1 Write equivalent fractions using the same denominator.

$$\frac{43}{16} = \frac{16 + 16 + 11}{16}$$

$$= \frac{16}{16} + \frac{16}{16} + \frac{11}{16}$$

$$= 1 + 1 + \frac{11}{16}$$

$$= 2\frac{11}{16}$$

Method 2 Divide the numerator by the denominator. If there is a remainder, use it as the numerator.

$$\frac{43}{16} = 16\overline{)43} = 2\frac{11}{16}$$
$$\phantom{\frac{43}{16} = 16)}\underline{-32}$$
$$\phantom{\frac{43}{16} = 16)\,\,}11$$

EXERCISES Change to fractions.

1. 6
2. 8
3. $3\frac{3}{4}$
4. $2\frac{3}{16}$
5. $4\frac{7}{8}''$
6. $7\frac{1}{16}''$
7. $2\frac{3}{32}''$
8. $4\frac{11}{32}''$

Simplify.

9. $\frac{24}{8}$
10. $\frac{64}{16}$
11. $\frac{29}{16}$
12. $\frac{55}{32}$
13. $\frac{15}{4}''$
14. $\frac{39}{8}''$
15. $\frac{65}{32}''$
16. $\frac{29}{11}''$

3-5 ADDING MIXED NUMBERS

A wire tunnel is to be constructed from sheet metal with the dimensions shown in Figure 3-6. In order to cut the strip of sheet metal to the correct width, we need to find the *perimeter*, or distance around the outer face.

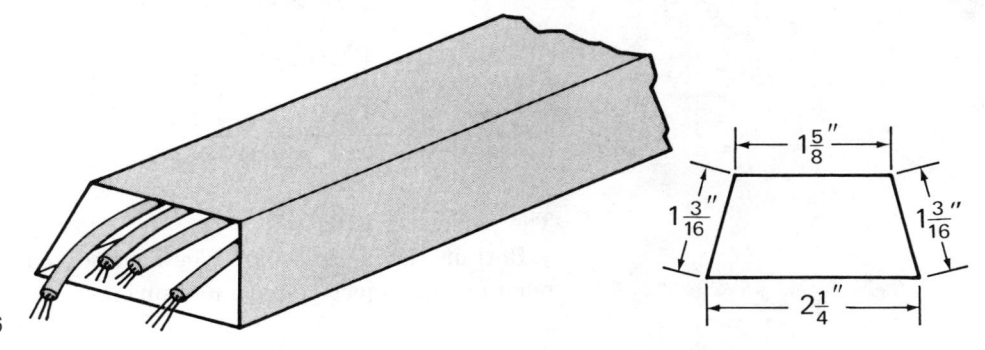

Figure 3-6

Using fractions

Perimeter = $2\frac{1}{4}'' + 1\frac{3}{16}'' + 1\frac{5}{8}'' + 1\frac{3}{16}''$

There are two ways of adding mixed numbers.

Method 1 Change the mixed numbers to fractions, then add. Simplify the answer.

Example 1

Find the perimeter of the wire tunnel in Figure 3-6.

Solution

$$P = 2\frac{1}{4} + 1\frac{3}{16} + 1\frac{5}{8} + 1\frac{3}{16}$$
$$= \frac{9}{4} + \frac{19}{16} + \frac{13}{8} + \frac{19}{16}$$
$$= \frac{36}{16} + \frac{19}{16} + \frac{26}{16} + \frac{19}{16}$$
$$= \frac{36 + 19 + 26 + 19}{16}$$
$$= \frac{100}{16}$$
$$= 6\frac{1}{4}$$

The perimeter is $6\frac{1}{4}''$.

Method 2 Add the whole numbers and the fractions separately, then combine to find the answer.

Example 2

Find the perimeter of the wire tunnel in Figure 3-6.

Solution

Perimeter $\begin{cases} 2\frac{1}{4} = 2 + \frac{1}{4} = 2 + \frac{4}{16} \\ 1\frac{3}{16} = 1 + \frac{3}{16} = 1 + \frac{3}{16} \\ 1\frac{5}{8} = 1 + \frac{5}{8} = 1 + \frac{10}{16} \\ + 1\frac{3}{16} = 1 + \frac{3}{16} = 1 + \frac{3}{16} \\ \hline 5 + \frac{20}{16} = 5 + 1 + \frac{4}{16} = 6\frac{1}{4} \end{cases}$

The perimeter is $6\frac{1}{4}''$.

Both methods are useful. The method you use will usually depend on the object you are measuring.

EXERCISES Find the sums.

1. $5\frac{1}{2} + 4\frac{3}{8}$
2. $2\frac{5}{8} + 7\frac{3}{32}$
3. $6\frac{7}{10} + 12\frac{47}{100}$
4. $9\frac{27}{32} + 3\frac{53}{64}$
5. $\frac{7}{8} + 10\frac{9}{10}$
6. $2\frac{1}{4}'' + 4\frac{1}{8}'' + 5\frac{3}{16}''$
7. $8\frac{7}{64}'' + 1\frac{9}{32}'' + 5\frac{5}{8}''$
8. $3\frac{3}{10}'' + \frac{3}{4}'' + 4\frac{1}{2}''$
9. $8\frac{15}{16}'' + 2\frac{7}{8}'' + 5\frac{3}{4}''$
10. $5\frac{5}{16}'' + 4\frac{7}{8}'' + 7\frac{3}{4}'' + 9\frac{1}{2}''$

11. Find the total length of the transmission gears.

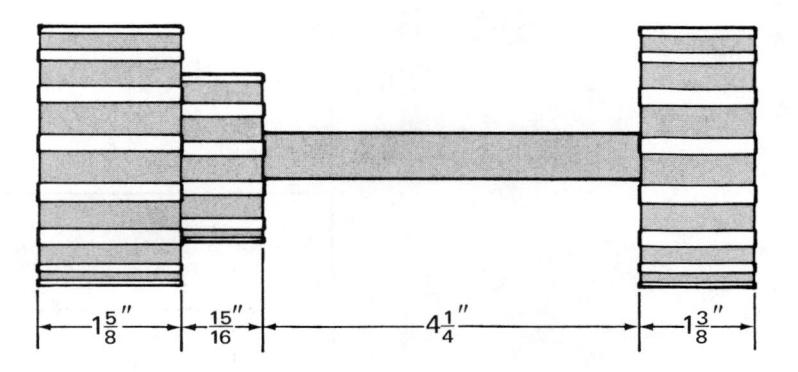

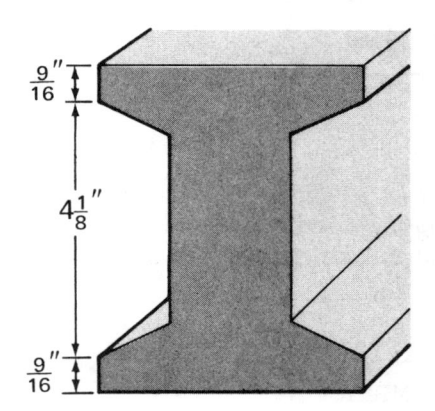

12. Find the height of the I beam.

13. How wide is the kitchen chopping block?

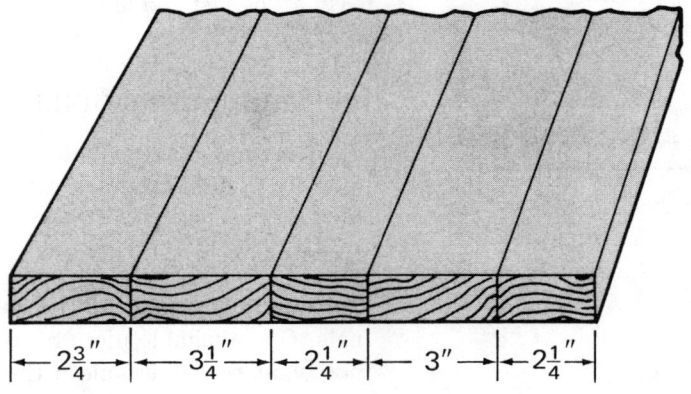

14. How long is the hammerhead?

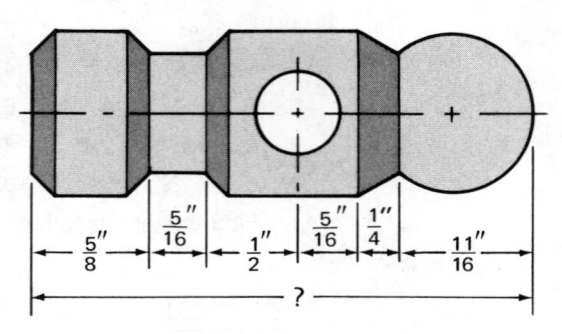

15. Find lengths A and B on the light switch plate.

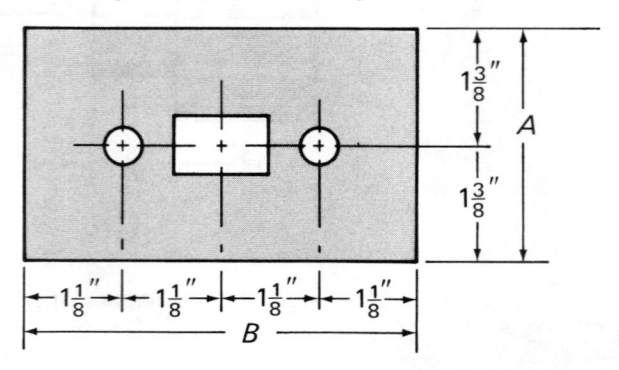

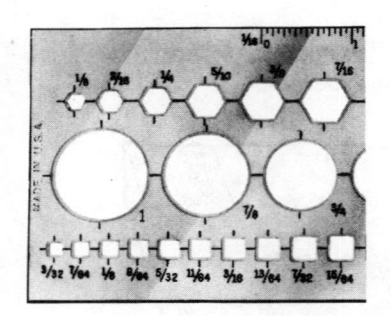

16. Drafters use a template when they draw common shapes in various sizes. Part of a drafting template is shown here. If the distance between each shape is $\frac{3}{32}''$, what is the distance between

   a. the first and last square?   b. the first and last circle?

3-6 SUBTRACTING MIXED NUMBERS

Subtracting mixed numbers is another skill that is used frequently in industrial work.

Example 1

An electrician needs $12\frac{1}{2}''$ of insulated wire to connect two terminals. One available piece measures $27\frac{3}{4}''$, but the electrician does not want to cut it unless there will be at least $15\frac{1}{8}''$ left to do another job. Should the electrician cut the wire?

66   Technical and industrial computations

Solution

To answer this question, we need to subtract $12\frac{1}{2}$ from $27\frac{3}{4}$, as shown here.

$$27\frac{3}{4} = 27\frac{3}{4}$$
$$-12\frac{1}{2} = -12\frac{2}{4}$$
$$\overline{\phantom{-12\frac{1}{2}}\ \ 15\frac{1}{4}}$$

Since there will be $15\frac{1}{4}''$ left, the electrician can make the cut.

To subtract mixed numbers, we change the fractions to equivalent fractions with the same denominator. Then we subtract the fractions and the whole numbers. Finally, we add the difference of the fractions to the difference of the whole numbers. (Remember: $15 + \frac{1}{4} = 15\frac{1}{4}$)

Example 2

In modifying an automobile engine the mechanic bores the cylinder to $4\frac{1}{16}''$ (see Figure 3-7). The original bore was $3\frac{7}{8}''$. By how much has the bore been enlarged?

Solution

To find the difference in the size of the bore before and after the machining, we must subtract $3\frac{7}{8}$ from $4\frac{1}{16}$, as shown here.

$$4\frac{1}{16} = 4\frac{1}{16} = 3\frac{17}{16}$$
$$-3\frac{7}{8} = -3\frac{14}{16} = -3\frac{14}{16}$$
$$\overline{\phantom{-3\frac{7}{8} = -3\frac{14}{16} =\ }\ \frac{3}{16}}$$

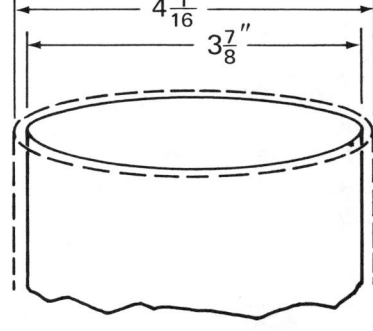

Figure 3-7

The bore was enlarged by $\frac{3}{16}''$.

Notice in Example 2 that in order to complete the subtraction, we must change $4\frac{1}{16}$ to $3\frac{17}{16}$. We can do this because

$$4\frac{1}{16} = 4 + \frac{1}{16} = 3 + 1 + \frac{1}{16} = 3 + \frac{16}{16} + \frac{1}{16} = 3 + \frac{17}{16}$$

Using fractions 67

EXERCISES Subtract.
1. $5\frac{7}{8} - \frac{3}{4}$
2. $3\frac{7}{16} - \frac{3}{8}$
3. $6\frac{15}{16} - \frac{5}{8}$
4. $9\frac{11}{16} - 8\frac{5}{8}$
5. $1\frac{5}{32} - \frac{1}{16}$
6. $2\frac{9}{16} - 1\frac{1}{4}$
7. $3\frac{5}{16}'' - 1\frac{1}{8}''$
8. $4\frac{3}{8}'' - 3\frac{1}{4}''$
9. $6\frac{7}{16}'' - 3\frac{3}{8}''$
10. $4 - 2\frac{3}{8}$
11. $3 - 1\frac{5}{16}$
12. $9\frac{1}{8} - 3\frac{3}{4}$
13. $2\frac{1}{4}'' - 1\frac{7}{8}''$
14. $16\frac{5}{8}'' - 12\frac{13}{16}''$
15. $17\frac{3}{8}'' - 15\frac{9}{16}''$

16. A plumber has a piece of plastic tubing $18\frac{1}{2}''$ long and plans to use a piece $12\frac{3}{4}''$ long. How many inches of unused tubing will be left?

17. To check the layout of the bracket shown below, a machinist measured distances *A* and *B*. Find *A* and *B*.

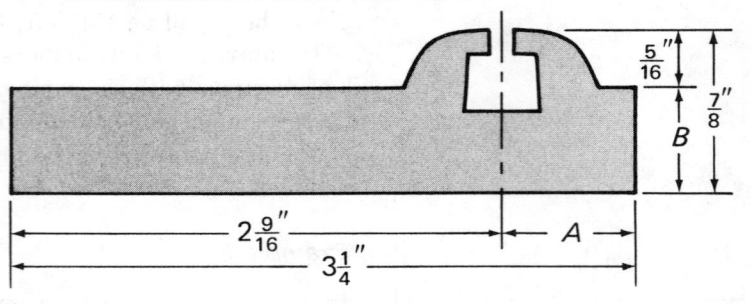

18. A sheet worker has to drill a hole in a metal plate according to the diagram shown. The worker centerpunched $22\frac{7}{8}''$ from the left edge and $1\frac{3}{8}''$ from the upper edge of the plate and then, to check the accuracy of the punch, measured distance *C* and distance *D*. What should these measurements be?

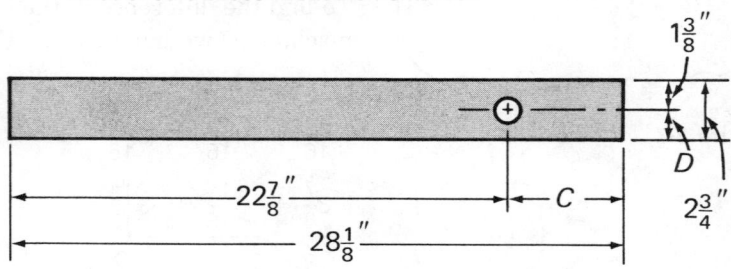

Dimensions in feet and inches

3-7 CHANGING INCHES TO FEET AND INCHES

So far the dimensions that you have studied have been given in inches only. In industry most lengths up to 72″ are stated in inches only. Lengths greater than 72″ are usually expressed in feet and inches. For example, a strip of plastic may measure 21″ wide (*not* 1¾′), but a door frame may be 7′2″ high (*not* 86″). However, those who plan to work in industry should be able to make calculations with dimensions in either form.

Example 1

A welder must butt-weld three sheets of sheet metal. The lengths of the sheets are 22″, 18″, and 41″. What will be the total length of the welded sheet?

Solution

22″ + 18″ + 41″ = 81″

Since 81″ is greater than 72″, we normally express this length as 6′9″. Therefore, the welded sheet will measure 6′9″ in length.

Example 2

A section of extruded plastic tubing $17\frac{1}{2}$″ long is cut from a piece $6'4\frac{3}{4}$″ long. What is the length of the remaining piece of tubing?

Solution

In order to subtract $17\frac{1}{2}$″ from $6'4\frac{3}{4}$″, we change $6'4\frac{3}{4}$″ to $76\frac{3}{4}$″.

$$\begin{array}{r} 76\frac{3}{4}'' \\ -17\frac{1}{2}'' \\ \hline 59\frac{1}{4}'' \end{array}$$

Thus, a piece of tubing $59\frac{1}{4}$″ long is left.

When most of the dimensions are given in both feet and inches, it is usually easier to express all of them in feet and inches. In this case we work with both units at once.

Example 3

In installing an automatic car wash, a plumber joins three sections of pipe. The sections are 6′8″, 11′6″, and 8′4″ in length. What is the total length of the pipe?

Solution

To make the work easier, we keep the numbers in columns.

$$\begin{array}{r} 6'\ 8'' \\ 11'\ 6'' \\ +\ \ 8'\ 4'' \\ \hline 25'18'' = 25' + 1' + 6'' \\ = 26'6'' \end{array}$$

The total length is 26'6".
Do you see why we change 18" to 1'6" and add it to 25'?

Example 4

A bookbinding machine used 237'3" of thread from a new spool to finish a job. If a spool holds 1000' of thread, how much thread is left on the spool?

Solution

In order to subtract 237'3" from 1000', we change 1000' to 999'12".
Thus,

$$\begin{array}{rl} 1000' = & 999'12'' \\ -\ 237'3'' = & -237'3'' \\ \hline & 762'9'' \end{array}$$

There are 762'9" of thread left.

EXERCISES Add or subtract. Express answers in inches or in feet and inches.

1. $14'' + 3\frac{3}{4}'' + 6''$
2. $19'' + 27\frac{1}{2}''$
3. $38\frac{3}{8}'' + 46\frac{3}{4}''$
4. $6'4'' - 26''$
5. $9'8\frac{3}{8}'' - 24\frac{1}{4}''$
6. $7'2'' - 10\frac{1}{2}''$
7. $13' + 6'3\frac{3}{4}'' + 5\frac{1}{4}''$
8. $9' + 2'4\frac{1}{2}'' + 5\frac{1}{2}''$
9. $8'4'' - (32'' + 6\frac{1}{2}'')$
10. $13'\frac{1}{2}'' - (36'' + 12\frac{1}{4}'')$

11. A plumber needs pieces of copper tubing in the following lengths: 8'6", 12'4", 9'8", and 7'3". How much pipe does the plumber need in all?

12. The inventory sheet shows that the hospital stockroom has only three pieces of clear plastic tubing. The lengths are 16'4$\frac{1}{2}$", 24'5$\frac{3}{4}$", and 8'. What is the combined length of the three pieces?

13. For a new industrial center, 350' of pipe must be laid. So far 195'8" of pipe has been laid. How much more pipe must be put in place?

14. The following lengths of rudder cable were required in an airplane: 10'3", 13'9", 7'9", and 11'5". Find the total length needed:

15. A piece 13′6″ long is cut from a 50′ plastic hose. How much of the hose is left?

16. From a sheet of fiberglass 12′ long, three pieces are to be cut. The pieces are to be 11″, 6′3¾″, and 17″ long. What will be the length of the remaining piece?

17. A spool of copper wire contains 25′ of wire. Four pieces are cut off. Their lengths are 12½″, 14¾″, 20⅝″, and 36 9/16″. How much wire remains?

3-8 CONSERVING MATERIALS

Because raw materials are expensive, industries try to minimize the amount of waste. Sometimes the waste material is converted into another product that can be sold. For example, cardboard is made from tree bark at a paper mill. In other branches of industry waste material is recycled. Old newspapers, for example, are broken down and converted into other paper products.

Planning how best to use waste material is an important part of industry. It is also a habit that you should try to develop now.

Example

Two wedge-shaped metal brackets are to be made according to the specifications in Figure 3-8.

Figure 3-8

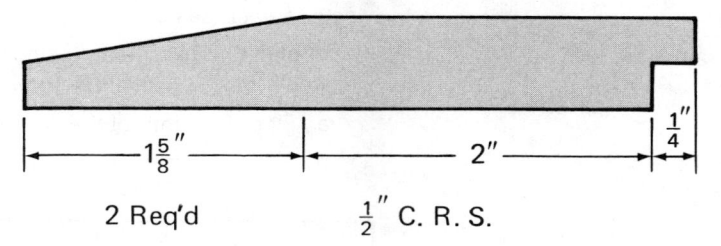

|← 1⅝″ →|← 2″ →| ¼″

2 Req'd   ½″ C. R. S.

Solution

The machinist would find three important facts about the braces in Figure 3-8:

1. The length: 1⅝″ + 2″ + ¼″
2. The number to be made: 2 (2 Req'd)
3. The material: ½″ cold-rolled steel (CRS)

Next the machinist would make a sketch showing all dimensions and allowances. He or she will allow ⅛″ for finishing each end. He or she will also allow 1/16″ for separating the pieces. The final drawing would look like Figure 3-9.

Figure 3-9

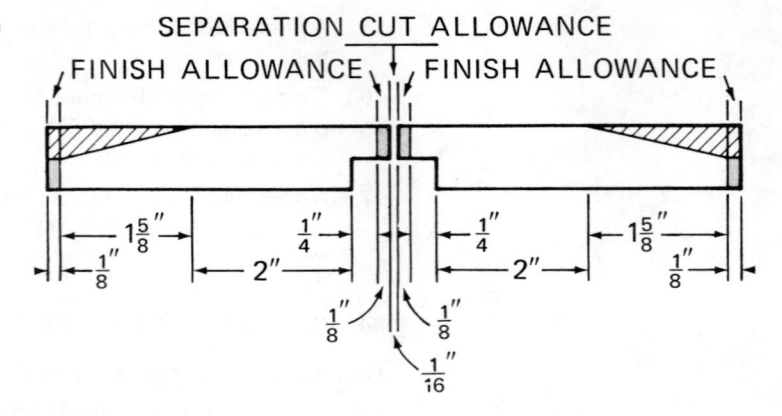

SEPARATION CUT ALLOWANCE
FINISH ALLOWANCE   **FINISH ALLOWANCE**

[dimensions: $1\frac{5}{8}"$, $\frac{1}{8}"$, $2"$, $\frac{1}{4}"$, $\frac{1}{4}"$, $2"$, $1\frac{5}{8}"$, $\frac{1}{8}"$, $\frac{1}{8}"$, $\frac{1}{8}"$, $\frac{1}{16}"$]

From the drawing the machinist can determine the length of the materials. Can you?

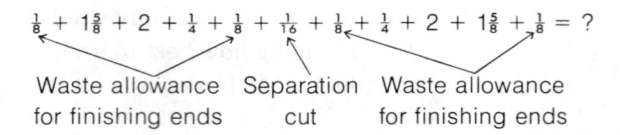

$$\frac{1}{8} + 1\frac{5}{8} + 2 + \frac{1}{4} + \frac{1}{8} + \frac{1}{16} + \frac{1}{8} + \frac{1}{4} + 2 + 1\frac{5}{8} + \frac{1}{8} = ?$$

Waste allowance for finishing ends   Separation cut   Waste allowance for finishing ends

Do you see how careful planning can reduce the amount of waste?

EXERCISES Answer the questions. Discuss how planning reduces the amount of waste in each case.

1. In planning to cut some stock, a printer found that the only waste would be the two strips shown.

    a. Find the width of the waste strip along the bottom.
    b. Find the width of the waste strip along the side.

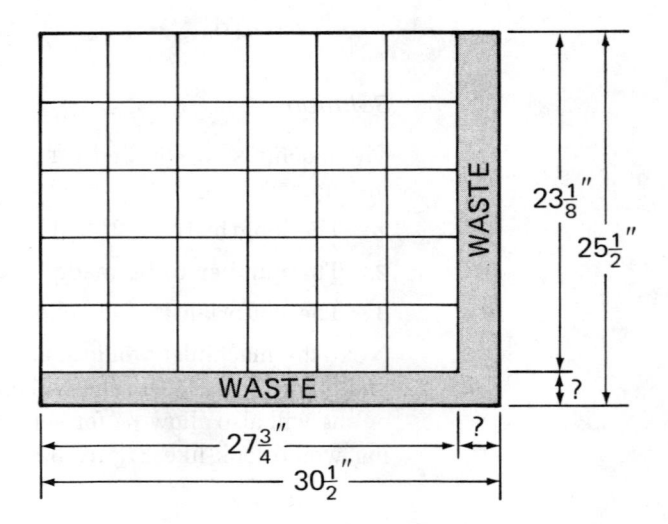

[diagram: rectangle with WASTE strips on right side and bottom; dimensions $23\frac{1}{8}"$, $25\frac{1}{2}"$, $27\frac{3}{4}"$, $30\frac{1}{2}"$, with ? marks]

72 Technical and industrial computations

2. What length of steel stock is needed to make the U bracket? Allow $\frac{3}{16}''$ for each bend and $\frac{1}{4}''$ on each end for finishing.

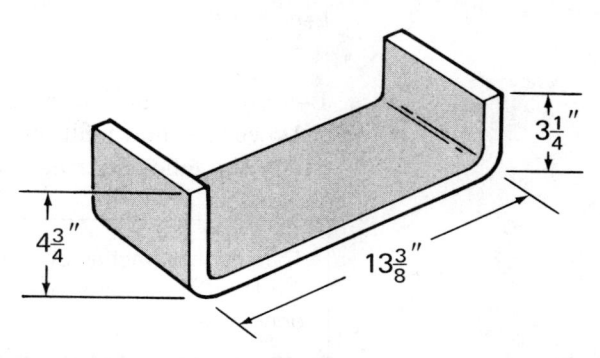

3. An electronics kit contains a spool of insulated wire 18′ long. The following lengths are to be cut: $12\frac{5}{8}''$, $33\frac{3}{16}''$, $4\frac{7}{8}''$, $25\frac{1}{2}''$. The instruction manual suggests allowing $\frac{1}{16}''$ for each cut.

   a. Draw a diagram to show how to cut the wire.
   b. What length will be cut from the spool?
   c. How much wire will remain on the spool after cutting?

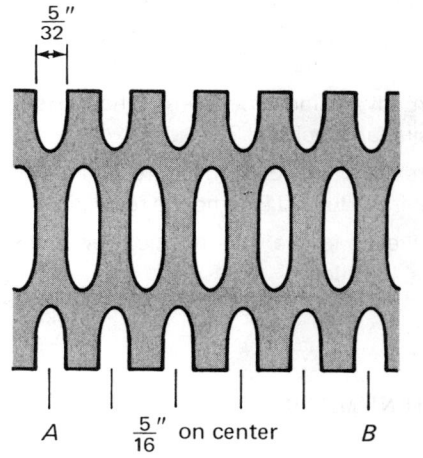

Figure 3-10

Multiplying fractions

3-9 MULTIPLYING A FRACTION AND A WHOLE NUMBER

Figure 3-10 shows a part of metal grille that covers the fan in an air conditioner. A grille like this is made in an automatic perforating machine according to specified dimensions.

To find the distance from A to B, we could add $\frac{5}{16}''$ five times. For example,

$$\text{Length } AB = \frac{5}{16} + \frac{5}{16} + \frac{5}{16} + \frac{5}{16} + \frac{5}{16} = \frac{25}{16} = 1\frac{9}{16}''$$

However, we usually find it easier and faster to multiply. For example,

$$\text{Length } AB = 5 \times \frac{5}{16} = \frac{5 \times 5}{16} = \frac{25}{16} = 1\frac{9}{16}''$$

Example

Suppose there were 47 slots between A and B in Figure 3-10. Find length AB.

Solution

Length $AB = 47 \times \dfrac{5}{16} = \dfrac{47 \times 5}{16} = \dfrac{235}{16} = 14\dfrac{11}{16}$

Length AB would measure $14\dfrac{11}{16}''$.

Do you see how multiplication helps us find the answer quickly? Study the following rule.

> To multiply a fraction by a whole number, multiply the numerator of the fraction by the whole number and place the product over the denominator.

EXERCISES Multiply. Write the answers in lowest terms.

1. $7 \times \dfrac{1}{8}$
2. $5 \times \dfrac{3}{4}$
3. $\dfrac{3}{10} \times 6$
4. $\dfrac{5}{16} \times 8$
5. $\dfrac{9}{64} \times 2$
6. $\dfrac{3}{100} \times 25$
7. $36 \times \dfrac{3}{64}$
8. $2 \times \dfrac{3}{32}$
9. $100 \times \dfrac{5}{8}''$
10. $17 \times \dfrac{7}{10}''$
11. $\dfrac{7}{8}'' \times 8$
12. $\dfrac{7}{16}'' \times 16$

13. What is the total weight of 7 cast-iron pulleys, if each weighs $\dfrac{3}{4}$ lb?

14. A surveyor marks off 12 adjacent lots. The frontage of each lot measures $\dfrac{1}{10}$ mi. What is the total frontage in miles?

15. A printer sets the type for an advertising brochure in $\dfrac{3}{4}$ hour. How long would it take to set 5 similar brochures?

16. The distance between the threads of the screw in Figure 3-11 is $\dfrac{3}{32}''$. How far does the screw travel if it is turned 8 complete revolutions?

17. A tile mason uses $\dfrac{7}{8}''$ square tiles to form a wall mosaic. How wide is the wall if the mason uses 184 tiles?

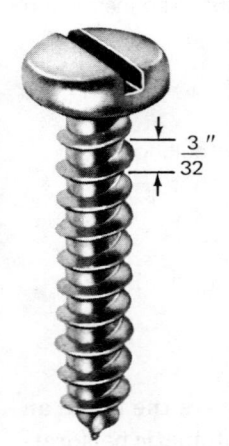

Figure 3-11

3-10 MULTIPLYING A MIXED NUMBER AND A WHOLE NUMBER

Airplanes undergo a great deal of stress when in flight. To provide strength in parts like the wing or body, aerospace manufacturers use rivets during assembly.

Figure 3-12 shows a portion of a wing panel with holes for the rivets. A "Caution" stripe is to be painted as shown. How long will the stripe be?

To find AB, we multiply $1\dfrac{3}{8}''$ by 7, or $7 \times 1\dfrac{3}{8}$. There are two methods we can use.

Figure 3-12

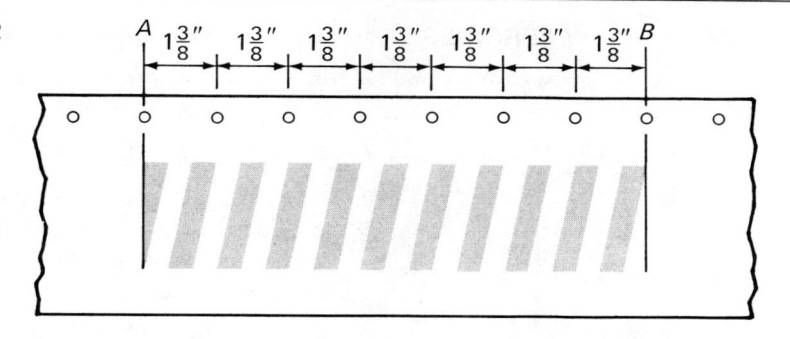

Method 1 The mixed number $1\frac{3}{8}$ can be changed to a fraction, $1\frac{3}{8} = \frac{11}{8}$. Now we multiply:

$$7 \times 1\frac{3}{8} = 7 \times \frac{11}{8} = \frac{77}{8} = 9\frac{5}{8}$$

The stripe between A and B will be $9\frac{5}{8}''$.

Method 2 Since $1\frac{3}{8} = 1 + \frac{3}{8}$, we can multiply 1 by 7 and $\frac{3}{8}$ by 7, then add the two results. Thus,

$$7 \times 1\frac{3}{8} = (7 \times 1) + \left(7 \times \frac{3}{8}\right)$$
$$= 7 + \frac{21}{8}$$
$$= 7 + 2\frac{5}{8}$$
$$= 9\frac{5}{8}$$

Again, the stripe will be $9\frac{5}{8}''$.

Would you agree with the following rule?

> To multiply a mixed number by a whole number, we can work in two ways:
> 1. a. Change the mixed number to a fraction; then
>   b. multiply by the whole number. Or
> 2. a. write the mixed number as the sum of a whole number and a fraction; then
>   b. multiply each part by the whole number; and
>   c. add the products.

Using fractions

EXERCISES Multiply. Reduce the answers to lowest terms.

1. $1\frac{7}{8} \times 4$
2. $2\frac{7}{8} \times 5$
3. $3\frac{5}{16} \times 10$
4. $8 \times 2\frac{7}{8}$
5. $22\frac{1}{2} \times 11$
6. $17 \times 12\frac{1}{8}$
7. $35\frac{2}{4}'' \times 8$
8. $20 \times 11\frac{7}{8}''$
9. $5 \times 7\frac{15}{16}''$

10. A furniture refinisher is calculating the amount of material needed to repair the 4 legs of a stool. The refinisher finds that, allowing $\frac{1}{4}''$ at each end for waste, the repair will need a piece of lumber $13\frac{3}{8}''$ long for each leg.
    a. How many inches of lumber does the refinisher need for 4 legs?
    b. Give the answer in feet and inches.

11. The height, or riser, of a single step in a flight of stairs is $7\frac{1}{2}''$. There are 14 steps in the flight.
    a. What is the height of the flight of stairs in inches?
    b. What is the height of the flight of stairs in feet and inches?

12. Find the height and width of the storage area of the small-parts chest.

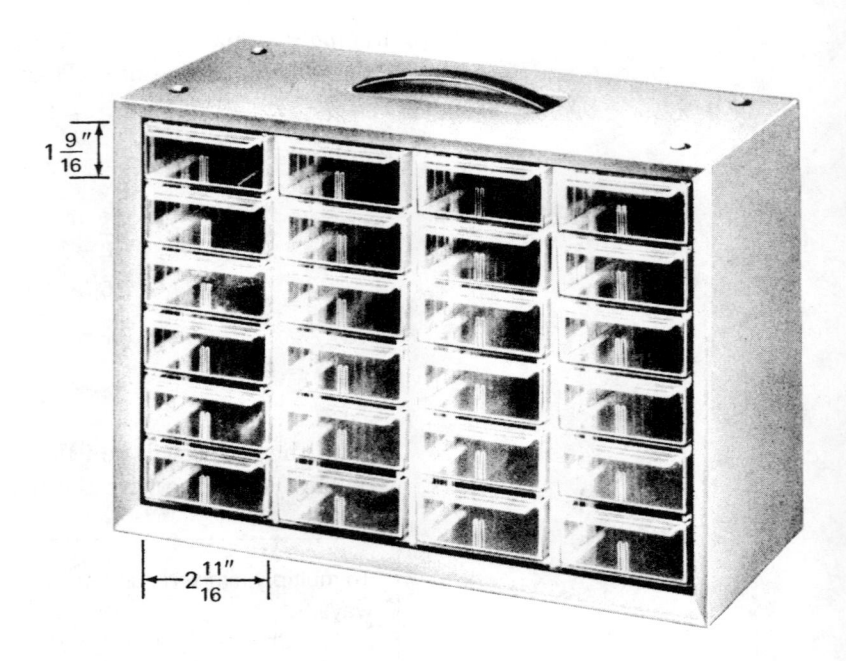

13. *L* is the length of the side of a dovetailed drawer.
    a. Find *L*.
    b. Can a matching piece be made from stock that is $9\frac{3}{8}''$ wide?

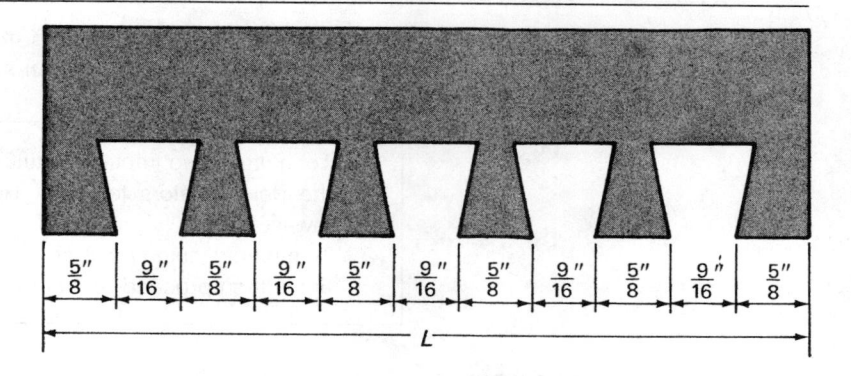

3-11 MULTIPLYING TWO FRACTIONS

Many situations in industry involve multiplying two fractions. Study the following examples.

Example 1

One turn of the handle of a bench vise in Figure 3-13 closes the jaws by $\frac{3}{16}''$. How much do the jaws close if we give the handle only a half-turn?

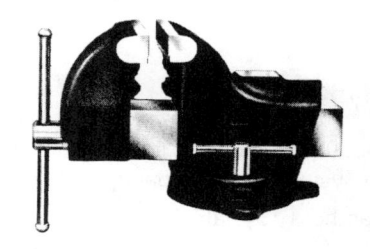

Figure 3-13

Solution

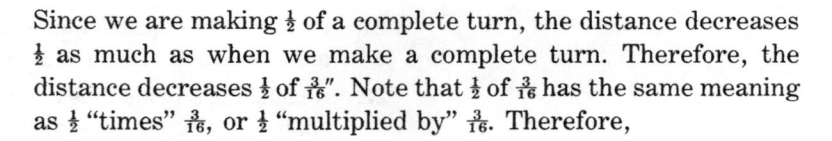

Since we are making $\frac{1}{2}$ of a complete turn, the distance decreases $\frac{1}{2}$ as much as when we make a complete turn. Therefore, the distance decreases $\frac{1}{2}$ of $\frac{3}{16}''$. Note that $\frac{1}{2}$ of $\frac{3}{16}$ has the same meaning as $\frac{1}{2}$ "times" $\frac{3}{16}$, or $\frac{1}{2}$ "multiplied by" $\frac{3}{16}$. Therefore,

$$\frac{1}{2} \text{ of } \frac{3}{16} = \frac{1}{2} \times \frac{3}{16} = \frac{1 \times 3}{2 \times 16} = \frac{3''}{32}$$

A half-turn of the vise handle closes the jaws by $\frac{3}{32}''$.

Example 2

If a printing press uses $4\frac{3}{4}$ ounces of ink in 1 hour, how much will it use during a $3\frac{1}{2}$-hour run?

Solution

Since the press would use $3\frac{1}{2}$ times more ink than in 1 hour, we can find the amount of ink used by multiplying $3\frac{1}{2} \times 4\frac{3}{4}$. Thus,

$$3\frac{1}{2} \times 4\frac{3}{4} = \frac{7}{2} \times \frac{19}{4} = \frac{7 \times 19}{2 \times 4} = \frac{133}{8} = 16\frac{5}{8}$$

The printing press will use $16\frac{5}{8}$ oz of ink.

From these examples we can state the following rules.

> 1. To multiply two fractions, multiply the numerators together and the denominators together. Reduce the resulting fraction to lowest terms.
> 2. To multiply mixed numbers, change the mixed numbers to fractions, then proceed as above.

EXERCISES

Multiply. Write the answers in lowest terms.

1. $\frac{1}{2} \times \frac{1}{4}$
2. $\frac{5}{16} \times \frac{14}{16}$
3. $\frac{1}{3} \times \frac{7}{8}$
4. $\frac{9}{10} \times \frac{4}{5}$
5. $\frac{22}{7} \times \frac{3}{8}$
6. $\frac{7}{2} \times \frac{5}{3}$
7. $3\frac{1}{2} \times 2\frac{5}{8}''$
8. $3\frac{1}{2}'' \times 3\frac{1}{5}$
9. $3\frac{1}{10} \times 2\frac{1}{10}$
10. $\frac{7}{9} \times \frac{2}{3}$
11. $\frac{3}{5} \times \frac{3}{4}$
12. $\frac{15}{18} \times \frac{5}{3}$
13. $\frac{7}{3} \times \frac{21}{5}$
14. $6\frac{7}{8}'' \times 3\frac{4}{5}$
15. $7\frac{1}{5} \times 2\frac{1}{2}''$

Find each of the following:

16. $2\frac{1}{2}$ times 4
17. Half of $4\frac{7}{8}$
18. $1\frac{1}{4}$ times $\frac{7}{8}''$
19. A third of $\frac{11}{64}''$

20. The weight of 1 gal of water is about $8\frac{1}{4}$ lb. What is the weight of the water in an automobile cooling system with a capacity of $4\frac{1}{2}$ gal?
21. An automobile gasoline tank can hold $15\frac{1}{2}$ gal. How many gallons are in the tank when it is $\frac{3}{4}$ full?
22. An automatic bottler has a capacity of $23\frac{1}{2}$ gal per min. How much soda can it bottle in $3\frac{1}{2}$ min?

TRICKS OF THE TRADE

Here is a method you can use to help you subtract lengths quickly. Suppose you want to cut a piece $2\frac{7}{8}''$ wide from a board $5\frac{3}{8}''$ wide. You may think like this:

$$5\frac{3}{8} - 2\frac{7}{8} = \underline{\quad ? \quad}$$

$$\left(5\frac{3}{8} + \frac{1}{8}\right) - \left(2\frac{7}{8} + \frac{1}{8}\right) = \underline{\quad ? \quad}$$

$$5\frac{1}{2} - 3 = 2\frac{1}{2}$$

The piece left is $2\frac{1}{2}''$ wide.

Can you see why this method works?

Technical and industrial computations

23. If it costs $1\frac{3}{4}$¢ per kilowatt-hour of electricity, what is the cost for $8\frac{1}{2}$ kWh?

24. The weight of a sheet of aluminum is $16\frac{3}{4}$ lb. If this sheet is cut and $\frac{2}{3}$ is put back in stock, what is the weight of the piece that is used?

25. Using a machinist's rule, an engineer found the dimensions for a sheet metal patch. The patch must be $9\frac{3}{10}''$ by $7\frac{7}{10}''$.

    a. What is the area of the patch in square inches? (Area = length × width)

    b. If it costs $3\frac{1}{2}$¢ per square inch to rustproof the material, what will it cost to treat this patch?

3-12 CANCELLATION

If one bundle of shingles weighs $23\frac{3}{4}$ lb, what do $6\frac{1}{5}$ bundles weigh? To find the answer, we multiply $23\frac{3}{4}$ by $6\frac{1}{5}$.

$$6\frac{1}{5} \times 23\frac{3}{4} = \frac{31}{5} \times \frac{95}{4} = \frac{31 \times 95}{5 \times 4} = \frac{2945}{20} = 147\frac{1}{4}$$

Thus, $6\frac{1}{5}$ bundles will weigh $147\frac{1}{4}$ lb.

Sometimes we can save time in multiplying fractions by using a short cut. In the example above, notice that

$$\frac{31}{5} \times \frac{95}{4} = \frac{31 \times 95}{5 \times 4}$$

We can reduce $\frac{31 \times 95}{5 \times 4}$ before we multiply by dividing both the numerator and denominator by 5 because

$$\frac{31 \times (19 \times 5)}{5 \times 4} = \frac{31 \times 19 \times 5}{5 \times 4}$$

$$= \frac{31 \times 19}{4} \times \frac{5}{5}$$

$$= \frac{31 \times 19}{4} \times 1$$

$$= \frac{589}{4} = 147\frac{1}{4}$$

Here is a short way to show our work.

$$6\frac{1}{5} \times 23\frac{3}{4} = \frac{31}{\cancel{5}_1} \times \frac{\cancel{95}^{19}}{4} = \frac{31 \times 19}{4} = \frac{589}{4} = 147\frac{1}{4}$$

Using fractions

This short cut is called *cancellation*. Here is a helpful rule to follow when you cancel.

> To cancel when multiplying fractions, divide the numerator and denominator of a fraction by the same number.
> *Caution:* Cancellation may be used *only* when multiplying fractions. It must *not* be used when adding or subtracting fractions.

Example

The water in a public swimming pool rises at a rate of $3\frac{1}{8}''$ an hour during filling. What will be the height after $9\frac{3}{5}$ hours?

Solution

To find the answer, we multiply $3\frac{1}{8}$ by $9\frac{3}{5}$.

$$3\frac{1}{8} \times 9\frac{3}{5} = \frac{25}{8} \times \frac{\overset{5}{\cancel{48}}}{\underset{1}{\cancel{5}}} \quad \text{(divide by 5)}$$

$$= \frac{5}{\underset{1}{\cancel{8}}} \times \frac{\overset{6}{\cancel{48}}}{1} \quad \text{(divide by 8)}$$

$$= 30$$

The water will be 30″ deep.

Notice that we can cancel twice. In actual practice we usually show all the cancellations in one step, as below.

$$3\frac{1}{8} \times 9\frac{3}{5} = \frac{\overset{5}{\cancel{25}}}{\underset{1}{\cancel{8}}} \times \frac{\overset{6}{\cancel{48}}}{\underset{1}{\cancel{5}}} = 30$$

Do you see how cancelling can shorten your work?

EXERCISES Multiply, using cancellation wherever possible. Write the answers in lowest terms.

1. $\frac{5}{6} \times \frac{9}{20}$
2. $\frac{3}{8} \times \frac{8}{9}$
3. $\frac{2}{5} \times \frac{7}{10}$
4. $\frac{24}{10} \times \frac{5}{8}$
5. $\frac{15}{36} \times \frac{9}{10}$
6. $\frac{27}{64} \times \frac{4}{9}$
7. $\frac{20}{9} \times \frac{5}{6}$
8. $4 \times \frac{5}{16}$
9. $24 \times \frac{3}{4}$
10. $2\frac{1}{16} \times 1\frac{1}{3}$
11. $10\frac{1}{8} \times 7\frac{1}{9}$
12. $5\frac{5}{8}'' \times \frac{4}{9}$
13. $3\frac{5}{8}'' \times 8$
14. $3\frac{3}{4}'' \times 16$
15. $1\frac{1}{3} \times \frac{3}{4}''$

Technical and industrial computations

Dividing fractions

3-13 RECIPROCALS

For every fraction there is another fraction such that when we multiply the two fractions, the result is 1. For example, if we multiply $\frac{2}{3}$ by $\frac{3}{2}$, we get $\frac{2}{3} \times \frac{3}{2} = \frac{6}{6} = 1$. The fractions $\frac{2}{3}$ and $\frac{3}{2}$ are called *reciprocals*.

Whole numbers have reciprocals too. Recall that 5 can be written as the fraction $\frac{5}{1}$. Since $\frac{5}{1} \times \frac{1}{5} = \frac{5}{5} = 1$, we see that $\frac{1}{5}$ and 5 are reciprocals.

We use reciprocals to solve common industrial problems that involve dividing fractions.

EXERCISES Give the reciprocal of each number. Then show that the product of the number and its reciprocal is 1.

1. $\frac{3}{4}$ 2. $\frac{2}{9}$ 3. $\frac{7}{2}$ 4. $\frac{8}{5}$
5. $\frac{22}{7}$ 6. $\frac{15}{16}$ 7. $\frac{31}{32}$ 8. $\frac{10}{3}$
9. 53   10. 117   11. $8\frac{2}{3}$ 12. $3\frac{7}{8}$

3-14 DIVIDING A FRACTION BY A WHOLE NUMBER

When a machinist has to make several identical small parts, he or she begins with a large piece of stock. The first thing done is the detail work like drilling holes and knurling. After this, the machinist cuts the stock into the required number of pieces. Dividing fractions occurs frequently in this kind of operation.

Example

Three rectangular insulators must be made from a nylon strip $8\frac{1}{4}''$ long (see Figure 3-14). How long will each be?

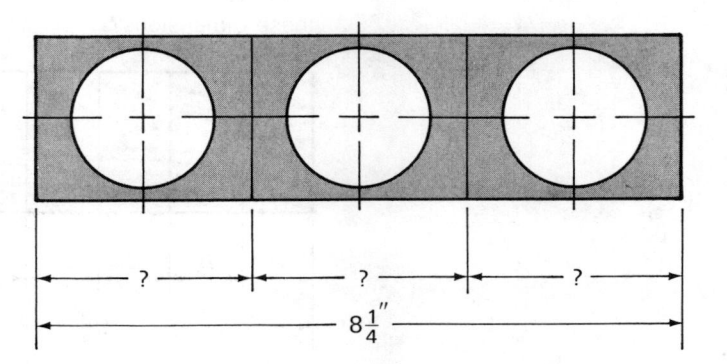

Figure 3-14

Using fractions

Solution

Since $8\frac{1}{4} = \frac{33}{4}$, we wish to divide $\frac{33}{4}$ by 3. Notice that

$$\frac{33}{4} = \frac{11 + 11 + 11}{4} = \frac{11}{4} + \frac{11}{4} + \frac{11}{4}$$

Do you see that when $\frac{33}{4}$ is divided into 3 parts, the result is $\frac{11}{4}$? The width of each insulator must be $\frac{11}{4}$, or $2\frac{3}{4}''$.

An easier way to show the division is to multiply $\frac{33}{4}$ by $\frac{1}{3}$, the reciprocal of 3. In this case we get

$$\frac{33}{4} \times \frac{1}{3} = \frac{\overset{11}{\cancel{33}}}{4} \times \frac{1}{\underset{1}{\cancel{3}}} = \frac{11 \times 1}{4 \times 1} = \frac{11}{4} = 2\frac{3}{4}$$

Thus, $\quad \dfrac{33}{4} \div 3 = \dfrac{33}{4} \times \dfrac{1}{3} = \dfrac{11}{4} = 2\dfrac{3}{4}$

Notice that dividing a quantity by 3 is the same as *multiplying* by $\frac{1}{3}$. Since we treat division as multiplication by the reciprocal, we can cancel.

Here's a helpful rule to remember.

To divide a fraction by a whole number, multiply the fraction by the reciprocal of the whole number.

EXERCISES Divide. Cancel where possible. Give the answers in lowest terms.

1. $\frac{9}{10} \div 3$ 2. $\frac{4}{5} \div 2$ 3. $\frac{16}{35} \div 4$
4. $\frac{3}{7} \div 8$ 5. $\frac{15''}{16} \div 3$ 6. $\frac{27''}{64} \div 9$
7. $\frac{63''}{100} \div 7$ 8. $\frac{5''}{16} \div 7$ 9. $3\frac{1}{3} \div 5$
10. $1\frac{23}{25} \div 4$ 11. $5\frac{1}{16}'' \div 27$ 12. $9\frac{3}{8}'' \div 25$

13. To make a six-sided stereo speaker cabinet, a piece of mahogany must be sawed so that *A, B, C, D, E,* and *F* are equal. What are these dimensions?

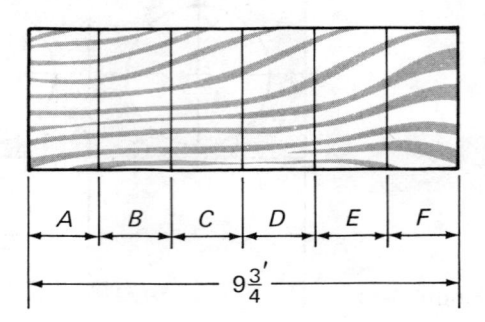

82   Technical and industrial computations

14. An earthmover removed $64\frac{1}{2}$ cu yd of gravel in 4 hours. What is its rate per hour?

15. A strip of aluminum $26\frac{1}{4}"$ long is to be sheared into 5 pieces of equal length. What is the length of each piece?

16. Three $8\frac{1}{2}"$ by $11"$ photographs are to be mounted on background stock. The photos will be equally spaced "east-west" and centered "north-south."

    a. Find dimensions A, B, C, and D.
    b. Find dimensions E and F.

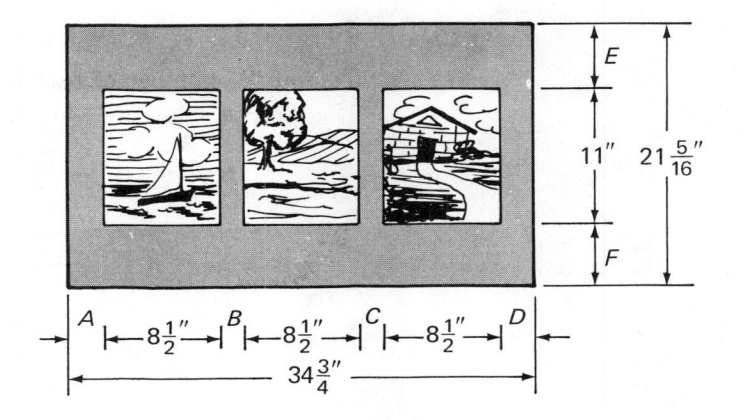

17. The cinder block in Figure 3-15 was formed with the dimensions given below. The three mortar holes are the same size.

    a. Find the width of the holes.
    b. Find the length of the holes.

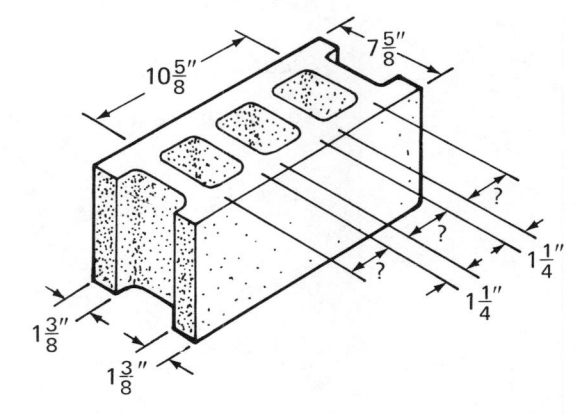

Figure 3-15

Using fractions 83

3-15 DIVIDING BY A FRACTION

Frequently, industrial calculations involve dividing by fractions. Study the following examples.

Example 1

The coffee table in Figure 3-16 is to be built with the top 30" wide. The plans call for finished pine boards $1\frac{7}{8}$" wide. How many boards will be needed to construct the top?

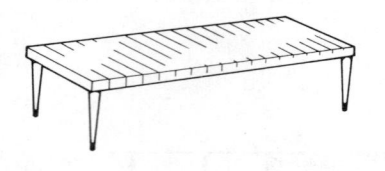

Figure 3-16

Solution

To find the number of boards needed, we have to divide 30" by $1\frac{7}{8}$". We can show the work like this:

$$30 \div 1\frac{7}{8} = \frac{30}{1} \div \frac{15}{8}$$

$$= \frac{\overset{2}{\cancel{30}}}{1} \times \frac{8}{\underset{1}{\cancel{15}}}$$

$$= \frac{2 \times 8}{1 \times 1}$$

$$= 16$$

Thus, 16 boards are needed.

Example 2

A numerical-control lathe automatically advances the cutting tool $1\frac{3}{8}$" every minute. At this rate, how long would it take to turn an iron rod $24\frac{3}{4}$" long?

Solution

To find the time, we must divide $24\frac{3}{4}$ by $1\frac{3}{8}$.

$$24\frac{3}{4} \div 1\frac{3}{8} = \frac{99}{4} \div \frac{11}{8}$$

$$= \frac{\overset{9}{\cancel{99}}}{\underset{1}{\cancel{4}}} \times \frac{\overset{2}{\cancel{8}}}{\underset{1}{\cancel{11}}}$$

$$= \frac{9 \times 2}{1 \times 1}$$

$$= 18$$

It would take 18 minutes to turn the rod.

These examples suggest the following rule.

> To divide a fraction by a fraction, multiply the first fraction by the reciprocal of the other.
>
> In division involving mixed numbers, change the mixed numbers to fractions, then proceed as above.

EXERCISES Divide. Follow the flow chart in Figure 3-17 if you need help.

1. $12 \div \frac{6}{7}$ 2. $18 \div \frac{3}{16}$ 3. $7 \div \frac{4}{9}$
4. $4 \div \frac{8}{9}$ 5. $\frac{4}{9} \div \frac{8}{9}$ 6. $\frac{5}{16} \div \frac{5}{8}$
7. $\frac{22}{7} \div \frac{4}{3}$ 8. $\frac{32}{5} \div \frac{1}{5}$ 9. $8 \div 4\frac{4}{5}$
10. $46'' \div 1\frac{7}{16}$ 11. $\frac{3''}{4} \div 2\frac{1}{2}$ 12. $8\frac{3''}{4} \div 2\frac{1}{2}$

13. A sheet of vinyl plastic is $25\frac{1}{2}''$ wide. How many strips, each $2\frac{5}{8}''$ wide, can be cut from the sheet?

14. A piece of rubber weather stripping $26\frac{1}{4}''$ long is required along the bottom of a storm window. How many strips can be cut from a roll $43'3''$ long?

15. One turn of the handle moves the tailstock of a lathe $\frac{3}{32}''$. How many turns must be made to move the tailstock $1\frac{1}{2}''$?

16. The adjusting nut on a plane lowers the blade $\frac{5}{64}''$ for each turn. How many times must the nut be turned to lower the blade $\frac{5}{32}''$?

17. How many roofing boards $1\frac{3}{16}''$ thick are there is a stack $2'2''$ high?

18. How many pieces of solid copper wire $2\frac{1}{2}''$ long can be cut from a piece $15'6''$ long?

19. Draw a flow chart to show how to find a reciprocal. Use it to find the reciprocals of:

    a. $\frac{1}{3}$ b. 3 c. $\frac{7}{8}$ d. $2\frac{1}{4}$ e. $\frac{19}{8}$.

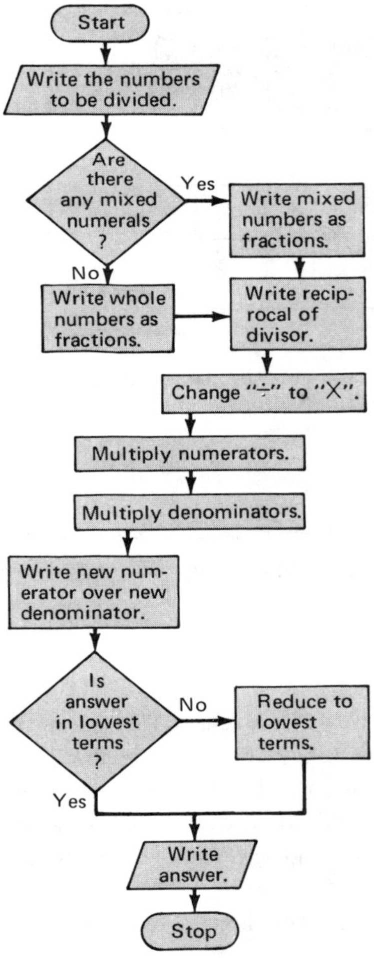

Combining operations with fractions

3-16 ESTIMATING

The daily operation of some industries is so large that exact figures for important information cannot be given. For example, it would be almost impossible to know how many cartons are in a storage warehouse at a certain time. Instead, industries often rely on estimates to help them plan their operations.

Estimating is a way of making a "reasonable guess." Estimates can give us an idea of the amount of material that is available or of

Figure 3-17

how much is needed. Estimates can also help us see whether our calculations are accurate.

Example 1

What is the total width of 5 pieces of $\frac{7}{8}''$ strapping?

Solution

Step 1 Since $\frac{7}{8}''$ is slightly less than 1", we expect the answer to be less than $5 \times 1''$, or 5". Since $\frac{7}{8}''$ is slightly more than $\frac{3}{4}''$, we expect the answer to be more than $5 \times \frac{3}{4}''$ or $3\frac{3}{4}''$. We, therefore, estimate the answer to be between $3\frac{3}{4}''$ and 5".

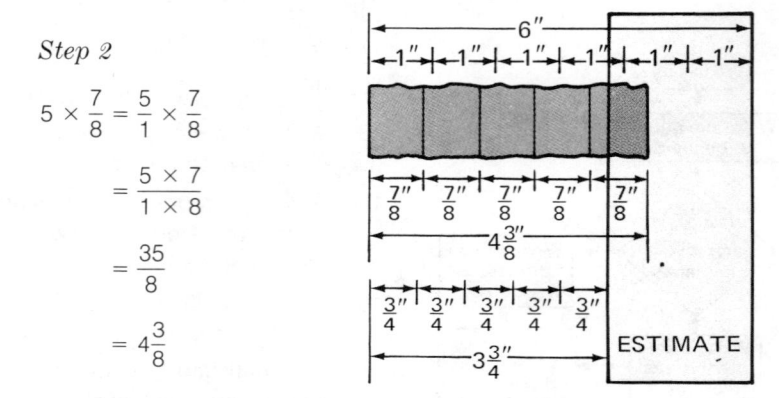

Step 2

$$5 \times \frac{7}{8} = \frac{5}{1} \times \frac{7}{8}$$

$$= \frac{5 \times 7}{1 \times 8}$$

$$= \frac{35}{8}$$

$$= 4\frac{3}{8}$$

The total width is $4\frac{3}{8}''$. This answer is reasonable, since $4\frac{3}{8}$ is between $3\frac{3}{4}$ and 5.

Example 2

An iron plate 65" long must be cut to form 6 pieces of equal width. Allowing $\frac{1}{8}''$ for each of 5 saw cuts, about how long will each piece be?

Solution

Step 1 Since 65" is slightly less than 66", a reasonable guess for the upper limit would be about 11" (66" ÷ 6). Considering the 5 saw cuts, each of which is less than an inch, we can allow at the most 5" for the saw cuts. Then the plate, less the saw cuts, would be about 60". The lower limit for each piece would be about 10" (60" ÷ 6).

Therefore, a reasonable guess for the length of each piece would be between 10" and 11".

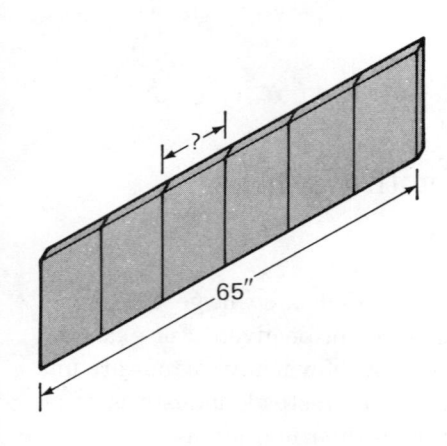

86 Technical and industrial computations

Step 2 Amount of material after cut allowance:

$$65 - \left(5 \times \frac{1}{8}\right) = 65 - \frac{5}{8} = 64\frac{3}{8}$$

Width of each piece:

$$64\frac{3}{8} \div 6 = \frac{515}{8} \times \frac{1}{6}$$

$$= \frac{515 \times 1}{8 \times 6}$$

$$= \frac{515}{48}$$

$$= 10\frac{35}{48}, \text{ or } 10\frac{3}{4} \text{ (approx.)}$$

Each plate will be about $10\frac{3}{4}''$ wide. This answer is reasonable, since $10\frac{3}{4}$ is between 10 and 11.

EXERCISES Estimate the answers. Then use your estimates to check the answers.

1. In making a binding post $\frac{3}{4}''$ long, $\frac{1}{8}''$ of stock is wasted. How many binding posts can be made from a rod 25" long?

2. How many thumb screws $\frac{7}{8}''$ long can be cut from a bar of cold-rolled steel 36" long, if $\frac{3}{16}''$ is wasted in each cut?

3. A plank $11\frac{3}{4}''$ wide is to be cut lengthwise into strips $2\frac{3}{4}''$ wide. Allowing $\frac{1}{8}''$ for each saw cut, find the number of $2\frac{3}{4}''$ strips that can be cut.

4. A garment worker cut 7 pieces of equal length from a strip of cloth 50" long. Allowing $\frac{1}{8}''$ scrap border for each of the 6 cuts, find the length of each piece.

5. Weight bars are used to press tablet sheets together before the glue is applied. How many bars $\frac{9}{16}''$ wide can be placed side by side in a space $8\frac{1}{2}''$ wide?

6. A board 11' long is to be cut into partitions for displaying stereo tapes. How many partitions $6\frac{3}{4}''$ long can be cut if $\frac{1}{8}''$ is allowed for each saw cut?

7. A street 340' long is to be curbed with blocks of granite, each 30" long. If $\frac{1}{4}''$ is allowed between the blocks for expansion, how many blocks are needed for the curb?

8. A 145' conveyor belt moves cartons from the storage area to the loading platform. Each carton is 18" wide. If a $\frac{3}{4}''$ space is left between cartons, what is the maximum capacity of the conveyor belt?

3-17 SOLVING PROBLEMS

Besides being able to estimate, tomorrow's skilled industrial worker must be able to solve problems involving several operations with fractions.

Example 1

A printer is going to ship 44 used plates in a wooden crate. Each plate weighs $1\frac{3}{4}$ lb. If the crate can safely hold 75 lb, can the printer ship the plates in the crate?

Solution

Step 1 Estimate the total weight of the plates.

44 × 1 lb = 44 lb

44 × 2 lb = 88 lb

The total weight is between 44 and 88 lb.

Step 2 Find the total weight of the plates.

$$44 \times 1\frac{3}{4} = \frac{44}{1} \times \frac{7}{4}$$

$$= \frac{44 \times 7}{1 \times 4}$$

$$= \frac{11 \times 7}{1}$$

$$= 77$$

The plates weigh 77 lb.

Step 3 Compare with the safe weight.

77 is greater than 75

Therefore, the crate cannot be used to ship the plates.

Example 2

A tie rack is to be manufactured from a piece of clear plastic according to the plan in Figure 3-18. Equally spaced holes must be drilled so that plastic pegs can be inserted. How many holes can be drilled?

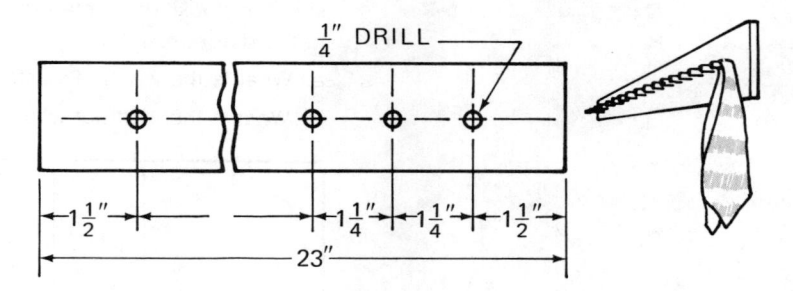

Figure 3-18

Solution

Step 1 Estimate the number of holes.

23" ÷ 1" = 23

23" ÷ 2" = 11½, rounded to 12

There will be between 12 and 23 holes.

Step 2 Find the length of the space where the holes will be drilled.

23 − (1½ + 1½) = 23 − 3 = 20

The space measures 20".

Step 3 Find the number of holes.

$$20 \div 1\frac{1}{4} = \frac{20}{1} \div \frac{5}{4}$$

$$= \frac{\overset{4}{\cancel{20}}}{1} \times \frac{4}{\underset{1}{\cancel{5}}}$$

$$= 16$$

There will be 16 spaces. Therefore, there will be 17 holes.

Using fractions 89

EXERCISES Solve the problems. Use estimates to check your work.

1. A printer's assistant must cut cards $4\frac{1}{4}''$ by $8\frac{1}{8}''$ from a 17" by 22" sheet of stock. The preferred method of cutting the stock is shown in the diagram.

    a. What is the width of waste strip A?

    b. What is the width of waste strip B?

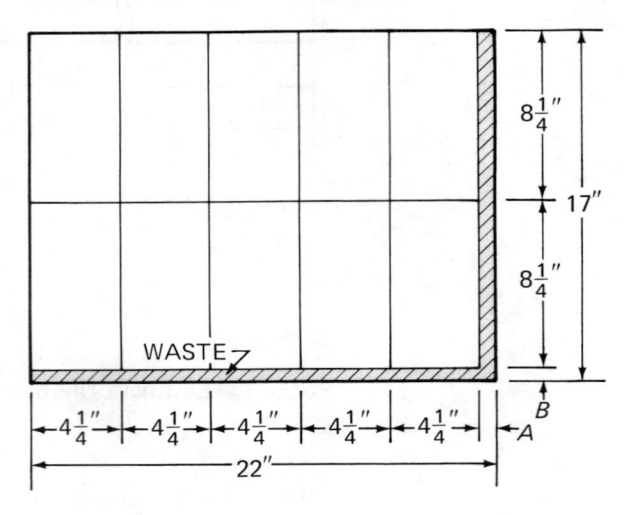

2. There must be 25 brackets with the dimensions shown stamped from a strip of no. 20–gauge sheet metal.

    a. How long is the bracket?

    b. How wide is the bracket?

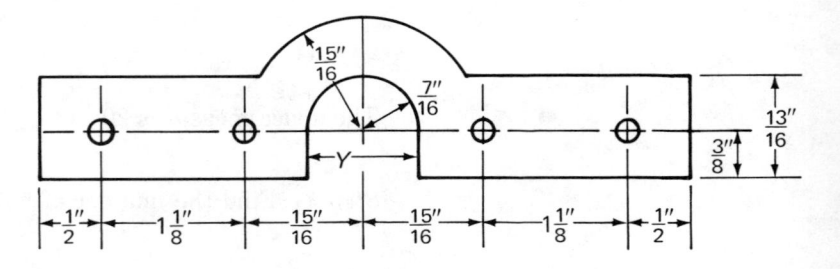

3. In making a bolt $2\frac{5}{8}''$ long, the allowance for waste is $\frac{1}{8}''$. How many feet of stock are used in making 36 bolts?

4. How many display shelves 2'3" long can be cut from a board 14' long if $\frac{1}{8}''$ is allowed for each saw cut?

5. A pattern for the cork sleeve on a woodburning pen is shown in Figure 3-19. The allowance for waste and cutting for each piece is $\frac{1}{4}''$. How many cork sleeves can be made from a cork rod 24" long?

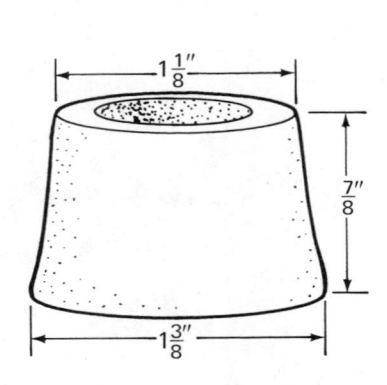

Figure 3-19

90 Technical and industrial computations

6. A service technician needs pieces of copper tubing 8¼", 9⅞", and 14½" long to install the cooling unit in an air conditioner. Allowing 1/16" for each cut, how much tubing does the technician need to assemble 16 such units? Give the answer in feet and inches.

7. A technical drawing 6¾" wide by 8⅛" high is to be centered on a sheet of 12" by 18" paper. How much of a margin should be left on each side of the drawing?

8. A tenon ¾" by 2⅜" is cut in the center of a piece 1½" by 3¼".
   a. How wide are the shoulders on the sides?
   b. How wide are the shoulders on the top and bottom?

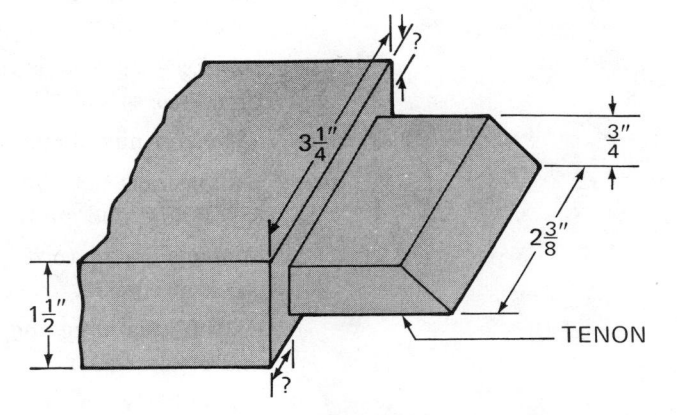

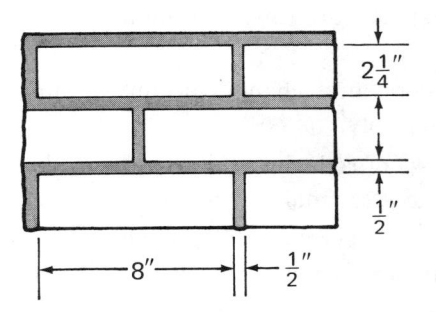

9. In building a brick wall 12' long and 6'5" high, bricks 8" long and 2¼" high are used. The mortar between the bricks is ½" thick.
   a. How many bricks are needed for one layer? (Count a fraction of a brick as a whole brick.)
   b. How many layers of brick are there in the completed wall?

10. Eight T rails are to be cut from a piece of stock 20' long. In cutting each piece, 8/16" of stock is wasted.
    a. How much stock is used in cutting the eight pieces?
    b. What is the length of the piece of unused stock?
    c. If the price of the stock is 62¢ a foot, what is the cost of the 20' piece of stock?

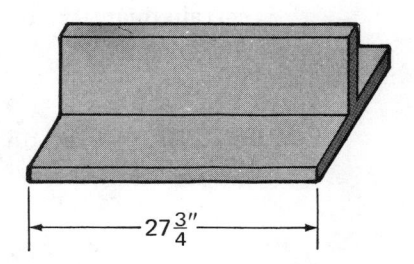

Using fractions 91

Taking inventory

1. To add (subtract) fractions with the same denominator, add (subtract) the numerators and use the same denominator. (p. 54)
2. To add (subtract) fractions with different denominators, change the fractions to equivalent fractions with a common denominator, then add (subtract). (p. 55)
3. The *least common denominator (LCD)* is the smallest number that can be divided evenly by all the denominators of the fractions that are to be added or subtracted. (p. 57)
4. A *mixed number* is made up of a whole number and a fraction. (pp. 60–63)
5. Whole or mixed numbers can be written as fractions. (p. 60)
6. Dimensions not shown on technical drawings can be found by adding and subtracting fractions. (pp. 63–67)
7. Dimensions up to 72″ are usually given in inches only. Those greater than 72″ are usually given in feet and inches. (p. 68)
8. Careful planning and accurate measurements help conserve materials. (p. 71)
9. To multiply fractions, multiply the numerators together and the denominators together. Reduce the resulting fraction to lowest terms. (p. 78)
10. To multiply mixed numbers, first change the mixed numbers to fractions, then multiply. (p. 78)
11. *Cancellation* is the process of reducing to lowest terms before multiplying instead of after multiplying. (p. 80)
12. The *reciprocal* of a number is another number which when multiplied by the first number gives 1 as the product. (p. 81)
13. To divide fractions, multiply the first fraction by the reciprocal of the second fraction. (p. 85)
14. An *estimate* is a "reasonable guess" as to the amount of a certain quantity. (p. 85)

Measuring your skills

Add or subtract. Give the answer in lowest terms. (3-1, 3-2)

1. $\frac{17}{64} + \frac{13}{64}$ 2. $\frac{2}{32} + \frac{11}{32} + \frac{9}{32}$

3. $\frac{13}{16}'' - \frac{5}{16}''$ 4. $\frac{5}{8}'' - \frac{3}{8}''$
5. $\frac{3}{8} + \frac{1}{4}$ 6. $\frac{1}{8} + \frac{3}{16} + \frac{5}{32}$
7. $\frac{3}{4}'' - \frac{1}{8}''$ 8. $\frac{7}{32}'' - \frac{3}{64}''$

Find the lowest common denominator for the fractions. (3-3)

9. $\frac{1}{3}, \frac{1}{4}, \frac{1}{5}$ 10. $\frac{3}{4}, \frac{7}{8}, \frac{3}{10}$

11. Change $17\frac{5}{8}$ to a fraction. (3-4)
12. Simplify $\frac{106}{4}$. (3-4)
13. Simplify $\frac{56}{8}$. (3-4)

Add or subtract. Give each answer in lowest terms. (3-5, 3-6)

14. $4\frac{1}{8} + 5\frac{1}{4} + 4\frac{3}{8}$ 15. $8\frac{5}{8} - 2\frac{1}{4}$
16. $3\frac{31}{32} - 2\frac{11}{16}$ 17. $7\frac{3}{4} + 1\frac{1}{8} + 9\frac{1}{16}$
18. $1\frac{9}{16}'' + 5\frac{5}{8}''$ 19. $4\frac{5}{16}'' - 2\frac{3}{4}''$
20. $5'' - 2\frac{1}{8}''$ 21. $12\frac{1}{2}'' + 5\frac{1}{4}'' + 6\frac{1}{16}''$

Multiply. Give the answer in lowest terms. (3-9, 3-10, 3-11)

22. $13 \times \frac{1}{4}$ 23. $4 \times \frac{4}{9}$ 24. $\frac{7}{8} \times 5$
25. $\frac{1}{7} \times 3\frac{3}{4}$ 26. $\frac{2}{5} \times 2\frac{1}{3}$ 27. $\frac{3}{8} \times 2\frac{1}{8}$
28. $\frac{3}{4} \times \frac{3}{10}$ 29. $\frac{7}{3} \times \frac{5}{4}$ 30. $\frac{5}{8} \times \frac{4}{9}$

Multiply, using cancellation. (3-12)

31. $\frac{4}{5} \times \frac{3}{16}$ 32. $\frac{1}{8}'' \times \frac{4}{7}$ 33. $\frac{5}{8} \times \frac{2}{3} \times \frac{12}{25}$

Give the reciprocals. (3-13)
34. 3   35. $\frac{1}{5}$ 36. $\frac{4}{7}$

Divide. Give the answers in lowest terms. (3-14, 3-15)

37. $\frac{15}{32} \div 3$ 38. $\frac{3}{4} \div 15$ 39. $\frac{18}{25} \div 9$
40. $\frac{2}{9} \div \frac{1}{5}$ 41. $\frac{4''}{9} \div \frac{3}{8}$ 42. $3\frac{5}{8}'' \div 2\frac{1}{3}$

43. A contractor wants to add a 42″ section of stockade fence to an existing fence. If the fence now measures 72′4″, how long will it be with the addition? (3-7)

44. In a warehouse, cartons move along a conveyor belt that is 423′6″ long. Only 117′ of the conveyor belt is at ground level; the rest is overhead. How much of the belt is overhead? (3-7)

45. Three sections of insulated telephone cable are cut from a 1000′ reel. The pieces measure 37′4″, 68′2″, and 92′8″. How much cable is left on the reel? (3-7)

46. A 36′ roll of wall covering must be cut into strips to cover a wall 7′2″. At the end of each piece, 4″ are allowed for matching the pattern. (3-8)

   a. Draw a diagram to show how the wall covering will be cut.

Using fractions   93

b. How many strips can be cut from the roll?

c. How much material will be wasted?

Estimate first, then find the answer. (3-16, 3-17)

47. To make one engine head bolt, $4\frac{3}{4}''$ of stock is required, including all allowances. How much stock is needed to make 48 head bolts? Give the answer in feet and inches.

48. An airplane uses gasoline at the rate of $15\frac{3}{4}$ gal per hr. How much fuel does it use in a flight of $5\frac{1}{2}$ hr?

49. Radio speaker grilles $3\frac{3}{4}''$ by $5\frac{3}{4}''$ are stamped from a strip of lightweight material 4" wide. A $\frac{3}{8}''$ waste piece is allowed between grilles. How long a strip is needed to manufacture 36 grilles?

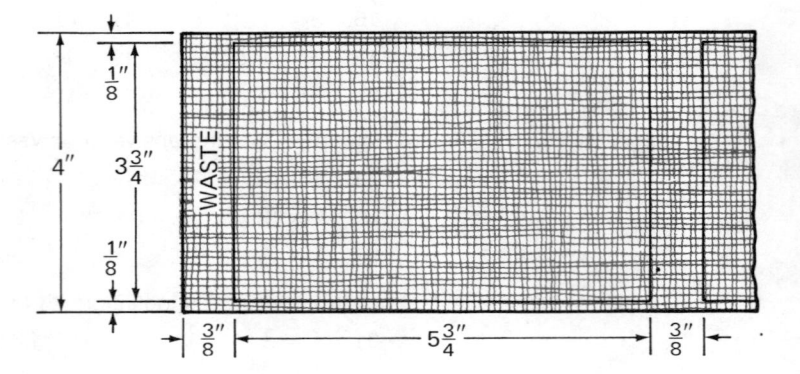

50. A panel of an airplane wing measures $9'3\frac{1}{2}''$ by $54\frac{1}{2}''$. The rivet holes are drilled $1\frac{3}{16}''$ apart, and the end holes were located $1\frac{1}{8}''$ from the ends of the section. How many holes are needed

a. along the length of the panel?

b. along the width of the panel?

c. around the perimeter of the panel?

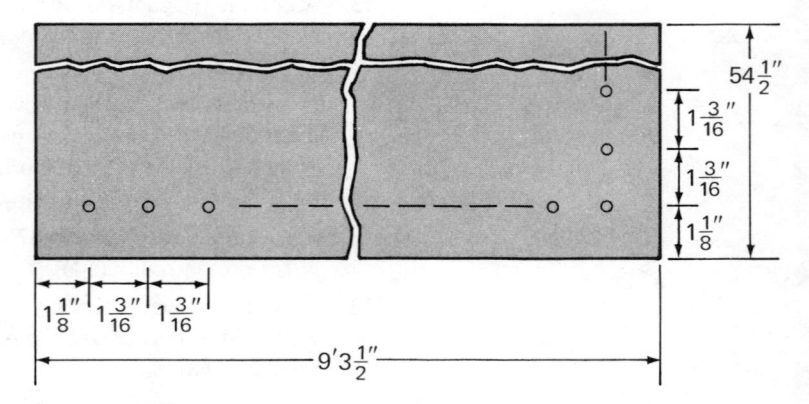

4 Decimals, percents

After completing this chapter, you should be able to:
1. Write equivalent fractions, decimals, and percents.
2. Add, subtract, multiply, and divide decimals.
3. Use a micrometer to take precise measurements.
4. Use percents to determine discounts, tolerances, and efficiencies.

Decimals and fractions

4-1 READING AND WRITING DECIMALS

In Chapter 1 you learned about the machinist's scale (see Figure 4-1). Recall that this scale is divided into hundredths of an inch.

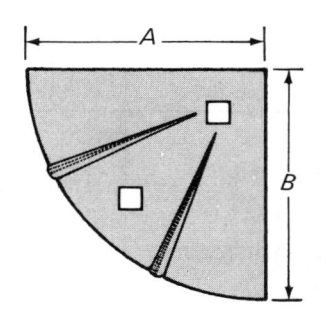

Figure 4-2

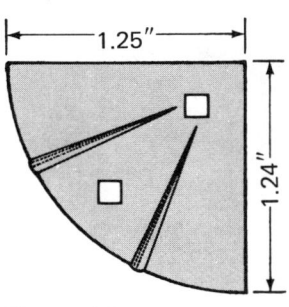

Figure 4-3

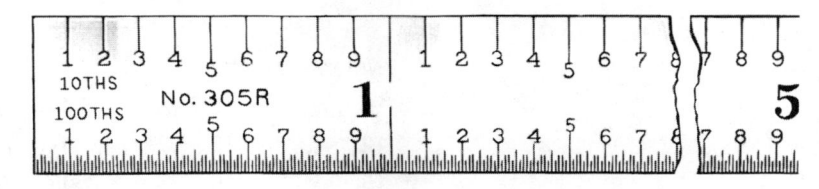

Figure 4-1

If a machinist's scale is available, measure lengths A and B on the bicycle kickstand mounting shown in Figure 4-2. Do you see that A is $1\frac{24}{100}''$ and B is $1\frac{25}{100}''$? The pattern for manufacturing this kickstand mounting might look like Figure 4-3. Notice how the dimensions $1\frac{24}{100}''$ and $1\frac{25}{100}''$ are shown in Figure 4-3 as 1.24" and 1.25". Such numbers are called *decimals*.

Decimals are special fractions whose denominators are 10, or 100, or 1000, and so on. Instead of writing the denominator of decimal fractions, we write only the numerator, preceded by a *decimal point*. Below are several more examples of decimals.

Fraction	Words	Decimal
$\frac{3}{10}$	three-tenths	.3
$\frac{37}{100}$	thirty-seven hundredths	.37
$\frac{217}{1000}$	two hundred seventeen thousandths	.217
$\frac{45}{10,000}$	forty-five ten-thousandths	.0045

The pattern in the decimals above suggests the following idea.

In a decimal, the number of digits after the decimal point indicates the number of zeros in the denominator of the fraction. For example,

$$.325 = \frac{325}{1000}$$

3 digits — 3 zeros

EXERCISES Write each as a common fraction and as a decimal.

1. four-tenths
2. seven-tenths
3. seven hundredths
4. thirty-six hundredths
5. ninety-three hundredths
6. nine thousandths
7. one hundred fifty-five thousandths
8. three hundred five thousandths

Write as decimals.

9. $\frac{2}{10}$
10. $\frac{9}{100}$
11. $\frac{6}{1000}$
12. $\frac{125}{1000}$
13. $\frac{900}{1000}$
14. $\frac{95}{10,000}$
15. $\frac{18}{1000}$
16. $\frac{83}{100}$
17. $\frac{5}{10}$

Write as fractions.

18. .1
19. .12
20. .322
21. .27
22. .125
23. .2737
24. .015
25. .0725
26. .0069

---

96 Technical and industrial computations

Using a machinist's scale, measure the lengths indicated in the diagrams. Write the dimensions as decimals.

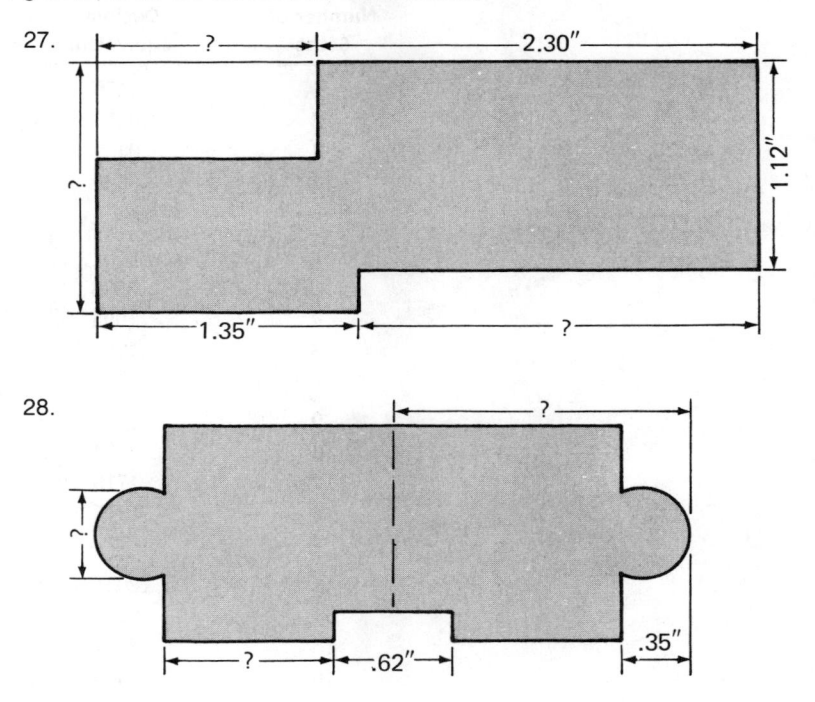

27.

28.

4-2 EQUIVALENT DECIMALS AND FRACTIONS

The dimensions of the bolt in Figure 4-4 are given in fractions. To machine a bolt like this on a metal lathe that is calibrated in hundredths of an inch, we need to know the dimensions as decimals. Usually we can find the most common equivalents in a table, as in Table 4-1.

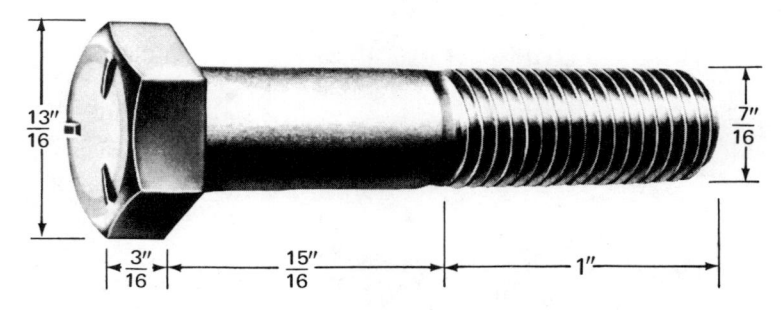

Figure 4-4

Decimals, percents 97

Table 4-1 *Decimal equivalents (four-place approximations)*

Number of 64ths	Decimal equivalent	Number of 64ths	Decimal equivalent
1	.0156	33	.5156
2	.0313	34	.5313
3	.0469	35	.5469
4	.0625	36	.5625
5	.0781	37	.5781
6	.0938	38	.5938
7	.1094	39	.6094
8	.125	40	.625
9	.1406	41	.6406
10	.1563	42	.6563
11	.1719	43	.6719
12	.1875	44	.6875
13	.2031	45	.7031
14	.2188	46	.7188
15	.2344	47	.7344
16	.25	48	.75
17	.2656	49	.7656
18	.2813	50	.7813
19	.2969	51	.7969
20	.3125	52	.8125
21	.3281	53	.8281
22	.3438	54	.8438
23	.3594	55	.8594
24	.375	56	.875
25	.3906	57	.8906
26	.4063	58	.9063
27	.4219	59	.9219
28	.4375	60	.9375
29	.4531	61	.9531
30	.4688	62	.9688
31	.4844	63	.9844
32	.5	64	1.

Example 1

Change $\frac{5}{8}''$ to a decimal.

Solution

First, change $\frac{5}{8}$ to an equivalent fraction with 64 as denominator.

$$\frac{5}{8} \times \frac{8}{8} = \frac{40}{64}$$

Now find the numerator 40 in the column headed "Number of 64ths." Next to it is the decimal equivalent, .625″.

Example 2

Change $\frac{5}{16}''$ to a decimal.

Solution

$\frac{5}{16} = \frac{20}{64}$. Next to 20 in the column headed "Number of 64ths" is the decimal .3125. Thus $\frac{5}{16}'' = .3125''$.

From time to time we may also need to find an equivalent fraction for a decimal. We can also use Table 4-1 to do this.

Example 3

An optical lens must be precision-ground to a thickness of .0469″. What is this dimension as a fraction?

Solution

Find .0469 in the column headed "Decimal equivalent." In the column to the left, read the numerator 3. Thus, $.0469'' = \frac{3}{64}''$.

Sometimes in your work you may not have a table close at hand. At other times the dimensions you are dealing with may not appear in the table. In these instances, you will need to find equivalent decimals and fractions. These rules can be helpful.

---

1. To change a fraction to its decimal equivalent, divide the numerator by the denominator.
2. To change a decimal to a fraction,
   a. count the number of decimal places; then
   b. write the fraction using that number of zeros in the denominator; and
   c. reduce the fraction to lowest terms.

Example 4

Change $\frac{5}{8}$ to a decimal.

Solution

$$\frac{5}{8} = 5 \div 8 = 8\overline{)5.000}$$

$$\begin{array}{r} .625 \\ 8\overline{)5.000} \\ -4\ 8 \\ \hline 20 \\ -16 \\ \hline 40 \\ -40 \\ \hline 0 \end{array}$$

Therefore, $\frac{5}{8} = .625$.

Example 5

Change 4.375 to a fraction.

Solution

$$4.375 = 4\frac{375}{1000} \quad \text{divide by 125}$$

$$= 4\frac{3}{8}$$

Thus, $4.375 = 4\frac{3}{8}$.

EXERCISES Use Table 4-1 to find the decimal equivalent of each fraction.

1. $\frac{3}{4}$ 2. $\frac{9}{32}$ 3. $\frac{7}{8}$ 4. $\frac{41}{64}$
5. $\frac{9}{16}$ 6. $\frac{3}{16}$ 7. $\frac{29}{32}$ 8. $\frac{1}{4}$

Use Table 4-1 to find the fraction whose value is nearest each of the following decimals.

9. .62 10. .525 11. .925 12. .435
13. .280 14. .385 15. .775 16. .651

Give the decimal equivalent of each.

17. $\frac{1}{5}$ 18. $\frac{2}{3}$ 19. $\frac{7}{8}$ 20. $\frac{3}{16}$
21. $\frac{5}{16}$ 22. $\frac{15}{16}$ 23. $\frac{3}{32}$ 24. $\frac{5}{32}$
25. $\frac{9}{32}$ 26. $1\frac{1}{4}''$ 27. $2\frac{3}{4}''$ 28. $1\frac{5}{16}''$

Write each decimal as a mixed number or fraction in lowest terms.

29. .25 30. .4 31. .6 32. .35
33. .75 34. .125 35. .875 36. .3125
37. .4375 38. 1.625″ 39. 3.15″ 40. 5.75″

41. An automotive-parts supplier lists the dimensions of all parts in decimals. What size piston rings must a mechanic buy to replace rings that are $3\frac{7}{8}″$ in diameter?

42. The thickness of a hexagon-shaped nut is $\frac{7}{8}$ the diameter of the bolt, reduced to the nearest sixty-fourth of an inch. Find the thickness of hex nuts for bolts with the following diameters:

   a. 1.125″   b. .625″   c. .6875″   d. 1.1875″

43. Write the dimensions of the bicycle shift lever in Figure 4-5 as fractions.

44. Find the fractional equivalent of each decimal to the nearest sixteenth.

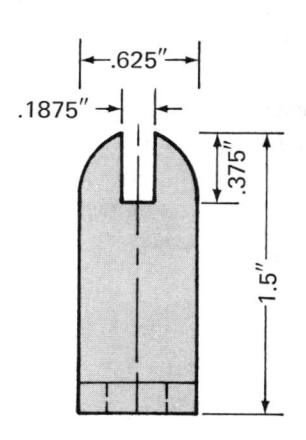

   a. .4   b. .15   c. .125   d. 1.625

Figure 4-5

Using decimals

4-3 ADDING AND SUBTRACTING DECIMALS

To find the length of the carpenter's brace shown in Figure 4-6, we must add several dimensions.

When adding decimals, we write them in a column, with the decimal points in a line.

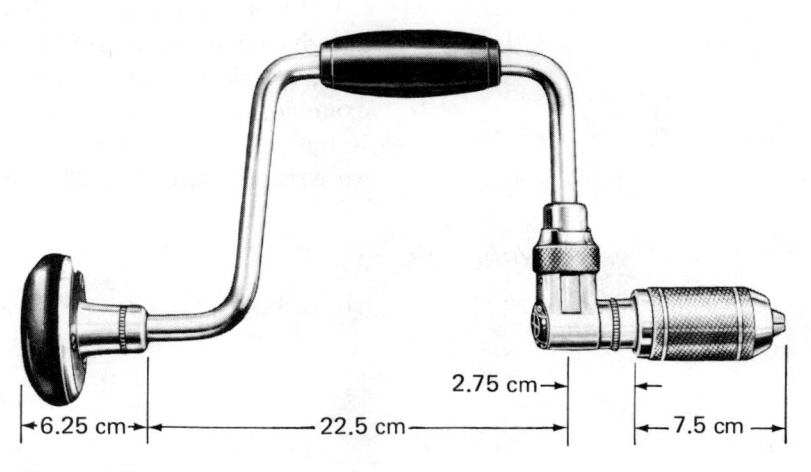

Figure 4-6

Decimals, percents 101

	Do this!	6.25 cm	Not this!	6.25 cm
		22.5 cm		22.5 cm
		2.75 cm		2.75 cm
		+ 7.5 cm		+ 7.5 cm
		39.00 cm		?? cm

If we wish to find the length of the brace excluding the handle, we need to subtract. Again we line up the decimal points. A helpful idea is to add enough zeros after the decimal point to give each decimal the same number of places. Thus, to subtract 6.25 from 39, we write:

```
  39.00 cm
-  6.25 cm
  32.75 cm
```

Placing zeros after the last digit to the right of the decimal point does not change the value of the decimal. The numbers 39, 39.0, and 39.00 are the same, except for the indicated degree of accuracy.

We can summarize our work this way.

> To add or subtract decimals, write them in a column, making sure the decimal points are in a line. Then proceed as with whole numbers.

Did you notice that the dimensions of the brace in Figure 4-6 are given in centimeters? Because the metric system is based on tens, metric dimensions are usually written as decimals. For example, the shank of the brace measures 22 cm + 5 mm. Since 10 mm = 1 cm, 5 mm = $\frac{1}{2}$ cm, or as a decimal, .5 cm. Therefore, we write the length as 22.5 cm.

EXERCISES   Add.

1. 4.03
   +2.91

2. 43.6
   +29.5

3. 16.82
   + 4.73

4. 53.39 cm
   +14.52 cm

102 Technical and industrial computations

5.   14.54 cm
      + 2.98 cm

6.   18.304
      + 7.683

7. 8.46 + 5.2 + .845 + .234
8. 10.76 + .18 + 4.2654 + 88.005
9. 68 + .056 + .0072 + 3.4
10. 1.65 + 2.0735 + .484

Subtract.

11.   3.9
     −2.7

12.   4.6 mm
     −3.2 mm

13.   5.93 cm
     −2.61 cm

14.   17.8 cm
     − 8.2 cm

15.   5.932
     −2.811

16.   16.38
     −11.97

17.   6.
     −3.71

18.   4.3
     − .582

19.   125.83
     − 14.7

U.S. system

20. The smaller end of a tapered pin is 1.125″ in diameter. The diameter of the other end is .385″ larger. Find the diameter of the larger end.

Metric system

23. Find the inside diameter CD of the extruded plastic tubing. The outside diameter and the thickness are shown in the cross-section.

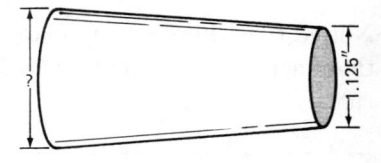

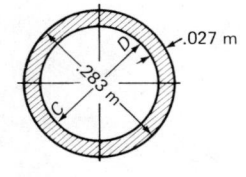

21. Find dimension A in the routed wood panel.

24. Find the distance between the centers of the holes in the drill jig.

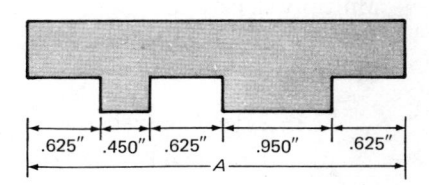

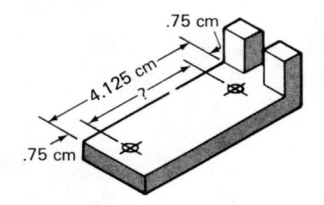

Decimals, percents   103

22. Find dimensions *A* and *B* in the step bracket.

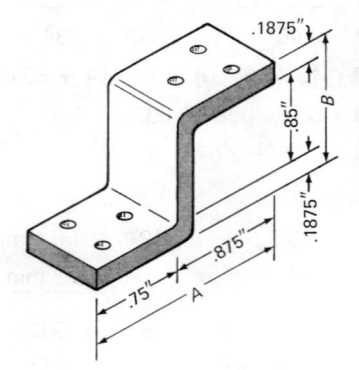

25. Two holes must be drilled in a lock template. Find dimensions *A* and *B*.

4-4 MULTIPLYING DECIMALS

The wood jointer in Figure 4-7 is set to cut at .125″. If a board is passed through the jointer 3 times, by how much is the board planed down?

To answer this question, we must multiply the amount planed each time, .125″, by the number of cuts, 3. We show our work as follows:

```
   .125″ ← 3 decimal places
×     3
   .375″ ← 3 decimal places
```

Notice that there are as many decimal places in the answer as there are in the decimal fraction. This fact suggests the following rule.

The number of decimal places in a product is the sum of the number of decimal places in the factors.

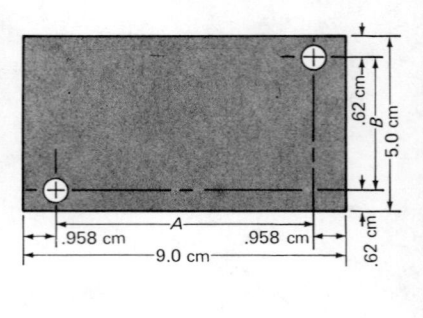

Figure 4-7

Example 1

Multiply .27 by .6.

Solution

```
   .27 ← 2 decimal places
×  .6  ← 1 decimal place
  .162 ← 3 decimal places
```

104 Technical and industrial computations

Example 2

Multiply 4.3 by .005.

Solution

$$\begin{array}{r} 4.3 \leftarrow \text{1 decimal place} \\ \times .005 \leftarrow \text{3 decimal places} \\ \hline 215 \leftarrow \text{??} \end{array}$$

We should have 4 decimal places in the answer, but we only have 3 digits in the answer. In this case, it is necessary to insert a zero to the left of the digit 2. Then we locate the decimal point. Thus, the product is .0215.

It is a good practice to make an estimate before beginning your work. See how the estimate helps you locate the decimal point in Example 3.

Example 3

A strip of heat-treated steel 21.4 cm by 3.21 cm is needed to replace a defective piece. What is the area of this piece?

Solution

Step 1   Estimate.

21.4 is about 21.

3.21 is about 3.

21 × 3 = 63

The strip has an area of about 63 sq cm.

Step 2   Solve.

$$\begin{array}{r} 21.4 \\ \times 3.21 \leftarrow \text{three significant figures} \\ \hline 214 \\ 4\ 28 \\ 64\ 2 \\ \hline 68.694 \end{array}$$

Therefore, the area is 68.7 sq cm, rounding off to three significant figures.

EXERCISES Locate the decimal point in the answer.
1. .6 × .7 = 42
2. .4 × 7 = 28
3. .7 × .21 = 147
4. 2.5 × 25 = 625
5. .7 × .866 = 6062
6. .3 × 1.414 = 4242
7. .007 × 1.8 = 126
8. .08 × .003 = 24
9. .003 × .004 = 12

Multiply. Check the answer by estimating.
10. 1.732 × 5
11. .007 × 130
12. .6 × .6
13. .15 × .15
14. 1.155 × 8.05
15. .063 × .0047
16. 57.3 × 35.2
17. 46.25 × 8.4
18. 68.8 × 1.43
19. 4.276 × 24
20. 4.39 × 3.4
21. 130.2 × 5.006

U.S. system

22. Each pound of a brass alloy used for stampings contains .58 lb of copper, .405 lb of zinc, and .015 lb of lead. A bar of this alloy weighs 8.5 lb. Find the amounts of copper, zinc, and lead in the bar.

23. In a square the side S is always about .707 times the length of the diagonal D. Find S when D is 1.75" long.

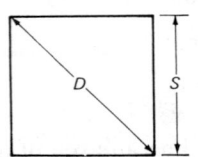

24. A countersunk-head rivet has a diameter D of .5". Dimension A is .425 times D. Dimension B is 1.85 times D. Find dimensions A and B. (See next page.)

Metric system

26. If 1 m of a steel rod 2 cm in diameter weighs 1.61 kilograms (kg), what is the weight of a rod of this steel that is 8.25 m long?

27. The distance F across the flats of a hexbolt is .866 times D. Find F when D is 4.4 cm.

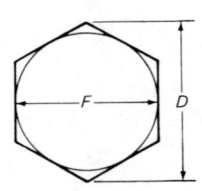

28. A mechanic must bore a hole large enough to seat the head of a bolt. The diameter of the hole must be 1.414 times the side of the bolt head. If the bolt is 3.2 cm on each side, what must be the diameter of the hole? (See next page.)

106 Technical and industrial computations

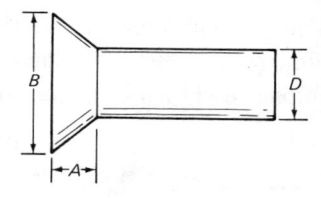

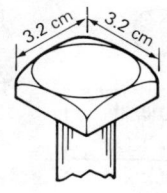

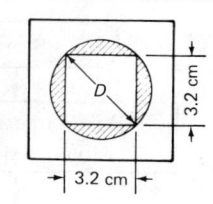

25. The thickness of a sheet of no. 1-gauge sheet metal is .28125″. How high is a stack containing 48 sheets?

29. Using the diagram in exercise 23, find S when D is 3.6 cm.

4-5 MICROMETERS

The *micrometer* is an instrument used in making precise measurements. Some micrometers can measure accurately to .001″. Others are calibrated to give accurate measurements in metric units.

The important parts of a micrometer are shown in a cut-away view in Figure 4-8.

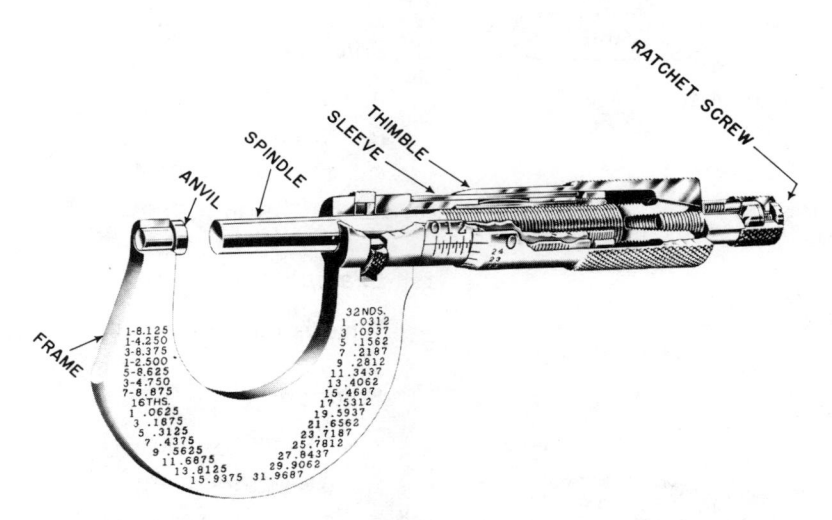

Figure 4-8

A micrometer is calibrated according to two scales—one on the sleeve, the other on the thimble. On a micrometer for measuring in the U.S. system, each numeral on the sleeve indicates .100″.

Decimals, percents 107

Each of the subdivisions on the sleeve is .025″. There are 25 divisions on the thimble. Each division equals .001″. These relationships are shown in Figure 4-9.

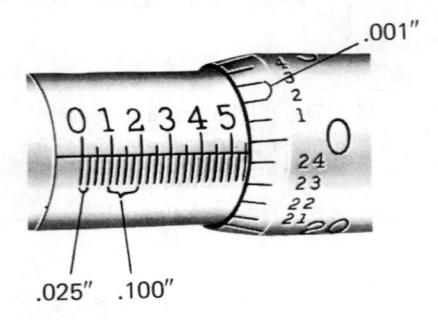

.025″   .100″

Figure 4-9

The flow chart in Figure 4-10 will help you read the setting on a micrometer.

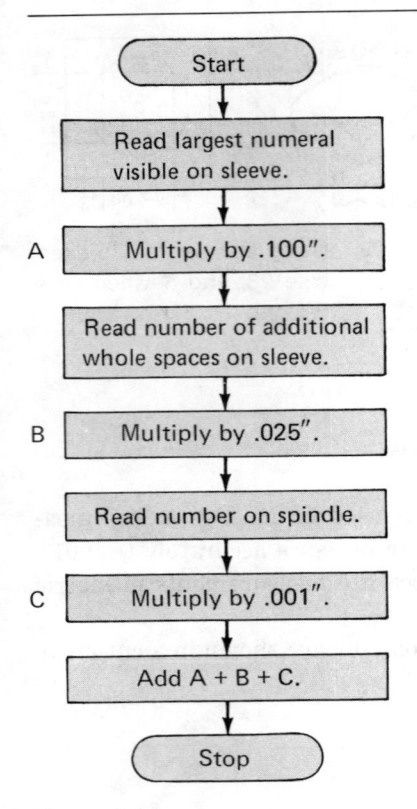

Example 1

Read the micrometer.

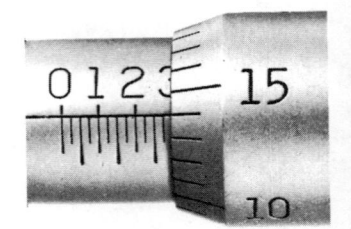

Solution

A. 2 × .100″ = .200″
B. 3 × .025″ = .075″
C. 14 × .001″ = .014″
    Sum = .289″

Example 2

What is the reading?

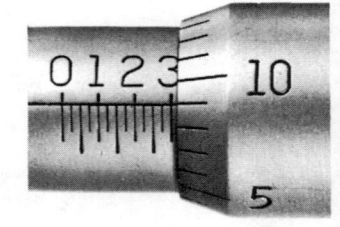

Solution

A. 3 × .100″ = .300″
B. 0 × .025″ = .000″
C. 9 × .001″ = .009″
    Sum = .309″

Metric micrometers also have a scale on the sleeve and a scale on the thimble. Figure 4-11 shows the basic units on these scales.
The flow chart in Figure 4-12 can be used to read the settings on a metric micrometer.

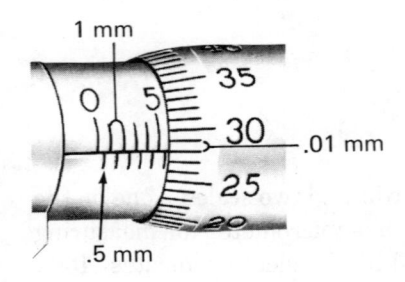

A — Read largest numeral visible on sleeve. → Multiply by .100″.
B — Read number of additional whole spaces on sleeve. → Multiply by .025″.
C — Read number on spindle. → Multiply by .001″.
Add A + B + C.

Figure 4-10

1 mm
.5 mm
.01 mm

Figure 4-11

108   Technical and industrial computations

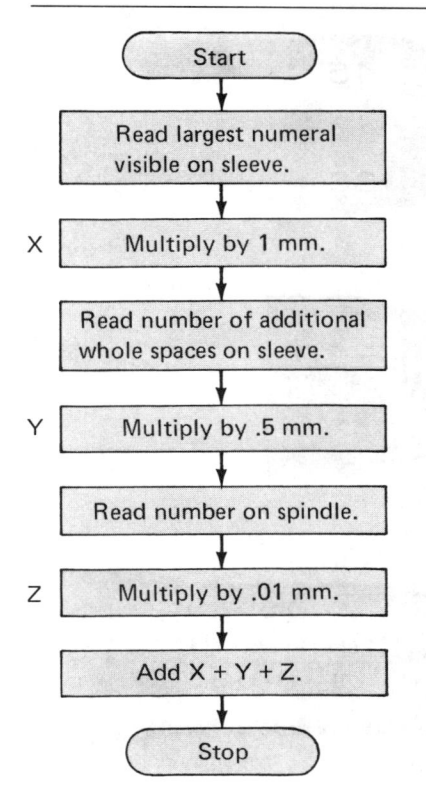

Figure 4-12

Start → Read largest numeral visible on sleeve. → X Multiply by 1 mm. → Read number of additional whole spaces on sleeve. → Y Multiply by .5 mm. → Read number on spindle. → Z Multiply by .01 mm. → Add X + Y + Z. → Stop

Example 3

Read the micrometer.

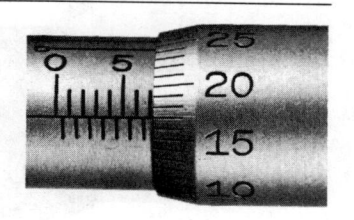

Solution

X. 5×1 mm = 5 mm
Y. 4×0.5 mm = 2.0 mm
Z. $17 \times .01$ mm = .17 mm
         Sum = 7.17 mm

Example 4

What is the reading?

Solution

X. 5×1 mm = 5 mm
Y. $7 \times .5$ mm = 3.5 mm
Z. $29 \times .01$ mm = .29 mm
         Sum = 8.79 mm

EXERCISES   *U.S. system*   *Metric system*

Read the micrometers.

1.

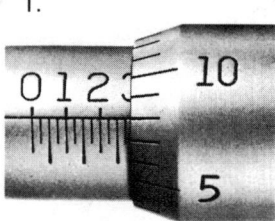

2.

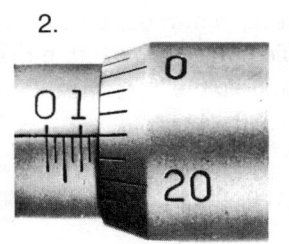

7.

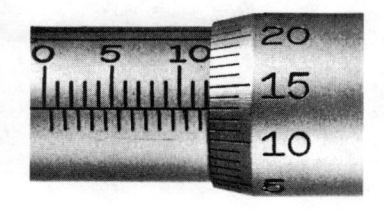

8.

Decimals, percents 109

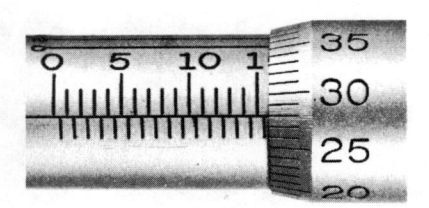

3. 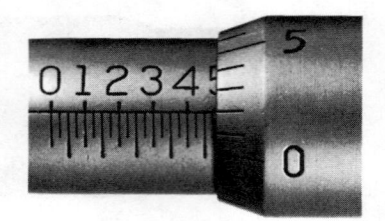
4. 
5. 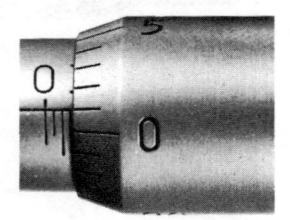
6. 
9. 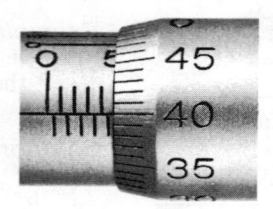
10. 
11. 
12.

TRICKS OF THE TRADE

To multiply a number by 10, move the decimal point 1 place to the *right*. To multiply by 100, move the decimal point 2 places to the *right*, and so on.

.234 × 10 = 2.34
.234 × 100 = 23.4

4-6 DIVIDING DECIMALS

Frequently it is necessary to divide decimals in making calculations. Do you recall how to divide decimals? If not, look at the examples below.

Example 1

Figure 4-13 shows the cross-section of a laminated helicopter rotor blade. Find the thickness of each sheet of laminating material if the five layers in the blade measure 6.485″.

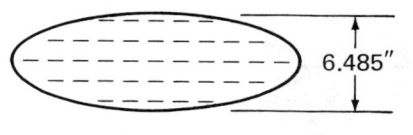

Figure 4-13

Solution

To find the answer, we need to divide 6.485 by 5. We can show the work in this way:

```
        1.297
    5) 6.485
       -5
        1 4
       -1 0
          48
         -45
          35
         -35
           0
```

Each sheet of laminating material is 1.297″ thick.

Notice that the decimal points are lined up when we divide decimals.

Example 2

It takes 13.5 volts to activate an electric relay. How many 1.5-volt batteries, connected in series, are needed to activate the relay?

Solution

To find the number of batteries, we must divide 13.5 by 1.5. The flow chart in Figure 4-14 can help us.

Decimals, percents 111

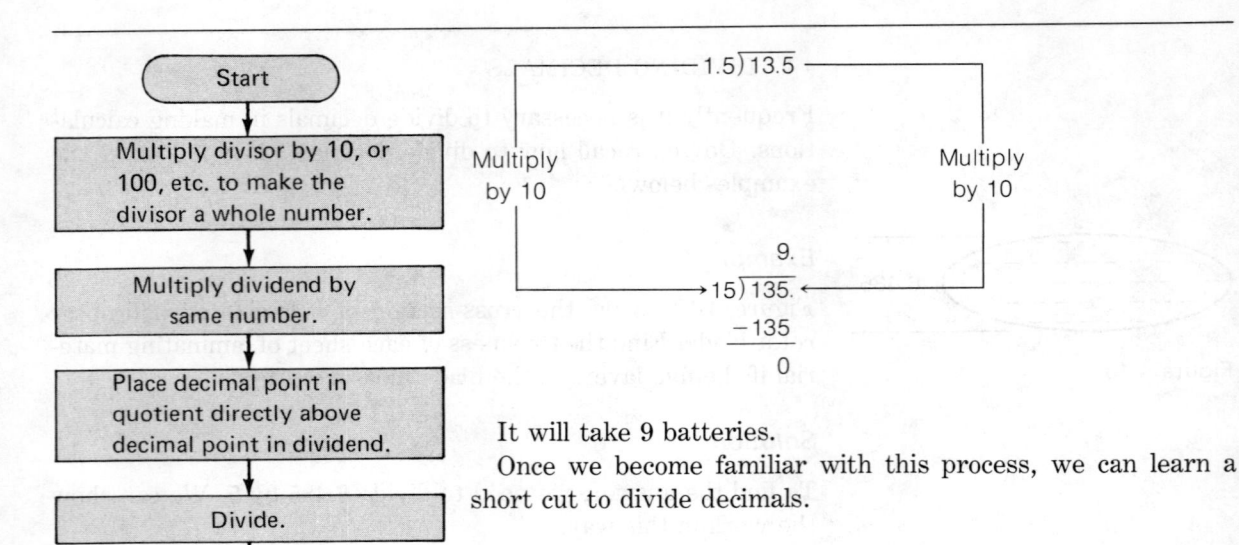

```
        Start
          ↓
  Multiply divisor by 10, or
  100, etc. to make the
  divisor a whole number.
          ↓
  Multiply dividend by
     same number.
          ↓
  Place decimal point in
  quotient directly above
  decimal point in dividend.
          ↓
        Divide.
          ↓
         Stop
```

Figure 4-14

```
              1.5) 13.5
   Multiply                    Multiply
   by 10                        by 10
                   9.
              → 15) 135. ←
                 −135
                    0
```

It will take 9 batteries.

Once we become familiar with this process, we can learn a short cut to divide decimals.

Example 3

A rectangular drip pan is placed under a printing press to keep ink and solvent off the floor. The pan covers an area of 1.725 sq m. If it is .75 m wide, how long is it?

Solution

We must divide 1.725 by .75.

```
              2.3
       .75) 1.725
           −1 50
              225
             −225
                0
```

The pan measures 2.3 m long.

We multiply both the divisor and dividend by 100. We show this by moving the decimal point 2 places to the right. How many places to the right would you move the decimal point if you were multiplying the divisor and dividend by 1000?

These examples suggest the following rule.

112 Technical and industrial computations

> To divide decimals, follow these steps:
> a. If the divisor is a decimal, multiply both the divisor and dividend by 10, or 100, or 1000, and so on, to make the divisor a whole number (this is the same as moving the decimal point to the right); then
> b. place the decimal point directly above the decimal point in the dividend; and
> c. divide as with whole numbers.

EXERCISES Divide.

1. 4.2 ÷ 2
2. .248 ÷ 8
3. 78.84 ÷ 9
4. .4632 ÷ 4
5. 184.25 ÷ 5
6. 12.465 ÷ 3
7. 28.8 ÷ 12
8. .064 ÷ 16
9. 1.275 ÷ 75
10. 11.25 ÷ 25
11. .665 ÷ 19
12. 69.16 ÷ 28
13. 4.5 ÷ .5
14. .48 ÷ .6
15. 19.8 ÷ 6
16. 4.96 ÷ 4
17. 10.71 ÷ .07
18. .225 ÷ .15
19. .125 ÷ .25
20. .9 ÷ .003
21. .805 ÷ .0035

U.S. system

22. To find the outside diameter of a pipe, we divide the circumference of the pipe by 3.14. What is the outside diameter of a pipe whose circumference is 15.7"?

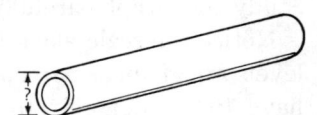

Metric system

27. The inside diameter of a pipe is equal to the inside circumference divided by 3.14. Find the inside diameter of a pipe with an inside circumference of 23.608 cm.

TRICKS OF THE TRADE

To divide a number by 10, move the decimal point 1 place to the *left*. To divide by 100, move the decimal point 2 places to the *left*, and so on.

758 ÷ 10 = 75.8

758 ÷ 100 = 7.58

Decimals, percents 113

23. An airplane propeller 4.25" thick is made of five layers of uniformly thick laminating material. Find the thickness of one layer of this material.

24. A stack of 55 sheets of watch-spring brass is 3.520" high. What is the thickness of each sheet?

25. Eight complete turns of a nut advance it 1.00". How far does it advance on each full turn?

26. A 250' roll of TV antenna lead costs $8.75. What is the cost per foot?

28. Rustproofing is applied to automobile frames by dipping them in vats. After four treatments, the coating on the metal surface measured 3.69 mm. About how much coating was applied each time?

29. If 1 sq m of no. 18-gauge sheet steel weighs .97 kg, what is the area in square meters of a piece of this steel weighing 7.954 kg?

30. A metric wood screw advances 1.8 mm for each complete turn. How many turns will advance it 12.6 mm?

31. It costs $3770 to fuel an atomic reactor with 4.4 g of radioactive material. What is the cost per gram?

Percents

4-7 MEANING OF PERCENT

In Chapter 1 you studied industrial graphs like Figure 4-15. Study this graph carefully.

Notice the scale along the side of the graph. The air-pollution levels are given as percents. Percents are special fractions. They have 100 as denominator. The word *percent* means "parts per hundred." Thus, 35 percent means 35 parts per hundred, and this can be written $\frac{35}{100}$, .35, or 35%.

In your work you may have to change a percent to a decimal or a fraction. Here's how to do it.

To change a percent to a decimal:
a. move the decimal point two places to the *left;* and
b. drop the percent sign.

TIMETABLE FOR CLEANING UP AUTO ENGINES

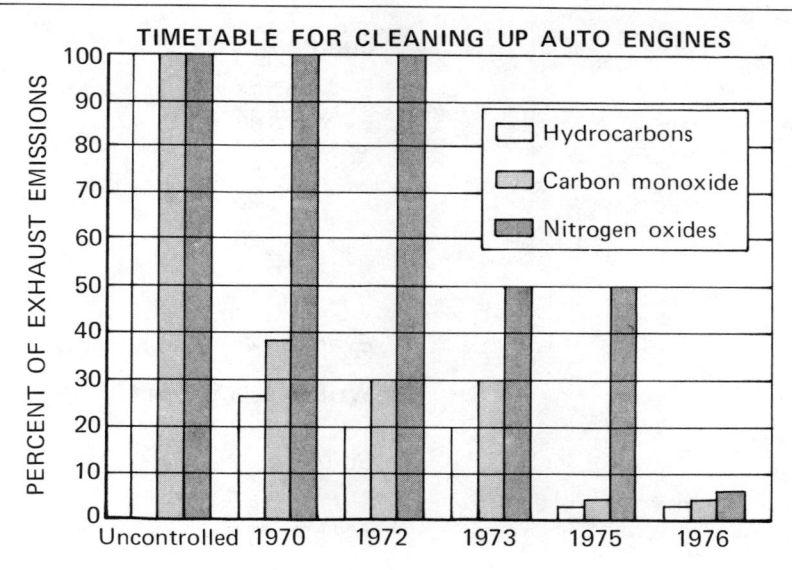

Figure 4-15

Example 1

Write 35% as a decimal.

Solution

.35.% → .35

Example 2

Write 6% as a decimal.

Solution

.06.% → .06

To change a fraction to a percent:
a. change the fraction to a decimal; then
b. move the decimal point two places to the *right;* and
c. add a percent sign.

Example 3

Write ¾ as a percent.

Decimals, percents

Solution

$$\frac{3}{4} = 4\overline{)\begin{array}{r} .75 \\ 3.00 \\ -2\,8 \\ \hline 20 \\ -20 \\ \hline 0 \end{array}} = .75 = 75\%$$

Example 4

Write $\frac{7}{8}$ as a percent.

Solution

$$\frac{7}{8} = 8\overline{)\begin{array}{r} .875 \\ 7.000 \\ -6\,4 \\ \hline 60 \\ -\,56 \\ \hline 40 \\ -\,40 \\ \hline 0 \end{array}} = .875 = 87.5\%, \text{ or } 87\frac{1}{2}\%$$

EXERCISES Write the percents as decimals.

1. 3%   2. 4%   3. 78%   4. 93%
5. 25.3%   6. 16.5%   7. .2%   8. 108.2%

Write the percents as fractions in lowest terms.

9. 25%   10. 50%   11. 75%   12. 40%
13. 60%   14. 90%   15. $12\frac{1}{2}$%   16. $37\frac{1}{2}$%
17. $62\frac{1}{2}$%   18. 100%   19. .5%   20. 1.5%

Write the decimals as percents.

21. .14   22. .63   23. 1.45   24. .062
25. .032   26. .017   27. .4   28. 1.25

Write the fractions as percents.

29. $\frac{1}{2}$ 30. $\frac{3}{4}$ 31. $\frac{2}{5}$ 32. $\frac{3}{5}$
33. $\frac{3}{8}$ 34. $\frac{5}{8}$ 35. $\frac{7}{8}$ 36. $\frac{7}{20}$
37. $\frac{13}{20}$ 38. $\frac{4}{25}$ 39. $\frac{7}{50}$ 40. $\frac{43}{80}$

For exercises 41 through 44, refer to the graph in Figure 4-15.

41. From 1972 to 1973 nitrogen oxides dropped to 50%. Write 50% as a fraction.
42. In 1976 exhaust contained no more than 4% hydrocarbons. Write 4% as a fraction.
43. With no pollution controls, emissions would be 100%. Write 100% as a decimal.
44. By 1976 hydrocarbons dropped to ¼ the amount present in 1972. What is ¼ as a decimal? as a percent?

4-8 DISCOUNTS

A dealer who buys an item from a manufacturer usually pays a *discount* price, called the *net price*. Discount rates are often stated as percents. Since you may have to work with discounts on your job, you will need to study them now.

Example

A $13.00 set of calipers is put on sale at 15% off the regular price. What is the sale price?

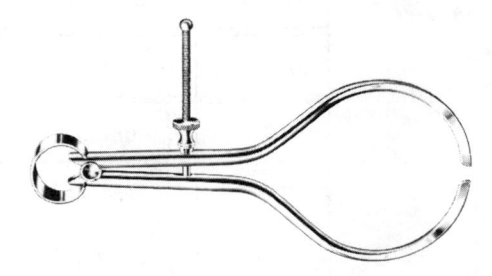

Solution

Percent means "hundredths." Therefore, 15% = .15.

Discount = .15 of $13.00 = .15 × $13.00

$$= \$1.95$$

$$\text{Sale price} = \$13.00 - \$1.95$$

$$= \$11.05$$

The set of calipers will sell for $11.05.

EXERCISES   Find the following amounts.

1. 20% of $840
2. 30% of $560
3. 25% of $84
4. 12½% of $80
5. 6% of $402
6. 15% of $90

Find the missing items in this purchase order.

7.

Quan.	Item	Unit cost	Cost	% discount	Net cost
5	hinges	.35	1.75	20%	1.40
7	locks	.40	?	25%	?
2	sockets	1.15	?	30%	?
4	switches	2.35	?	15%	?
				Total net cost	?

Copy and complete this purchase order.

8.

ACME PLUMBING SUPPLY CO.

Purchase Order No. 4279

Date: 4/5
For: Ortega's Plumbing   Phone: 443-2921
Address: 1434 Ridge Blvd.   City: Waco, Texas

QUANTITY	ITEM	UNIT COST	COST	DISCOUNT	NET COST
5	$\frac{3}{4}''$ unions	.85	4.25	20%	3.40
8	$\frac{3}{4}''$ 45° ells	.65		30%	
4	$\frac{3}{4}''$ valves	2.50		25%	
18	$\frac{3}{4}'' \times 6''$ nozzles	.95		15%	
5	$\frac{3}{4}''$ faucets	2.00		20%	
12	$\frac{3}{4}''$ 90° ells	.75		30%	
18	$\frac{3}{4}''$ fittings	.35		35%	
8	$\frac{3}{4}''$ T-joints	.55		20%	

Total _____
Less 2% Cash Discount _____
Net Total _____

118   Technical and industrial computations

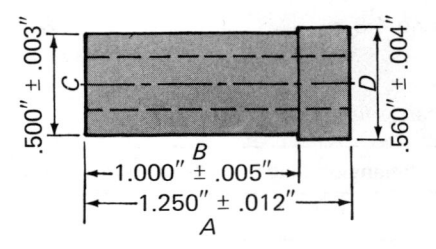

Figure 4-16

4-9 MACHINE TOLERANCE

The bushing in Figure 4-16 must be machined to certain limits of precision. Notice that dimension A must be $1.250'' \pm .012''$. (The symbol $\pm$ is read "plus or minus.") These figures show that the basic size of A is $1.250''$. However, to allow for errors in manufacturing, the length of A can vary from $1.238''$ to $1.262''$ and still be acceptable. The measurements $1.238''$ and $1.262''$ are called the *dimension limits*. The difference between the dimension limits is called the *tolerance*. The tolerance of dimension A is $1.262'' - 1.238'' = .024''$.

Example 1

Find the dimension limits and tolerance for dimension B in Figure 4-16.

Solution

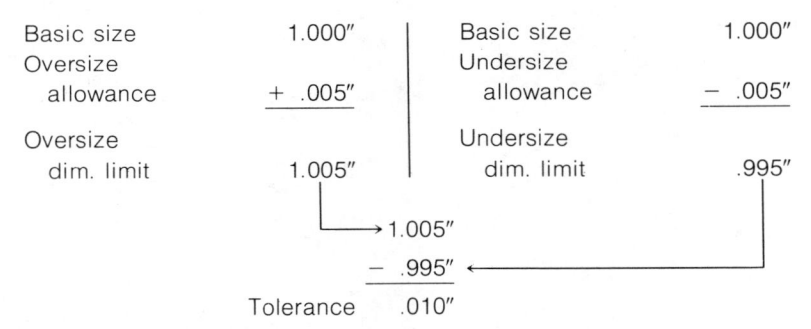

Basic size	1.000″	Basic size	1.000″
Oversize allowance	+ .005″	Undersize allowance	− .005″
Oversize dim. limit	1.005″	Undersize dim. limit	.995″

$$1.005''$$
$$- .995''$$
Tolerance .010″

The specifications for some machined parts require very high levels of precision. This means that the percent of *relative error* must be very small, 1% or less. To determine the percent of relative error, we can use this formula.

$$\text{Relative error} = \frac{1}{2} \times \frac{\text{Tolerance}}{\text{Basic size}}$$

Example 2

What is the percent of relative error for dimension A in 4-16?

Solution

Relative error = $\frac{1}{2} \times \frac{.024}{1.250} = \frac{1}{2} \times .0192 = .0096$

Percent of relative error = $.0096 \times 100 = .96\%$, or 1% (approx.)

Decimals, percents

EXERCISES Copy and complete Tables 4-2 and 4-3.

Table 4-2 *Table of dimension limits, U.S. system*

	Basic size	Tolerance	Dimension limits	% of relative error
1.	2.500″	±.125″	?	?
2.	4.375″	±.005″	?	?
3.	2.445″	±.002″	?	?
4.	.828″	±.005″	?	?
5.	1.875″	±.004″	?	?

Table 4-3 *Table of dimension limits, metric system*

	Basic size	Tolerance	Dimension limits	% of relative error
6.	322 mm	±.5 mm	?	?
7.	45 mm	±.5 mm	?	?
8.	625 mm	±.3 mm	?	?
9.	83 mm	±.2 mm	?	?
10.	39 mm	±.25 mm	?	?

11. Draw two tables like Tables 4-2 and 4-3. Complete the tables for dimensions A, B, C, D, and E in Figures 4-17 and 4-18.

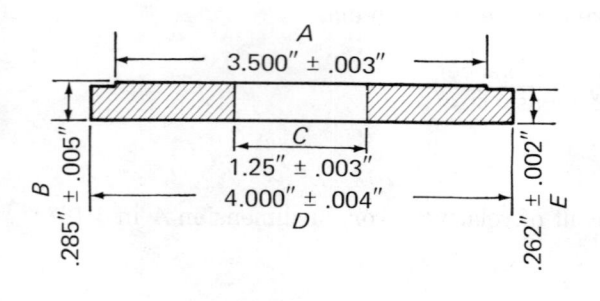

Figure 4-17

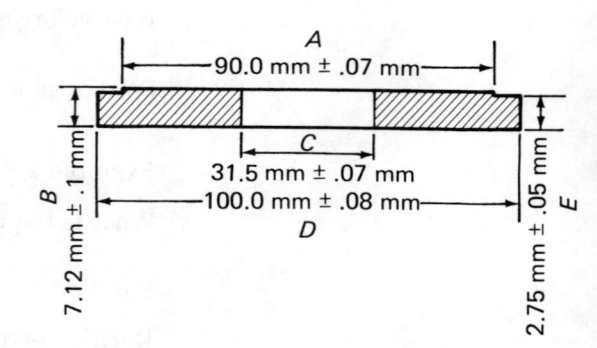

Figure 4-18

120 **Technical and industrial computations**

12. The cylinder bore on a certain engine is 4.062" ± .005". Find the dimension limits, the tolerance, and the percent of relative error.

13. The specifications for an economy automobile engine show the bore as 856 mm ± .5 mm. Find the dimension limits, the tolerance, and the percent of relative error.

4-10 ELECTRICAL TOLERANCE

Electronic resistors like the one in Figure 4-19 are used in radios and television sets to control the flow of electric current. Resistors are rated in *ohms*, the standard unit of electrical resistance.

Resistors have colored bands to indicate their ohms rating and their tolerance. Study the sample resistor and code shown in Table 4-4.

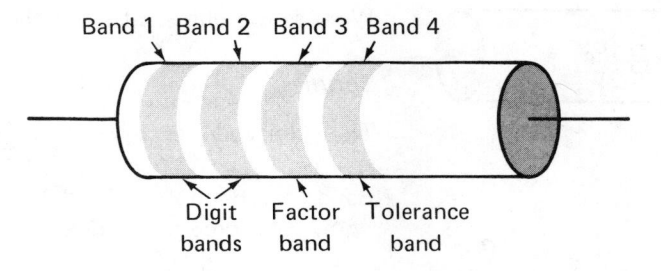

Figure 4-19

Table 4-4 *Resistor color code*

Color	Band 1	Band 2	Band 3	Band 4
Brown(BR)	1	1	× 10	None ± 20%
Red(R)	2	2	× 100	Silver(S) ± 10%
Orange(O)	3	3	× 1000	Gold(GD) ± 5%
Yellow(Y)	4	4	× 10,000	Red(R) ± 2%
Green(GN)	5	5	× 100,000	
Blue(BL)	6	6	× 1,000,000	
Violet(V)	7	7	Grey(GR) ÷ 100	
Grey(GY)	8	8	Gold(GD) ÷ 10	
White(W)	9	9	Black(BK) ÷ 1	
Black(BK)	0	0		

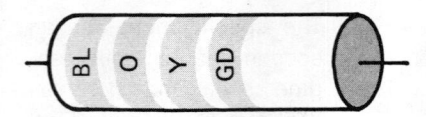

Figure 4-20

Example 1

Read the basic ohms rating of the resistor in Figure 4-20.

Solution

The color code is:

Band 1	Band 2	Band 3	Band 4
blue	orange	yellow	gold
6	3	× 10,000	±5%

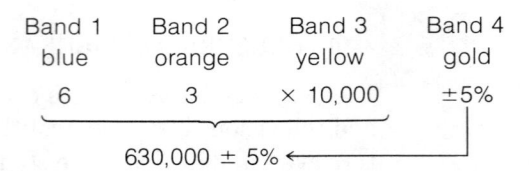

$$630,000 \pm 5\%$$

The resistance is 630,000 ohms, ± 5%.

Example 2

What is the ohms rating of the resistor in Figure 4-21?

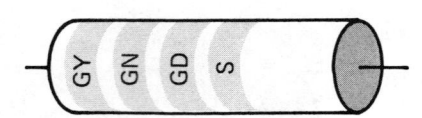

Figure 4-21

Solution

The color code is:

Band 1	Band 2	Band 3	Band 4
grey	green	gold	silver
8	5	÷ 10	± 10%

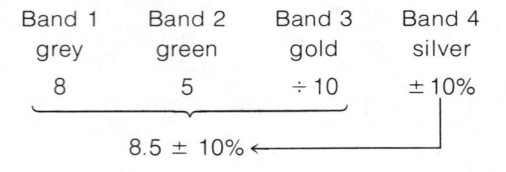

$$8.5 \pm 10\%$$

The resistance is 8.5 ohms, ± 10%.

Example 3

What are the tolerance limits of the resistor in Figure 4-21?

Solution

Band 4 is silver. The tolerance is ± 10% of the basic ohms rating, or .85 ohm. Thus,

```
   8.50              8.50
 + .85             - .85
   9.35              7.65
    ↑                 ↑
    └─── Tolerance limits ───┘
```

122 **Technical and industrial computations**

EXERCISES Find the basic ohms rating of the resistors. Use the color code on page 121.

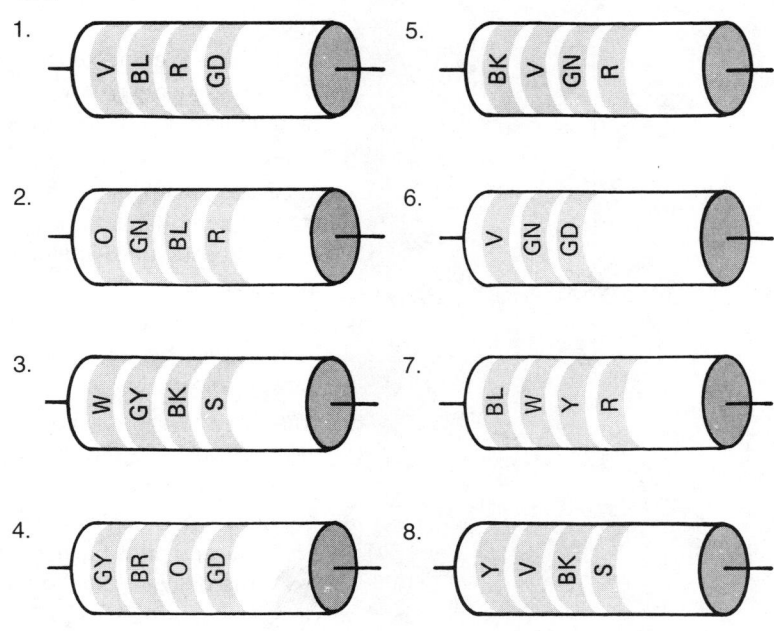

1. V BL R GD
2. O GN BL R
3. W GY BK S
4. GY BR O GD
5. BK V GN R
6. V GN GD
7. BL W Y R
8. Y V BK S

4-11 EFFICIENCY

When power is converted from one form to another, some of the power is wasted. The *efficiency* of a conversion is the ratio of power produced to power used. Efficiency is usually expressed as a percent.

Example 1

A turbine in a hydroelectric plant converts power with an efficiency rate of 60%. If the power of the water that drives the turbine is rated at 750,000 horsepower, how much electrical power is generated by the turbine?

Solution

60% of 750,000 = .60 × 750,000

= 450,000

The electrical output is about 450,000 horsepower.

EXERCISES

1. An automobile engine rated at 150 horsepower delivers only 120 horsepower to the transmission. What is the efficiency of the engine?

2. An improved, computerized welding machine that used to complete 75 pieces per day now completes 90 pieces each day. By what percent did its efficiency increase?

3. A metal machine company replaced an old lathe with a numerically controlled machine. With the old lathe, a machinist could produce 124 finished items in one day. With the new machine, production has increased to 284 items a day. What is the percent of increase in production?

4. A torque converter behind a 222-horsepower engine loses 36 horsepower because of fluid transmission. What percent of the power is lost?

5. The input to a generator powered by a nuclear reactor is 15,000 horsepower. The efficiency of the generator is 63%. What horsepower is delivered?

6. An automobile engine develops 316 horsepower. The transmission and rear-end drive have a combined efficiency of 68%. What horsepower is delivered to the rear wheels?

7. A duplicating machine can make 9 copies every 15 seconds. It is replaced by a newer model that can make 48 copies per minute. How much more efficient is the newer machine in copies per minute? in efficiency percentage?

Taking inventory

1. *Decimals* are fractions whose denominators are 10, or 100, or 1000, and so on. (p. 96)

2. To add or subtract decimals, write the numbers in a column with the decimal points in line. Then proceed as with whole numbers. (p. 102)

3. The number of decimal places in the product of two decimals equals the sum of the number of decimal places in the factors. (p. 104)

4. *Micrometers* are instruments used for making precise measurements. (p. 107)

5. When decimal fractions are divided, the divisor should first be changed to a whole number. (p. 113)

6. A *percent* is a decimal fraction whose denominator is understood to be 100. (p. 114)

7. A *discount* is a reduction in the marked price of an item. The discount price is called the *net price*. (p. 117)
8. *Tolerance* is the difference between the greatest and smallest acceptable dimensions or specifications. (p. 119)
9. A tolerance is usually given in specifying the size of a machined item or in specifying the ohms rating of electrical resistors. (pp. 119, 121)
10. The *efficiency* of a machine is the ratio of the power produced (output) to the power used (input). (p. 123)

Measuring your skills

Write as decimals. (4-1, 4-2)

1. $\frac{9}{10}$ 2. $\frac{67}{100}$ 3. $\frac{3}{8}$

Write as fractions. (4-1, 4-2)

4. .25 5. .4 6. .65

U.S. system

Add or subtract. (4-3)

7. 3.72"
   +5.69"

8. 19.637"
   − 8.298"

9. Find the missing dimensions.

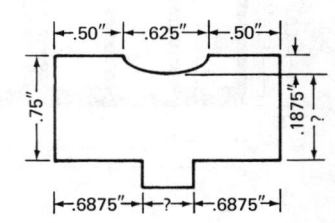

Multiply. (4-4)

10. 3.778"
    × .4

11. 29.73"
    × 13.4

Metric system

Add or subtract. (4-3)

17. 3.217 cm
    +4.693 cm

18. 14.09 mm
    − 6.97 mm

19. Find the missing dimensions.

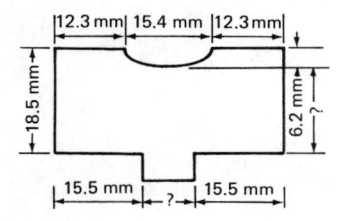

Multiply. (4-4)

20. 8.11 cm
    × 1.5

21. 13.45 cm
    × 17.3

Decimals, percents 125

12. Gold leaf has a thickness of .03". What is the thickness of seven sheets of gold leaf?

Read the micrometer. (4-5)

13.

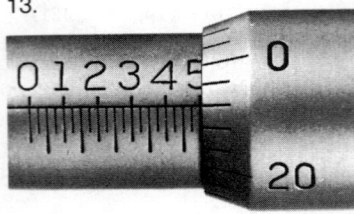

Divide. Give your answers to the nearest thousandth. (4-6)

14. 3.7" ÷ .58
15. 9.12" ÷ 3.9
16. A punch press stamps lantern reflectors from a sheet of aluminum, as shown below. Dimension A shows how much is used for each piece. How many pieces can be stamped from the length indicated? (4-6)

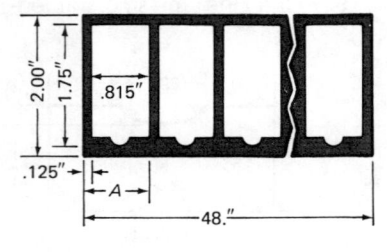

22. A strip of copper used in making printed circuits is .4 mm thick. Find the thickness of nine layers of this material.

Read the micrometer. (4-5)

23.

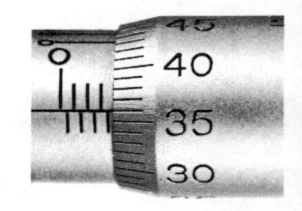

Divide. (4-6)

24. 2.1 mm ÷ 5.1
25. 13.07 mm ÷ .08
26. Solve Exercise 16, using the figure below. (4-6)

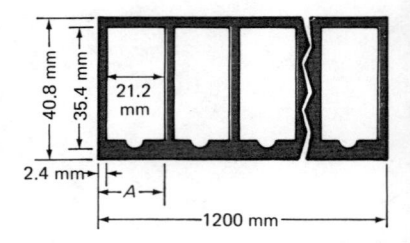

Write the percents as fractions. (4-7)

27. 16%   28. 65%   29. 113%

Write the percents as decimals. (4-7)

30. 40%   31. 25%   32. 120%

Write as percents. (4-7)

33. .6   34. 3.5   35. $\frac{4}{5}$ 36. $\frac{5}{12}$

37. An offset printing press retails for $7395. Find the net price if a 5% discount is allowed for full payment within 90 days. **(4-8)**

38. Find the dimension limits, tolerance, and percent of relative error. **(4-9)**

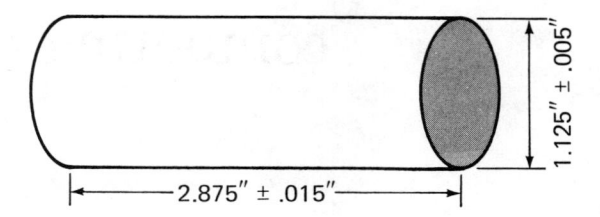

Give the ohms readings and the tolerance of the resistors. **(4-10)**

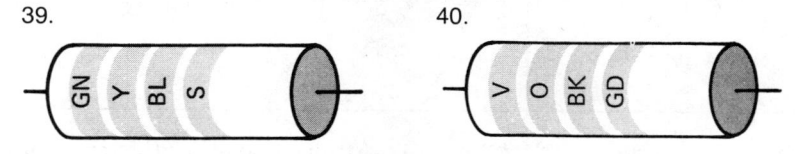

39.

40.

41. The input for an electric motor is 8000 watts. The output is 7200 watts. What is the percent efficiency? **(4-11)**

5 Using hand-held calculators and other computing devices

After completing this chapter, you should be able to:
1. Use your calculator to check arithmetic.
2. Use your calculator to solve problems.
3. Use your calculator to make computations.
4. Identify other computing devices and know where they are used.

Hand-held calculators

5-1 TYPES OF CALCULATORS

All the mathematics presented in the first four chapters is essential for work in every technical field. To help you use the mathematics you have learned, you may wish to use a hand-held calculator. Figures 5-1 through 5-4 show several different kinds.

Figure 5-1 shows a basic calculator that you can use to compute all arithmetic problems. Figure 5-2 shows a calculator that you can use in more advanced and scientific work. After you have completed Part 2 of this book, this type of calculator will be useful. Figure 5-3 shows a calculator that can help you with business computations. The calculator in Figure 5-4 differs from the others only in that the information is printed on a tape.

All calculators operate in one of two basic ways: Reverse Polish notation or algebraic logic. You will have to read the instructions supplied with your calculator to learn how to use it.

Figure 5-1

Figure 5-2

Example 1

Use your calculator to check the following addition problem.

56 in. + 8 in. + 43 in. + 29 in. = 136 in.

Example 2

Use your calculator to check the following subtraction problem.

981 m − 546 m = 435 m

Figure 5-3

Figure 5-4

Using hand-held calculators and other computing devices   129

EXERCISES Use your calculator to check the following problems.

1.   26 yd   2.   457 rpm
     +72 yd   −126 rpm
      98 yd   331 rpm

3.   156   4. $8\overline{)328}$ = 41
    ×   7
    1092

5. 12 tons + 121 tons + 56 tons + 7 tons + 205 tons = 401 tons
6. 234 cc − 78 cc = 156 cc
7. 325 × 14 = 4550
8. 7260 ÷ 121 = 60

Use your calculator to check the following problems. Circle the problems with the correct answer.

9.   846 ft   10.   4472 rpm
     209 ft   −2593 rpm
      98 ft   1879 rpm
     127 ft
     556 ft
   +342 ft
   2278 ft

11.   10179   12. $109\overline{)72594}$ = 676
    ×   253
    2575287

5-2 MAKING CALCULATIONS

Your calculator makes addition, subtraction, multiplication, and division of whole numbers very easy. You can also perform these same operations on integers very easily when you use your calculator. Make sure you know how to express signed numbers on your calculator.

Example 1

Use your calculator to find which of the following expressions gives the same answer as (23 + 58) + 91.

a. (23 + 91) + (58 + 91)

b. 71 + 91

c. 23 + (58 + 91)

Solution

Using your calculator, you find that 23 + 58 = 81 and 81 + 91 = 172. Also, 58 + 91 = 149 and 23 + 149 = 172. Therefore, 23 + (58 + 91) is the correct answer.

Example 2

Estimate the answer to the following multiplication problem. Then use your calculator to find the exact answer.

$$\begin{array}{r} 326 \\ \times\ 48 \\ \hline \end{array}$$

Solution

Rounding off 326 to 300 and 48 to 50, we have 300 × 50 = 15,000. Using your calculator, you find that 326 × 48 = 15,648.

Example 3

Add the following integers: $^-9 + {}^+11 + {}^+8 + {}^-7$.

Solution

Using your calculator, enter $^-9$ and add 11 to get 2; add 8 to get 10; enter $^-7$ and add to get 3.

$^-9 + {}^+11 + {}^+8 + {}^-7 = 3$

Example 4

Subtract: $^-22 - {}^-13$.

Solution

When you use your calculator, you do not need to change the sign of the second number and then proceed as in addition. Simply enter $^-22$ and $^-13$, then push the subtract button to obtain $^-22 - {}^-13 = {}^-9$.

EXERCISES Use your calculator to do the following problems.

1. 56 in. + 8 in. + 43 in. + 29 in.
2. 596 m − 321 m
3. 956
   × 72
4. 15,876 ÷ 24
5. $^{+}10 + {^{+}4} + {^{-}8}$
6. $^{+}100 - {^{-}80} + {^{-}50} - {^{-}35}$
7. $^{-}212$
   $-{^{-}114}$
8. $^{-}2756$
   × $^{-}23$
9. $^{+}12 \times {^{-}3} \times {^{-}20}$
10. $^{-}576 \div {^{+}4}$

Estimate the answer to each of the following problems. Then use your calculator to find the exact answer.

11. 596 ft
    207 ft
    +113 ft
12. 457 rpm
    −126 rpm
13. 784
    × 32
14. 109)72594
15. $^{-}410$
    $^{+}375$
    $^{+}512$
    $+{^{-}276}$
16. $\dfrac{^{+}5280}{^{-}50}$
17. $^{-}504$
    ×$^{+}103$

Use your calculator to do the following mixed calculations.

18. $(^{+}10 - {^{-}5} \times {^{+}12})/(^{+}12 + {^{+}2})$
19. $(^{-}12 \times {^{-}15} \times {^{+}2})/(^{-}3 \times {^{+}4} \times {^{-}2})$
20. $(5 + 5 + 5 + 5 + 5 + 5)/(6 + 6 + 6 + 6 + 6)$

5-3 SOLVING PROBLEMS

The hand-held calculator can speed up much of our work in solving problems. As long as we know which operations to perform when solving a problem, the calculator will give us the answer much faster than we can get it using paper and pencil. In this respect, the calculator is a very useful tool in all technical work.

Example 1

Find the length shown by the letter A on the diagram.

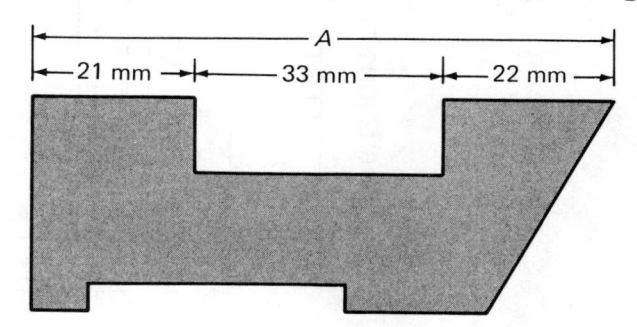

Solution

The length of A is found by adding 21 mm, 33 mm, and 22 mm. Using your calculator, you find that 21 mm + 33 mm + 22 mm = 76 mm.

Example 2

The sales receipts of a printing firm indicated the following sales for the first six months of this year:

January	$ 98,524.00
February	102,671.00
March	100,159.00
April	75,778.00
May	87,234.00
June	95,872.00

What were the average monthly sales for the first six months?

Solution

To find the average, you must add the sales figures and divide by 6. Using your calculator, you find that the total sales were $560,238.00. Dividing by 6, 560,238 ÷ 6 = 93,373. The average monthly sales for the first six months were $93,373.00.

APPLIANCE RETAIL SALES

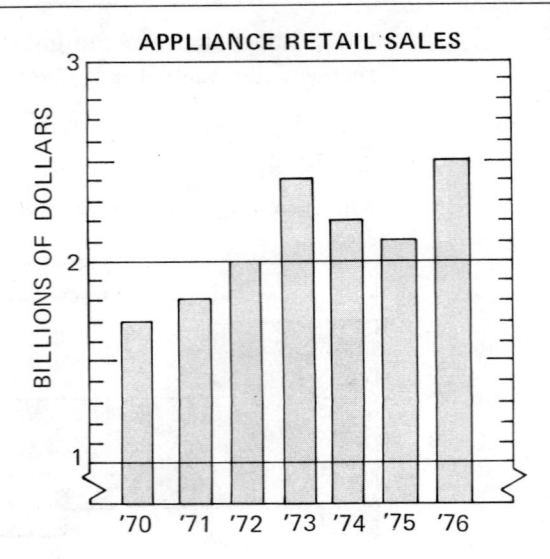

Figure 5-5

EXERCISES Use Figure 5-5 and your calculator to answer the following questions.

1. What were the total appliance sales, in billions of dollars, for the seven years shown in the graph?
2. By how much did 1976 sales exceed the sales in 1970?
3. The appliance industry expects to double its 1976 sales figure by 1980. What is the projected 1980 sales figure?
4. What is the average amount of sales for the years shown on the graph?
5. Use your calculator to find the lengths shown by letters on the diagram.

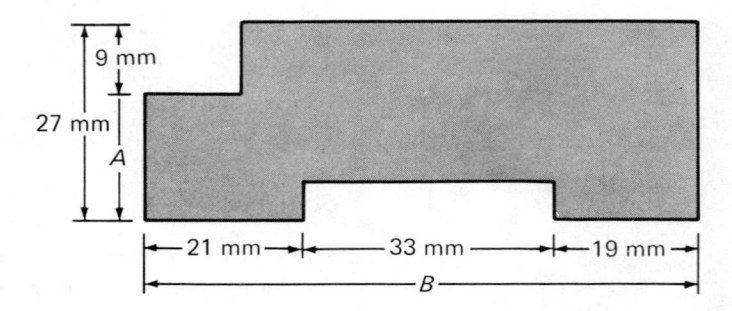

Use your calculator to solve the following problems involving integers.

6. Find the sum of the voltages shown in the following diagram.

134 Technical and industrial computations

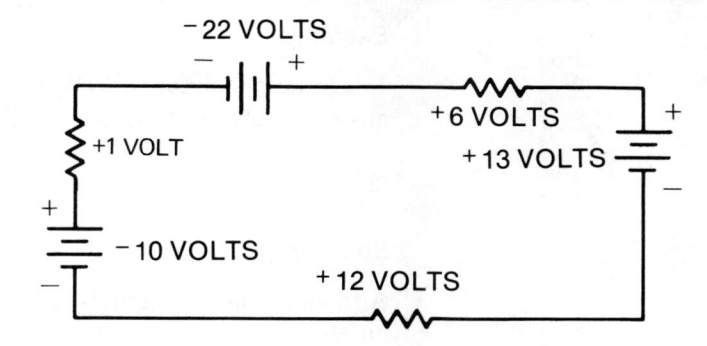

```
        −22 VOLTS
       −  ||  +
   ┌───┤|├──────────w──────────┐
   │                  +6 VOLTS  │ +
   ≷ +1 VOLT          +13 VOLTS ═
   │                            │ −
 + │                            │
   ═ −10 VOLTS                  │
 − │          +12 VOLTS         │
   └──────────w─────────────────┘
```

7. The temperature of a supercooled chemical was recorded as $^-124°$. Gradually the chemical was heated to $^+32°$. What was the change in temperature of the chemical?

8. A metal rod expands 1 mm for each 5° between 0°C and 10°C, and 2 mm for each 5° from 10°C to 20°C. If the rod was 34 mm in length at 0°C, what will its length be when the temperature reaches 15°C?

9. In order to find the profit in dollars on a certain construction project, it is necessary to find the answer to the problem

$$\frac{20{,}000 + (^-5 \times {}^+3200)}{^+16}$$

Find the profit.

Advanced calculations

5-4 LONG DIVISION

Long-division problems usually require a great deal of work and time. Using a calculator makes division as easy as any other operation. Most division problems have a remainder. Calculators give the remainder as a decimal, usually to 8 or 9 places. We normally round off the decimals to 2 or 3 places.

Example 1

Use your calculator to find the answer to 32 ÷ 7. Round your answer to the nearest hundredth.

Solution

Your calculator may give the answer 32 ÷ 7 = 4.571428571. Rounding to hundredths, 32 ÷ 7 = 4.57.

Example 2

Use your calculator to solve the following problem. Round your answer to the nearest hundredth.

29)59.13

Solution

Although the dividend is a decimal, you can easily place the number in your calculator. Then divide by 29. The answer is 2.038965517. Rounded off to hundredths, the answer is 2.04.

EXERCISES Use your calculator to do the division problems.

1. 84 ÷ 6
2. 144 ÷ 12
3. 8)328
4. 23)529
5. 15,876 ÷ 24
6. 109)72594

Use your calculator to do the division problems, and give all answers rounded to the nearest hundredth.

7. 48 ÷ 17
8. 59.8 ÷ 50.3
9. 380)120
10. 23)68.9

Use your calculator to place the decimal point in each answer.

11. 259 / 25)64.83
12. 867 / .54)4.684
13. 975 / 1.3)12.685
14. 97357 / .07)68.15
15. 2.1 ÷ 120 = 175

5-5 WORKING WITH FRACTIONS AND DECIMALS

Calculators do not work fraction problems directly. All fractions must be converted to decimals. Using your calculator to work with decimals is as easy as using it to work with whole numbers. You must remember to place the decimal point in the calculator for each problem. When you use your calculator, you avoid having to find common denominators.

Example 1

Convert $\frac{3}{4}$ to a decimal.

Solution

Use your calculator to divide 3 by 4.

$\frac{3}{4} = 3 \div 4 = .75$

Example 2

Convert $\frac{7}{9}$ to a decimal, and round off to four places.

Solution

$\frac{7}{9} = 7 \div 9 = .777777777$

$\phantom{\frac{7}{9} = 7 \div 9} = .7778$

Example 3

Convert $3\frac{5}{16}$ to a decimal.

Solution

$3\frac{5}{16} = 3 + \frac{5}{16}$

$\phantom{3\frac{5}{16}} = 3 + (5 \div 16)$

$\phantom{3\frac{5}{16}} = 3 + .3125$

$\phantom{3\frac{5}{16}} = 3.3125$

Example 4

Use your calculator to convert the mixed numbers to decimals, then multiply. Round all decimals to three places.

$3\frac{5}{8} \times 12\frac{2}{3}$

Solution

$3\frac{5}{8} = 3 + \frac{5}{8} = 3 + .625 = 3.625$

$12\frac{2}{3} = 12 + \frac{2}{3} = 12 + .667 = 12.667$

$3\frac{5}{8} \times 12\frac{2}{3} = 3.625 \times 12.667$

$\phantom{3\frac{5}{8} \times 12\frac{2}{3}} = 45.917875$

$\phantom{3\frac{5}{8} \times 12\frac{2}{3}} = 45.918$

EXERCISES Convert the fractions to decimals and round the decimals to four places.

1. $\frac{7}{8}$ 2. $\frac{9}{16}$ 3. $\frac{17}{32}$ 4. $\frac{7}{10}$
5. $\frac{15}{8}$ 6. $\frac{8}{3}$ 7. $1\frac{3}{8}$ 8. $7\frac{7}{16}$

Use your calculator to work the decimal problems. Round your answers to hundredths.

9.   35
    ×1.6
    ─────

10.   5.49
    ×3.96
    ─────

11.   9.08
    ×.035
    ─────

12. 9.8/.5   13. 81.7/.52   14. 138/.32

Convert the fractions to decimals, work the problems, and round the answers to three places.

15. $27\frac{1}{3}$
     $38\frac{3}{4}$
    $+14\frac{1}{8}$
    ─────

16. $21\frac{3}{10}$
    $- 6\frac{9}{16}$
    ─────

17. $19\frac{1}{8}$
    $\times 7\frac{3}{5}$
    ─────

18. $39\frac{7}{10} \div 12\frac{2}{9}$

Use your calculator to solve the following problems. Round the answers to three places.

19. Find the missing dimensions in the following diagram.

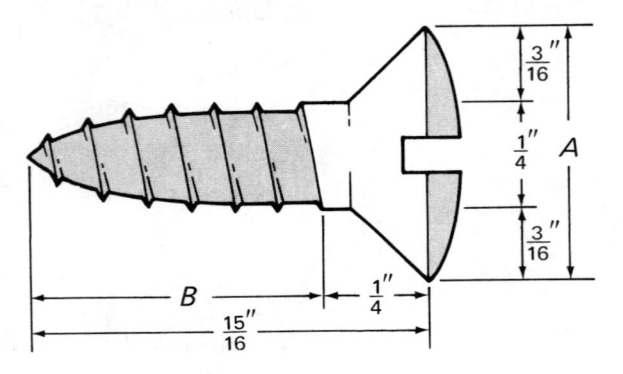

20. Find lengths A and B on the light switch plate.

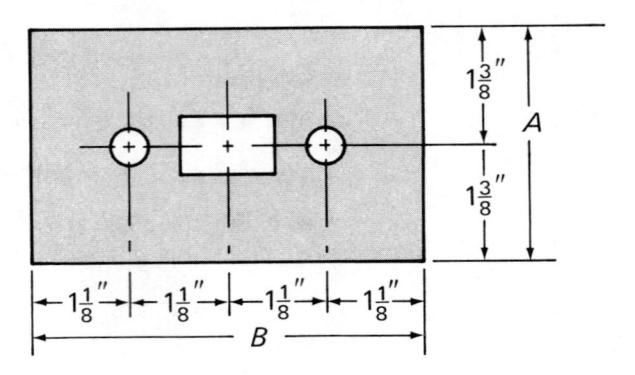

21. A spool of copper wire contains 50′ of wire. Four pieces are cut off. Their lengths are $12\frac{1}{8}''$, $14\frac{3}{4}''$, $20\frac{5}{8}''$, and $36\frac{9}{16}''$. How much wire remains?

22. Using a machinist's rule, an engineer found the dimensions for a sheet metal patch. The patch must be $9\frac{3}{10}''$ by $7\frac{7}{10}''$.

    a. What is the area of the patch in square inches? (Area = length × width)

    b. If it costs $3\frac{1}{2}$¢ per square inch to rustproof the material, what will it cost to treat this patch?

23. How many pieces of solid copper wire $2\frac{1}{2}''$ long can be cut from a piece 15′6″ long?

24. Find the inside diameter CD of the extruded plastic tubing. The outside diameter and the thickness are shown in the cross-section.

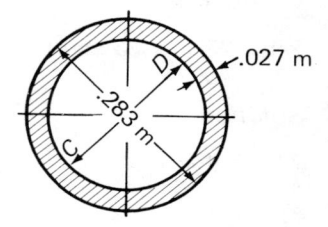

25. A mechanic must bore a hole large enough to seat the head of a bolt. The diameter of the hole must be 1.414 times the side of the bolt head. If the bolt is 3.2 cm on each side, what must be the diameter of the hole?

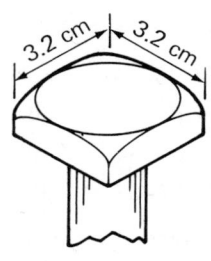

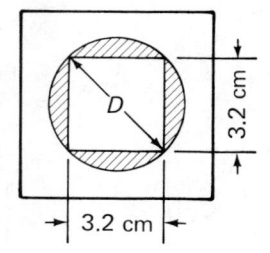

5-6 WORKING WITH PERCENTS

Your calculator can handle percents in one of two ways. If it has a percent button, then you can find 12% of 125 by multiplying 125 by 12 and pushing the percent button.

125 × 12% = 15

If your calculator does not have a percent button, then you must change 12% to a decimal before multiplying by 125.

$$12\% = .12$$
$$125 \times .12 = 15$$

Example 1

Convert 25.8% to a decimal.

Solution

$$25.8\% = .258$$

Example 2

Find 4% of 7.85 and round the answer to two places.

Solution

$$7.85 \times 4\% = .31$$

or

$$7.85 \times .04 = .31$$

EXERCISES

Convert the percents to decimals.

1. 78% 2. 3% 3. 25.3%
4. .2% 5. 108.2% 6. 93%

Find the following amounts.

7. 20% of $840   8. 30% of $560
9. 25% of $84   10. 12½% of $80
11. 6% of $402   12. 15% of $90

13. A turbine in a hydroelectric plant converts power with an efficiency of 62%. If the power of the water that drives the turbine is rated at 650,000 horsepower, how much electrical power is generated by the turbine?

14. An improved, computerized welding machine that used to complete 75 pieces per day now completes 90 pieces each day. By what percent did its efficiency increase?

15. An automobile engine develops 316 horsepower. The transmission and rear-end drive have a combined efficiency of 68%. What horsepower is delivered to the rear wheels?

Figure 5-6

5-7 DESK CALCULATORS AND COMPUTERS

People who do technical work often make their mathematical calculations while sitting at a desk. For these people, a desk calculator may be more convenient than a hand-held calculator. A desk calculator usually has larger keys, or buttons, and a larger display area than a hand-held calculator. Figure 5-6 illustrates a desk calculator with a display on which the results appear.

Some desk calculators also print the information and the results on a paper tape, which lets you check your calculations easily. Figure 5-7 shows a desk calculator that prints the results on paper tape and also flashes the results on the display.

People have used many kinds of calculators. One of the earliest was the abacus. Before the hand-held calculator, many people in technical work used a slide rule. All these instruments are now surpassed as calculators by the electronic computer. Computers can perform long and complex calculations much faster than calculators. They can also store information to be recalled later when it is needed in performing calculations. Computers can be designed and programmed to perform tasks based on mathematical calculations. These tasks may include running a sawmill and operating a stamping machine, as well as many others.

You need advanced knowledge to program a computer to perform various tasks. Once a computer is programmed, it can be as easy to operate as a calculator. Each year more and more technical work is being done with the aid of computers.

Figure 5-7

EXERCISES
1. Select any technical area. Go to the library and try to find information that will tell how the computer is used in that area. Report to your class.
2. Find out what advanced training you will need in order to operate a computer in your technical area.

Taking inventory

1. There are many kinds of hand-held calculators. (**pp. 128-129**)
2. Calculators can be used to check arithmetic. (**p. 129**)
3. Calculators make addition, subtraction, multiplication, and division of whole numbers and integers very easy. (**p. 130**)
4. When you know how to solve a problem, a calculator will help you get the answer rapidly. (**pp. 132-133**)
5. Calculators make long division very easy and fast. (**p. 135**)
6. Calculators change all fractions to decimals. (**p. 136**)
7. On calculators with a percent button, percent calculations are as simple as multiplication. (**p. 139**)
8. Computers work much faster and have more uses than calculators. (**p. 141**)

Measuring your skills

Use your calculator to check the following problems. (**5-1**)

1. 271 m
    235 m
    +418 m
    924 m

2. 321 km
    17 km
    106 km
    + 56 km
    500 km

3. $5761.00
    − 3975.00
    $1786.00

4. 5976
    × 105
    627480

5. 47
    225)10575

Estimate the answer to each of the following problems. Then use your calculator to find the exact answer. (**5-2**)

6. 56 ft
    109 ft
    1257 ft
    250 ft
    +1009 ft

7. 4795 km
    −1886 km

8. 395
    × 49

9. 96)3264

10. [$^+28 - (^-6 \times {}^+7)]/(^+12 - {}^+2)$

Use your calculator to solve the following problems. (5-3)

11. A contractor's payroll for the first six months was recorded as follows:

January:	$12,650.00
February:	10,725.00
March:	11,435.00
April:	13,600.00
May:	15,850.00
June:	21,450.00

    What was the contractor's average monthly payroll for the six months?

Use your calculator to do the following problems, and round your answer to the nearest hundredth. (5-4, 5-5)

12. .07)68.15

13. 7.52
     ×2.76

14. $14\frac{5}{8}$
     $- 6\frac{9}{16}$

15. $54\frac{7}{10} \div 12\frac{1}{9}$

16. What is the total length of a wire formed by splicing four pieces of wire of lengths $10\frac{1}{8}''$, $12\frac{3}{4}''$, $31\frac{5}{8}''$, and $33\frac{9}{16}''$?

Use your calculator to do the following problems involving percents. (5-6)

17. 12% of 560
18. 102% of 41
19. 10.5% of 120
20. $33\frac{1}{3}$% of 52

Measuring your progress

Read the meters. (1-1, 1-2, 1-3, 1-4)

1.

2.

Table 5-1 *Freight shipments*

Month	Tons
Jan.	34
Feb.	39
Mar.	36
Apr.	46
May	51
June	50

3. Show the information in Table 5-1 in a bar graph and a broken-line graph. **(1-5, 1-6, 1-7)**

4. Give the length of each line in sixteenths of an inch and in millimeters. **(1-8, 1-9, 1-10, 1-11)**

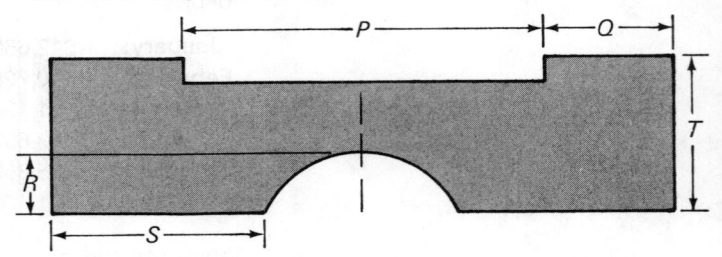

5. Copy the drawing. Measure the parts of the diagram that are not dimensioned. Show all the dimensions on your copy. **(1-12, 1-13, 1-14)**

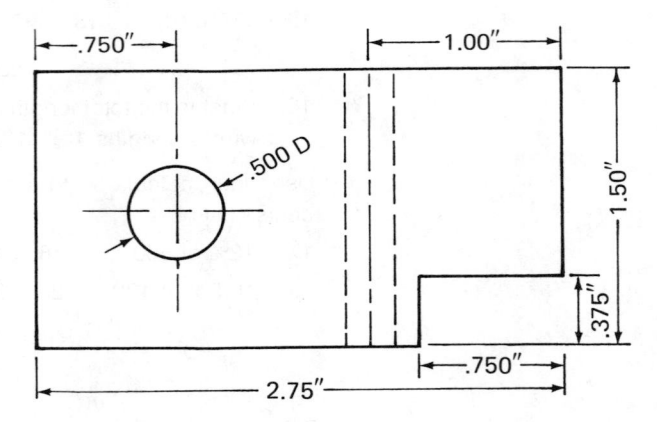

Add or subtract. **(2-1, 2-2, 2-3, 2-4, 2-5)**

6. 879
    +322

7. 172
    −158

8. $^+84 + {}^-27$

9. $^-54 + {}^+37$

10. $^-43 - {}^-13$

Multiply or divide. **(2-6, 2-7)**

11. 234
    × 25

12. $23\overline{)7843}$

13. $^+18 \times {}^-2$

144 Technical and industrial computations

14. $^-15 \times {}^-4$ 15. $^-125 \div {}^+25$

Add or subtract. (3-1, 3-2, 3-3)

16. $\frac{15}{32}'' - \frac{12}{32}''$ 17. $\frac{3}{8}'' + \frac{3}{16}'' + \frac{3}{32}''$ 18. $\frac{7}{16}'' - \frac{3}{8}''$

19. Find the missing dimensions on the C clamp. (3-4, 3-5, 3-6)

20. To fit between two fence posts, a split rail 8' long must be shortened by $15\frac{3}{4}''$. What will be the length of the rail after sawing? (3-7, 3-8)

Multiply. Cancel wherever possible. (3-9, 3-10, 3-11, 3-12)

21. $8 \times \frac{5}{16}$ 22. $3\frac{1}{8} \times 4$ 23. $1\frac{5}{8} \times 2\frac{1}{4}$

Divide. (3-13, 3-14, 3-15, 3-16, 3-17)

24. $\frac{3}{4} \div 9$ 25. $\frac{5}{8} \div \frac{7}{16}$ 26. $\frac{1}{4} \div 1\frac{3}{16}$

Write the fractions as decimals. (4-1, 4-2)

27. $\frac{3}{8}$ 28. $\frac{7}{20}$ 29. $\frac{5}{16}$

Write the decimals as fractions. (4-1, 4-2)

30. .3   31. .75   32. .375

Perform the indicated operation. (4-3, 4-4, 4-5, 4-6)

33. 3.172"
    +2.973"

34. 7.03 mm
    −1.73 mm

35. 4.273
    × .27

36. $.87 \overline{)27.84}$

37. During a sale a set of high-speed drills is marked down 20%. If the regular price is $9.95, how much is the sale price? (4-7, 4-8, 4-9, 4-10, 4-11)

38. What are the dimension limits and tolerance of A and B? (4-7, 4-8, 4-9, 4-10, 4-11)

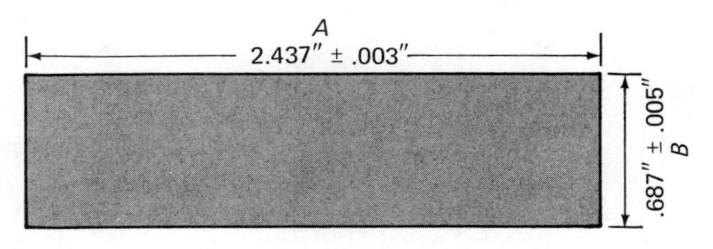

Use your calculator to perform the indicated operation. (5-1, 5-2, 5-3)

39. $^+212 + {}^-140 + {}^+20 + {}^-85$
40. $^+16 - {}^+10 + {}^-5 - {}^-35$
41. $^-256 \times {}^+42$ 42. $^+1288 \div {}^-56$

43. Use your calculator to find the lengths shown by letters on the diagram.

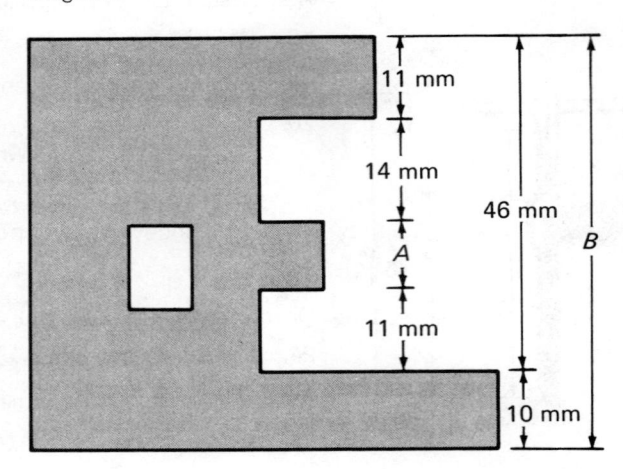

Use your calculator to do the following problems. Round the answers to three places. (5-4, 5-5, 5-6)

44. $144 \overline{)7289}$ 45. $1.05 \overline{)19.578}$

46.   18.21   47. $96\frac{1}{3}$
   $\times\ .035$ $12\frac{5}{8}$
                        $+33\frac{7}{16}$

48. How many pieces of telephone jack wire 4'6" long can be cut from a piece 50'0" long?
49. What is 15.2% of $1250?
50. A mechanic's hydraulic lift jack retails for $2595. Find the net price if a 5% discount is allowed for full payment within 90 days.

2 Technical and industrial formulas

In this section of the text you will learn the procedures for solving industrial problems involving algebraic, geometric, and trigonometric formulas.

Chapters 6 through 9 introduce the algebraic techniques you will need to solve mathematical problems related to industry. You will learn to solve equations; to write and use formulas to determine perimeters, areas, and volumes; and to make conversions within and between the United States system and the metric system of measurement. You will also learn to read, draw, and interpret graphs of linear and nonlinear equations.

Chapter 10 introduces the basic concepts of geometry that are related to industrial problems. Topics covered include angle relationships, parallel and perpendicular lines, congruent and similar triangles, and geometric constructions.

Chapter 11 continues the geometric theme combined with some algebraic concepts. This chapter introduces ratio and proportion and discusses their application to similar figures and scale drawings.

Part 2 concludes with an introduction to trigonometry. We define the sine, cosine, and tangent ratios, and we discuss their application in solving triangles and related industrial problems.

6 Equations, formulas

After completing this chapter, you should be able to:
1. Simplify number expressions.
2. Solve equations by using addition, subtraction, multiplication, and division.
3. Use formulas to solve typical industrial problems.

Components of equations

6-1 TERMS, EXPRESSIONS, AND EQUATIONS

A strip of stainless steel molding must go around the edge of the table in Figure 6-1. To cut the strip to the correct length, we need to find the perimeter of the table top. We can show the calculation as follows:

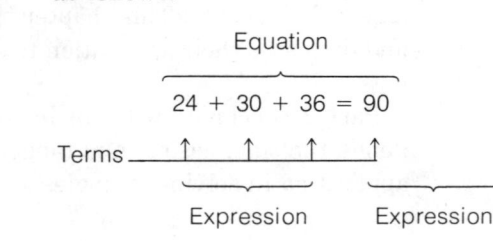

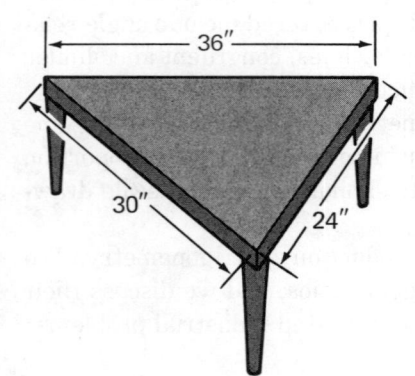

$$\underbrace{\underbrace{24 + 30 + 36}_{\text{Expression}} = \underbrace{90}_{\text{Expression}}}_{\text{Equation}}$$

with terms 24, 30, 36, and 90.

Figure 6-1

As you can see, an *equation* is a statement showing that two *expressions* are equal. An expression may consist of several *terms*, like 24 + 30 + 36. Or it may have just one term, like 90.

148

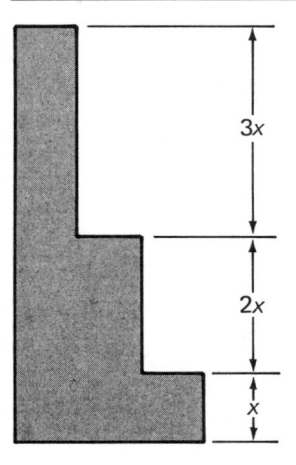

Figure 6-2

Sometimes in our work we deal with expressions that contain "unknown" terms. For example, Figure 6-2 shows the pattern for a step-block. Because step-blocks can be made in any size, we use the dimensions x, $2x$, and $3x$ to show the relative size of the steps. Thus, the steps could be 1 ft, 2 ft, and 3 ft, or 1 cm, 2 cm, and 3 cm.

The overall length of the step-block is $3x + 2x + x$. To make calculations easier, we can combine *similar terms*. Thus,

Length = 3x + 2x + x
 = (3 + 2 + 1) x
 = 6x

The overall length of the step-block would be 6 ft or 6 cm, depending on the unit of measure represented by x.

Study the following examples of combining terms.

Example 1

Combine $6m + 3m - 2m$.

Solution

6m + 3m − 2m = (6 + 3 − 2)m
 = 7m

Example 2

Combine $\ell + w + \ell + w$.

Solution

ℓ + w + ℓ + w = (ℓ + ℓ) + (w + w)
 = (1 + 1)ℓ + (1 + 1)w
 = 2ℓ + 2w

> Notice that we combine the ℓ terms with ℓ terms and the w terms with w terms. We cannot combine the ℓ and w terms because they are not similar terms.

Equations, formulas 149

Example 3

Write an expression for the length of the template.

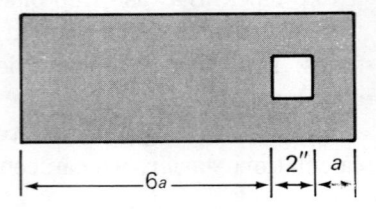

Solution

Length = 6a + 2 + a

= (6a + a) + 2

= (6 + 1)a + 2

= 7a + 2

Notice again that we can combine only the similar terms.

Example 4

Combine $8x - (3x + 4x)$.

Solution

8x − (3x + 4x)

= 8x − [(3 + 4)x]

= 8x − 7x

= (8 − 7)x

= x

Example 5

Write an expression for the perimeter (the sum of the lengths of the sides) of the cross-section of the V block on the next page.

Solution

x + y + .25y + .5x + .5x + .25y + y

= (x + .5x + .5x) + (y + .25y + .25y + y)

= (1 + .5 + .5)x + (1 + .25 + .25 + 1)y

= 2x + 2.5y

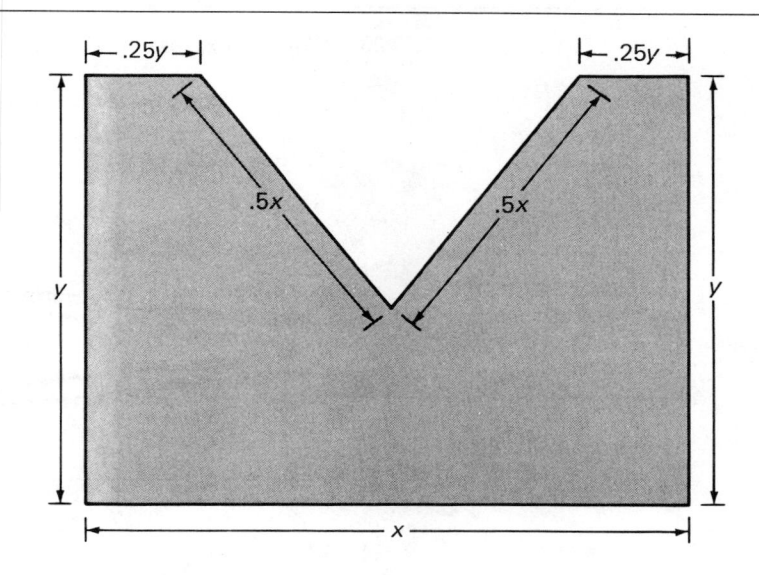

EXERCISES Combine the similar terms to simplify the expressions.

1. $4a + a$
2. $7c - 2c$
3. $3x + x$
4. $5t - 2t$
5. $x + x$
6. $3y + 2y$
7. $5m + 3m + 2m$
8. $2t - t$
9. $6s - 3s$
10. $5y + 2y + 9y$
11. $10r - 3r$
12. $2x + 2x + 2x$
13. $3x + 4x + 1$
14. $9y - 2y + 6$
15. $4t + 8 + 3t$
16. $2s + 2s + 5$
17. $3x + 2y + 4x + 2y$
18. $9a + 8b + 2b + 4a$
19. $5a - (3a + a)$
20. $12p - (7p + 2p)$

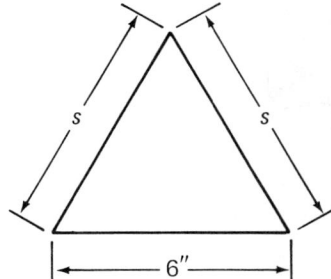

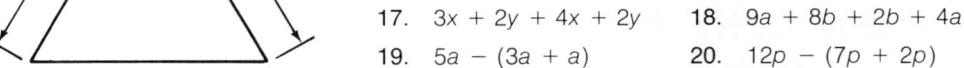

21. Write an expression for the perimeter of the triangle shown at the left.

22. A square shaft is milled from a round steel bar. Write an expression for the diameter of the bar.

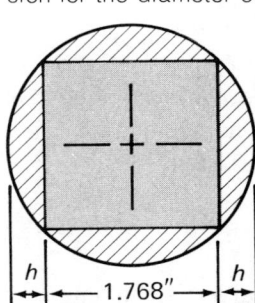

23. Write an expression for the length of the inside calipers.

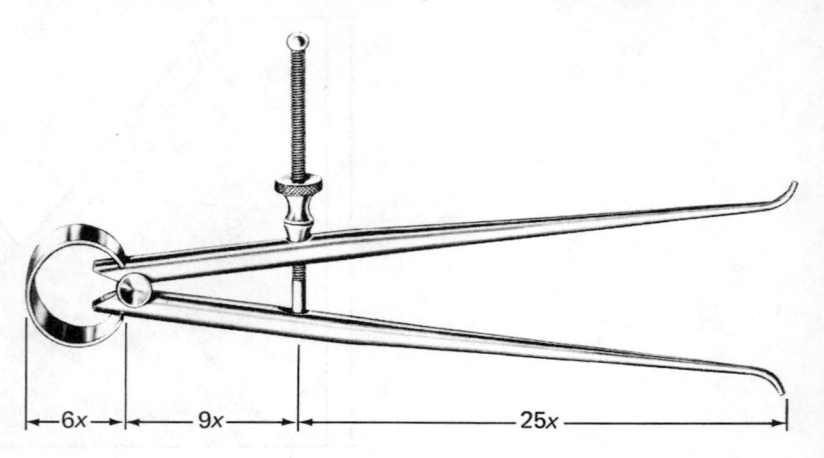

|←6x→|←9x→|←25x→|

24. Write an expression for the diameter of the largest pulley in the diagram.

6-2 EXPONENTS

The formula for the area of a square, as in Figure 6-3, can be written as

A = s × s

where s is the length of the side.

The formula for the volume of a cube, as in Figure 6-4, can be written as

V = e × e × e

where e is the length of the edge.

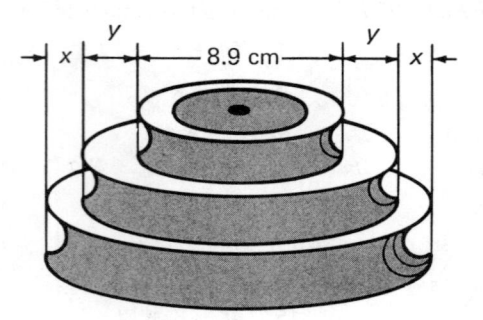

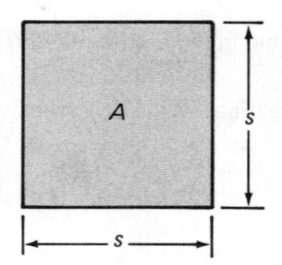

Figure 6-3

152 Technical and industrial formulas

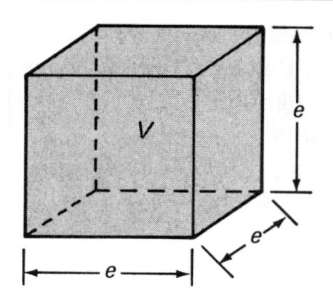

Figure 6-4

Sometimes formulas can be written with special symbols that help simplify them. One such symbol is the *exponent*. An *exponent* is a small number written above and to the right of another number. It indicates how many times the number is used as a factor. For example,

$$2^3 = 2 \times 2 \times 2 = 8$$

or $\quad 3^4 = 3 \times 3 \times 3 \times 3 = 81$

We can use an exponent to simplify the area and volume formulas. Thus,

$$A = s \times s = s^2$$

and $\quad V = e \times e \times e = e^3$

Many of the formulas that you will use in industry and the trades will contain exponents. Study these examples.

Example 1

What is the area of a square microscope slide that has a side 2.7 cm long?

Solution

Area $= s^2 = (2.7)^2 = 2.7 \times 2.7 = 7.29$
The area is 7.29 sq cm.

Example 2

Find the volume of a cube-shaped laboratory container that measures 6.3 cm along an edge.

Solution

Volume $= e^3 = (6.3)^3 = 6.3 \times 6.3 \times 6.3 = 250.047$
The volume is 250 cu cm (approx.).

It is often convenient to write very large or very small numbers as powers of 10.

$\quad\quad 1,000 = 10^3$

$1,000,000 = 10^6$

$2,000,000 = 2 \times 10^6$

Equations, formulas   153

The distance around the earth is approximately 25,000 mi, or 2.5×10^4 mi. One light year, the distance that light travels in one year, is approximately 9,460,000,000,000 km or 9.46×10^{12} km.

For very small numbers we use negative exponents.

$$10^{-3} = \frac{1}{10^3} = \frac{1}{1000} = .001$$

$$10^{-8} = \frac{1}{10^8} = .00000001$$

$$.000037 = 3.7 \times 10^{-5}$$

We can use powers of 10 to simplify calculations using very large or very small numbers.

Example 3

Multiply 6,500,000 by 7,400,000.

Solution

$6,500,000 = 6.5 \times 10^6$

$7,400,000 = 7.4 \times 10^6$

$(6,500,000)(7,400,000)$

$\quad = (6.5 \times 10^6)(7.4 \times 10^6)$

$\quad = (6.5)(7.4)(10^6)(10^6)$

$\quad = (6.5)(7.4)(10^{6+6})$

$\quad = (6.5)(7.4)(10^{12})$

$\quad = 48.1 \times 10^{12}$

$\quad = 48,100,000,000,000$

Example 4

The distance from the earth to the moon is approximately 200,000 mi. A radio wave travels at 186,000 mi per sec. How long will it take the radio wave to reach the moon?

Solution

$$\frac{200,000}{186,000} = \frac{2 \times 10^5}{1.86 \times 10^5}$$

$$= \frac{2}{1.86} \times \frac{10^5}{10^5}$$

$$= 1.08 \text{ sec}$$

Example 5

Multiply .00037 by .000042.

Solution

$.00037 = 3.7 \times 10^{-4}$

$.000042 = 4.2 \times 10^{-5}$

$(.00037)(.000042)$

$\quad = (3.7)(10^{-4})(4.2)(10^{-5})$

$\quad = (3.7)(4.2)(10^{-5})(10^{-4})$

$\quad = (3.7)(4.2)(10^{-9})$

$\quad = 15.54 \times 10^{-9} = .00000001554$

EXERCISES Calculate.

1. 3^2
2. 5^2
3. 7^2
4. 10^3
5. 8^3
6. 4^2
7. 5^3
8. 6^3
9. 4^4
10. $(1.2)^2$
11. $(2.3)^2$
12. $(5.6)^2$
13. $(3.9)^3$
14. $(7.9)^2$
15. $(5.4)^3$
16. $(21.7)^2$
17. $(3.04)^2$
18. $(12.6)^3$

Write using powers of 10.

19. 1,000,000
20. 16,000
21. 6,700,000
22. .0001
23. .000045
24. .00000236

Calculate.

25. $6,900,000 \times 3000$
26. $.00005 \times .00025$
27. $640,000 \div 1600$
28. $.00016 \div .04$

U.S. system

29. Find the area of a square bench plate that has sides $8\frac{1}{4}''$ long. Use the formula $A = s^2$.

30. Find the volume of a cube-shaped storage bin that has an edge $2\frac{1}{8}'$ in length. Use the formula $V = e^3$.

Metric system

32. What is the area of a square pane of glass that is 22.5 cm on each side? Use the formula $A = s^2$.

33. A clear plastic cube is used to display photos. Find the volume of the cube if the edge is 12.2 cm long. Use the formula $V = e^3$.

31. The formula for the area of a circle is $A = 3.14r^2$, where r represents the radius of the circle. Find the area of a circular flower bed having a radius of 7'.

34. The radius of a circle on a bicycle-safety test course is 14 m. Find the area of the circular region. Use the formula $A = 3.14r^2$.

6-3 THE ORDER OF OPERATIONS

During your work you may come across an expression like $(2 \times 7) + [6 \div (3 - 1)]$. How would you simplify this?

To help us simplify expressions like this, we can use the steps in the flow chart in Figure 6-5.

Example 1

Simplify $2 \times 7 + 6 \div (3 - 1)$.

Solution

Step A $2 \times 7 + 6 \div (3 - 1)$

Step B $2 \times 7 + 6 \div 2$

Step C $14 + 3$

17

Example 2

Simplify $4 \times (9 + 2) \div 2 + 2$.

Solution

Step A $4 \times (9 + 2) \div 2 + 2$

Step B $4 \times 11 \div 2 + 2$

Step B $44 \div 2 + 2$

Step C $22 + 2$

24

156 Technical and industrial formulas

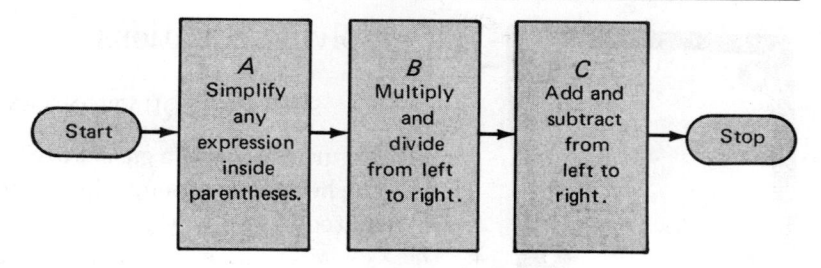

Figure 6-5

Example 3

Simplify $10^3 \div 25 - (3 \times 6)$.

Solution

Step A $10^3 \div 25 - \underline{(3 \times 6)}$

Step B $\underline{10^3} \div 25 - 18$

$\underline{1000 \div 25} - 18$

Step C $\underline{40 - 18}$

22

EXERCISES Simplify each expression.

1. $4 \times 5 + 7$
2. $18 \div 3 + 36$
3. $3 + 4 \times 6$
4. $(15 \div 3) \div 5$
5. $21 - 6 \div 3$
6. $60 - 8 \div 2$
7. $(3 \times 15) \div 5$
8. $(20 \div 4) \times 3$
9. $5 \times 6 + 8$
10. $25 - 3 \times 6$
11. $8 \times 4 \div 2$
12. $27 \div 9 + 3$
13. $8 \times (7 - 4) + 11$
14. $[11 \times (3 + 2)] \div 5$
15. $23 + [18 \div (5 + 1)]$
16. $12 \times 8 + (12 \times 2)$
17. $(6 \div 2) + 1 \times 4$
18. $18 \div 9 + (3 \times 3)$
19. $(12 \times 2) \div 1$
20. $(7 + 4 + 3) \div (2 + 1)$
21. $2 \times 3^2 - 5$
22. $4^2 \div 8 + (3 \times 2)$

Equations, formulas

Solving equations

6-4 SOLVING EQUATIONS BY SUBTRACTION

Figure 6-6 shows a gasket used in sealing the rocker-arm cover to the head of an automobile engine. The length of the gasket can be stated in two ways:

A. $2 + x + 2$, or $x + 4$, represents the length.
B. The length is 17".

The equation $x + 4 = 17$ indicates that the two quantities $x + 4$ and 17 are equal.

To solve the equation, we must find the number which x represents. We can solve this equation by using the following rule.

> Subtracting the same number from both sides of an equation does not change the equality of the equation.

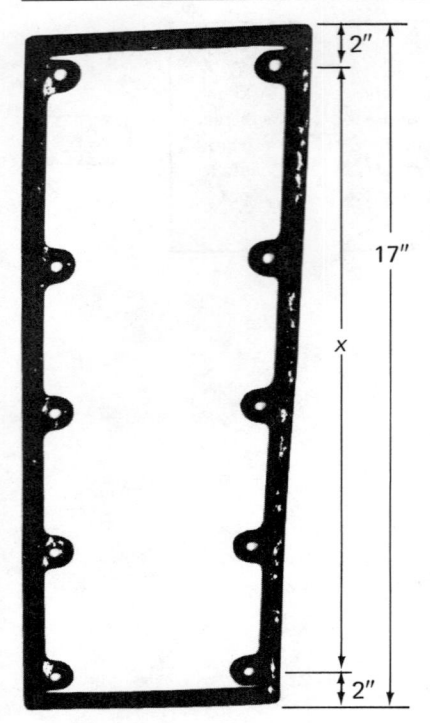

Figure 6-6

Thus,
$$\begin{aligned} x + 4 &= 17 \\ -4 &= -4 \quad \text{(subtracting 4)} \\ \hline x &= 13 \end{aligned}$$

Therefore, the center portion of the gasket is 13" long.

Study these other examples of solving equations by using subtraction.

Example 1

Solve $a + \frac{4}{9} = \frac{2}{3}$.

Solution

$$a + \frac{4}{9} = \frac{2}{3}$$

$$a + \frac{4}{9} - \frac{4}{9} = \frac{2}{3} - \frac{4}{9} \quad \left(\text{subtracting } \frac{4}{9}\right)$$

$$a + 0 = \frac{2}{9}$$

$$a = \frac{2}{9}$$

158   Technical and industrial formulas

Example 2

A 2.7-cm hole must be enlarged to allow a cable connector to pass through. The cable is 3.6 cm in diameter. By how much must the hole be enlarged?

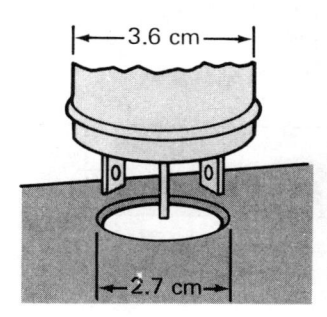

Solution

Let y represent the amount that the hole must be enlarged. Then,

$$y + 2.7 = 3.6$$
$$y + 2.7 - 2.7 = 3.6 - 2.7$$
$$y - 0 = .9$$
$$y = .9$$

Therefore, the hole must be widened by .9 cm, or 9 mm.

EXERCISES Use the subtraction rule to solve the equations.

1. $x + 7 = 11$
2. $y + 9 = 14$
3. $a + 11 = 20$
4. $b + 7 = 15$
5. $c + 9 = 26$
6. $a + 37 = 198$
7. $b + 15 = 71$
8. $c + 31 = 103$
9. $69 + y = 311$
10. $84 + z = 99$
11. $a + \frac{1}{2} = 4\frac{3}{4}$
12. $b + \frac{3}{8} = 2\frac{7}{8}$
13. $c + \frac{5}{16} = 7\frac{9}{16}$
14. $d + .35 = 2.89$
15. $y + 1.2 = 47.9$
16. $x + \frac{5}{6} = 3\frac{1}{6}$
17. $y + \frac{4}{5} = 9\frac{1}{5}$
18. $z + 1\frac{1}{4} = 8$
19. $89.01 + y = 205.44$
20. $z + 3.4 = 21.4$

For exercises 21-28, write the equation for the unknown dimension. Then solve the equation.

U.S. system

21. Find length A on the metal plate.
22. Find length B.

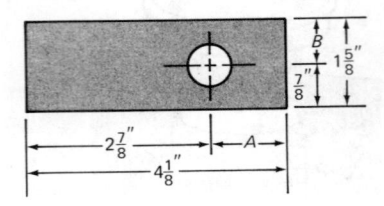

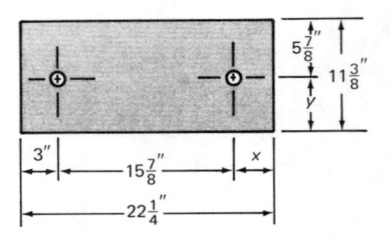

Metric system

25-26. Use the diagram below to solve exercises 21 and 22.

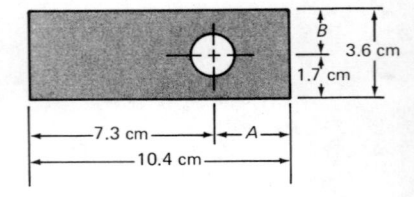

23. Find dimension x for spacing the knobs on the drawer.
24. Find dimension y.

27-28. Use the diagram below to solve exercises 23 and 24.

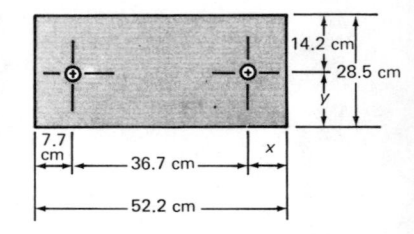

TRICKS OF THE TRADE

To multiply any number by 5, multiply half the number by 10.

$5 \times 54 = \frac{54}{2} \times 10 = 27 \times 10 = 270$

Odd numbers work the same way.

$5 \times 23 = \frac{23}{2} \times 10 = 11.5 \times 10 = 115$

An easier way to multiply odd numbers by 5 is to halve the number that is 1 less than the given number. Then affix a final digit 5 to your answer.

$5 \times 135 \rightarrow 2\overline{)134}^{\,67} \rightarrow 675$

Technical and industrial formulas

6-5 SOLVING EQUATIONS BY ADDITION

Figure 6-7 shows the seal plate of a ship's porthole. The distance from the center of the glass to the inner seal is $6\frac{1}{4}''$. The seal plate itself is $1\frac{7}{8}''$ wide. What is the distance R from the center of the glass to the outer seal? What is the overall diameter of the seal plate?

We can write the following equation based on the information in Figure 6-7:

$$R - 1\tfrac{7}{8} = 6\tfrac{1}{4}$$

To find the value of R that solves this equation, we can use the following rule.

> Adding the same number to both sides of an equation does not change the equality of the equation.

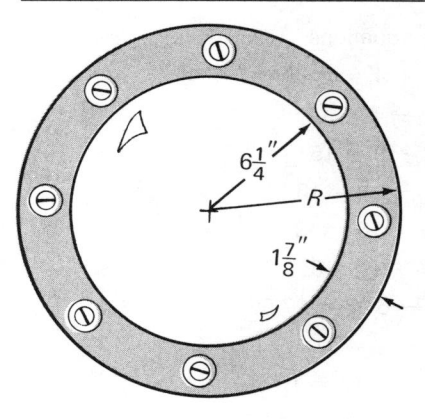

Thus,
$$\begin{array}{rl} R - 1\tfrac{7}{8} = & 6\tfrac{1}{4} \\ + 1\tfrac{7}{8} = & + 1\tfrac{7}{8} \quad \text{(adding } 1\tfrac{7}{8}\text{)} \\ \hline R = & 8\tfrac{1}{8} \end{array}$$

Figure 6-7

Therefore, the distance from the center of the glass to the outer seal is $8\frac{1}{8}''$. Since the diameter of a circle is twice the radius, the overall diameter of the seal plate is $2 \times 8\frac{1}{8}''$, or $16\frac{1}{4}''$.

Here is another example of an equation that is solved by using addition. Study it carefully.

Example

Solve $x - 14.3 = 17.9$.

Solution

$$\begin{array}{r} x - 14.3 = 17.9 \\ x - 14.3 + 14.3 = 17.9 + 14.3 \quad \text{(adding 14.3)} \\ x + 0 = 32.2 \\ x = 32.2 \end{array}$$

Do you see how adding 14.3 to each side keeps the equation "balanced"?

Equations, formulas

EXERCISES Use the addition rule to solve the equations.

1. $x - 6 = 20$
2. $y - 4 = 7$
3. $a - 8 = 19$
4. $b - 6 = 15$
5. $c - 11 = 15$
6. $x - 8 = 15$
7. $y - 5 = 23$
8. $z - 11 = 18$
9. $a - 21 = 32$
10. $b - 16 = 19$
11. $y - \frac{2}{3} = \frac{1}{3}$
12. $d - 3\frac{1}{2} = 6\frac{3}{4}$
13. $x - 9.3 = 21.7$
14. $f - 6.33 = 27.96$
15. $z - 3.5 = 7.9$
16. $z - \frac{1}{4} = 31$
17. $h - 3\frac{1}{4} = 9\frac{7}{8}$
18. $y - 2.18 = 1.78$
19. $d - .065 = 8.183$
20. $x - 1.784 = 14.693$

For exercises 21–28, write the equation for the unknown dimension. Then solve the equation.

U.S. system

21. What is the size of a rod if .023" must be ground off in order to have the proper thickness of 1.375"?

22. Find the original width of a piece of aluminum if, after a $\frac{3}{32}$" strip is cut off, it is $\frac{17}{32}$" wide.

23. A walnut panel needed for a stereo cabinet must be $22\frac{1}{2}$". The piece is made by cutting off $1\frac{3}{4}$" from a longer piece of stock. What is the length of the stock before cutting?

24. After .19" is milled off, the thickness of a steel plate is 1.67". What was the thickness before the milling?

Metric system

25. Find the outside diameter of a piece of plastic tubing that has an inside diameter of 4.31 cm and a thickness of .51 cm.

26. In order to have the proper length of 22.62 cm, a machinist mills a connecting rod down by .11 cm. What is the length of the rod before milling?

27. Find the outside diameter of a pipe that has an inside diameter of 23.4 mm and a thickness of 4.6 mm.

28. After 1.3 cm is planed off the top, a block of wood is 6.28 cm thick. How thick was the block of wood before planing?

6-6 SOLVING EQUATIONS BY DIVISION

The hallway in Figure 6-8 contains 28 sq ft of floor space. An asphalt floor is to be laid in the hallway. If the flooring comes in rolls 4′ wide, how long a piece is needed for the job?

The area of a rectangle is equal to the length times the width. Thus, the area of the hall is $4 \times \ell$ or simply 4ℓ. We can show the equation as

$4\ell = 28$

To solve an equation like this, we use the following rule.

> Dividing both sides of an equation by the same number (except zero) does not change the equality of the equation.

Using this division rule, we solve the equation in this way:

$4\ell = 28$

$\dfrac{4\ell}{4} = \dfrac{28}{4}$ (dividing by 4)

$\ell = 7$

The length of flooring needed is 7′.

Do you see how the division rule allows us to find the unknown quantity? Here are other examples that use the division rule.

Example 1

Solve $9m = 38$.

Solution

$9m = 38$

$\dfrac{9m}{9} = \dfrac{38}{9}$ (dividing by 9)

$m = \dfrac{38}{9}$

$m = 4\dfrac{2}{9}$

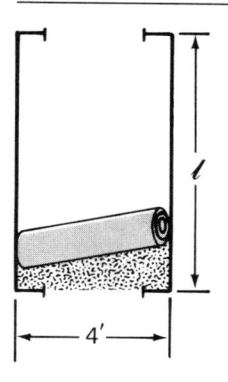

Figure 6-8

Equations, formulas

Example 2

Solve $.7x = .35$.

Solution

$.7x = .35$

$\dfrac{.7x}{.7} = \dfrac{.35}{.7}$ (dividing by .7)

$x = .5$

In Examples 1 and 2 we have the product and one factor. Whenever we must find the other factor, we use the division rule.

EXERCISES Use the division rule to solve the equations.

1. $6a = 24$
2. $12d = 60$
3. $8x = 32$
4. $9y = 63$
5. $4t = 44$
6. $25w = 125$
7. $14d = 42$
8. $4y = 39$
9. $9c = 71$
10. $6n = 23$
11. $.6y = .54$
12. $.2a = 3.8$
13. $.4p = 36$
14. $7m = 4.9$
15. $12y = .048$
16. $.07r = 5.6$
17. $6b = .18$
18. $1.3k = .065$

For exercises 19–26, write the equation for the unknown quantity. Then solve the equation.

U.S. system

19. The area of an appliance-store showroom is 9600 sq ft. If the width is 80 ft, find the length of the showroom.

20. A temporary spacer made of three washers on a bolt measures $1\tfrac{7}{32}''$ in length. How thick is each washer?

21. A standard sheet of printing paper has an area of 374 sq in. The paper measures 17 in. wide. What is the length of the sheet?

Metric system

23. A rectangular airplane wing panel is 7200 sq cm in area. The width of the panel is 60 cm. What is the length?

24. Four table legs can be turned from a piece of pine 422 cm long. How long is each leg?

25. A dining room table has three extra leaves. The leaves extend the length of the table 43.5 cm. What is the width of each leaf?

164   Technical and industrial formulas

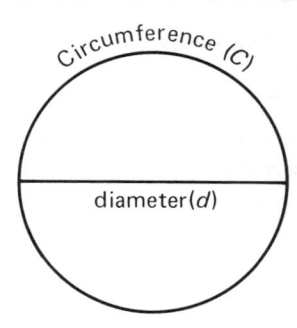

22. To find the circumference (distance around) of a circle, we use the equation $C = \pi d$. Find the diameter d of a circular swimming pool if $C = 63.55'$. Use 3.14 for π.

26. Solve exercise 22, using 20.5 m for C.

6-7 SOLVING EQUATIONS BY MULTIPLICATION

The drill press in Figure 6-9 uses pulleys and pulley belts to transmit power. To vary the speed of rotation, the belt can be changed from one pulley to another.

Figure 6-9

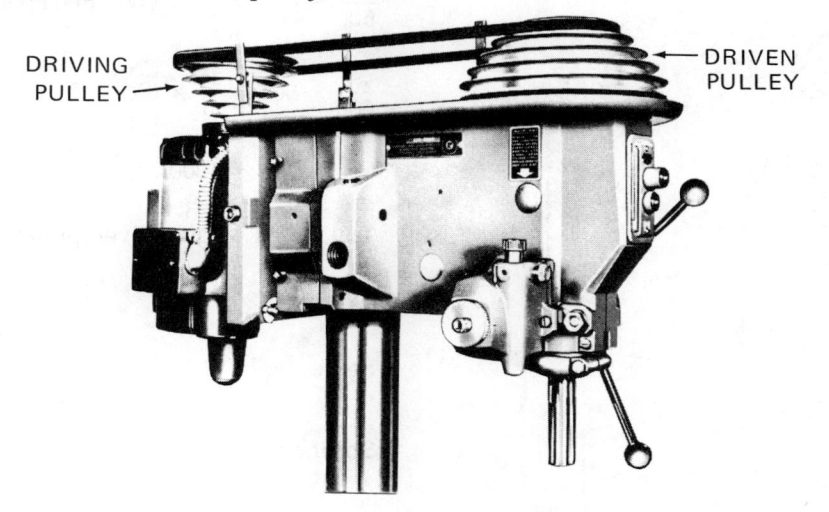

The relationship between the diameters of the pulleys and the speeds of the pulleys is given as follows:

$$\frac{\text{Speed of driven pulley }(s)}{\text{Speed of driving pulley }(S)} = \frac{\text{Dia. of driving pulley }(D)}{\text{Dia. of driven pulley }(d)}$$

Example 1

A driving pulley with a diameter of 15" and a speed of 600 rpm is connected to a driven pulley that has a diameter of 5". What is the speed of the driven pulley?

Solution

By substituting the known values in the formula, we get

$$\frac{s}{600} = \frac{15}{5}$$

Equations, formulas 165

To solve an equation like this, we can use the following rule.

> Multiplying both sides of an equation by the same number (except zero) does not change the equality of the equation.

We solve the equation in the following way:

$$\frac{s}{600} = \frac{15}{5}$$

$$\frac{s}{600} \times 600 = \frac{15}{5} \times 600 \quad \text{(multiplying by 600)}$$

$$\frac{s}{\cancel{600}} \times \cancel{600} = \frac{\cancel{15}^{3}}{\cancel{5}} \times 600$$

$$s = 1800$$

The speed of the driven pulley is 1800 rpm.

Example 2

Solve $\dfrac{x}{3.2} = .8$.

Solution

$$\frac{x}{3.2} = .8$$

$$\frac{x}{3.2} \times 3.2 = .8 \times 3.2 \quad \text{(multiplying by 3.2)}$$

$$x = 2.56$$

EXERCISES Use the multiplication rule to solve the equations.

1. $\dfrac{x}{5} = 6$ 2. $\dfrac{y}{7} = 3$ 3. $\dfrac{z}{3} = 12$

4. $\dfrac{a}{15} = 11$ 5. $\dfrac{c}{17} = 9$ 6. $\dfrac{x}{12} = 8$

7. $\dfrac{d}{5} = 14$ 8. $\dfrac{x}{3} = 18$ 9. $\dfrac{w}{7} = \dfrac{2}{3}$

10. $\dfrac{y}{7} = \dfrac{7}{8}$ 11. $\dfrac{z}{3} = \dfrac{5}{8}$ 12. $\dfrac{x}{2} = \dfrac{3}{16}$

13. $\dfrac{x}{1.3} = 2.6$ 14. $\dfrac{h}{7} = 1.2$ 15. $\dfrac{d}{1.9} = 14$

16. $\dfrac{y}{5} = .83$ 17. $\dfrac{x}{4.3} = 0.5$ 18. $\dfrac{h}{1.9} = \dfrac{6}{2.3}$

For exercises 19–24, write the equation for the unknown quantity. Then solve the equation.

U.S. system

19. The relationship between dimensions *d* and *F* of the hex nut is given by the equation $d = \dfrac{F}{.866}$. If $d = .34''$, find *F*.

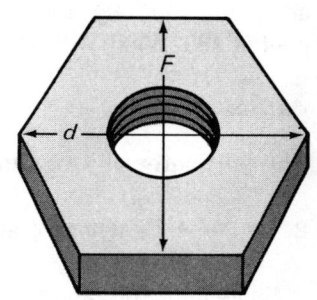

20. A 6″ driving pulley is connected to a 2″ driven pulley by a V belt. If the driving pulley turns at 450 rpm, what is the speed of the driven pulley?

21. The resistance of 250′ of electrical wire is 8.9 ohms. Find the resistance of 600′ of the wire. Use the following equation:

$$\dfrac{600}{250} = \dfrac{r}{8.9}$$

Metric system

22. The relationship between *F* and *d* is the same for a metric hex nut as for a U.S. hex nut. This relationship is given by the formula $d = \dfrac{F}{.866}$. Find the dimension *F* for each of the following size metric hex nuts:

  a. $d = 5.53$ mm
  b. $d = 1.2$ cm
  c. $d = 1.7$ cm

23. A table saw blade is attached to a 9.2-cm pulley. The motor has a 12.4-cm pulley. If the motor pulley turns at 1850 rpm, what is the speed of the saw blade in rpm?

24. The resistance of 30 cm of copper wire is 1.3 ohms. What is the resistance of 160 cm of the wire? Use the following equation:

$$\dfrac{30}{160} = \dfrac{1.3}{r}$$

6-8 SOLVING EQUATIONS BY SEVERAL METHODS

In your work you will often need to solve equations, using several of the methods you have just studied. The flow chart in Figure 6-10 will help organize your method of solving such equations.

Figure 6-10

Example 1

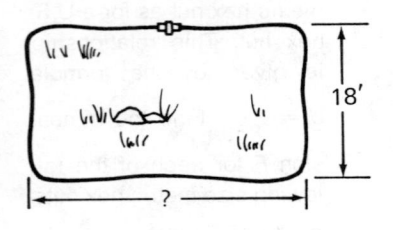

To get an idea of the size of a swimming pool that he wants to build, Mr. Phelps takes some rough measurements. He forms a rectangular loop on the ground with his 100′ garden hose. After adjusting the length and width, he finds the best width for the pool is 18′. What will be the length?

Solution

The perimeter of a rectangle is given by the equation

$2\ell + 2w = P$, where ℓ and w are the length and width.

Since $P = 100$, and $w = 18$, we can substitute these measures in the equation. Thus,

$2\ell + 2(18) = 100$

$2\ell + 36 = 100$

To solve this equation, we use the subtraction rule first, then the division rule.

$2\ell + 36 - 36 = 100 - 36$ (subtracting 36)

$2\ell + 0 = 64$

or $2\ell = 64$

Then $\dfrac{2\ell}{2} = \dfrac{64}{2}$ (dividing by 2)

$\ell = 32$

Therefore, the pool will be 32′ long.

168   Technical and industrial formulas

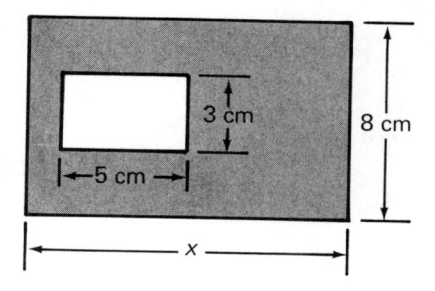

Example 2

After a 3-cm by 5-cm part has been stamped from a piece of sheet metal, the scrap material, shown by shading, is thrown away. The scrap piece is 8 cm wide and contains 65 sq cm of sheet metal. What is the length of the scrap piece?

Solution

The area of the scrap piece is equal to the difference in the areas of the large and small rectangles.

Area of the large rectangle = $8x$

Area of the small rectangle = 5×3, or 15

Thus, $8x - 15 = 65$

To solve the equation, we use the addition rule first, then the division rule.

$$8x - 15 = 65$$
$$8x - 15 + 15 = 65 + 15 \quad \text{(adding 15)}$$
$$8x + 0 = 80$$

or $\quad 8x = 80$

Then $\quad \dfrac{8x}{8} = \dfrac{80}{8} \quad$ (dividing by 8)

$$x = 10$$

Therefore, the length of the scrap is 10 cm.

Notice in these examples that we use the rules for solving equations in the order shown in the flow chart in Figure 6-10.

EXERCISES Solve the equations.

1. $2x + 13 = 35$
2. $8x + 41 = 105$
3. $5x - 32 = 53$
4. $9a - 48 = 6$
5. $\dfrac{x}{2} + 1 = 9$
6. $\dfrac{y}{3} + 5 = 29$
7. $3t + 7 = 25$
8. $9y + 8 = 17$
9. $4p - 13 = 27$
10. $8y - 3 = 61$

Equations, formulas

11. $\frac{a}{5} - 3 = 2$ 12. $\frac{b}{8} - 5 = 1$

13. $.3y \div 1.8 = 6.5$

14. $1.2z + 4.3 = 115.7$

15. $7y - \frac{4}{3} = \frac{5}{6}$

16. $7t - \frac{5}{8} = \frac{11}{16}$

17. $1.4x + 3 = 7$

18. $.5x - 4 = 1.3$

19. $3a + \frac{3}{4} = \frac{7}{8}$

20. $5y + \frac{1}{6} = \frac{2}{3}$

For exercises 21–28, write the equation for the unknown quantity. Then solve the equation.

U.S. system

21. The perimeter of the triangle below is 26″. The sides marked *s* are equal in length. Find *s*.

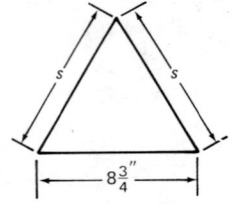

    (triangle with two sides marked *s* and base $8\frac{3}{4}″$)

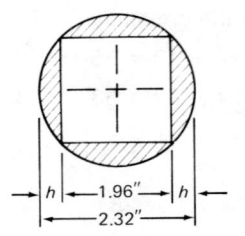

Metric system

25. The lengths of the sides of the triangle are given below. The perimeter of the triangle is 46.6 mm. Find the lengths of sides *AB* and *BC*.

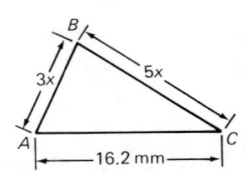

    (triangle with AB = 3x, BC = 5x, AC = 16.2 mm)

22. A square shaft is milled from a round bar, as shown in the diagram. Find the depth of the cut *h*.

    (round bar with square, dimensions: h, 1.96″, h; overall 2.32″)

26. Solve exercise 22, using the dimensions given on this figure.

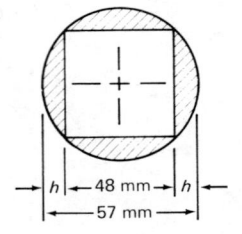

    (round bar with square, dimensions: h, 48 mm, h; overall 57 mm)

170   **Technical and industrial formulas**

23. Find length *x* on the light switch plate.

27. Solve exercise 23, using the dimensions given on this figure.

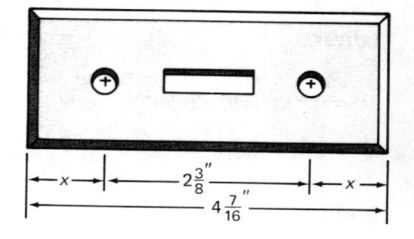

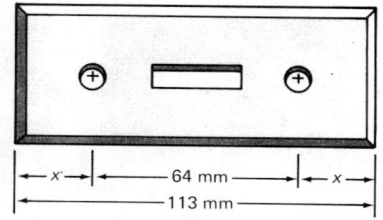

24. Find dimensions *x* and *y* on the die pattern.

28. Solve exercise 24, using the dimensions given on this figure.

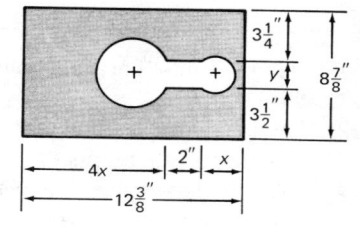

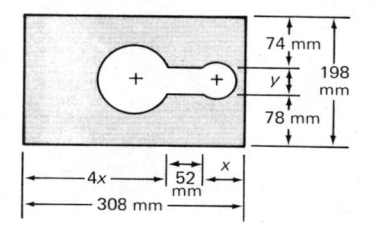

Working with formulas

6-9 WRITING FORMULAS FROM RULES

Many of the common relationships and processes used in industry can be stated as *rules*. For example, the appliance industry depends heavily on this rule:

> The power in watts used by an electrical appliance is equal to the product of the voltage and the number of amperes.

Usually an equation representing the rule will be used in performing calculations and measurements. The equation, or the

Equations, formulas 171

formula, for the rule just given can be expressed this way:

$P = V \times A$

where

P = the power in watts

V = the voltage in volts

and

A = the current in amperes

We can use the following general procedure for writing formulas from rules.

---

1. Read the rule, carefully noting the words that represent quantities.
2. Select a letter to stand for the unknown value of each quantity.
3. Determine which operations are used to relate the quantities.
4. Write the letters, together with the operation signs, to make the formula agree with the rule.

Example 1

Write a formula for this rule: The number of square centimeters is equal to 10,000 times the number of square meters.

Solution

Step 1 The important expressions are "the number of square centimeters" and "the number of square meters."

Step 2 Let c = the number of square centimeters.
Let m = the number of square meters.

Step 3 The quantities are related by multiplication.

Step 4 The formula is $c = 10,000\ m$.

Example 2

The number of screw threads per inch is equal to 1 divided by the pitch of the screw.

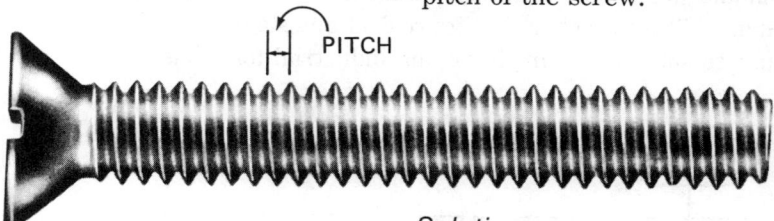

Solution

Step 1   The important expressions are "the number of screw threads per inch" and "the pitch of the screw."

Step 2   Let t = the number of screw threads per inch.
Let p = the pitch of the screw.

Step 3   The quantities are related by division.

Step 4   The formula is $t = 1 \div p$, or $t = \dfrac{1}{p}$.

EXERCISES

Write a formula for each rule.

1. The number of inches is equal to 12 times the number of feet.
2. The number of meters is equal to .01 times the number of centimeters.
3. The number of kilograms is equal to .454 times the number of pounds.
4. The net cost is equal to the marked price less the discount.
5. The area of a sheet of plywood is equal to the length times the width.
6. The number of dollars equals the number of cents divided by 100.
7. The diagonal of a square is equal to 1.414 times the side of the square.
8. To find the circumference of a circle, multiply the diameter by 3.1416.
9. The distance traveled by an automobile equals the number of hours it travels multiplied by the speed.
10. The number of tons is equal to the number of pounds divided by 2000.

6-10 SOLVING PROBLEMS WITH FORMULAS

Every field of industry uses formulas in its daily operation. You should begin now to develop skill in solving problems with formulas. The flow chart in Figure 6-11 gives a method that you can use to solve problems involving industrial formulas.

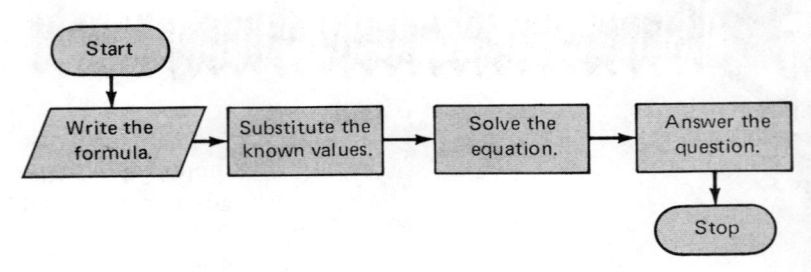

Figure 6-11

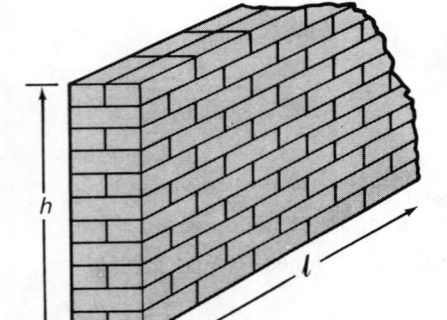

Example 1

The number of standard bricks needed to build a wall is about 21 times the volume of the wall in cubic feet. How many bricks will it take to build a wall 6" wide, 8' high, and 14' long?

Solution

We can write the formula for building the wall as follows:

$n = 21V$

where n = the number of bricks

and V = the volume of the wall in cubic feet

Since the wall is a rectangular solid, its volume in cubic feet can be found by multiplying the length by the width by the height. Thus we can write the formula as:

$n = 21\ell wh$

where n = the number of bricks required

ℓ = the length of the wall in feet

w = the width of the wall in feet

and h = the height of the wall in feet

174   **Technical and industrial formulas**

By substituting the known values in the formula, we get

$n = 21\ell wh$

$n = 21 \times 14 \times \dfrac{1}{2} \times 8$

Thus, $n = 1176$

Therefore, about 1176 bricks are needed.

In Example 1 we can solve the equation easily once we substitute the known quantities into the formula. In the following example we will need to use the multiplication and addition rules to solve the equation.

Example 2

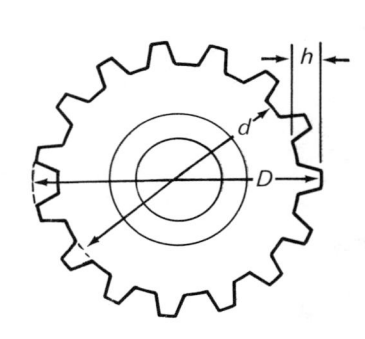

The depth h of a gear tooth is found by dividing the difference between the major diameter D and the minor diameter d by 2. Find the major diameter if the minor diameter is 5.25 cm and the depth of the gear tooth is .25 cm.

Solution

We write the formula for the relationship described above as

$$h = \dfrac{D - d}{2}$$

We then substitute the known values and solve the equation as follows:

$$h = \dfrac{D - d}{2}$$

$$.25 = \dfrac{D - 5.25}{2}$$

$$.25 \times 2 = \dfrac{D - 5.25}{2} \times 2 \quad \text{(multiplying by 2)}$$

$$.25 \times 2 = \dfrac{(D - 5.25) \times \overset{1}{\cancel{2}}}{\underset{1}{\cancel{2}}}$$

$$.50 = D - 5.25$$

$$.50 + 5.25 = D - 5.25 + 5.25 \quad \text{(adding 5.25)}$$

$$5.75 = D + 0$$

or $D = 5.75$

Thus, the major diameter is 5.75 cm.

Do you see how the method in the flow chart (Figure 6-11) helps us plan our solutions?

Example 3

If $H = \dfrac{P}{2} + .01$ and $H = .26$, find P.

Solution

$$H = \dfrac{P}{2} + .01$$

Substitute $H = .26$ into the formula.

$$.26 = \dfrac{P}{2} + .01$$

Solve the resulting equation for P.

$$.26 - .01 = \dfrac{P}{2} + .01 - .01 \quad \text{(subtract .01)}$$

$$.25 = \dfrac{P}{2}$$

$$.25 \times 2 = \dfrac{P}{2} \times 2 \quad \text{(multiply by 2)}$$

$$.5 = P$$

EXERCISES Substitute the values into the formula and solve the resulting equation.

1. $C = 3.14\, d$
   a. Find C, if $d = .6$.
   b. Find d, if $C = 10.048$.

2. $F = .866\, D$
   a. Find F, if $D = 1.5$.
   b. Find D, if $F = 1.1258$.

3. $d = 1.41s$
   a. Find d, if $s = 8.2$.
   b. Find s, if $d = 12.831$.

4. $P = 4s$
   a. Find P, if $s = 3\frac{5}{8}$.
   b. Find s, if $P = 58\frac{3}{4}$.

5. $P = 2\ell + 2w$
   a. Find P, if $\ell = 16$ and $w = 12$.
   b. Find ℓ, if $P = 48$ and $w = 9\frac{1}{2}$.
   c. Find w, if $P = 36.5$ and $\ell = 12.75$.

6. $A = \frac{1}{2}bh$
   a. Find A, if $b = 7$ and $h = 18$.
   b. Find b, if $A = 24\frac{3}{8}$ and $h = 7\frac{1}{2}$.
   c. Find h, if $A = 8.14$ and $b = 4.4$.

U.S. system

7. Using the formula from Example 1, find the number of bricks needed to build a wall 20′ long, 8′ high, and 4″ thick.

8. How many bricks are needed for a wall 30′ long, 4′ high, and 6″ thick?

9. How many bricks are in a wall 48′4″ long, 8′ high, and 8″ thick?

10. The diagonal d of a square is related to the side s of the square by the formula $d = 1.414s$. Find the length of the diagonal of a square table whose side is 32″.

11. Find the length of the diagonal cross-beam of a square shed whose side is 9′ long.

Metric system

12. The minor diameter of a gear is 8.2 cm, and the depth of the gear tooth is .35 cm. Using the formula from Example 2, find the major diameter of the gear.

13. The minor diameter of a clock gear is 15.6 cm, and the depth of the gear tooth is .83 cm. Find the major diameter.

14. The minor diameter of a watch gear is 7.3 mm, and the major diameter is 7.8 mm. What is the depth of the gear tooth?

15. The formula for finding the circumference C of a circle is $C = 3.1416d$, where d is the diameter of the circle. Find the circumference of a circular fish pond that has a diameter of 14 m.

16. Find the diameter of a round birdbath that has a circumference of 61.7 cm.

Taking inventory

1. An *equation* is a statement that indicates that two number expressions are equal. (p. 148)

2. An *exponent* is a symbol that indicates how many times a number is used as a factor. (p. 153)

3. To simplify an expression involving several operations:
   a. Do the operations inside the parentheses.
   b. Do the multiplications and divisions from left to right.
   c. Do the additions and subtractions from left to right. (p. 157)
4. The same number can be subtracted from or added to both sides of an equation without changing the equality of the equation. (pp. 158, 161)
5. Both sides of an equation may be divided by or multiplied by the same number (except zero) without changing the equality of the equation. (pp. 163, 166)
6. A *formula* is an equation that uses symbols to represent a relationship between numbers. (p. 172)

Measuring your skills

Simplify the expressions. (6-1)
1. $7a + 9a + 4a$
2. $6b - 2b$

Calculate. (6-2)
3. 3^4
4. $\left(\frac{1}{2}\right)^2$
5. $(1.7)^3$

Simplify the expressions. (6-3)
6. $17 + 5 \times (6 - 2)$
7. $(27 - 3) + (5 - 1)$

Solve the equations. (6-4, 6-5, 6-6, 6-7, 6-8)
8. $x + 8 = 23$
9. $t + 15 = 17$
10. $9y = \frac{3}{5}$
11. $\frac{y}{8} = \frac{3}{16}$
12. $a - 8 = 13$
13. $b - 1.5 = 4.3$
14. $5x = 3.5$
15. $\frac{m}{9} = 6$
16. $6y - 8 = 28$
17. $\frac{a}{3} + .04 = .85$

Write a formula for the statements. (6-9)
18. The amount of work done is equal to the product of the force and the distance.
19. The area of a triangle is equal to half the product of the base and the height.

Use the formulas to find the unknown values.

U.S. system

20. The volume of a rectangular container: $V = \ell \times w \times h$. Use $V = 36$ cu in., $\ell = 6$ in., $w = 2$ in. to find the value of h.

21. Find the depth of the screw thread, using the formula $H = \dfrac{P}{2} + .01$.

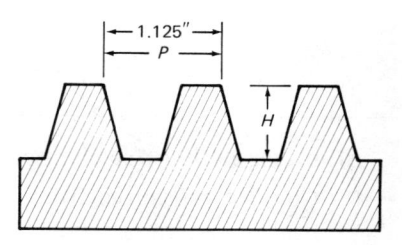

Metric system

22. The volume of a cylindrical container: $V = 3.14\ r^2 h$. Use $r = 3.6$ cm, $h = 7.2$ cm to find the value of V.

23. Find the speed of the driven pulley, using the formula $\dfrac{s}{S} = \dfrac{D}{d}$.

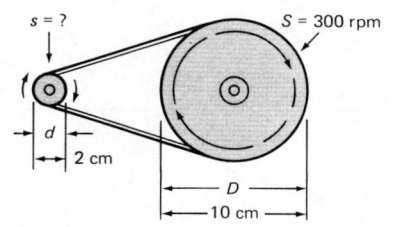

7 Length, area, volume

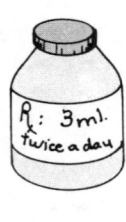

After completing this chapter, you should be able to:

1. Convert measures from one unit to another in the United States system.
2. Convert measures from one unit to another in the metric system.
3. Convert measures in the United States system to measures in the metric system, and vice versa.
4. Find the perimeters of rectangles and the circumferences of circles.
5. Find the area of parallelograms, triangles, trapezoids, and circles.
6. Find the surface area of cylinders, cones, and spheres.
7. Find the volume of rectangular solids, cylinders, cones, pyramids, and spheres.
8. Convert from Celsius to Fahrenheit, and vice versa.
9. Solve problems using the metric and United States systems.

Length

7-1 LINEAR MEASURES

Labels or signs like those in Figure 7-1 will no doubt be commonplace before very long. It is therefore important for you to learn to think in the metric system. Since you may be "on the job"

Figure 7-1

before the changeover is complete, you will need to feel at home with whichever system, United States or metric, is used.

The metric system is used all over the world because it is so easy to use. It is based on the decimal system. To convert from one unit to another, you simply multiply by 10 or 100 or 1000 or their reciprocals, .1 or .01 or .001, and so on.

There are three basic units in the metric system, the *meter*, the *kilogram*, and the *liter*.

A *meter* is about 3½″ longer than a yard. Figure 7-2 shows the difference between a meter and a yard.

Figure 7-2

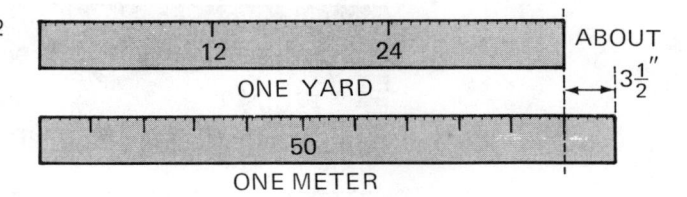

The kilogram (1000 grams) equals about 2.2 lb. You will see measures in grams on tire-balancing weights.

The *liter* is a little more than a quart. You would buy gas for your car by the liter. Figure 7-3 compares the size of a liter and a quart.

Figure 7-3

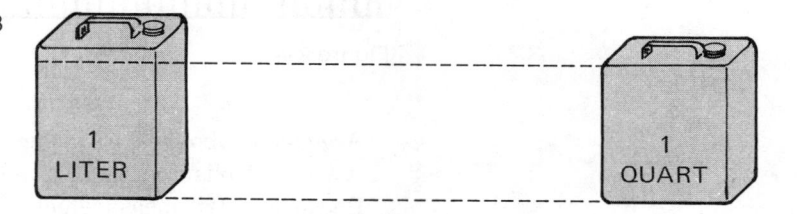

Units larger or smaller than these basic units are named by using prefixes. For example, "kilo" is used with "gram" to form "kilogram," "centi" with "meter" to form "centimeter," and so on. The most common prefixes, with their meanings, are the following:

milli: one-thousandth (.001)
centi: one-hundredth (.01)
deci: one-tenth (.1)
kilo: one thousand times (1000)

The following table shows the relationships between the commonly used linear metric units. Some of these relationships should be familiar to you by now.

Length, area, volume 181

Table 7-1 *Metric linear units*

10 *milli*meters (mm)	=	1 *centi*meter (cm)	or	1 mm = .1 cm
10 *centi*meters	=	1 *deci*meter (dm)	or	1 cm = .1 dm
10 *deci*meters	=	1 meter (m)	or	1 dm = .1 m
1000 meters	=	1 *kilo*meter (km)	or	1 m = .001 km

Earlier you studied the relative size of the centimeter and millimeter compared to the inch. Recall that it takes 2.54 cm to make an inch (see Figure 7-4).

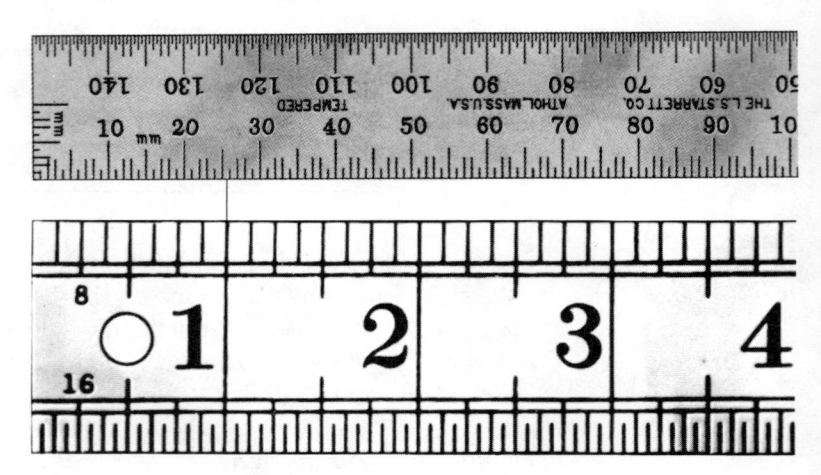

Figure 7-4

A *meter* is about 39.37 inches, or a little more than a yard.
A *kilometer* is about $\frac{5}{8}$ of a mile.

Because metric measurements are used widely in industry, you should become familiar with the different units. Study the following examples.

Example 1

3.5 m = __?__ cm

Solution

Since 100 cm = 1 m, the *conversion factor* for changing meters to centimeters is *100*.

3.5 m = (100 × 3.5) cm

= 350 cm

Technical and industrial formulas

Example 2

11.3 km = ____?____ m

Solution

Since 1000 m = 1 km, the conversion factor for changing kilometers to meters is 1000.

11.3 km = (1000 × 11.3) m

= 11,300 m

Example 3

A rectangular piece of plywood is 120 cm wide and 240 cm long. What are the dimensions in meters?

Solution

Since 100 cm = 1 m, the conversion factor is .01; that is, we divide by 100 or multiply by .01.

Width: 120 cm = (120 × .01) m = 1.2 m
Length: 240 cm = (240 × .01) m = 2.4 m
Dimensions: 1.2 m by 2.4 m

In Example 3, why is the conversion factor .01 instead of 100 as in Example 1? The reason is that you are changing from a smaller unit to a larger one. Example 1 is just the opposite—you are changing from a larger unit to a smaller one.

EXERCISES

1. Read the ruler at the points indicated.

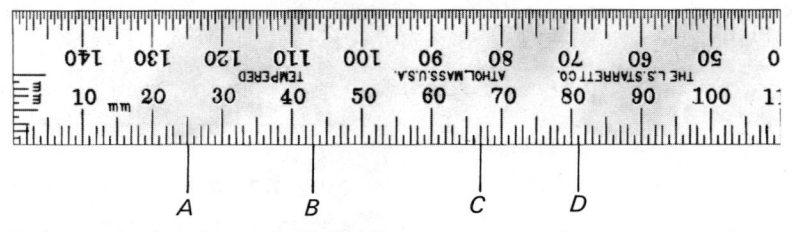

Estimate the lengths of the following segments to the nearest centimeter. Check your estimates with your ruler.

2. ├─────────────────────────────┤
3. ├──────────────────────────────────┤
4. ├──────────────┤

Length, area, volume 183

Complete.

5. 152 cm = __?__ m
6. 14 m = __?__ cm
7. 14 m = __?__ mm
8. 3 km = __?__ m
9. 3 km = __?__ cm
10. 6.7 m = __?__ mm
11. 1784 m = __?__ km
12. 14.3 m = __?__ cm

Copy and complete the table.

	mm	cm	dm	m
13.	?	1	?	?
14.	?	?	1	?
15.	1	?	?	?
16.	?	?	?	1

17. A micrometer = .000001 meter. How many micrometers are there in 12.5 mm? (*Micro* means "$\frac{1}{1,000,000}$ part of.")

18. A TV station broadcasts at a frequency of 50 megahertz. (*Mega* means 1,000,000 times.) State the frequency in kilohertz.

7-2 LINEAR CONVERSIONS

For the three-step pulley in Figure 7-5, the diameters are 11.5 cm, 10.2 cm, and 8.9 cm. To replace the assembly with a pulley with measurements in the United States system, what approximate measurements would you use?

This problem is typical of the kind of problem that a change-over to the metric system will bring. To solve this problem, we need a conversion table. Table 7-2 shows the United States system equivalents of the common metric linear units. To convert measures, multiply by the appropriate conversion factor.

In your work with the United States and metric systems, remember this important point.

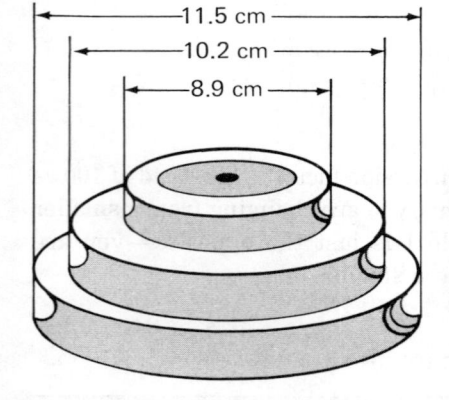

Figure 7-5

> In changing measurements from one system to another, your answers are approximate whenever the equivalents are approximate.

Table 7-2 *Approximately equivalent linear units*

1 in. = 25.4 mm	1 mm = .0394 in.
1 in. = 2.54 cm	1 cm = .3937 in.
1 ft = 3.05 dm = .305 m	1 dm = .328 ft
1 yd = .9144 m	1 m = 1.0936 yd
1 mi = 1.61 km	1 km = .621 mi

184 Technical and industrial formulas

Example 1

12 in. = __?__ cm

1 in. = 2.54 cm

Solution

The conversion factor is 2.54. An estimate for the answer is "somewhere between 24 and 36," since $2 \times 12 = 24$ and $3 \times 12 = 36$.

12 in. = (12 × 2.54) cm

= 30.48 cm

Is the answer reasonable?

Example 2

For the pulley assembly in Figure 7-5, convert the diameters to inches. Tell whether a three-step pulley with diameters of $3\frac{1}{2}$ in., 4 in., and $4\frac{1}{2}$ in. is a reasonable replacement.

Solution

1 cm = .3937 in. The conversion factor is .3937.

11.5 cm = (11.5 × .3937) in.

= 4.52755 in., or 4.5 in.

10.2 cm = (10.2 × .3937) in.

= 4.01574 in., or 4.0 in.

8.9 cm = (8.9 × .3937) in.

= 3.50393 in., or 3.5 in.

Is the replacement part a reasonable one to use?

Example 3

A road sign lists the speed limit as 50 miles per hour. If you are driving a European car that shows speed in kilometers per hour, what is the speed limit?

Length, area, volume

Solution

1 mi = 1.61 km. The conversion factor is 1.61.

50 miles = (50 × 1.61) km

= 80.5 km, or 81 km

The speed limit is 81 km per hr.

EXERCISES Decide which is the greater length.

1. 4 cm or 1 in.
2. 1 ft or 1 m
3. ½ in. or 1 cm
4. ½ km or 1 mi

Arrange the following lengths from the longest to the shortest.

5. 1 ft, ½ m, ½ yd
6. 55 cm, 1 ft, ½ m

Complete the conversions, rounding to the nearest tenth.

7. 3.8 mi = __?__ km
8. 15 km = __?__ mi
9. 4 in. = __?__ cm
10. 25 mm = __?__ in.
11. 100 yd = __?__ m
12. 1760 yd = __?__ m
13. 1 mi = __?__ m
14. 76.7 cm = __?__ ft

15. A set of metric wrenches has the following sizes: 6 mm, 8 mm, 10 mm, 12 mm, and 14 mm. Change these measurements to the United States system. Give your answer to the nearest thousandth of an inch.

16. The gap of the spark plug in Figure 7-6 is 2 mm. What is the gap in inches?

17. A micrometer measures the diameter of the wrist watch in Figure 7-7 to be 3 cm. What is the diameter of the watch in millimeters? in inches?

18. A builder must use glass at least 3/16" thick for a construction job. To the nearest millimeter, how thick must the glass be?

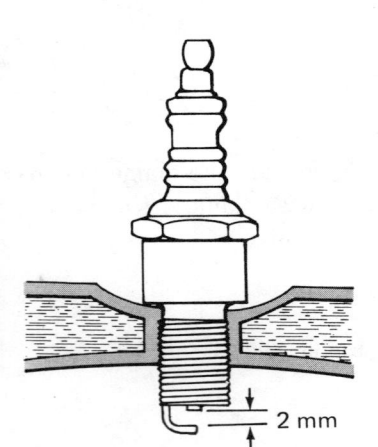

Figure 7-6

Figure 7-7

7-3 PERIMETER OF A RECTANGLE OR SQUARE

The *perimeter* of any figure is the distance around the figure. You can find the perimeter of any shape by adding the lengths of the sides.

Figure 7-8 shows a rectangle that has a length of 4 cm and a width of 3 cm. Because we have two pairs of sides with the same

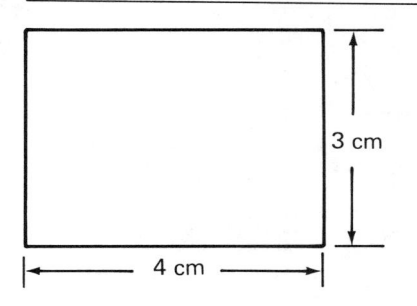

Figure 7-8

measure, we can take the sum of double the length and double the width. So the formula for the perimeter of a rectangle that has a length ℓ and a width w is

$P = 2\ell + 2w$

For a square with a side s, we can find the perimeter by this formula.

$P = 4s$

Example 1

Find the perimeter of the triangle shown.

Solution

$25 + 28 + 22 = 75$

The perimeter is 75 mm.

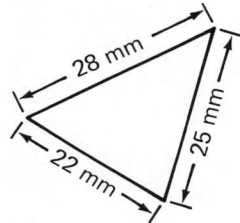

Example 2

Find the perimeter of a rectangle with a length of 7 ft and a width of 54 in.

Solution

We must convert all the dimensions to the same unit.

54 in. $= \frac{54}{12}$ ft $= 4\frac{1}{2}$ ft

$P = 2\ell + 2w$

  $= (2 \times 7) + (2 \times 4\frac{1}{2})$

  $= 14 + 9$

  $= 23$ ft

EXERCISES

U.S. system

1. Find the perimeter of a triangle with sides of 3", 6", and 5".

Metric system

8. Find the perimeter of a triangle with sides of 12 mm, 18 mm, and 21 mm.

Length, area, volume 187

2. Find the perimeter of the figure.

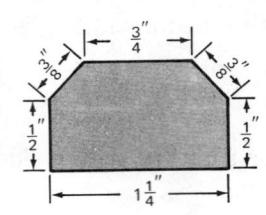

3. Find the perimeter of a rectangle with a length of 5 ft and a width of 39 in.

4. Find the perimeter of a square with sides of 54 in.

5. Find the perimeter of a rectangle with a length of 4 ft 6 in. and a width of 27 in.

6. A room has the shape and dimensions shown. How much floor molding would be needed for the room?

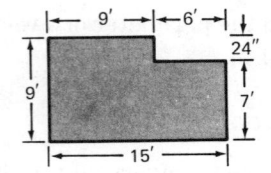

7. The figure shows the shape of a parking lot. How much fencing is needed to enclose it?

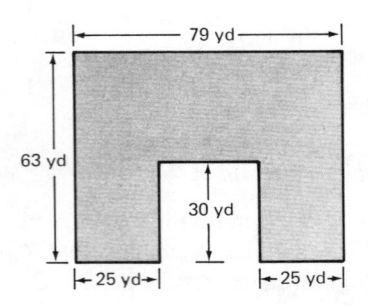

9. Find the perimeter of the figure.

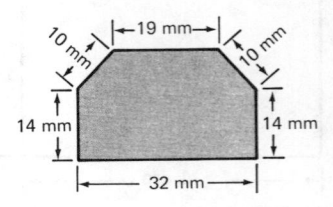

10. Find the perimeter of a rectangle with a length of 230 mm and a width of 560 mm.

11. Find the perimeter of a square with sides of 95 mm.

12. Find the perimeter of a rectangle with a length of 562 mm and a width of 385 mm.

13. A room has the shape and dimensions shown. How much floor molding would be needed for the room?

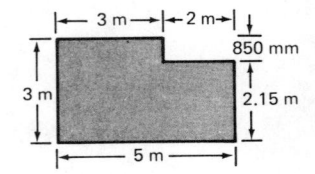

14. The figure shows the shape of a parking lot. How much fencing is needed to enclose it?

188   Technical and industrial formulas

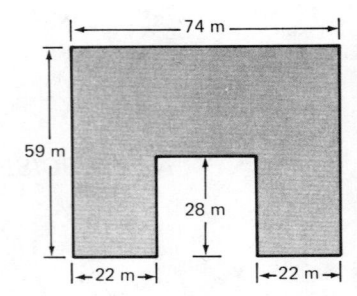

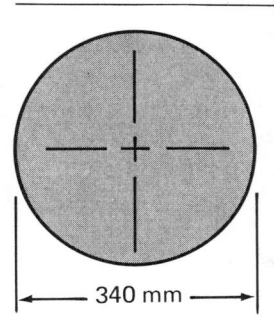

Figure 7-9

7-4 CIRCUMFERENCE

Many parts of machines and tools are circular. Wheels, rings, and pulleys are some of the many circular objects commonly used in industry and the trades. It is important to know the relationships among the measurements of a circle.

Figure 7-9 represents a circular table. Suppose we want to know how much metal edging is needed to go around the table. We are finding the *circumference* of the circle, or the distance around it. We know the *diameter* is 340 mm. So we can use the formula:

$$C = \pi d$$

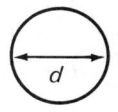

where C represents the circumference

d represents the diameter

π is approximately 3.14

The number π can thus be defined as the ratio c/d, where c and d represent the circumference and diameter of any circle. While π is an irrational number, in our calculations we can use various rational (fractional) approximations to π. The most common approximations used in calculations are:

$$\pi = 3.14 \qquad \pi = 3.1416 \qquad \pi = \frac{22}{7}$$

Practical reality dictates which approximation to use.

We can find the amount of edging needed for the circular table in Figure 7-9 as follows:

$C = \pi d$

$ = 3.14 \times 340$

$C = 1067.60$ mm

We should buy 1.068 m of edging.

Example 1

Find the circumference of a circle with a diameter of 11 in.

Length, area, volume 189

Solution

$C = \pi d$

$ = 3.14 \times 11$

$C = 34.54$ in.

Since the diameter is given to the nearest inch, we write the circumference to the nearest inch: $C = 35$ in.

EXERCISES

U.S. system

Find the circumference of a circle with the given diameter.

1. $d = 5$ in.
2. $d = 17$ ft
3. $d = 36$ yd
4. $d = 4\frac{3}{8}$ in.
5. $d = 7.22$ in.
6. How much edging would it take for the table represented? 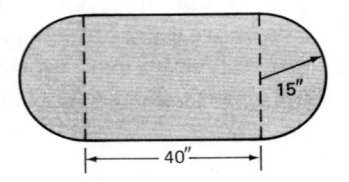
7. Find the circumference of pulley A in Figure 7-10.
8. Find the circumference of pulley B in Figure 7-10.

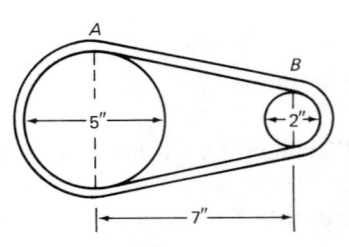

Metric system

Find the circumference of a circle with the given diameter.

10. $d = 5$ cm
11. $d = 1.3$ m
12. $d = 42$ mm
13. $d = 44.3$ m
14. $d = 5.42$ m
15. How much metal edging would it take for the table represented? 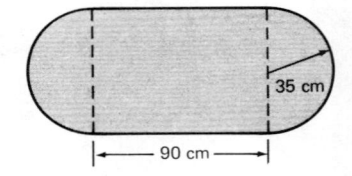
16. Find the circumference of pulley A in Figure 7-11.
17. Find the circumference of pulley B in Figure 7-11.

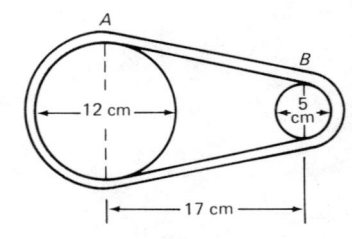

Figure 7-10

Figure 7-11

9. The length of belt needed for the pulleys in Figure 7-10 is approximated by the formula:

$$L = .52(C_A + C_B) + 2\ell$$

where C_A = circumference of pulley A

C_B = circumference of pulley B

ℓ = distance between the centers of the pulleys

L = length of the belt

Find L for the pulleys in Figure 7-10.

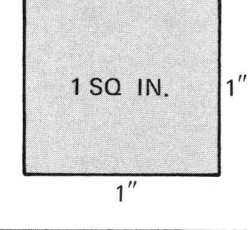

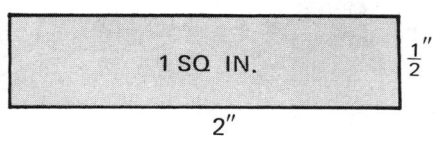

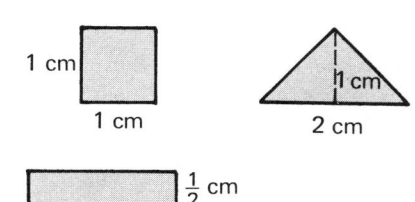

18. The length of belt needed for the pulleys in Figure 7-11 is approximated by the formula:

$$L = .52(C_A + C_B) + 2\ell$$

where C_A = circumference of pulley A

C_B = circumference of pulley B

ℓ = distance between the centers of the pulleys

L = length of the belt

Find L for the pulleys in Figure 7-11.

Area

7-5 AREA MEASUREMENT

Any area that is equivalent to the area of a square one inch on a side is a *square inch*. Each diagram above is 1 sq in. in area.

Can you check that each diagram below has an area of 1 sq cm?

To convert an area measurement to a different unit of measure, multiply by the appropriate conversion factor. The following tables show the common conversion units for area in both systems.

Table 7-3 *United States area measure*

144 sq in. = 1 sq ft	1 sq in. = $\frac{1}{144}$ sq ft
9 sq ft = 1 sq yd	1 sq ft = $\frac{1}{9}$ sq yd

Table 7-4 *Metric area measure*

100 sq mm = 1 sq cm	1 sq mm = .01 sq cm
100 sq cm = 1 sq dm	1 sq cm = .01 sq dm
100 sq dm = 1 sq m	1 sq dm = .01 sq m

The flow chart in Figure 7-12 tells how to change a measure from one unit to another.

Length, area, volume

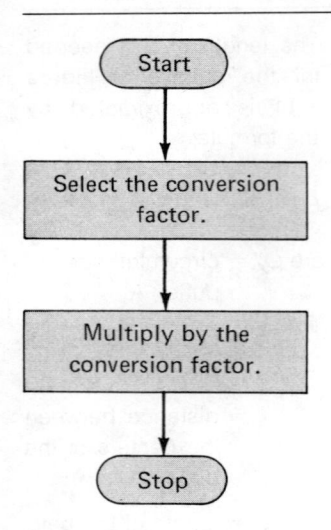

Figure 7-12

[Flowchart: Start → Select the conversion factor. → Multiply by the conversion factor. → Stop]

Example 1

54 sq ft = ___?___ sq yd

[Flowchart: Start → Conversion factor is $\frac{1}{9}$. → Multiply by $\frac{1}{9}$. → Stop]

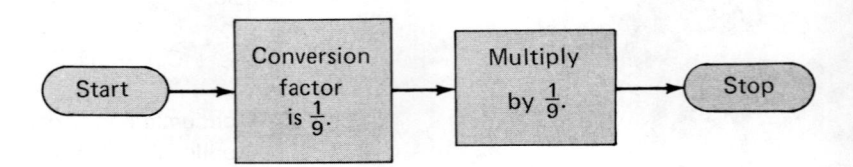

Solution

54 sq ft = (54 × $\frac{1}{9}$) sq yd

= 6 sq yd

Example 2

56,400 sq cm = ___?___ sq m

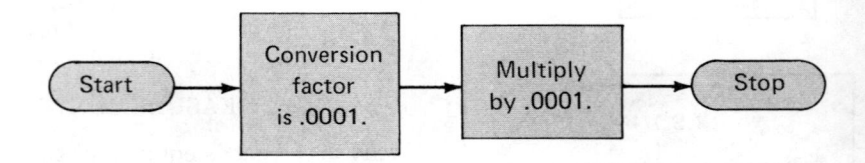

[Flowchart: Start → Conversion factor is .0001. → Multiply by .0001. → Stop]

Solution

1 sq cm = .01 sq dm

1 sq dm = .01 sq m

Therefore,

1 sq cm = (.01 × .01) sq m

= .0001 sq m

56,400 sq cm = (56,400 × .0001) sq m

= 5.64 sq m

Example 3

A house contains approximately 1040 sq ft of living space. What is the area in square yards? in square meters?

192   Technical and industrial formulas

Solution

Step 1

1040 sq ft = (1040 × 1/9) sq yd

= 116 sq yd (approx.)

Step 2 From Table 7-2,

1 ft = .305 m

1 sq ft = (.305)² sq m

= (.305 × .305) sq m

= .09 sq m (approx.)

1040 sq ft = (1040 × .09) sq m

= 93.6 sq m

EXERCISES   What factor do we use to make the following conversions?
1. Square inches to square feet
2. Square yards to square inches
3. Square millimeters to square centimeters
4. Square meters to square centimeters
5. Square feet to square yards

Complete the following statements.
6. 152 sq ft = __?__ sq yd
7. 3.5 sq cm = __?__ sq mm
8. 9½ sq yd = __?__ sq ft
9. 3.6 sq m = __?__ sq dm
10. 300 sq in. = __?__ sq ft
11. 15½ sq ft = __?__ sq in.
12. A parking lot measures 2520 sq ft in area. How much would it cost to blacktop the lot at a cost of $5.60 per sq yd?
13. A storage shed contains approximately 108 sq ft. How much would it cost to carpet the room at $9.75 per sq yd?
14. The press room at a lithograph company has a length of 60 ft and a width of 40 ft. What is the area in square yards?
15. How much would it cost to install a suspended ceiling in a room measuring 9 ft by 27 ft? The ceiling tiles are 12" square and cost $.27 each.
16. The blueprint of a house includes a 4 m by 5 m patio. What would be the cost of laying a flagstone patio, if the flagstone costs $6.95 per sq m?

7-6 AREA OF PARALLELOGRAMS

A *parallelogram* is any four-sided, closed figure with opposite pairs of sides parallel. Can you see that the rectangle and square shown in Figure 7-13 are parallelograms?

PARALLELOGRAM   RECTANGLE   SQUARE

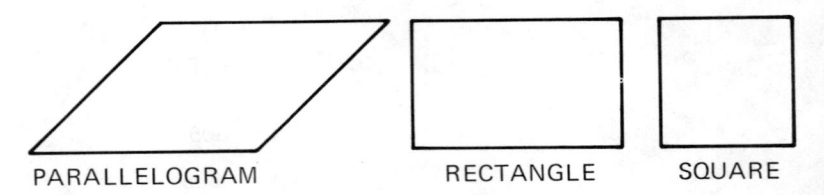

Figure 7-13

A *rectangle* is a parallelogram with four right angles. A *square* is a rectangle that has four sides of the same length.

The *area* of, or space inside, the rectangle in Figure 7-14 is 12 units, each a square. We always use *square units* to measure area.

To find the area of the parallelogram shown in Figure 7-15, imagine that you cut off the triangular region DAF and paste it over the triangular region CBE. Can you see that the area of parallelogram $ABCD$ is 5×3, or 15 square units? The distance AF is called the *height* of the parallelogram, and it is perpendicular to the *base*, AB.

So to find the area of any parallelogram, we can use the formula

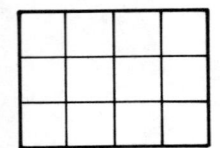

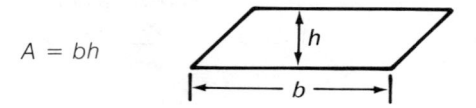

$$A = bh$$

Figure 7-14

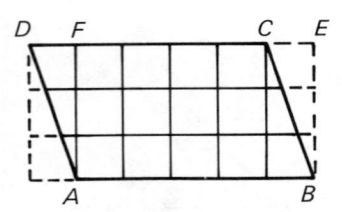

Figure 7-15

For a rectangle, the area formula is usually written

$$A = \ell w$$

And for the area of a square, we have the formula

$$A = s \times s$$
$$A = s^2$$

194   Technical and industrial formulas

Example

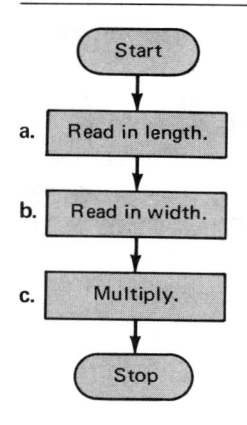

Find the area of a rectangle with a length of 14.3 cm and a width of 6.1 cm.

Solution

a. 14.3

b. 6.1

c. 14.3 × 6.1 = 87.23

d. 87.23 sq cm

Flowchart:
- Start
- a. Read in length.
- b. Read in width.
- c. Multiply.
- Stop

EXERCISES

U.S. system

Copy and complete the table. Find the area of a rectangle with the length and width given.

	Length	Width	Area
1.	18 ft	8 ft	?
2.	27 yd	16 yd	?
3.	6 ft	42 in.	?
4.	$3\frac{1}{8}$ in.	$2\frac{7}{8}$ in.	?

Find the area of a square with the sides given.

5. $s = 16$ ft
6. $s = \frac{3}{4}$ in.

Copy and complete the table. Find the area of a parallelogram with the base and height given.

	Base	Height	Area
7.	9 ft	3 ft	?
8.	$5\frac{1}{2}$ in.	4 in.	?
9.	14.2 ft	6.7 ft	?

10. A rectangular driveway measures 51 yd by 4 yd. How much would it cost to blacktop the driveway at $7.28 per square yard?

Metric system

Copy and complete the table. Find the area of a rectangle with the length and width given.

	Length	Width	Area
11.	16 m	9 m	?
12.	36 m	24 m	?
13.	5 cm	1.4 cm	?
14.	3.4 m	6.72 m	?

Find the area of a square with the sides given.

15. $s = 15$ cm
16. $s = 12.3$ m

Copy and complete the table. Find the area of a parallelogram with the base and height given.

	Base	Height	Area
17.	14 cm	9 cm	?
18.	16.3 cm	5.6 cm	?
19.	12.3 m	4.12 m	?

20. A rectangular driveway measures 54 m by 4 m. How much would it cost to blacktop the driveway at $7.28 per square meter?

Length, area, volume

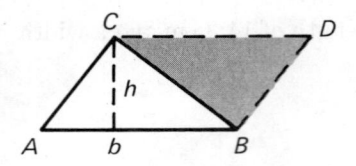

Figure 7-16

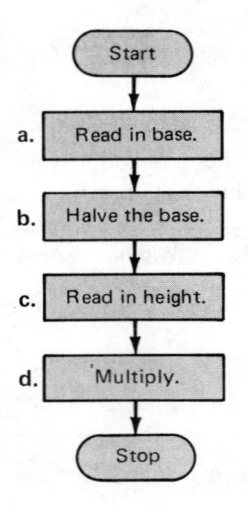

a. Read in base.
b. Halve the base.
c. Read in height.
d. Multiply.

7-7 AREA OF TRIANGLES AND TRAPEZOIDS

The area of a *triangle* is related to the area of a parallelogram. Figure 7-16 shows triangle ABC, with a *base* (b) and a *height* (h) perpendicular to the base. Does region ABC appear to be the same size and shape as region BCD? Together, these regions make up parallelogram $ABCD$. So the area of triangle ABC is $\frac{1}{2}$ of the area of $ABCD$.

So we have the formula for the area of a triangle:

$A = \frac{1}{2} bh$

Example 1

Find the area of a triangle that has a base of 14 cm and a height of 8 cm.

Solution

a. 14
b. $\frac{1}{2} \times 14 = 7$
c. 8
d. $8 \times 7 = 56$
e. 56 sq cm

A *trapezoid* is a four-sided figure that has exactly one pair of parallel sides. Two different trapezoids are shown in Figure 7-17. All trapezoids have two different length bases. We refer to them as b_1 and b_2.

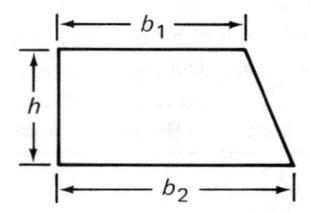

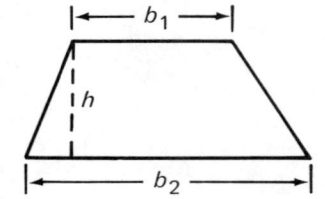

Figure 7-17

The area of a trapezoid can be found by using the formula

$A = \frac{1}{2} h (b_1 + b_2)$

196 Technical and industrial formulas

Flowchart

a. Read in height.
b. Halve the height.
c. Read in b_1.
d. Read in b_2.
e. Add $b_1 + b_2$.
f. Multiply by half the height.

Example 2

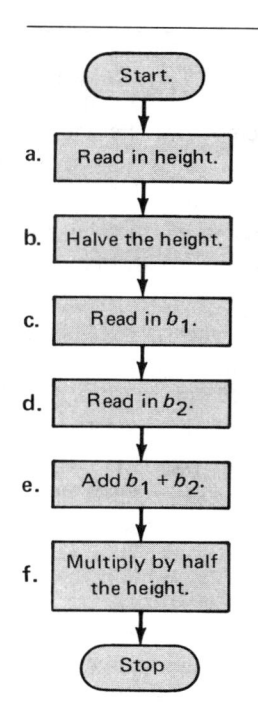

Find the area of a trapezoid that has bases of 14 cm and 18 cm and a height of 6 cm.

Solution

a. 6
b. $\frac{1}{2} \times 6 = 3$
c. 14
d. 18
e. $14 + 18 = 32$
f. $32 \times 3 = 96$
g. 96 sq cm

EXERCISES

U.S. system

Find the area of a triangle with the height and base given.

1. $b = 14$ in., $h = 10$ in.
2. $b = 7\frac{1}{2}$ in., $h = 5$ in.
3. $b = 120$ ft, $h = 80$ ft
4. $b = \frac{5}{8}$ in., $h = \frac{3}{4}$ in.

Copy and complete the table. Find the area of a trapezoid with the bases and height given.

	b_1	b_2	h	A
5.	12 in.	16 in.	10 in.	?
6.	5 in.	3 in.	$7\frac{1}{2}$ in.	?
7.	120 ft	110 ft	80 ft	?
8.	$\frac{3}{4}$ in.	1 in.	$\frac{5}{8}$ in.	?

Metric system

Find the area of a triangle with the height and base given.

11. $b = 14$ cm, $h = 10$ cm
12. $b = 165$ mm, $h = 85$ mm
13. $b = 1.2$ m, $h = .5$ m
14. $b = 110$ cm, $h = 50$ cm

Copy and complete the table. Find the area of a trapezoid with the bases and height given.

	b_1	b_2	h	A
15.	16 cm	12 cm	5 cm	?
16.	180 mm	240 mm	80 mm	?
17.	1.2 m	.5 m	.25 m	?
18.	2.32 cm	2.82 cm	1.6 cm	?

Length, area, volume

9. A trapezoidal table top has bases of 30 in. and 60 in. The height is 25½ in.
   a. How many square inches of vinyl are needed to cover the table top?
   b. If the vinyl is cut from a rectangular piece that is 30 in. by 60 in., how much vinyl is wasted?

10. A metal plate is made using the dimensions shown in the figure.
    a. Find the area of the trapezoid.
    b. Find the area of the rectangle that is cut out of the trapezoid.
    c. Find the area of the plate.

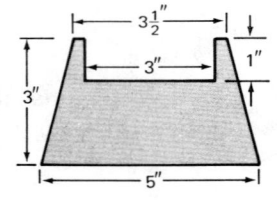

19. A trapezoidal table top has bases of 75 cm and 150 cm. The height is 67.5 cm.
    a. How many square centimeters of vinyl are needed to cover the table top?
    b. If the vinyl is cut from a rectangular piece that is 75 cm by 150 cm, how much is wasted?

20. A metal plate is made using the dimensions shown in the figure.
    a. Find the area of the trapezoid.
    b. Find the area of the rectangle that is cut out of the trapezoid.
    c. Find the area of the plate.

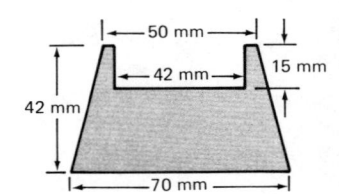

7-8 AREAS OF SPECIAL REGIONS

Sometimes we use more than one formula to find the area of a region.

For example, to find the area of the sheet metal plate shown in Figure 7-18, we find the area of the large rectangle and then subtract from it the area of the square.

Area of rectangle = 20 × 10 = 200

Area of square = 3 × 3 = 9

Area of plate = Area of rectangle − Area of square

= 200 − 9

= 191 sq mm

Figure 7-18

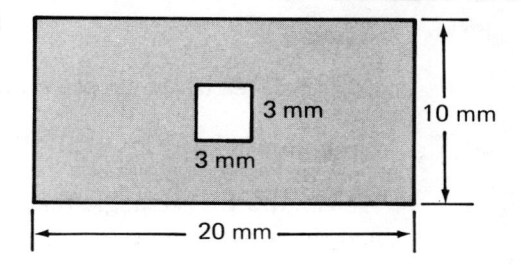

Sometimes we add to find the area of a region. Study the example below.

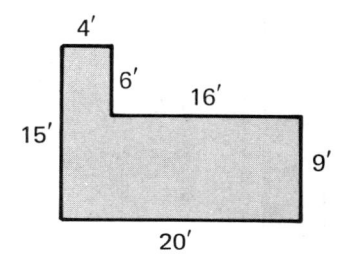

Example 1

Find the area of the L-shaped patio shown.

Solution 1

We can divide the region into two rectangles, and then combine the two areas.

Area of I = 4 × 6
= 24 sq ft
Area of II = 9 × 20
= 180 sq ft
Total area = 24 + 180
= 204 sq ft

Solution 2

We can find the area of the large rectangle and subtract the area of rectangle A.

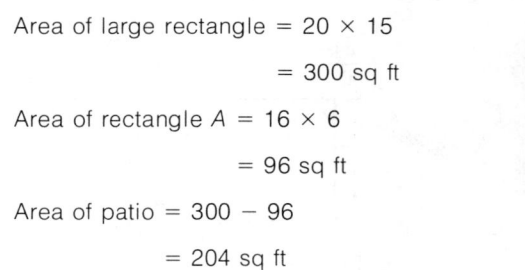

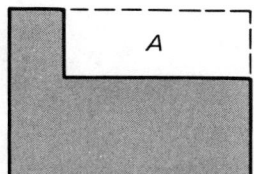

Area of large rectangle = 20 × 15
= 300 sq ft
Area of rectangle A = 16 × 6
= 96 sq ft
Area of patio = 300 − 96
= 204 sq ft

Length, area, volume 199

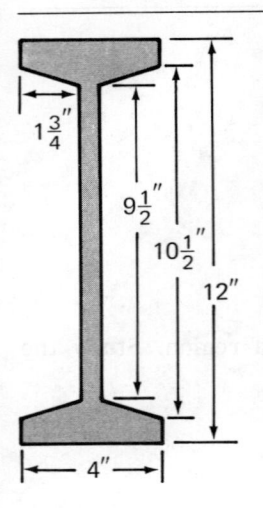

Example 2

Find the area of the cross-section of the I beam shown.

Solution

We can draw a rectangle that encloses the region. Then we can find the area of the two trapezoids and subtract them.

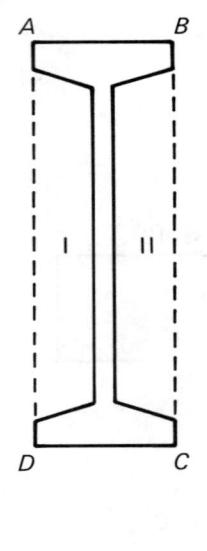

Area of ABCD = 4 × 12

= 48

Area of I = $\frac{1}{2}h(b_1 + b_2)$

$= \frac{1}{2} \times 1\frac{3}{4} (9\frac{1}{2} + 10\frac{1}{2})$

$= \frac{1}{2} \times \frac{7}{4} \times \frac{20}{1}$

$= \frac{1}{2} \times \frac{7}{\underset{1}{\cancel{4}}} \times \frac{\overset{5}{\cancel{20}}}{1}$

$= \frac{7 \times 5}{2}$

$= 17\frac{1}{2}$ sq in.

Area of II = $17\frac{1}{2}$ sq in.

Area of beam = 48 − ($17\frac{1}{2} + 17\frac{1}{2}$)

= 48 − 35

= 13 sq in.

EXERCISES

U.S. system

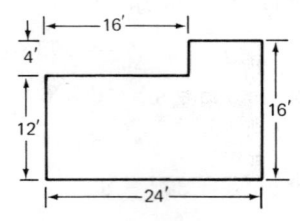

Figure 7-19

Metric system

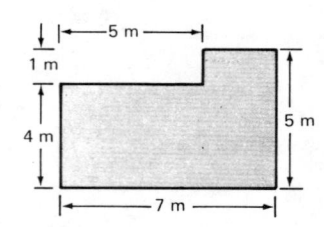

Figure 7-20

1. Find the area of the L-shaped patio shown in Figure 7-19.
2. Find the area of the cross-section of the concrete wall and footing shown. 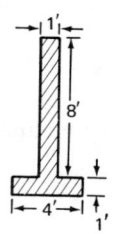
3. Find the area of the side of the building shown in the figure. 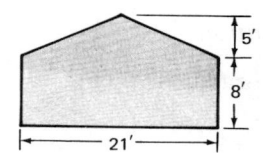
4. Find the area of the template shown in the figure. 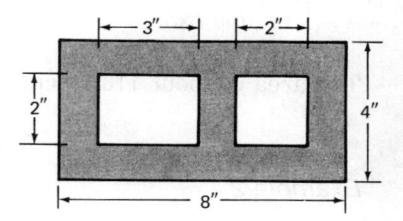
5. Find the area of the cross-section of the concrete highway support shown in the figure. 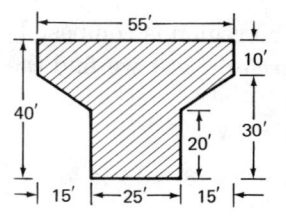
6. Find the area of the L-shaped patio shown in Figure 7-20.
7. Find the area of the cross-section of the concrete wall and footing shown. 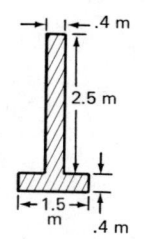
8. Find the area of the side of the building shown in the figure. 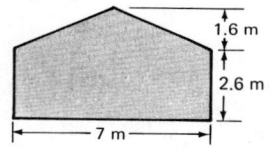
9. Find the area of the template shown in the figure. 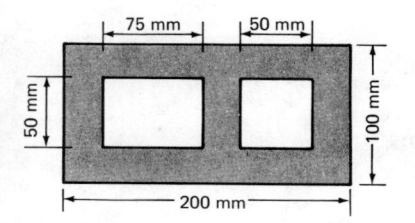
10. Find the area of the cross-section of the concrete highway support shown in the figure.

Length, area, volume 201

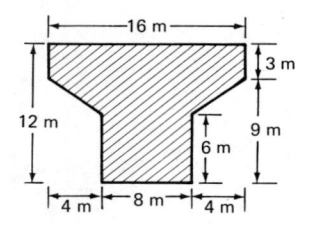

7-9 AREAS OF CIRCLES

Recall from Section 7-4 that to find the circumference of a circle, we need to use the number π, which is approximately equal to 3.14.

To find the area of a circle we also need to use this famous Greek number. The formula for the area of a circle is:

$$A = \pi \times r \times r \text{ or } A = \pi r^2$$

Remember that r^2 means to multiply the radius times itself, not times 2.

Example 1

Find the area of the circle shown.

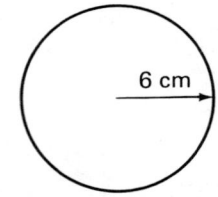

Solution

The radius is 6 cm.

$A = \pi r^2$
$A = 3.14 \times 6 \times 6$
$ = 3.14 \times 36$
$A = 113 \text{ (approx.)}$

The area is about 113 sq cm.

Example 2

Find the area of the region shown.

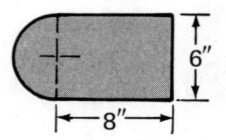

Solution

The region is composed of a rectangle and a semicircle (half of a circle). First we find the area of the rectangle.

$A = \ell w$
$A = 8 \times 6 = 48 \text{ sq in.}$

Now we find the area of the semicircle. To do this we find half the area of a circle whose radius is 3″. (Note that 6″, shown in the diagram, is the diameter, not the radius.)

$A = \pi r^2$

$ = 3.14 \times 3 \times 3$

$A = 28.26$ sq in.

Now to find the area of the semicircle, divide 28.26 by 2. 28.26 ÷ 2 = 14.13 sq in.

Now we add the area of the rectangle to the area of the semicircle: 14.13 + 48 = 62.13

The area of the region is about 62 sq in.

EXERCISES

U.S. system

Find the area of a circle with the radius given.

1. 5 in.
2. 8 ft
3. 15 yd
4. 7.3 in.
5. $4\frac{1}{2}$ in.
6. .25 in.
7. Find the area of the top of a circular tank with a diameter of 12 ft.
8. Find the area of the metal blank shown.

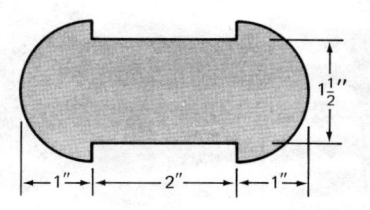

Metric system

Find the area of a circle with the radius given.

10. 6 mm
11. 100 mm
12. 1.3 m
13. 3.7 cm
14. 35 mm
15. 1.02 m
16. Find the area of a camera lens that has a diameter of 50 mm.
17. Find the area of the metal blank shown.

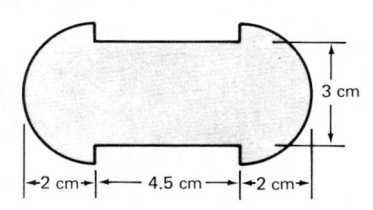

Length, area, volume   203

9. A cross-section of a pipe is shown. Find the area as follows:
   a. Find the area of the outer circle.
   b. Find the area of the inner circle and subtract it.

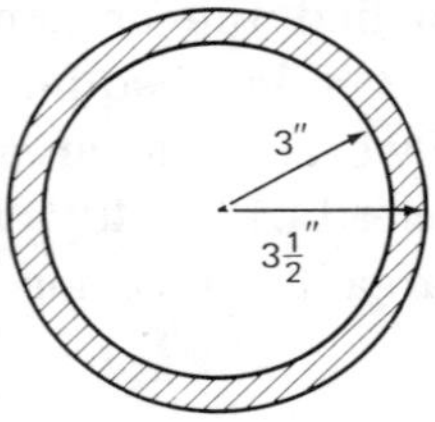

18. A cross-section of a pipe is shown. Find the area as follows:
    a. Find the area of the outer circle.
    b. Find the area of the inner circle and subtract it.

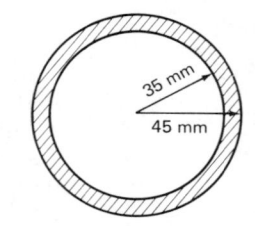

7-10 AREAS OF CYLINDERS, SPHERES, AND CONES

Suppose we need to know the area of the metal used to make the pipes shown in Figure 7-21. Each pipe has the shape of an open-ended *cylinder*.

When we find the area of the metal used for the pipes, we are finding the lateral area of a cylinder.

The formula for the *lateral area (L)* of a cylinder is:

$L = 2\pi rh$

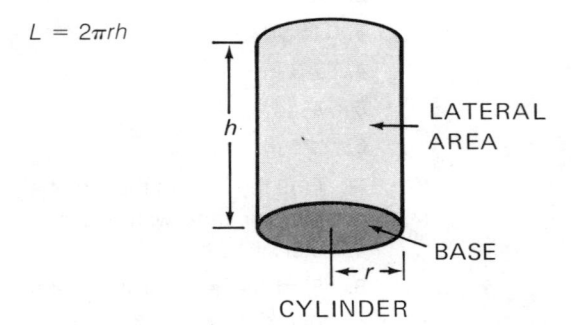

Figure 7-21

204 Technical and industrial formulas

So if a pipe is 6 feet long and has a radius of 6 inches, we can find the area of metal used as follows:

$L = 2\pi rh$

$= 2 \times 3.14 \times \frac{1}{2} \times 6$ (changing 6" to $\frac{1}{2}$')

$L = 18.84$ sq ft

Sometimes we may wish to find the total surface area of a cylinder. To the lateral area, we must add the area of the two bases, which are circular. So the formula for the *total surface area* (*S*) of a cylinder is the following:

S = Lateral area + 2 × (area of one base)

$S = 2\pi rh + 2\pi r^2$

Example 1

Find the total surface area of a cylindrical can that has a radius of 3 cm and a height of 8 cm.

Solution

$S = 2\pi rh + 2\pi r^2$

$= (2 \times 3.14 \times 3 \times 8) + (2 \times 3.14 \times 3 \times 3)$

$= 150.72 + 56.52$

$S = 207.24$ sq cm

To find the surface area of the spherical fuel tanks shown in Figure 7-22, we use the formula for the *surface area of a sphere*:

$S = 4\pi r^2$

Example 2

Find the surface area of the spherical fuel tank if it has a radius of 10 m.

Solution

$S = 4\pi r^2$

$= 4 \times 3.14 \times 10 \times 10$

$S = 1256$ sq m

Figure 7-22

Sometimes we want to calculate the surface area of a *cone*. For example, the funnel at the bottom of the exhaust collector shown in Figure 7-23 has the shape of a cone.

The *lateral area* (*L*) of a cone can be found by using the formula:

$$L = \pi r s$$

where *r* = the radius of the base

s = the slant height

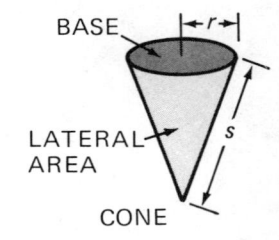

To find the *total surface area* (*S*) of a cone, we add the area of the base to the lateral area.

$$S = \pi r^2 + \pi r s$$

Figure 7-23

Example 3

Find the surface area of a cone with a radius of 5 cm and a slant height of 8 cm.

Solution

$S = \pi r^2 + \pi r s$

$= (3.14 \times 5 \times 5) + (3.14 \times 5 \times 8)$

$= 78.50 + 125.60$

$S = 204.10$ sq cm

EXERCISES

U.S. system

Find the total surface area of a cylinder with the radius and height given.

1. $r = 5$ in., $h = 10$ in.
2. $r = 8$ in., $h = 12$ in.
3. $r = 1\frac{1}{2}$ in., $h = 3$ in.
4. $r = 3.4$ in., $h = 1.1$ in.

Find the surface area of a sphere with the radius given.

5. $r = 4$ in.
6. $r = 6$ ft

Metric system

Find the total surface area of a cylinder with the radius and height given.

13. $r = 6$ cm, $h = 10$ cm
14. $r = 10$ mm, $h = 25$ mm
15. $r = 1.1$ m, $h = .5$ m
16. $r = 40$ mm, $h = 80$ mm

Find the surface area of a sphere with the radius given.

17. $r = 5$ cm
18. $r = 40$ mm

206 Technical and industrial formulas

Find the surface area of a cone with the radius and slant height given.

7. $r = 2$ in., $s = 5$ in.
8. $r = 10$ ft, $s = 12$ ft
9. $r = 80$ yd, $s = 150$ yd
10. How many square inches of sheet metal are needed to make the can shown?

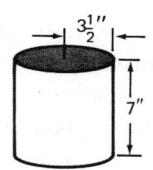

11. How many square inches of metal are needed to make a cone-shaped funnel like the one shown?

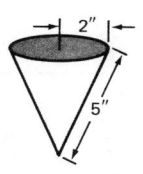

12. A spherical gas tank has a diameter of 20 ft. How many gallons of paint will it take to paint the tank if one gallon covers 200 sq ft?

Find the surface area of a cone with the radius and slant height given.

19. $r = 50$ mm, $s = 80$ mm
20. $r = 9$ m, $s = 14$ m
21. $r = 3.5$ m, $s = 6$ m
22. How many square millimeters of sheet metal are needed to make the can shown?

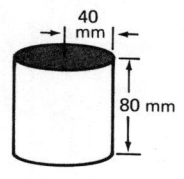

23. How many square millimeters of metal are needed to make a cone-shaped tunnel like the one shown?

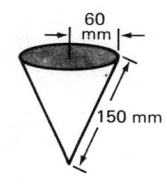

24. A spherical gas tank has a diameter of 6 m. How many cans of paint will it take to paint the tank if one can covers 18 sq m?

Volume

7-11 VOLUME MEASUREMENT

Any volume equivalent to the volume of a cube that has an edge one inch in length is called a *cubic inch*.

Similarly, any volume equivalent to the volume of a cube that has an edge one centimeter long is a *cubic centimeter*.

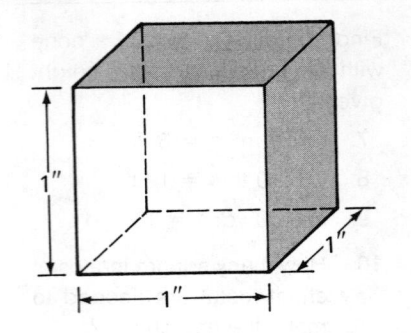

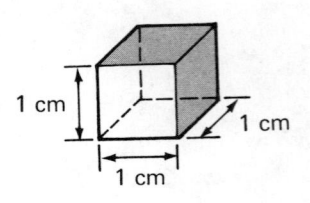

Figure 7-24

Figure 7-24 shows the relative size of the basic United States unit of volume (the cubic inch) and the basic metric unit of volume (the cubic centimeter).

Some of the common conversion factors for volume units are given in the tables below.

Note: Another abbreviation for 1 cu cm is 1 cc (read "one c-c"). Still another is 1 cm³. You should be familiar with all three.

To convert a particular volume measure to a different unit of measurement, multiply by the appropriate conversion factor.

Table 7-5 *United States volume measure*

1728 cu in. = 1 cu ft	1 cu in. = $\frac{1}{1728}$ cu ft
27 cu ft = 1 cu yd	1 cu ft = $\frac{1}{27}$ cu yd

Table 7-6 *Metric volume measure*

1000 cu mm = 1 cu cm	1 cu mm = .001 cu cm
1000 cu cm = 1 cu dm	1 cu cm = .001 cu dm
1000 cu dm = 1 cu m	1 cu dm = .001 cu m

Example 1

$3\frac{1}{2}$ cu yd = ___?___ cu ft

Solution

The conversion factor is 27.

$3\frac{1}{2}$ cu yd = ($3\frac{1}{2}$ × 27) cu ft

= $94\frac{1}{2}$ cu ft

208   Technical and industrial formulas

Example 2

35,200 cu mm = ___?___ cu cm

Solution

The conversion factor is .001.

35,200 cu mm = (35,200 × .001) cu cm

= 35.2 cu cm

Example 3

12,960 cu in. = ___?___ cu ft

Solution

The conversion factor is $\frac{1}{1728}$.

12,960 cu in. = ($\frac{1}{1728}$ × 12,960) cu ft

= 7.5 cu ft

Example 4

A small-parts bin measures 6 cm by 5 cm by 3 cm. What is the volume (a) in cubic centimeters? (b) in cubic inches?

Solution

a. The formula is $V = \ell \times w \times h$.

V = (6 × 5 × 3) cu cm

= 90 cu cm

b. From Table 7-2, 1 cm = .39 in.

1 cu cm = (.39)³ cu in.

= (.39 × .39 × .39) cu in.

= .06 cu in. (approx.)

V = 90 cu cm = (90 × .06) cu in.

= 5.4 cu in.

Length, area, volume

EXERCISES What is the conversion factor for each of the following:

1. Cubic feet to cubic inches?
2. Cubic yards to cubic inches?
3. Cubic millimeters to cubic centimeters?
4. Cubic meters to cubic centimeters?
5. Cubic feet to cubic meters?

Complete.

6. $2\frac{1}{4}$ cu ft = __?__ cu in.
7. 2.1 cu cm = __?__ cu mm
8. 3.7 cu m = __?__ cu dm
9. $9\frac{1}{4}$ cu yd = __?__ cu ft
10. 90 cu ft = __?__ cu yd
11. $17\frac{1}{3}$ cu yd = __?__ cu ft

12. A railway roadbed requires 110,000 cu yd of gravel. The trucks used on the project can haul 229.5 cu ft per load. How many truckloads of gravel will it take to complete the project?
13. The platform of a rapid transit station requires 1440 cu ft of concrete. How much will the concrete cost at $9.55 per cu yd?
14. A concrete batching plant produces $3\frac{1}{2}$ cu yd of concrete at a time. How many loads of concrete does it take to fill a truck with a capacity of 189 cu ft?
15. How many cubic yards of concrete are needed to lay a 1'-deep foundation for a garage measuring 25 ft by 30 ft?
16. If a tank has a volume of 40 cu yd, how many gallons will it hold? (1 cu ft = 7.5 gal, approximately)
17. In building a swimming pool, a contractor excavates an area measuring 18' by 36' to an average depth of 4'. If the excavated soil is used as land fill, how many cubic yards does the excavation yield?
18. Skylab, an experimental space station, contains a cylindrical workshop 48 ft long and 21.6 ft in diameter. Express its measurements in meters (to the nearest tenth).
19. In exercise 18, what is the space available in the workshop, to the nearest cubic meter?

7-12 VOLUMES OF RECTANGULAR SOLIDS

To find the *volume* of, or amount of space inside, a rectangular solid, we use *cubic measure*. Whereas in measuring area we used square units, in measuring volume we use cubic units. For example, to find the volume of the solid shown in Figure 7-25, we count the number of cubes that will fit in the solid.

You can see that there are 4 × 2, or 8 units in one layer. There

Figure 7-25

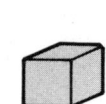

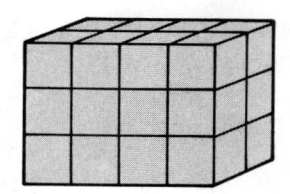

are 3 layers, so there are 4 × 2 × 3 cubic units in all. The volume is 24 cubic units.

To find the volume of any rectangular solid whose length, width, and height we know, we can use the formula:

$V = \ell w h$

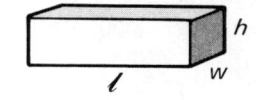

Example

How many cubic yards of dirt must be excavated to form a basement 35 ft × 20 ft × 8 ft?

Solution

$V = \ell w h$

$\quad = 35 \times 20 \times 8$

$V = 5600$ cu ft

But we need our answer in cubic yards.

To find how many cubic feet are in one cubic yard, look at Figure 7-26. You can see that there are 3 × 3 × 3, or 27 cubic feet in 1 cubic yard.

So we can divide 5600 by 27 to find the number of cubic yards.

$5600 \div 27 = 207.4$ (approx.)

About 207 cubic yards of dirt must be removed.

3 FT
3 FT
3 FT
1 CUBIC YARD

Figure 7-26

EXERCISES

U.S. system

Find the volume of a rectangular solid with the dimensions given.

1. $\ell = 8$ in., $w = 6$ in., $h = 5$ in.
2. $\ell = 15$ ft, $w = 9$ ft, $h = 20$ ft

Metric system

Find the volume of a rectangular solid with the dimensions given.

11. $\ell = 5$ cm, $w = 9$ cm, $h = 4$ cm
12. $\ell = 20$ mm, $w = 35$ mm, $h = 50$ mm

Length, area, volume   211

3. $\ell = 9\frac{1}{2}''$, $w = 4\frac{1}{2}''$, $h = 3''$

4. $\ell = 5\frac{3}{4}''$, $w = 1\frac{3}{8}''$, $h = \frac{5}{8}''$

5. A swimming pool is to be excavated. Its dimensions are 40 ft × 20 ft × 8 ft. How many cubic yards of dirt must be removed?

6. How many cubic yards of concrete are needed to construct a driveway 50 ft long, 10 ft wide, and 8 in. deep?

7. A rectangular trough measures 30 ft × 3 ft × 2 ft. How many gallons of water will it hold if 1 cu ft = $7\frac{1}{2}$ gal?

8. What volume of concrete is needed to build a roadway 45 ft wide, 1 ft thick, and 10 mi long?

9. If 1 cu in. of cast aluminum weighs .093 lb, how much would a bar 1 in. thick, 2 in. wide, and 3 ft long weigh?

10. An 8-ft-deep basement is to be excavated for the L-shaped building shown. How many cubic yards of dirt must be removed?

13. $\ell = 3.5$ m, $w = 2.8$ m, $h = 5$ m

14. $\ell = 2.6$ m, $w = 3.55$ m, $h = 4$ m

15. A swimming pool is to be excavated. Its dimensions are 12 m × 6.2 m × 3 m. How many cubic meters of dirt must be removed?

16. How many cubic meters of concrete are needed to construct a driveway 17 m long, 3.1 m wide and 200 mm deep?

17. A rectangular trough measures 9 m × .75 m × .6 m. How many liters of water will it hold if 1 cu m = 1000 l?

18. What volume of concrete is needed to build a roadway 13.5 m wide, .3 m thick, and 8000 m long?

19. If 100 cu mm of cast aluminum weigh 2.58 grams, how much would a bar of cast aluminum 25 mm thick, 50 mm wide, and 900 mm long weigh?

20. A basement 2.4 m deep is to be excavated for the L-shaped building shown. How many cubic meters of dirt must be removed?

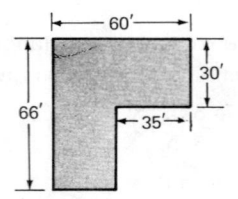

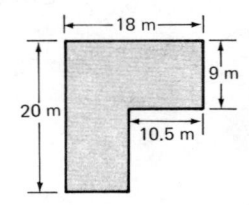

212   Technical and industrial formulas

7-13 VOLUMES OF CYLINDERS

Many everyday objects have the shape of a cylinder. Pipes and cans are the most common examples. Figure 7-27 shows how the cylinder for a lightweight steel beverage can is manufactured.

Figure 7-27

We find the volume of a cylinder in much the same way as we find volumes of rectangular solids. In a rectangular solid we multiplied the area of the base, in this case a rectangle, by the height of the solid.

V = Area of base $\times h$

$V = \ell w h$

In a cylinder, the base is a circle. Recall that the formula for the area of a circle is πr^2. So the formula for the volume of a cylinder is parallel to the formula for the volume of a rectangular solid:

V = Area of base $\times h$

$V = \pi r^2 h$

Example

Find the volume of a cylinder with a radius of 1.5 cm and a height of 9 cm.

Length, area, volume

Solution

$V = \pi r^2 h$

$= 3.14 \times 1.5 \times 1.5 \times 9$

$V = 63.5850$

The volume is about 63.6 cu cm.

EXERCISES

U.S. system

Find the volume of a cylinder with the radius and height given.

1. $r = 10$ in., $h = 8$ in.
2. $r = 4$ in., $h = 18$ in.
3. $r = 2\frac{1}{2}$ in., $h = 5$ in.
4. $r = 180$ yd, $h = 120$ yd
5. $r = \frac{3}{8}$ in., $h = \frac{3}{4}$ in.
6. A cylinder tank has a diameter of 30 ft and a height of 40 ft. How many gallons will it hold if 1 cu ft will hold about 7.5 gal?
7. The half-round steel rod shown has a diameter of $\frac{1}{2}$ in. and a length of 14 in. Find the volume of the rod.

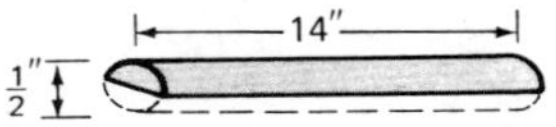

Metric system

Find the volume of a cylinder with the radius and height given.

8. $r = 10$ mm, $h = 20$ mm
9. $r = 9$ cm, $h = 15$ cm
10. $r = 2.8$ cm, $h = 3.5$ cm
11. $r = 3.5$ m, $h = 5.7$ m
12. $r = .85$ m, $h = 1.23$ m
13. A cylindrical tank has a diameter of 9 m and a height of 12 m. How many liters will it hold if 1 cu m will hold 1000 l?
14. The half-round steel rod shown has a diameter of 3 cm and a length of 25 cm. Find the volume of the rod.

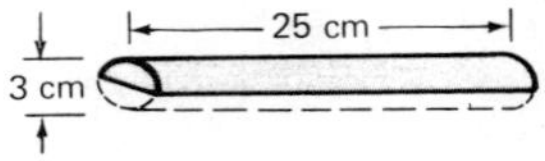

7-14 VOLUMES OF CONES, PYRAMIDS, AND SPHERES

A funnel is one of the most common examples of a design based on a *cone*.

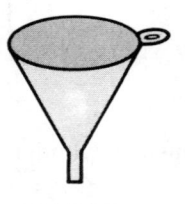

FUNNEL 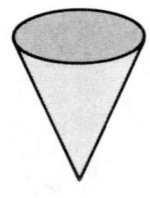 CONE

214 Technical and industrial formulas

Figure 7-28 shows a cone and a cylinder that have the same base and height. How do you think the volume of the cone compares to the volume of the cylinder?

Figure 7-28

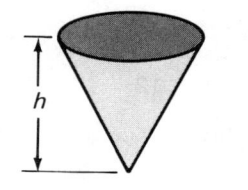

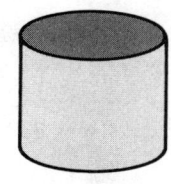

The volume of the cone is exactly one-third the volume of the cylinder. So the formula for the volume of a cone is:

$V = \frac{1}{3}\pi r^2 h$

Example 1

Find the volume of a cone with a radius of 3 in. and a height of 5 in.

Solution

$V = \frac{1}{3}\pi r^2 h$

$ = \frac{1}{3} \times 3.14 \times 3 \times 3 \times 5$

$V = 47.10$

The volume is about 47 cu in.

Compare the size of the *rectangular pyramid* in Figure 7-29 with the rectangular solid with the same base and height. You might guess that the volume of these two solid figures would compare in the same way as the volume of a cone and a cylinder with the same base and height.

Figure 7-29

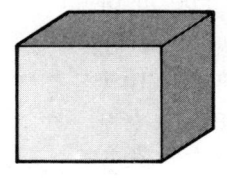

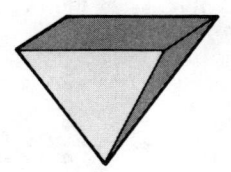

Indeed, the formula for the volume of a rectangular pyramid is:

$V = \frac{1}{3}\ell w h$

Length, area, volume   215

Example 2

Find the volume of the pyramid shown.

Solution

$V = \frac{1}{3}\ell wh$

$= \frac{1}{3} \times 10 \times 8 \times 12$

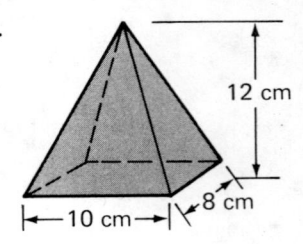

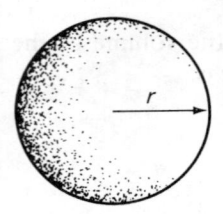

$V = 320$ cu cm

An interesting thing about spheres is that they have the greatest volume for their amount of surface area. It is for this reason that spherical tanks are commonly used to store gas and water.

The volume of a sphere can be found by using the formula:

$V = \frac{4}{3}\pi r^3$

Example 3

Find the volume of a sphere with a radius of 3 cm.

Solution

$V = \frac{4}{3}\pi r^3$

$= \frac{4}{3} \times 3.14 \times 3 \times 3 \times 3$

$V = 113.04$

The volume is about 113 cu cm.

EXERCISES

U.S. system

Find the volume of a cone with the radius and height given.

1. $r = 10$ in., $h = 6$ in.
2. $r = 18$ ft, $h = 12$ ft
3. $r = 1\frac{1}{2}$ in., $h = 5$ in.
4. $r = 50$ ft, $h = 20$ ft

Find the volume of a rectangular pyramid with the length, width, and height given.

5. $\ell = 11$ in., $w = 9$ in., $h = 5$ in.

Metric system

Find the volume of a cone with the radius and height given.

15. $r = 100$ mm, $h = 90$ mm
16. $r = 210$ mm, $h = 350$ mm
17. $r = 3.3$ m, $h = 7$ m
18. $r = .6$ m, $h = 1.2$ m

Find the volume of a rectangular pyramid with the length, width, and height given.

19. $\ell = 10$ mm, $w = 9$ mm, $h = 11$ mm

6. $\ell = 4$ ft, $w = 3$ ft, $h = 5$ ft
7. $\ell = 15$ in., $w = 18$ in., $h = 10$ in.
8. $\ell = 2\frac{1}{2}$ in., $w = 1\frac{1}{4}$ in., $h = 9$ in.

Find the volume of a sphere with the radius given.

9. $r = 10$ in.
10. $r = 6$ ft
11. $r = 300$ yd

12. A spherical water tank has a radius of 24 ft. How many gallons of water will it hold if 1 cu ft holds about 7.5 gal?

13. A cone-shaped container has a radius of 14 in. and a height of 10 in. How many gallons will it hold? (231 cu in. = 1 gal)

14. A concrete base for a light pole is constructed in the form of a square pyramid with the top section cut off, as shown. Find the volume of the light pole base.

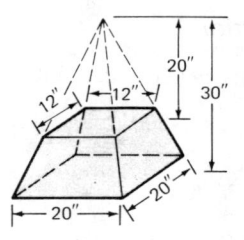

20. $\ell = 32$ cm, $w = 15$ cm, $h = 12$ cm
21. $\ell = 30$ mm, $w = 50$ mm, $h = 85$ mm
22. $\ell = 3.6$ m, $w = 2.1$ m, $h = .5$ m

Find the volume of a sphere with the radius given.

23. $r = 100$ mm
24. $r = 90$ m
25. $r = 3.3$ m

26. A spherical water tank has a radius of 7.2 m. How many liters of water will it hold if 1 cu m holds 1000 l?

27. A cone-shaped container has a radius of 35 cm and a height of 25 cm. How many liters will it hold? (1000 cu cm = 1 l)

28. A concrete base for a light pole is constructed in the form of a square pyramid with the top section cut off, as shown. Find the volume of the light pole base.

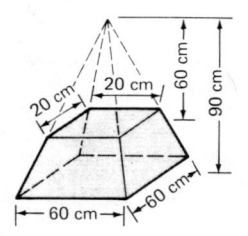

Weight, capacity, temperature

7-15 WEIGHT MEASUREMENT

The *weight* of an object is the *amount of force* that gravity exerts on the object. This force varies from place to place. The *mass* of an object is the *amount of material* of which the object is made.

For our purpose, we will use the terms interchangeably, since the objects we are studying are affected by the earth's gravity.

The basic unit of weight measurement in the United States system is the *pound* (*lb*). The basic unit of weight measurement in the metric system is the *kilogram*. The kilogram is the larger unit; 1 pound is approximately .454 kilogram. The relative size of these units is shown in Figure 7-30.

The tables below show the relationships among the units of weight.

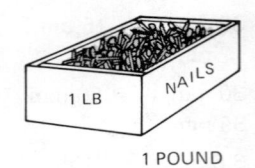

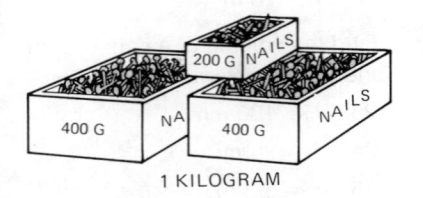

Figure 7-30

Table 7-7 United States weight measure

16 oz = 1 lb	1 oz = $\frac{1}{16}$ lb
2000 lb = 1 ton	1 lb = $\frac{1}{2000}$ ton = .0005 ton

Table 7-8 Metric weight measure

1000 milligrams (mg) = 1 g	1 mg = .001 g
1000 g = 1 kilogram (kg)	1 g = .001 kg
1000 kg = 1 metric ton (t)	1 kg = .001 t

Table 7-9 Approximately equivalent units of weight

1 oz = 28.350 g	1 g = .035 oz
1 lb = .454 kg	1 kg = 2.20 lb
1 ton = .907 t	1 t = 1.103 ton

As with the other units of measure, you will find that many industrial problems require changing from one unit of weight to another.

Example 1

$2\frac{1}{4}$ tons = ____?____ lb

Solution

$2\frac{1}{4}$ tons = ($2\frac{1}{4}$ × 2000) lb

= 4500 lb

Example 2

4500 g = ____?____ kg

Solution

4500 g = (4500 × .001) kg

= 4.5 kg

Example 3

.25 kg = ____?____ oz

218   Technical and industrial formulas

Solution

We first change kilograms to grams. The conversion factor is 1000.

.25 kg = (1000 × .25) g
= 250 g

We now can change grams to ounces. The conversion factor is .035.

250 g = (.035 × 250) oz
= 8.75 oz

Therefore, .25 kg = 8.75 oz

Example 4

If ½ kilogram of nails costs $.45, what is the cost per pound of the nails?

Solution

The nails cost $.90 per kg (2 × $.45).
Since 1 kg = 2.2 lb, we can say that 2.2 lb of nails cost $.90.
Thus, 1 lb of nails cost $.90/2.2 = $.41 (approx.).

EXERCISES

What conversion factor is used to change

1. pounds to tons?
2. metric tons to kilograms?
3. kilograms to grams?
4. kilograms to pounds?
5. pounds to metric tons?

Complete the weight conversions.

6. $24\frac{1}{4}$ lb = __?__ oz
7. 2.75 tons = __?__ lb
8. 22,500 lb = __?__ tons
9. 2700 kg = __?__ t
10. 1.72 kg = __?__ g
11. 454 g = __?__ mg
12. .151 t = __?__ kg
13. 5 lb = __?__ kg
14. 5.25 oz = __?__ g
15. 227 g = __?__ lb
16. 3.6 t = __?__ tons
17. 2.7 kg = __?__ lb
18. .75 kg = __?__ oz

19. A certain steel alloy weighs .28 lb per cu in. If a pipe contains 1200 cu in. of this metal, what is its weight in pounds? in kilograms?

Length, area, volume 219

20. A rectangular strip of metal is 500 cm long, 200 cm wide, and .5 cm thick. If the metal weighs .0077 kg per cu cm, find the weight of the metal strip in kilograms. What is the weight in pounds?
21. A commercial vehicle scale has a maximum capacity of 25,000 lb. What is its capacity in metric tons?
22. Test specifications for a certain metal container require it to withstand a force of 400 lb on the lid. If each container weighs 35 kg when full, how many may be safely stacked?
23. A radioactive material used in nuclear power plants costs $350 per gram. What would 1 oz of this material cost?
24. A tank has a volume of 1000 cu cm. How many kilograms of water can it hold? (1 cc weighs 1 g.)

7-16 CAPACITY MEASUREMENT

In the United States gasoline is purchased by the gallon. In most countries the price is quoted in liters. The *liter* (l) is the basic unit of capacity in the metric system.

A race car might be listed as having a 7-liter engine. A liter is equivalent to 1000 cu cm, so a 7-liter engine has a volume of 7000 cu cm.

To change this figure to cubic inches, we multiply by .06, the approximate number of cubic inches in a cubic centimeter. Can you see how to obtain this factor from Table 7-2?

7000 cu cm = (7000 × .06) cu in.

= 420 cu in.

So a 7-liter engine would be comparable to a 420 cu in. engine.

The tables that follow show the relationships among the common units of capacity.

Table 7-10 *United States capacity measure*

16 oz = 1 pt	1 oz = $\frac{1}{16}$ pt
2 pt = 1 qt	1 pt = $\frac{1}{2}$ qt
4 qt = 1 gal	1 qt = $\frac{1}{4}$ gal

Table 7-11 *Metric capacity measure*

1000 ml = 1 l	1 ml = .001 l
1000 l = 1 kiloliter (kl)	1 l = .001 kl

Table 7-12 *Approximately equivalent capacity units*

1 oz = 29.57 ml	1 ml = .034 oz
1 qt = .946 l	1 l = 1.057 qt
1 gal = 3.785 l	1 l = .264 gal

Example 1

2.25 l = __?__ ml

Solution

2.25 l = (1000 × 2.25) ml

= 2250 ml

Example 2

7.5 gal = __?__ l

Solution

7.5 gal = (3.785 × 7.5) l

= 28.3875 l, or 28.4 l

Example 3

If gasoline is listed at 42.9¢ per gallon, what is the cost per liter?

Solution

1 l = .264 gal

.264 × $.429 = $.113256, or 11.3¢ per liter

EXERCISES Perform the conversions.

1. 72 oz = __?__ qt
2. 50 pt = __?__ gal
3. 74.5 gal = __?__ qt
4. 1500 kl = __?__ l
5. 10.5 l = __?__ ml
6. 1752 l = __?__ kl
7. 25.33 ml = __?__ l
8. 500 ml = __?__ oz
9. 22.5 gal = __?__ l
10. 1 pt = __?__ ml
11. 28.7 l = __?__ gal
12. 2,500 ml = __?__ qt
13. A spherical tank has a radius of 6 m. How many liters of water will it hold? (Use π = 3.14.) (1 l = 1000 cu cm)
14. A cylindrical can has a volume of 525 cu cm. How many liters will it hold? (Use π = 3.14.) (1 ml = 1 cu cm)
15. If an automobile has a 450 cu in. engine, what is the capacity in liters?

16. To give a pleasant scent, 2 ml of perfume concentrate are used in 1 l of a liquid dishwashing detergent. If the detergent is packaged in liter containers, how much perfume concentrate must be used to scent 500 bottles of detergent?

17. If 1 l of a concentrated fertilizer is diluted with water to make 9 gal of fertilizer spray, how many milliliters of concentrate are in 1 gal of the diluted solution?

7-17 TEMPERATURE FORMULAS

Mr. Strong has a Fahrenheit thermometer in his shop, but the one outside the building has two scales, one in Fahrenheit and one in Celsius.

On the Fahrenheit thermometer the freezing point is 32°. The boiling point of water is 212°. On the Celsius thermometer 0° is the freezing point and 100° is the point at which water boils (see Figure 7-31).

Here are two formulas you can use to change the reading on one scale to the other.

$$C = \tfrac{5}{9}(F - 32) \qquad F = \tfrac{9}{5}C + 32$$

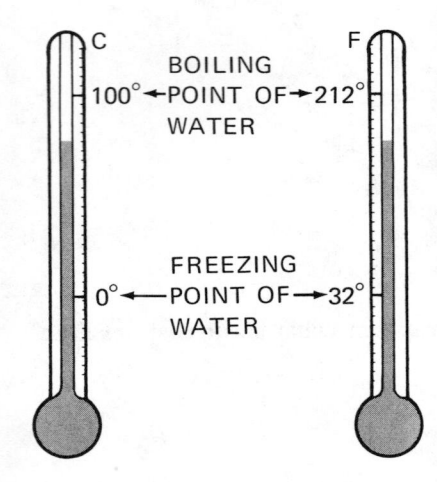

Figure 7-31

Example 1

20°C = __?__ °F

Solution

$F = \tfrac{9}{5}C + 32$

$= \tfrac{9}{5} \times 20 + 32$

$= 36 + 32 = 68.$ 68°F

Example 2

28°F = __?__ °C

Solution

$C = \tfrac{5}{9}(28 - 32)$

TRICKS OF THE TRADE

To multiply by 25, multiply $\tfrac{1}{4}$ of the number by 100.

$25 \times 68 = \tfrac{68}{4} \times 100 = 1700$

$25 \times 97 = 21.25 \times 100 = 2125$

If the second number in the parentheses is larger than the first, subtract the first from the second and use the minus sign, as $28 - 32 = -4$. What does the minus sign mean on a thermometer? Therefore,

$$C = \tfrac{5}{9}(28 - 32) = \tfrac{5}{9}(-4) = -\tfrac{20}{9} = -2\tfrac{2}{9}$$

$-2\tfrac{2}{9}°C$ means $2\tfrac{2}{9}°$ below zero on the Celsius scale.

EXERCISES Convert to Fahrenheit or Celsius readings as indicated.

1. 32°C = __?__ °F
2. 62°F = __?__ °C
3. 160°F = __?__ °C
4. 100°C = __?__ °F
5. 200°F = __?__ °C
6. 16°F = __?__ °C
7. 45°C = __?__ °F
8. 95°C = __?__ °F
9. The temperature fell from 80°C to 60°C in 1 hr. How much did the temperature drop in degrees on the Fahrenheit scale?
10. A soldering iron tip reaches a temperature of 830°F. What is this temperature in Celsius degrees?
11. Mercury freezes at −39°C. What is that temperature in degrees Fahrenheit?
12. Wrought iron melts at 1500°C. Express this in degrees Fahrenheit.

Taking inventory

1. The metric system and the United States system of measurement can be used for expressing measurements of length, area, volume, weight, and capacity. (p. 181)
2. The metric system is based on the decimal system; the conversion factors within the system are based on tens. (p. 181)
3. The basic units of the United States system are:
    a. the *yard* for length. (p. 181)
    b. the *gallon* for capacity. (p. 220)
    c. the *pound* for weight. (p. 218)
4. The basic units of the metric system are:
    a. the *meter* for length. (p. 182)
    b. the *liter* for capacity. (p. 220)
    c. the *kilogram* for weight. (p. 218)

5. The Celsius thermometer is scaled so that 0° corresponds to the freezing point of water and 100° its boiling point. These points on the Fahrenheit scale are 32° and 212°, respectively. (**p. 222**)

P = Perimeter, C = Circumference, A = Area,
L = Lateral area, S = Surface area, V = Volume

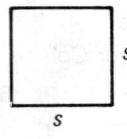

$P = 4s$ (**p. 187**)
$A = s^2$ (**p. 194**)

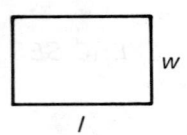

$P = 2\ell + 2w$ (**p. 187**)
$A = \ell w$ (**p. 194**)

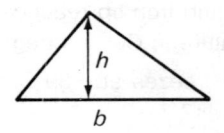

$A = \frac{1}{2}bh$ (**p. 196**)

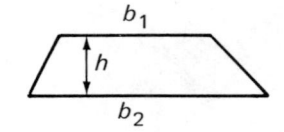

$A = \frac{1}{2}h(b_1 + b_2)$ (**p. 196**)

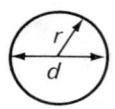

$C = \pi d$ (**p. 189**)
$A = \pi r^2$ (**p. 202**)

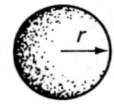

$S = 4\pi r^2$ (**p. 205**)
$V = \frac{4}{3}\pi r^3$ (**p. 216**)

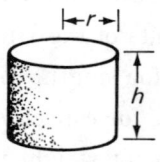

$L = 2\pi rh$ (**p. 204**)
$S = 2\pi rh + 2\pi r^2$ (**p. 205**)
$V = \pi r^2 h$ (**p. 213**)

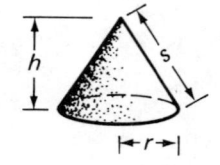

$L = \pi rs$ (**p. 206**)
$S = \pi r^2 + \pi rs$ (**p. 206**)
$V = \frac{1}{3}\pi r^2 h$ (**p. 215**)

224   Technical and industrial formulas

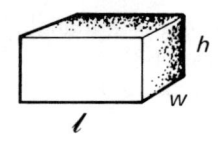

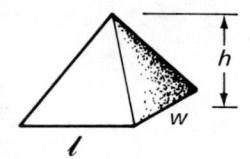

$V = \ell wh$ (p. 211)　　　$V = \frac{1}{3}\ell wh$ (p. 215)

Measuring your skills

1. 4.3 m = __?__ cm = __?__ mm (7-1)
2. .42 km = __?__ m = __?__ cm (7-1)
3. 1 ft = __?__ cm (7-2)
4. $4\frac{1}{2}$ mi = __?__ km (7-2)
5. 45 sq ft = __?__ sq yd (7-5)
6. 1500 sq m = __?__ sq km (7-5)
7. 10 sq m = __?__ sq yd (7-5)
8. If a rectangular patio floor is 24 ft long, 14 ft wide, and 4 in. thick, how many cubic meters of concrete will it take to construct the floor? (7-11)
9. 1.3 kg = __?__ g (7-15)　　10. 1.3 mg = __?__ g (7-15)
11. 3.5 lb = __?__ kg (7-15)　　12. 1460 ml = __?__ l (7-16)
13. 13.5 l = __?__ gal (7-16)　　14. 21.5 gal = __?__ qt (7-16)
15. If film developer costs $1.35 per gallon, what is the cost per liter? (7-16)
16. 85°C = __?__ °F (7-17)　　17. 85°F = __?__ °C (7-17)

U.S. system

18. Find the perimeter of a rectangle with a length of 8 in. and a width of 5 in. (7-3)
19. Find the circumference of a circle with a diameter of 5 in. (7-4)
20. Find the area of a parallelogram with a base of 18 in. and a height of 5 in. (7-6)

Metric system

28. Find the perimeter of a rectangle with a length of 12 cm and a width of 16 cm. (7-3)
29. Find the circumference of a circle with a diameter of 8 cm. (7-4)
30. Find the area of a parallelogram with a base of 25 mm and a height of 18 mm (7-6)

21. Find the area of a triangle with a base of 3½ in. and a height of 2 in. (7-7)

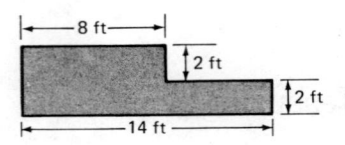

22. Find the area of the region shown. (7-8)

|←— 8 ft —→|
           ↕ 2 ft
                  ↕ 2 ft
|←——— 14 ft ———→|

23. Find the area of a circle with a radius of 4 ft. (7-9)

24. Find the total surface area of a cylinder with a radius of 6 in. and a height of 9 in. (7-10)

25. Find the volume of a rectangular solid with a length of 9 in., a width of 10 in., and a height of 3 in. (7-12)

26. Find the volume of a cylinder with a radius of 8 ft and a height of 12 ft. (7-13)

27. Find the volume of a cone with a radius of 9 in. and a height of 20 in. (7-14)

31. Find the area of a trapezoid with bases of 14 cm and 18 cm and a height of 9 cm. (7-7)

32. Find the area of the region shown. (7-8)

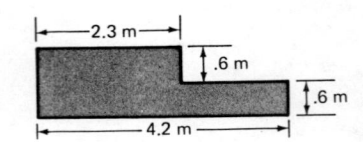

|←—2.3 m—→|
           ↕ .6 m
                  ↕ .6 m
|←——— 4.2 m ———→|

33. Find the area of a circle with a radius of 5 m. (7-9)

34. Find the surface area of a sphere with a radius of 30 mm. (7-10)

35. Find the volume of a rectangular solid with a length of 15 mm, a width of 20 mm, and a height of 8 mm. (7-12)

36. Find the volume of a cylinder with a radius of 25 mm and a height of 40 mm. (7-13)

37. Find the volume of a sphere with a radius of 10 mm. (7-14)

8 Polynomials

After completing this chapter, you should be able to:
1. Determine the degree of a monomial and a polynomial.
2. Add and subtract polynomials.
3. Multiply monomials and polynomials.
4. Factor trinomials and the difference of two squares.
5. Solve quadratic equations by factoring.
6. Solve quadratic equations using the quadratic formula.

Operations with polynomials

8-1 ADDING AND SUBTRACTING POLYNOMIALS

Terms like x^2, $5x$, and $4x^2y$ are called monomials. A *monomial* is a term that consists of a product of a constant and powers of variables. An indicated sum (or difference) of more than one monomial is called a *polynomial*.

Thus,

$7x^2 + 4x^3 + 2x - 5$

$15x + 7$

$14x^5 + 4x^4y + 3x^3y^2$

are polynomials.

The *degree of a monomial in a variable* is the exponent of the variable in the monomial. Thus, $4xy^7z^2$ is of degree 1 in x, 7 in y, and 2 in z. The *degree of a monomial* is the sum of the degrees in each of the variables. The monomial $4xy^7z^2$ has degree $1 + 7 + 2 = 10$.

The *degree of a polynomial* is the degree of the term that has the highest degree.

Thus, $7x^2 + 4x^3 + 2x - 5$ has degree 3, $15x + 7$ has degree 1, $14x^5 + 4x^4y + 3x^3y^2$ has degree 5.

To add polynomials, we add the similar terms.

Example 1

Add: $7x^3 - 4x^2 + 7x - 3$ and $x^3 + 2x^2 + 4$.

Solution

$$
\begin{array}{r}
7x^3 - 4x^2 + 7x - 3 \\
x^3 + 2x^2 + 4 \\
\hline
8x^3 - 2x^2 + 7x + 1
\end{array}
$$

Example 2

Find an expression for the triangle perimeter in Figure 8-1.

Figure 8-1

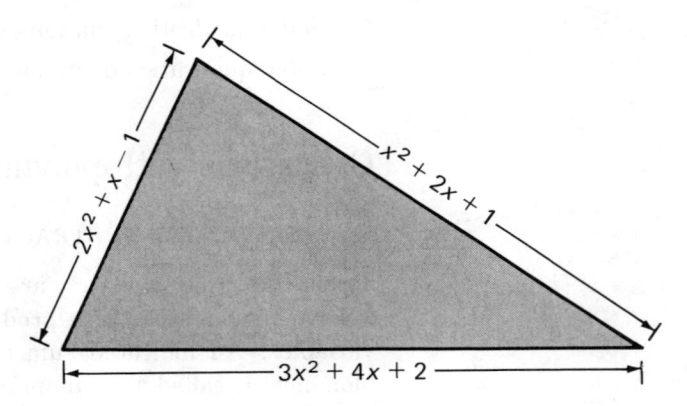

Solution

The perimeter is the sum of the three polynomials.

$$
\begin{array}{r}
x^2 + 2x + 1 \\
2x^2 + x - 1 \\
3x^2 + 4x + 2 \\
\hline
6x^2 + 7x + 2
\end{array}
$$

To subtract one polynomial from another, we add the negative of the polynomial to be subtracted to the other polynomial. Recall that the negative of an expression is obtained by changing the sign of each term in the expression. Thus, the negative of $-x^3$ is x^3, the negative of $x - 4$ is $-x + 4$, the negative of $-x^2 + 4x - 7$ is $x^2 - 4x + 7$.

Example 3

Subtract $x^2 - 4x + 6$ from $2x^2 - 3x + 4$.

Solution

The negative of $x^2 - 4x + 6$ is $-x^2 + 4x - 6$.

$$\begin{array}{r} 2x^2 - 3x + 4 \\ -x^2 + 4x - 6 \\ \hline x^2 + x - 2 \end{array}$$

Example 4

Find the area of the shaded part of the rectangle in Figure 8-2 if the area of the large rectangle is represented by the polynomial $4x^3 - 2x^2 + 7x$ and the area of the small rectangle is represented by the polynomial $2x^2 + 3x - 5$.

Figure 8-2

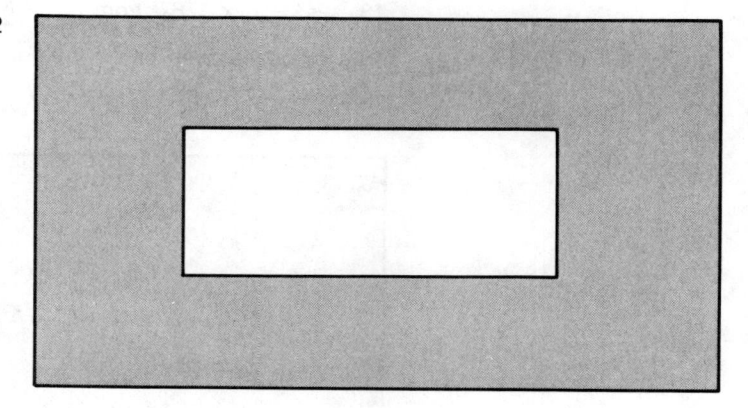

Solution

The negative of $2x^2 + 3x - 5$ is $-2x^2 - 3x + 5$. We add this polynomial to $4x^3 - 2x^2 + 7x$.

$$\begin{array}{r} 4x^3 - 2x^2 + 7x \\ -2x^2 - 3x + 5 \\ \hline 4x^3 - 4x^2 + 4x + 5 \end{array}$$

EXERCISES Determine the degree of each monomial.

1. $12x^3y^2$ 2. $11x^2y$ 3. $4x^4$ 4. $2xy$

Determine the degree of each polynomial.

5. $4x^3 - 2x^2 + 7x - 11$
6. $2xy + x$
7. $2x^3y + 3x^2y^2 + 4xy$
8. $5x^5 + 1$

Add the following polynomials.

9. $3x^2 + 4x + 6$
   $2x^2 + 2x - 1$

10. $5x^3 + 7x^2 - 2x + 7$
    $4x^2 + 3x - 2$

11. $4x^3y - 7x^2y^2 + 3y^3x$
    $2x^3y - 2x^2y^2 - 2y^3x$

12. $3.6x + 4.2y$
    $1.7x - 1.4y$

Write the negative of each polynomial.

13. $x^3 - 2x^2 - 5x + 7$
14. $x^5 - 1$

Subtract the following polynomials.

15. $2x^3 - 3x^2 + 7x - 5$ from $5x^3 + 4x^2 - 6x - 3$
16. $2x + 7$ from $-3x - 4$
17. $x^2 - x + 1$ from $x^3 - 1$
18. $2y^2 - 5yx + 6x^2$ from $7y^2 - 6x^2$
19. Find an expression for the perimeter of the rectangle in Figure 8-3.

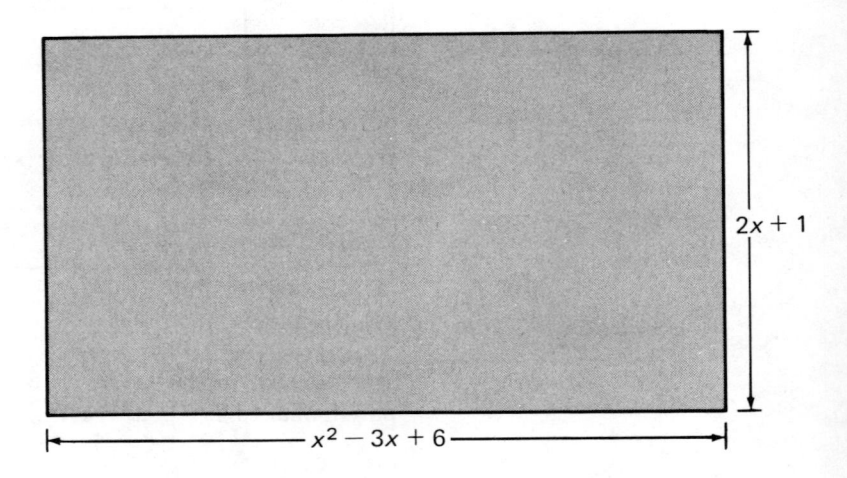

Figure 8-3

20. The fuel consumption of an engine is given by the expression $3.7x^2 + .4y + 7.2$. After a turbocharger is added to the engine, the consumption is $4.3x^2 + .3y + 7.2$. Write an expression for the increase in consumption caused by the turbocharger.

8-2 MULTIPLYING POLYNOMIALS

To multiply two monomials, we use the properties of real numbers and the properties of exponents.

Example 1

$x^2 \cdot x^3 = ?$

Solution

From the definition of an exponent,

$x^2 \cdot x^3 = (x \cdot x)(x \cdot x \cdot x)$

$= x \cdot x \cdot x \cdot x \cdot x$

$= x^5$

This example suggests the following rule.

$$x^a \cdot x^b = x^{a+b}$$

Example 2

$(3x^3y)(2x^2y^4) = ?$

Solution

$(3x^3y)(2x^2y^4) = (3 \cdot 2)(x^3 \cdot x^2)(y \cdot y^4)$

$= 6x^{3+2}y^{1+4}$

$= 6x^5y^5$

Note that the product of two monomials is a monomial.

To multiply a polynomial by a monomial, we multiply each term in the polynomial by the monomial.

Polynomials 231

Example 3

$(3x^2)(x^2 - 4x + 3) = ?$

Solution

$3x^2(x^2 - 4x + 3)$
$= (3x^2)(x^2) + (3x^2)(-4x) + (3x^2)(3)$
$= (3 \cdot x^2 \cdot x^2) + [3 \cdot (-4) \cdot x^2 \cdot x] + (3 \cdot 3 \cdot x^2)$
$= 3x^{2+2} - 12x^{2+1} + 9x^2$
$= 3x^4 - 12x^3 + 9x^2$

Example 4

$(2x^2y^2)(x^2 + 2xy - y^2) = ?$

Solution

$2x^2y^2(x^2 + 2xy - y^2)$
$= (2x^2y^2)(x^2) + (2x^2y^2)(2xy) + (2x^2y^2)(-y^2)$
$= (2 \cdot x^2 \cdot x^2 \cdot y^2) + (2 \cdot 2 \cdot x^2 \cdot x \cdot y^2 \cdot y) + [2(-1) \cdot x^2 \cdot y^2 \cdot y^2]$
$= 2x^{2+2}y^2 + 4x^{2+1}y^{2+1} - 2x^2y^{2+2}$
$= 2x^4y^2 + 4x^3y^3 - 2x^2y^4$

To multiply a polynomial by a second polynomial, we multiply each term in the first polynomial by each term in the second polynomial and combine the like terms.

Example 5

Multiply $2x - 3$ by $x + 6$.

Solution 1

$$
\begin{array}{r}
2x - 3 \\
x + 6 \\
\hline
2x^2 - 3x \\
12x - 18 \\
\hline
2x^2 + 9x - 18
\end{array}
$$

1. Multiply $2x - 3$ by x.
2. Multiply $2x - 3$ by 6.
3. Add like terms.

Solution 2

$(x + 6)(2x - 3)$

$= x(2x - 3) + 6(2x - 3)$

$= (2x^2 - 3x) + (12x - 18)$

$= 2x^2 + (-3x + 12x) - 18$

$= 2x^2 + 9x - 18$

An alternative procedure can be used when multiplying a pair of binomials of the form $(ax + b)(cx + d)$.

```
           Last term
   First   term
     ↓      ↓
   (ax + b)(cx + d)
     ↑  ↑   ↑
```
Inner and outer products

1. We obtain the first term in the product by multiplying the first terms of the binomials.

    $(ax)(cx) = acx^2$

2. The middle term of the product is the sum of the inner and outer products.

    $b(cx) + (ax)d = (bc + ad)x$

3. We obtain the last term of the product by multiplying the last terms of the binomials.

    $(b)(d) = bd$

Thus, $(ax + b)(cx + d) = acx^2 + (bc + ad)x + bd$

We now illustrate how to use this method to multiply the binomials of Example 5.

```
  ┌─2x²─┐
  │  ┌─-18┐
  │  │   │
 (x + 6)(2x - 3)
      ↑   ↑
  ─3x + 12x
```

1. The first term is $(x)(2x) = 2x^2$.
2. The middle term is $-3x + 12x = 9x$.
3. The last term is $(6)(-3) = -18$.

Thus, $(x + 6)(2x - 3) = 2x^2 + 9x - 18$.

Polynomials

Figure 8-4

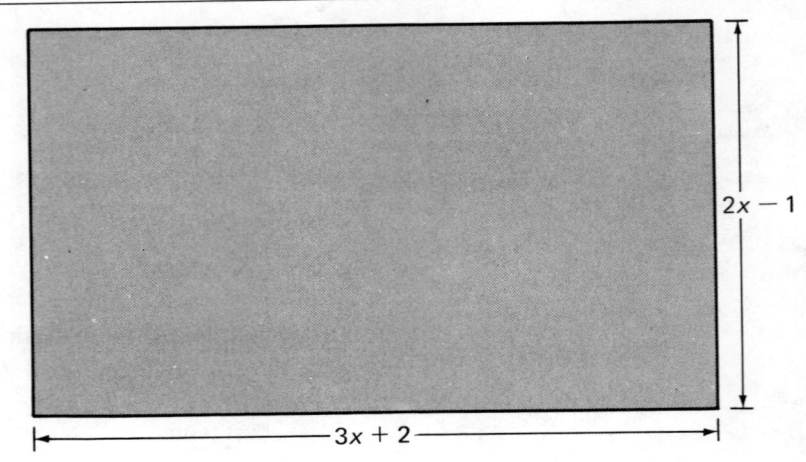

[Rectangle with width $2x-1$ and length $3x+2$]

Example 6

Find the area of the rectangle shown in Figure 8-4.

Solution

The length of the rectangle is $3x + 2$, and the width is $2x - 1$. Area = Length × Width = $(3x + 2)(2x - 1)$.

$$(3x + 2)(2x - 1)$$

with products: $6x^2$, -2, $4x$, $-3x$

$A = 6x^2 + (4x - 3x) - 2 = 6x^2 + x - 2$

Example 7

Multiply $a^2 + ab + b^2$ by $a - b$.

Solution 1

$$\begin{array}{l} a^2 + ab + b^2 \\ \underline{ a - b} \\ a^3 + a^2b + ab^2 \\ \underline{ -a^2b - ab^2 - b^3} \\ a^3 - b^3 \end{array}$$

1. Multiply $a^2 + ab + b^2$ by a.
2. Multiply $a^2 + ab + b^2$ by b.
3. Combine like terms.

234 Technical and industrial formulas

Solution 2

$(a - b)(a^2 + ab + b^2)$

$= a(a^2 + ab + b^2) - b(a^2 + ab + b^2)$

$= (a^3 + a^2b + ab^2) - (a^2b + ab^2 + b^3)$

$= a^3 + (a^2b - a^2b) + (ab^2 - ab^2) - b^3$

$= a^3 + 0 + 0 - b^3$

$= a^3 - b^3$

Example 8

A building with dimensions $AB = 2x + 1$ and $BC = x + 1$ is expanded to the dimensions $A'B = 3x + 2$, $BC' = x + 6$. By how much area is the original building expanded?

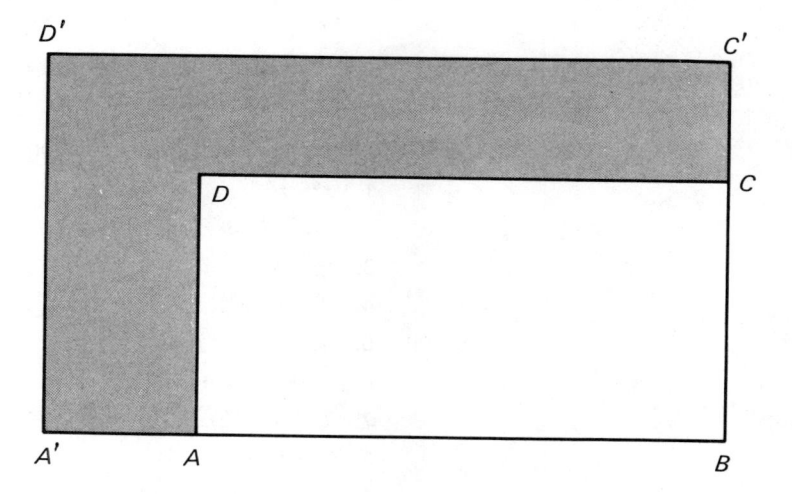

Solution

1. Compute the area of rectangle $ABCD$.

   $A_1 = (2x + 1)(x + 1)$

   $A_1 = 2x^2 + (x + 2x) + 1 = 2x^2 + 3x + 1$

Polynomials 235

2. Compute the area of rectangle $A'BC'D'$.

$$A_2 = (3x + 2)(x + 6)$$

$$A_2 = 3x^2 + (2x + 18x) + 12 = 3x^2 + 20x + 12.$$

3. Calculate $A_2 - A_1$.

$$A_2 - A_1 = (3x^2 + 20x + 12) - (2x^2 + 3x + 1)$$

$$\begin{array}{r} 3x^2 + 20x + 12 \\ -2x^2 - 3x - 1 \\ \hline \end{array}$$

$$A_2 - A_1 = x^2 + 17x + 11$$

EXERCISES Perform the indicated calculations.

1. $x^5 \cdot x^2$
2. $(4x^3y)(-3xy^2)$
3. $y^4 \cdot y$
4. $(2x^2y^4)(x^3)$
5. $(5x^3y^2)(\frac{1}{5}xy^2)$
6. $3(x^2 + 2x + 1)$
7. $5x(3x - 2)$
8. $4x^3(x^2 + x + 1)$
9. $-2x(x^3 - 5x^2 + 4x - 3)$
10. $3x^3(y^2 - 2y + 7)$
11. $(x + 3)(x - 2)$
12. $(2x + 3y)(x - 5y)$
13. $(x^2 + x + 1)(x - 1)$
14. $(5x - 7y)(3x - 2y)$
15. $(x^2 + y^2)(x - 2y)$
16. $(3x^2 + 2x + 1)(x + 1)$
17. $(4x^2 - 3y^2)(2x^2 + y^2)$
18. Find a polynomial expression for the area of the rectangle in Figure 8-5.

Figure 8-5

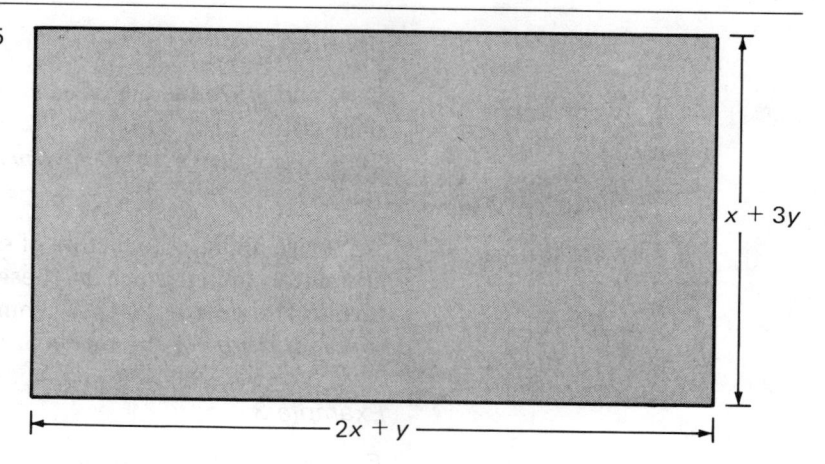

(Rectangle with width $2x + y$ and height $x + 3y$)

19. Find a polynomial expression for the volume of a rectangular solid with length $2x + 3$, width $2x - 1$, and height 5 ($V = \ell wh$).
20. Write an expression for the volume of a cylinder with radius r and height $2r - 1$ ($V = \pi r^2 h$).

Factoring

8-3 FACTORING POLYNOMIALS

Writing a polynomial as a product of two or more polynomials is called *factoring* the polynomial. We use the properties of numbers to find the factors.

The first step in factoring a polynomial is to determine the monomial, if any, that can be factored from the polynomial. This process is based on the property:

$ab + ac = a(b + c)$

Example 1

Factor $5x^2 + 10x$.

Solution

5 and x are factors of both terms of $5x^2 + 10x$. Thus, $5x^2 + 10x = 5x(x + 2)$, since $5x^2 = (5x)(x)$ and $10x = (5x)(2)$.

Example 2

Factor $2x^3y + 4x^2y^2 + 6xy^3$.

Solution

2, x, and y are factors of each of the three terms of the polynomial. Thus, $2x^3y + 4x^2y^2 + 6xy^3 = 2xy(x^2 + 2xy + 3y^2)$, since $2x^3y = (2xy)(x^2)$, $4x^2y^2 = (2xy)(2xy)$, and $6xy^3 = (2xy)(3y^2)$.

We are finding the factors of each of the terms in a polynomial and determining which of these factors are common to all the terms. The product of these common terms is called the *greatest common factor* of the terms of the polynomial.

Example 3

Factor out the greatest common factor for the polynomial $5x^2yz^3 + 10x^3y^2z^3 - 10x^2y^2z^2$.

Solution

$5x^2yz^3$ has the factors $5, x, x, y, z, z, z$. $10x^3y^2z^3$ has the factors $2, 5, x, x, x, y, y, z, z, z$. $10x^2y^2z^2$ has the factors $2, 5, x, x, y, y, z, z$. The common factors are $5, x, x, y, z, z$. The greatest common factor is

$$(5)(x)(x)(y)(z)(z) = 5x^2yz^2$$

since 5 appears once, x appears twice, y appears once, and z appears twice as a factor of each of the terms in the polynomial. Thus, $5x^2yz^3 + 10x^3y^2z^2 - 10x^2y^2z^2 = 5x^2yz^2(z + 2xy - 2y)$, since $5x^2yz^3 = (5x^2yz^2)(z)$, $10x^3y^2z^2 = (5x^2yz^2)(2xy)$, and $-10x^2y^2z^2 = (5x^2yz^2)(-2y)$.

Example 4

The formula for the total surface area of a cone is $S = \pi r^2 + \pi rs$. Write the formula with the right side of the equation in factored form.

Solution

Each of the terms on the right side of the equation has π and r as factors, since πr^2 has π, r, r as factors and πrs has π, r, s as factors. Thus, $S = \pi r(r + s)$.

We now illustrate the method of factoring *trinomials*, which are polynomials containing three terms and which are of degree 2 in one of the variables.

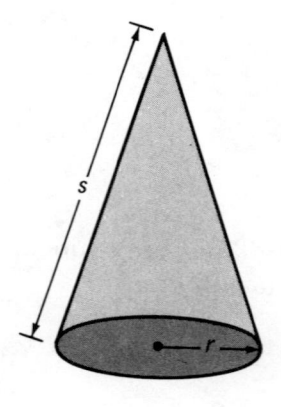

Example 5

Are $(x - 3)$ and $(x - 2)$ factors of the trinomial $x^2 - 5x + 6$?

Solution

We can answer this question by multiplying $(x - 3)$ by $(x - 2)$.

$$(x - 3)(x - 2) = x^2 + (-2x - 3x) + 6$$
$$= x^2 - 5x + 6$$

We can see that $x - 3$ and $x - 2$ are indeed factors of $x^2 - 5x + 6$.

In general, the factors (if any) of a trinomial are binomials. We use this fact to help us find these factors.

Example 6

Factor the trinomial $x^2 + 7x + 12$.

Solution

1. The factors (if any) are binomials.

   $x^2 + 7x + 12 = ($ $)($ $)$

2. The product of the first terms of the binomials must be x^2.

   $x^2 + 7x + 12 = (x$ $)(x$ $)$

3. The product of the last terms of the binomials must be 12. There are several possibilities: 12×1, 6×2, 4×3. However, we are looking for the pair of factors that has the sum 7. Thus, we use 4 and 3.

   $x^2 + 7x + 12 = (x + 4)(x + 3)$

Polynomials

4. We check the result by multiplying.

$$(x+3)(x+4)$$

with x^2 from first terms, 12 from last terms, and $3x + 4x$ from outer/inner.

$$(x+4)(x+3) = x^2 + (4x + 3x) + 12$$
$$= x^2 + 7x + 12$$

Example 7

Factor the trinomial $x^2 + 11x - 12$.

Solution

$$x^2 + 11x - 12 = (x \quad)(x \quad)$$

$$(x \quad)(x \quad)$$

with x^2 from first terms, -12 from last terms, and $11x$ from outer/inner.

The possible factors of -12 are the pairs $(-6)(2)$, $(6)(-2)$, $(-4)(3)$, $(4)(-3)$, $(12)(-1)$, $(-12)(1)$. We are looking for the particular pair that has the sum 11. The desired pair is $(12)(-1)$. Thus,

$$x^2 + 11x - 12 = (x + 12)(x - 1)$$

We check the result by multiplying.

$$(x+12)(x-1)$$

with x^2 from first terms, -12 from last terms, and $12x - x$ from outer/inner.

$$(x + 12)(x - 1) = x^2 + (12x - x) - 12$$
$$= x^2 + 11x - 12$$

Example 8

Factor the trinomial $6x^2 + 13xy + 7y^2$.

Solution

The possible factors of $6x^2$ are the pairs $(3x)(2x)$ and $(6x)(x)$. Suppose we try the pair $(3x)(2x)$:

$$6x^2 + 13xy + 7y^2 = (3x + \quad)(2x + \quad)$$

The factors of $7y^2$ are $7y$ and y.

$$(3x + 7y)(2x + y)$$
$$14xy + 3xy$$

$$(3x + y)(2x + 7y)$$
$$2xy + 21xy$$

Neither of these pairs of binomials gives $13xy$ for the middle term. We now try the pair $(6x)(x)$ as factors of the first term.

$$(6x + 7y)(x + y)$$
$$7xy + 6xy$$

$$(6x + y)(x + 7y)$$
$$xy + 42xy$$

The first pair of factors, $(6x + 7y)$ and $(x + y)$, gives the middle term $13xy$. Thus,

$$6x^2 + 13xy + 7y^2 = (6x + 7y)(x + y)$$

Example 9 shows how we combine the techniques of Examples 1 through 4 with the techniques of Examples 6 through 8.

Example 9

Factor the trinomial $3x^5 - 6x^4 + 3x^3$.

Solution

1. Factor out any monomial factor. The greatest monomial factor is $3x^3$.

$$3x^5 - 6x^4 + 3x^3 = 3x^3(x^2 - 2x + 1)$$

2. Factor the resulting trinomial.

$$3x^3(x^2 - 2x + 1) = 3x^3(x - 1)(x - 1)$$

Polynomials 241

Figure 8-6

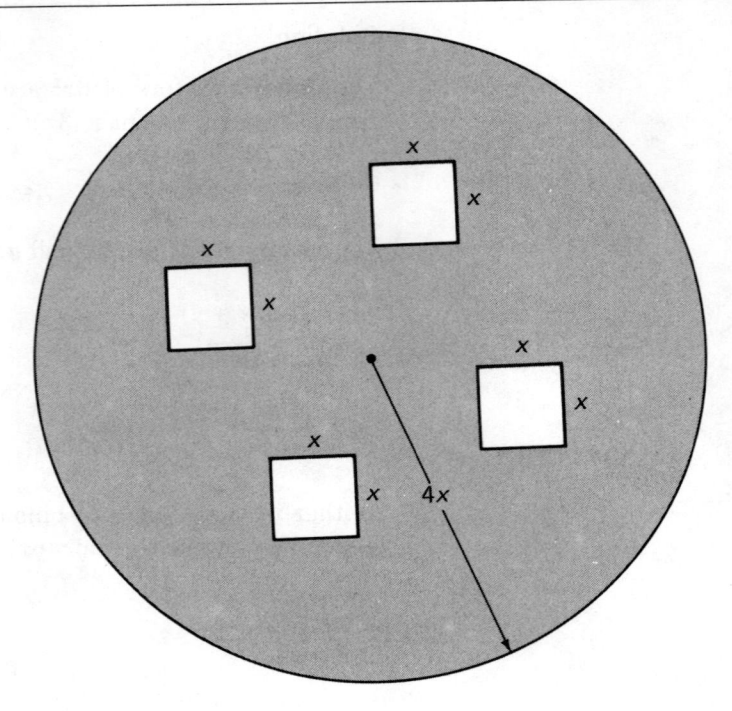

Example 10

Write an expression, in factored form, for the area of the shaded region in Figure 8-6.

Solution

1. The area of the circle is

$$A_1 = \pi(4x)^2 = 16\pi x^2$$

2. The area of the squares is $A_2 = 4x^2$. The area of the shaded region is

$$A = A_1 - A_2 = 16\pi x^2 - 4x^2$$
$$= 4x^2(4\pi - 1)$$

EXERCISES Factor the greatest monomial factor from each of the following polynomials.

1. $9x + 18$
2. $3x^2 - 9x$
3. $5ax + 5ay$
4. $a^2x^2 + ax$
5. $2x^4y + 4x^3y^2 + 8x^2y^3$
6. $6x^2yz - 12xy^2 + 21x^2y^2z^2$

Write each of the given trinomials as the product of a pair of binomials.

7. $x^2 + x - 6$
8. $x^2 - x - 72$
9. $x^2 + x - 20$
10. $4x^2 + 3x - 1$
11. $12y^2 + 2y - 14$
12. $3y^2 - 7y - 6$
13. $x^2 - 5x - 36$
14. $2x^2 - 13xy + 18y^2$
15. $x^2 + 14xy + 24y^2$
16. $15y^2 - 2y - 8$

For each of the following trinomials, factor out the greatest monomial factor, and write the resulting trinomial factor as the product of a pair of binomials.

17. $2x^2 + 20x + 50$
18. $xy^2 - 16xy + 64x$
19. $8x^2 + 8x + 2$
20. $15x^2y - 13xy + 2y$

Write an expression, in factored form, for the area of the shaded region.

21.

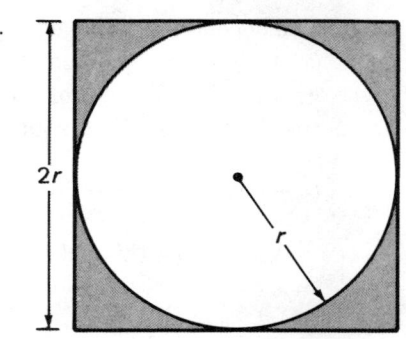

22.

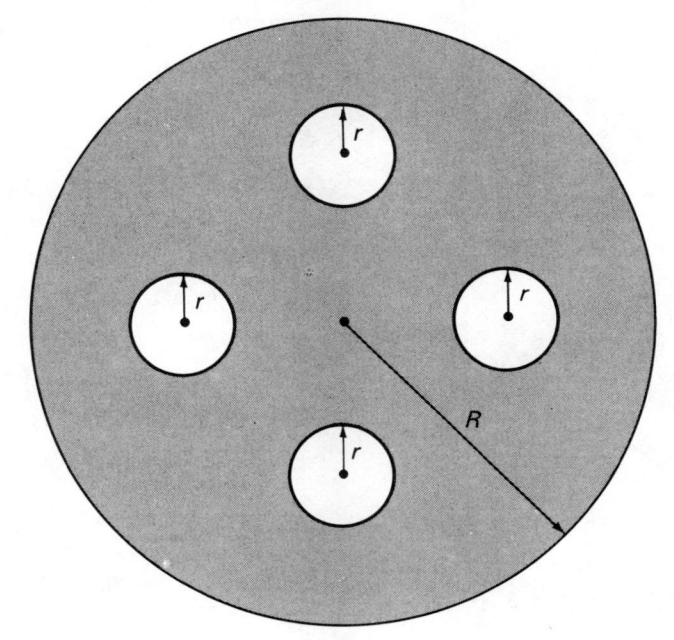

Polynomials 243

8-4 FACTORING THE DIFFERENCE OF TWO SQUARES

Certain polynomials occur frequently enough that they should be recognized as factorable. One such polynomial is a polynomial of the form $x^2 - y^2$, or *the difference of two squares*. Examine the three examples shown below.

$$
\begin{array}{ccc}
x + y & x - 7 & 3x + 2y \\
\underline{x - y} & \underline{x + 7} & \underline{3x - 2y} \\
x^2 + xy & x^2 - 7x & 9x^2 + 6xy \\
\underline{ - xy - y^2} & \underline{ + 7x - 49} & \underline{ - 6xy - 4y^2} \\
x^2 - y^2 & x^2 - 49 & 9x^2 - 4y^2
\end{array}
$$

The second and third of these products can be written as $(x - 7)(x + 7) = x^2 - 7^2$ and $(3x + 2y)(3x - 2y) = (3x)^2 - (2y)^2$. In general, the product of the sum and difference of two numbers is the difference of the squares of the numbers, or

$$(x + y)(x - y) = x^2 - y^2$$

We can use this rule to factor the difference of two squares.

Example 1

Factor $x^2 - 9$.

Solution

$x^2 - 9 = x^2 - 3^2 = (x + 3)(x - 3)$

Example 2

Factor $36x^2 - 49y^2$.

Solution

$36x^2 - 49y^2 = (6x)^2 - (7y)^2$
$ = (6x + 7y)(6x - 7y)$

Example 3

Factor $9x^2z - 36y^2z$.

Solution

1. We first factor out the monomial factor.

$$9z(x^2 - 4y^2)$$

2. We now factor the resulting binomial as the difference of two squares.

$$9z(x^2 - 4y^2) = 9z[x^2 - (2y)^2]$$
$$= 9z(x + 2y)(x - 2y)$$

Example 4

Figure 8-7 shows a plan for making a box without a cover from a square piece of metal by cutting out the corners. Write an expression, in factored form, for the area of the metal used in the box.

Solution

1. The area of the original square is

$$A_1 = x^2$$

2. The area of the metal that is cut out is

$$A_2 = 4y^2$$

3. The area of the metal used is

$$A = A_1 - A_2 = x^2 - 4y^2$$
$$= x^2 - (2y)^2$$
$$= (x + 2y)(x - 2y)$$

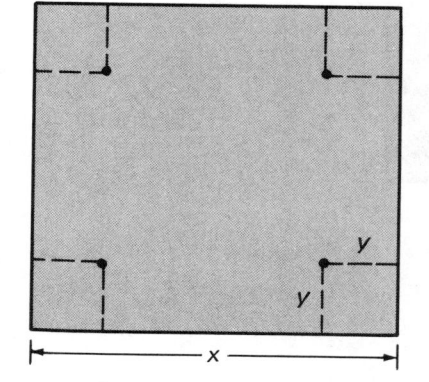

Figure 8-7

EXERCISES

Factor each of the expressions.

1. $x^2 - 9$
2. $y^2 - 121$
3. $4x^2 - 25$
4. $x^2 - 9y^2$
5. $16x^2 - 25y^2$

6. $x^2 - 144y^2$
7. $5x^2 - 125$
8. $48x^2 - 27y^2$
9. $18x^2 - 8$
10. $16x^2y^2 - z^2$

Write an expression, in factored form, for the area of the shaded part of each region.

11.

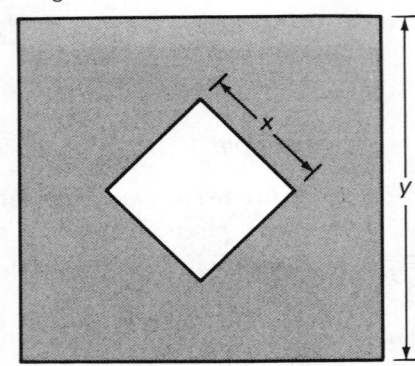

12.

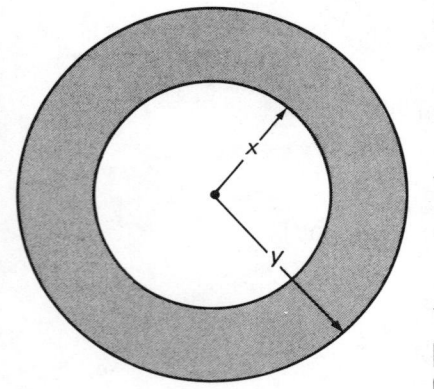

13.

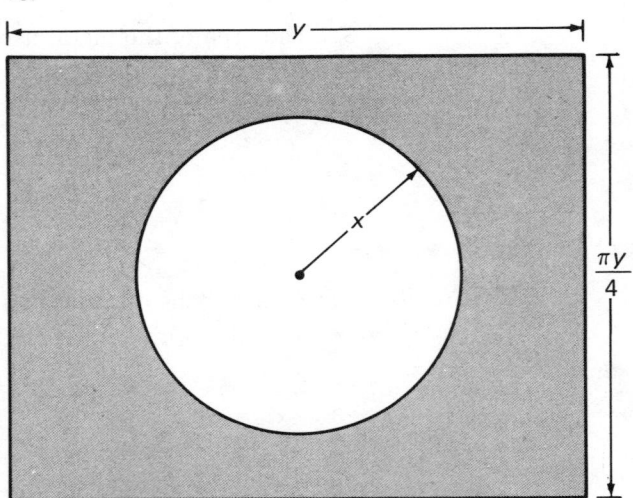

Quadratic equations

8-5 SOLVING QUADRATIC EQUATIONS BY FACTORING

In Sections 8-2 and 8-3, we learned to factor trinomial expressions and the difference of two squares. We now apply these techniques to solving *second-degree* or *quadratic equations* in a particular variable. The factoring method of solving equations is based on a property of real numbers.

> If a and b are real numbers and $a \cdot b = 0$, then at least one of a and b equals 0.

Example 1

Solve $x^2 - 7x + 12 = 0$.

Solution

1. Factor the left side of the equation.

   $(x - 3)(x - 4) = 0$

2. At least one of $x - 3$ and $x - 4$ is zero. If $x - 3 = 0$, then $x = 3$. If $x - 4 = 0$, then $x = 4$.

3. Check the numbers $x = 3$ and $x = 4$ in the equation $x^2 - 7x + 12 = 0$. For $x = 3$, $3^2 - 7(3) + 12 = 9 - 21 + 12 = 0$. For $x = 4$, $4^2 - 7(4) + 12 = 16 - 28 + 12 = 0$. Therefore, either $x = 3$ or $x = 4$.

Example 2

Solve $3x^2 + x - 10 = 0$.

Solution

1. Factor the left side of the equation.

   $(3x - 5)(x + 2) = 0$

2. At least one of $3x - 5$ and $x + 2$ is 0. If $3x - 5 = 0$, then $x = \frac{5}{3}$. If $x + 2 = 0$, then $x = -2$.

Polynomials 247

3. Check the numbers $x = \frac{5}{3}$ and $x = -2$ in the equation $3x^2 + x - 10 = 0$. For $x = \frac{5}{3}$, $3(\frac{5}{3})^2 + \frac{5}{3} - 10 = \frac{25}{3} + \frac{5}{3} - 10 = 0$. For $x = -2$, $3(-2)^2 + (-2) - 10 = 12 - 2 - 10 = 0$. Therefore, either $x = \frac{5}{3}$ or $x = -2$.

Example 3

A rectangular parking lot is 2 yd longer than it is wide. The area of the lot is 99 sq yd. Find the dimensions.

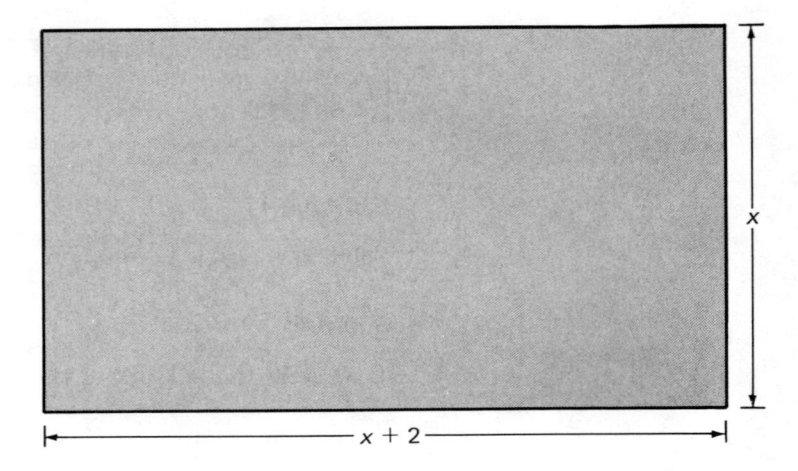

Solution

1. Write an equation that shows the relationship. Since $A = \ell \times w$, then $x(x + 2) = 99$, or $x^2 + 2x - 99 = 0$.
2. Solve the equation.

    $x^2 + 2x - 99 = 0$

    $(x + 11)(x - 9) = 0$

    If $x + 11 = 0$, then $x = -11$. If $x - 9 = 0$, then $x = 9$.
3. Interpret the results. We reject the solution $x = -11$, since the dimensions of a rectangle cannot be negative. If $x = 9$, then the width is 9 yd and the length is 11 yd. Do you see how we can check this conclusion?

Example 4

A piece of wire 44 cm long is cut into two pieces and bent to form two squares. If the sum of the areas of the squares is 61 sq cm, find the lengths of the sides of the squares.

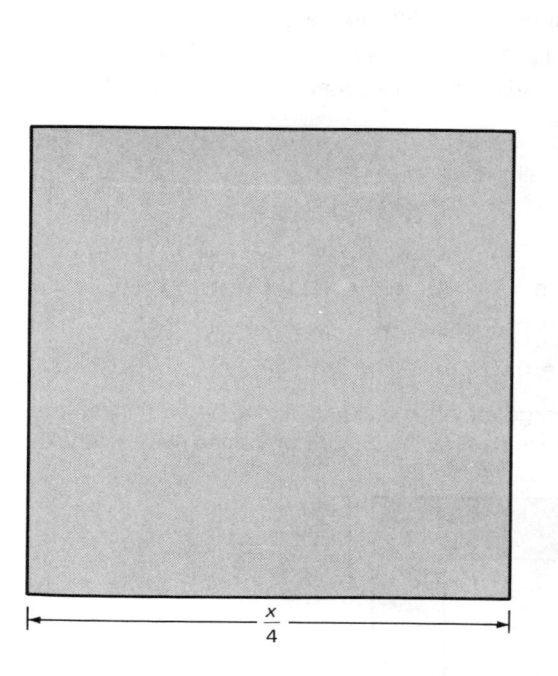

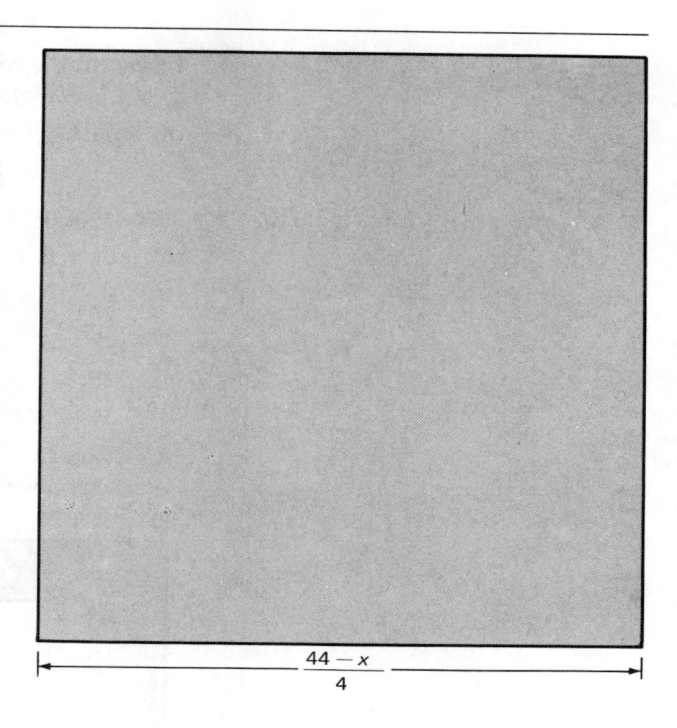

Solution

1. Write an equation that shows the relationship. Let x and $44 - x$ be the lengths of the two pieces of wire. Then $x/4$ and $(44 - x)/4$ are the lengths of the sides of the squares. The areas of the squares are represented by

$$\left(\frac{x}{4}\right)^2 + \left(\frac{44 - x}{4}\right)^2 = 61$$

2. Solve the equation.

$$\frac{x^2}{16} + \frac{(44 - x)^2}{16} = 61$$

$$x^2 + (44 - x)^2 = 976$$

$$x^2 + 1936 - 88x + x^2 = 976$$

$$2x^2 - 88x + 960 = 0$$

$$x^2 - 44x + 480 = 0$$

$$(x - 20)(x - 24) = 0$$

$$x = 20 \text{ or } x = 24.$$

Polynomials 249

3. Interpret the results. If $x = 20$, then $x/4 = 5$ and $(44 - x)/4 = 6$. If $x = 24$, then $x/4 = 6$ and $(44 - x)/4 = 5$. In either case the lengths of the sides of the two squares are 5 cm and 6 cm.

EXERCISES Solve the following equations.

1. $x^2 - 5x + 6 = 0$
2. $x^2 + 5x - 6 = 0$
3. $x^2 + x - 12 = 0$
4. $x^2 + 10x + 21 = 0$
5. $x^2 - 18x + 45 = 0$
6. $4x^2 + 4x + 1 = 0$
7. $x^2 - 36 = 0$
8. $3x^2 - 5x + 2 = 0$
9. $10x^2 - 17x + 3 = 0$
10. $x^2 + 5x - 84 = 0$

11. A rectangular storage area that is 30 units long and 20 units wide is to be doubled in area. Find the dimensions of the new rectangle.

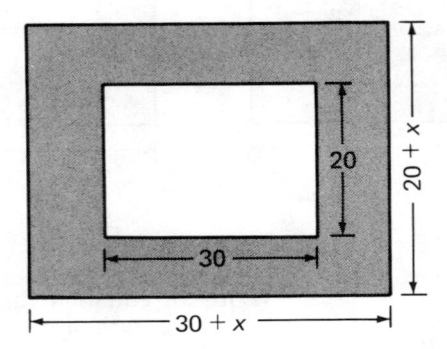

12. Find the missing dimensions for the plate in Figure 8-8 if the area of the shaded part is 105 square units.

Figure 8-8

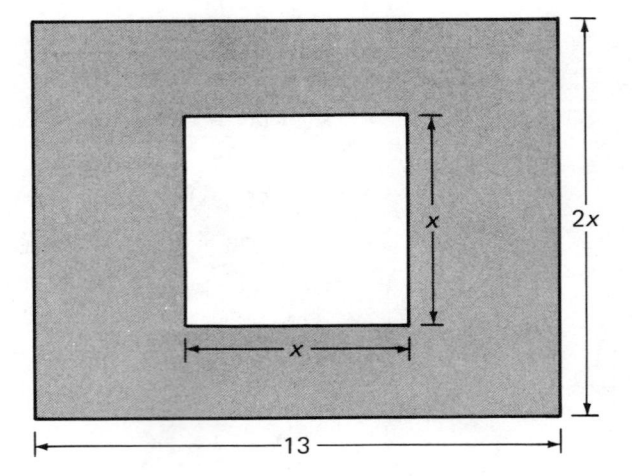

250 Technical and industrial formulas

13. An L-shaped brace is cut from the rectangular piece of metal in Figure 8-9. If the area of the brace is 4 square units, find the missing dimensions.

Figure 8-9

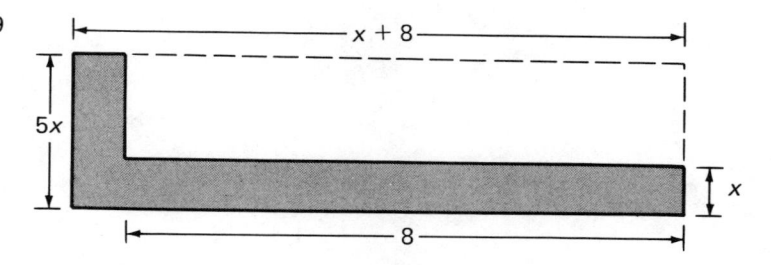

8-6 THE QUADRATIC FORMULA

Solving quadratic equations by factoring, as we did in Section 8-5, has limitations. Some trinomials cannot be factored using the methods we learned. To solve these equations, we use the *quadratic formula*, which is given in the following rule.

The solutions to the equation $ax^2 + bx + c = 0$ are

$$x_1 = \frac{-b + \sqrt{b^2 - 4ac}}{2a}$$

$$x_2 = \frac{-b - \sqrt{b^2 - 4ac}}{2a}$$

Example 1

Solve $x^2 - 2x - 2 = 0$.

Solution

Substitute $a = 1$, $b = -2$, $c = -2$ into the quadratic formula.

$$x_1 = \frac{-(-2) + \sqrt{(-2)^2 - 4(1)(-2)}}{2}$$

$$= \frac{2 + \sqrt{4 + 8}}{2}$$

$$= \frac{2 + \sqrt{12}}{2} = \frac{2 + 3.46}{2} = \frac{5.46}{2} = 2.73 \quad \text{(to the nearest hundredth)}$$

Polynomials 251

From the formulas, we can see that

$$x_2 = \frac{2 - \sqrt{12}}{2} = \frac{2 - 3.46}{2} = \frac{-1.46}{2} = -.73 \quad \text{(to the nearest hundredth)}$$

Example 2

Solve $-3x^2 + 4x + 8 = 0$.

Solution

$a = -3, b = 4, c = 8$.

$$x_1 = \frac{-4 + \sqrt{4^2 - 4(-3)(8)}}{2(-3)}$$

$$= \frac{-4 + \sqrt{16 + 96}}{-6}$$

$$= \frac{-4 + \sqrt{112}}{-6}$$

$$= \frac{-4 + 10.58}{-6}$$

$$= \frac{6.58}{-6} = -1.10 \quad \text{(to the nearest hundredth)}$$

$$x_2 = \frac{-4 - 10.58}{-6} = \frac{-14.58}{-6} = 2.43 \quad \text{(to the nearest hundredth)}$$

Example 3

A rectangular storage room is to be partitioned into two rooms, as shown in Figure 8-10. The original room had an area of 90 sq m. Find the dimensions of the two new rooms.

Figure 8-10

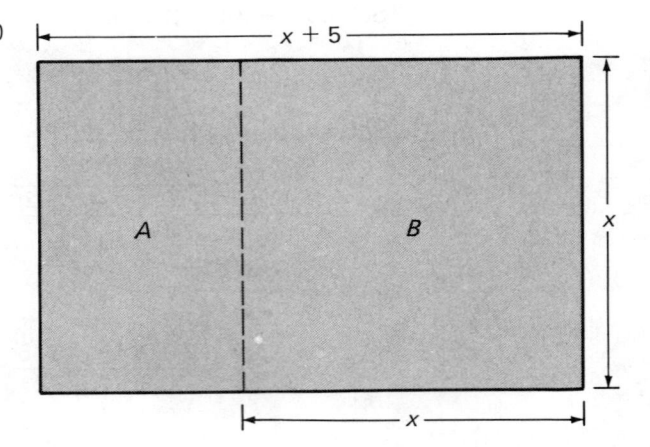

Solution

1. Write an equation that expresses the relationship. The area of the original room is 90 sq m. Thus, $x(x + 5) = 90$.
2. Solve the equation.

$$x^2 + 5x - 90 = 0$$

$$x_1 = \frac{-5 + \sqrt{5^2 - 4(1)(-90)}}{2(1)}$$

$$= \frac{-5 + \sqrt{25 + 360}}{2} = \frac{-5 + \sqrt{385}}{2}$$

$$= \frac{-5 + 19.62}{2} = \frac{14.62}{2} = 7.3 \text{ (approx.)}$$

$$x_2 = \frac{-5 - 19.62}{2} = \frac{-24.62}{2} = -12.3 \text{ (approx.)}$$

3. Interpret the results. We discard x_2 since it is negative. Room B is a square with 7.3-m sides. The dimensions of room A are:

Length = x = 7.3 m
Width = $x + 5 - x$ = 5 m

Example 4

A contractor agrees to provide 120 sq ft of outside concrete work with the purchase of a home. If the buyer decides on a rectangular patio that is 5 ft longer than it is wide, what are the dimensions of the patio?

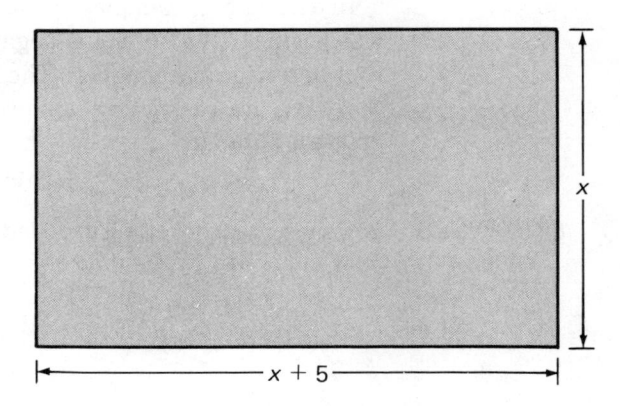

Polynomials

Solution

1. Write an equation for the relationship.

   $$x(x + 5) = 120$$

2. Solve the equation.

   $$x^2 + 5x = 120$$
   $$x^2 + 5x - 120 = 0$$
   $$a = 1, b = 5, c = -120$$
   $$x_1 = \frac{-5 + \sqrt{5^2 - (4)(1)(-120)}}{2}$$
   $$= \frac{-5 + \sqrt{25 + 480}}{2}$$
   $$= \frac{-5 + \sqrt{505}}{2}$$
   $$= \frac{-5 + 22.47}{2}$$
   $$= \frac{17.47}{2} = 8.7 \quad \text{(to the nearest tenth)}$$
   $$x_2 = \frac{-5 - 22.47}{2} = \frac{-27.47}{2} = -13.7$$

3. Interpret the results. We discard $x_2 = -13.7$ as a solution. The dimensions of the patio are:

   $w = x = 8.7$ ft

   $\ell = x + 5 = 13.7$ ft

In each of the examples in this section, the expression $b^2 - 4ac$ was positive. If $b^2 - 4ac$ is negative, we have to take the square root of a negative number. The solutions involve *complex numbers*. We are restricting our problems to those whose solutions are real numbers.

EXERCISES Solve the equations using the quadratic formula. Approximate the solutions to the nearest hundredth.

1. $x^2 - 4x + 2 = 0$
2. $4x^2 - 8x + 3 = 0$

3. $x^2 - 2x - 1 = 0$
4. $4x^2 + 4x - 11 = 0$
5. $x^2 - 2x - 3 = 0$
6. $3x^2 + 10x + 2 = 0$
7. $\dfrac{x^2}{2} - 2x - 1 = 0$
8. $-2x^2 + 12x - 13 = 0$
9. $3x^2 + 6x - 200 = 0$
10. $x^2 - 85x + 1200 = 0$
11. A rectangle is 5 units longer than it is wide. The area is 90 square units. Find the dimensions of the rectangle.
12. Find the dimensions of the right triangle in Figure 8-11. (Use the Pythagorean theorem.)

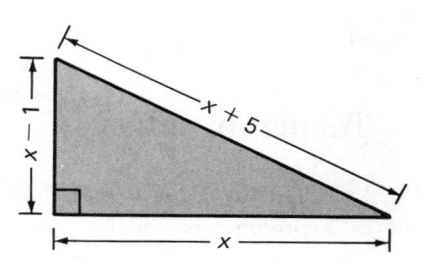

Figure 8-11

13. A parking lot 200 × 100 is to be tripled in area by adding the region shown in Figure 8-12. Find the value of x.

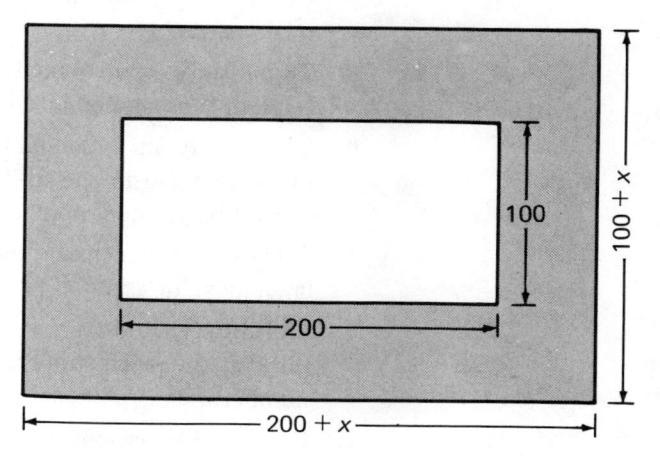

Figure 8-12

14. If the area of the cut-out part (A) of the rectangle in Figure 8-13 is 150 square units, find the dimensions of the original rectangle.

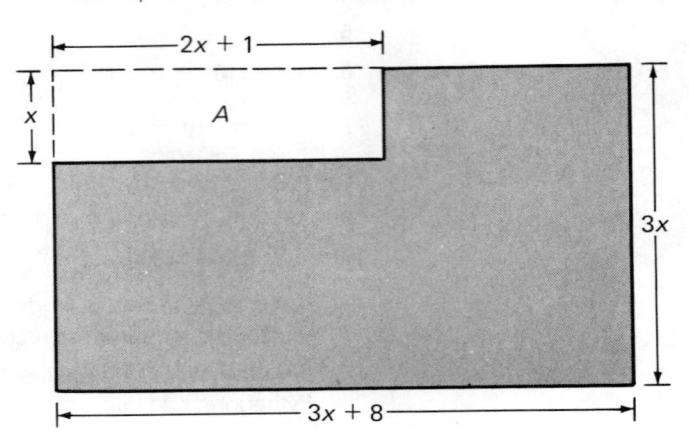

Figure 8-13

Taking inventory

1. The *degree of a monomial in a variable* is the exponent of the variable in the monomial. The *degree of a monomial* is the sum of the degrees of each of the variables. (p. 228)

2. The *degree of a polynomial* is the degree of the highest-degree term of the polynomial. (p. 228)

3. To add polynomials, we add the similar terms. (p. 228)

4. To subtract one polynomial from another, we add the negative of the polynomial to be subtracted to the other polynomial. (p. 229)

5. To multiply a polynomial by a monomial, we multiply each term in the polynomial by the monomial. (p. 231)

6. To multiply a polynomial by a second polynomial, we multiply each term in the first polynomial by each term in the second polynomial and combine like terms. (p. 232)

7. Some trinomials can be factored as the product of a pair of binomials. (p. 239)

8. The difference of the squares of two numbers can be factored as the product of the sum and the difference of the two numbers, $x^2 - y^2 = (x + y)(x - y)$. (p. 244)

9. Quadratic equations with the left side of the equation factorable are solved using this rule: If $a \cdot b = 0$, then either $a = 0$ or $b = 0$ or both. (p. 247)

10. Quadratic equations of the form $ax^2 + bx + c = 0$ that have real roots can be solved using the quadratic formula:

$$x_1 = \frac{-b + \sqrt{b^2 - 4ac}}{2a}$$

$$x_2 = \frac{-b - \sqrt{b^2 - 4ac}}{2a} \quad \text{(p. 251)}$$

Measuring your skills

Perform the indicated operations in each exercise. (8-1, 8-2)
1. $(2x^2 + 4x - 6) + (3x^2 - 2x + 1)$
2. $(5x^3 - 6x^2y^2 + 4y^3) + (3x^3 - 4y^3)$
3. $(2x^2 + 4x - 6) - (3x^2 - 2x + 1)$
4. $(5x^3 - 6x^2y^2 + 4y^3) - (3x^3 - 4y^3)$
5. $x^8 \cdot x^4$
6. $(3x^2y^2)(\frac{1}{3}xy^3)$
7. $(x - 5)(x + 7)$
8. $(2x - 3y)(x + 2y)$
9. $(x^2 - x + 1)(x + 1)$
10. $(3x^2 - 2y^2)(x^2 + 2y^2)$

Factor each of the polynomials. (8-3, 8-4)
11. $4x^2 - 16x$
12. $3x^2yz - 3xy^2 + 3x^2y^2z^2$
13. $x^2 - x - 6$
14. $x^2 + 5x - 36$
15. $2x^3 - 10x^2 - 72x$
16. $9x^2 - y^2$
17. $20x^2 - 125$

Solve each equation by factoring. (8-5)
18. $x^2 - x - 72 = 0$
19. $3x^2 - 7x - 6 = 0$
20. $x^2 - 5x - 36 = 0$

Solve each equation using the quadratic formula. (8-6)
21. $x^2 + 4x + 2 = 0$
22. $2x^2 - 11x - 6 = 0$
23. $x^2 - 6x - 1 = 0$
24. $x^2 - 16x + 64 = 0$

9 Graphing

After completing this chapter, you should be able to:
1. Locate points on a coordinate system.
2. Draw graphs of linear equations.
3. Solve pairs of linear equations graphically and algebraically.
4. Draw graphs of nonlinear equations.
5. Estimate maximum values and minimum values of nonlinear equations.

Linear equations

9-1 THE COORDINATE PLANE

In Chapter 7 we learned the formula $F = \frac{9}{5}C + 32$, which gives the relationship between temperatures on the Fahrenheit and Celsius scales. The following table shows some of the values that are obtained using the formula.

C	−100	0	40	100	200
F	−148	32	104	212	392

We can also list the data in the table as a set of *ordered pairs* of numbers, {(−100, −148), (0, 32), (40, 104), (100, 212), (200, 392)}, where the first number in each pair represents the Celsius temperature and the second number in each pair represents the corresponding Fahrenheit temperature. The term *ordered pair* is used to indicate a pair of numbers where the *order* in which the numbers are written is important.

Example 1

Write four ordered pairs (d, c), where d represents the diameter of a circle and c represents the circumference of the circle that corresponds to the given diameter.

Solution

Let $d = 1, 2, 3, 4$ be the diameters of circles. Using the formula $c = \pi d$ or $c = 3.14d$, the corresponding circumferences are 3.14, 6.28, 9.42, and 12.56. The set of ordered pairs is: {(1, 3.14), (2, 6.28), (3, 9.42), (4, 12.56)}.

To represent a set of ordered pairs on a graph, we introduce the idea of a *coordinate system*.

We make a coordinate system by drawing a pair of perpendicular lines, as in Figure 9-1, and putting a scale on each of the lines, beginning at the point where the lines intersect. We call this

Figure 9-1

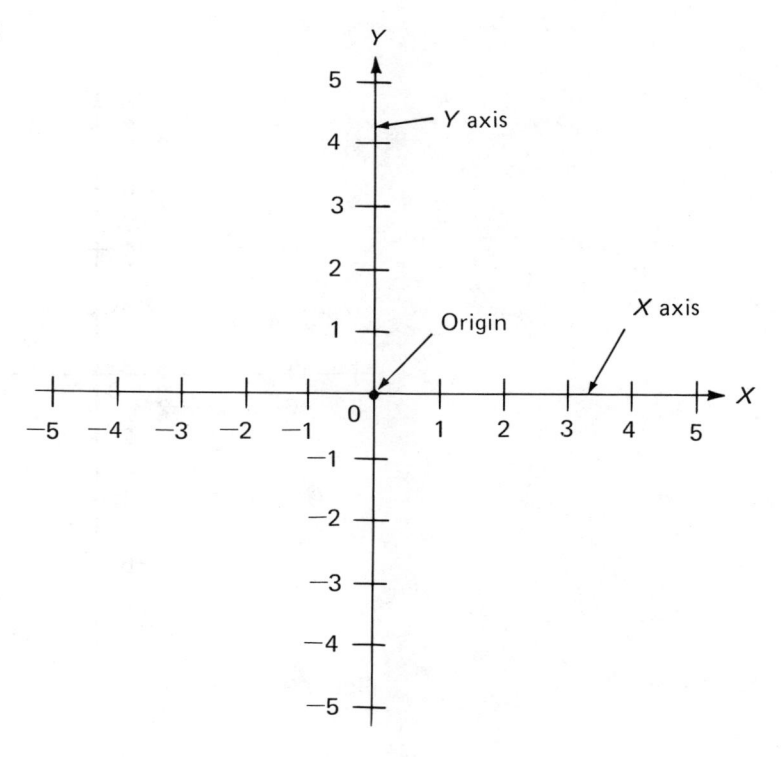

point the origin. The horizontal line is called the X *axis*, and the vertical line is called the Y *axis*. By custom, we scale the X axis from left to right, and we scale the Y axis from bottom to top. The size of the units used to scale the axes is completely arbitrary. We use whatever unit is most realistic. If the numbers in the ordered pairs are large, then it makes sense to use small units for the scales. If the numbers in the ordered pairs are small, then we use larger units for the scales. The examples in this section illustrate how to choose an appropriate scale.

Once we have drawn a coordinate system, we can plot the graph of any ordered pair of real numbers on this system, as in Figure 9-2. Point A represents the ordered pair (3, 2); point B, the ordered pair (6, 5); point C, the ordered pair (−4, 1); point D, the ordered pair (−3, −3); and point E, the ordered pair $(4, -\frac{5}{2})$.

Note that the first number in an ordered pair tells us how far the point is to the right or left of the Y axis, and the second number in the ordered pair tells us how far the point is above or below the X axis.

The ordered pairs of numbers are the *coordinates* of the point. The first number in the pair is the *x coordinate* of the point, and

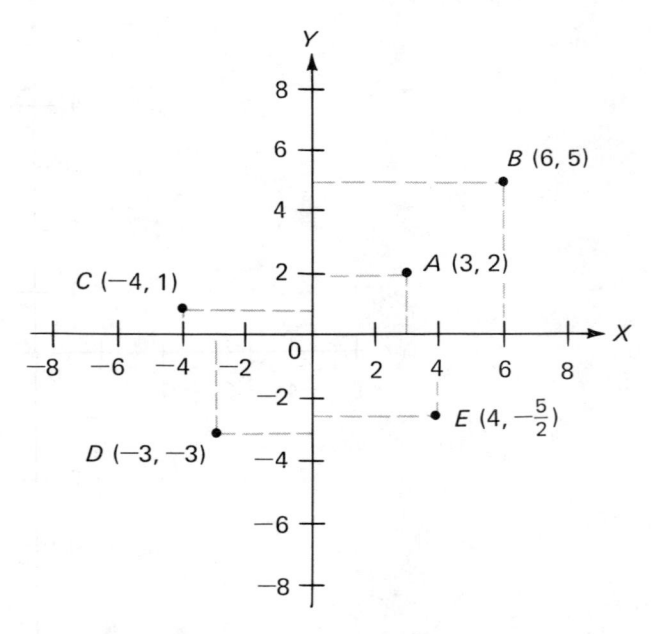

Figure 9-2

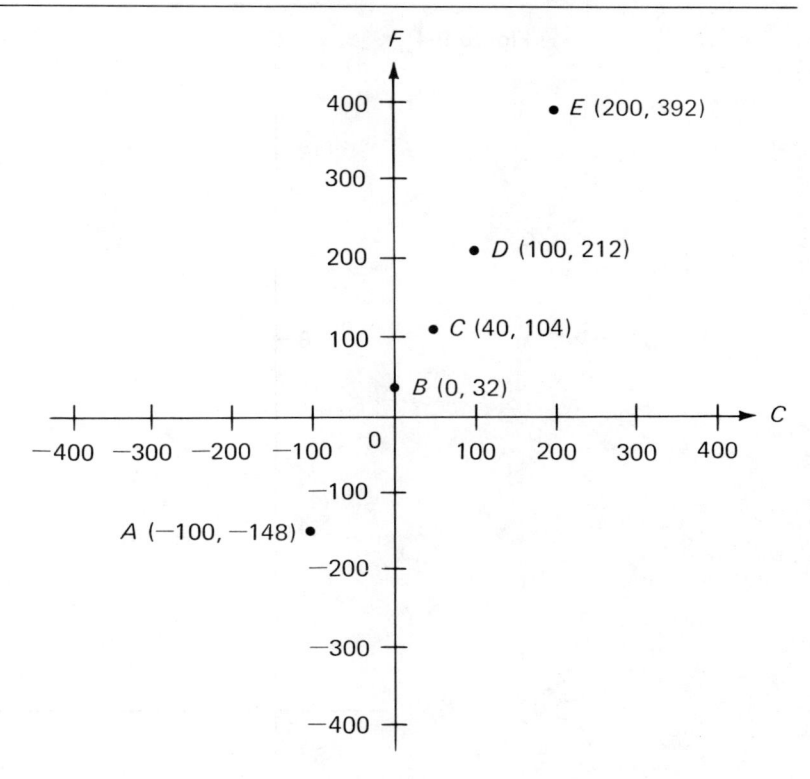

Figure 9-3

the second number in the pair is the *y coordinate* of the point. In the ordered pair (5, 8), the *x* coordinate is the first number, 5, and the *y* coordinate is the second number, 8.

Sometimes we need to adjust the scale on the axes to fit the conditions of the problem. Figure 9-3 shows a coordinate system that is appropriate for graphing the set of ordered pairs in the table of Celsius and Fahrenheit temperatures. Points *A*, *B*, *C*, *D*, and *E* represent the approximate positions of the points (−100, −148), (0, 32), (40, 104), (100, 212), and (200, 392) on the coordinate system. Do you see a pattern to the points?

Sometimes you do not need to include the parts of a coordinate system that have one or more negative coordinates. The context of Example 1 makes it unnecessary to consider negative values of *d* or *c*. Figure 9-4 shows the points *M* (1, 3.14), *N* (2, 6.28), *P* (3, 9.42), and *Q* (4, 12.56) plotted on an appropriate coordinate system.

Graphing

Figure 9-4

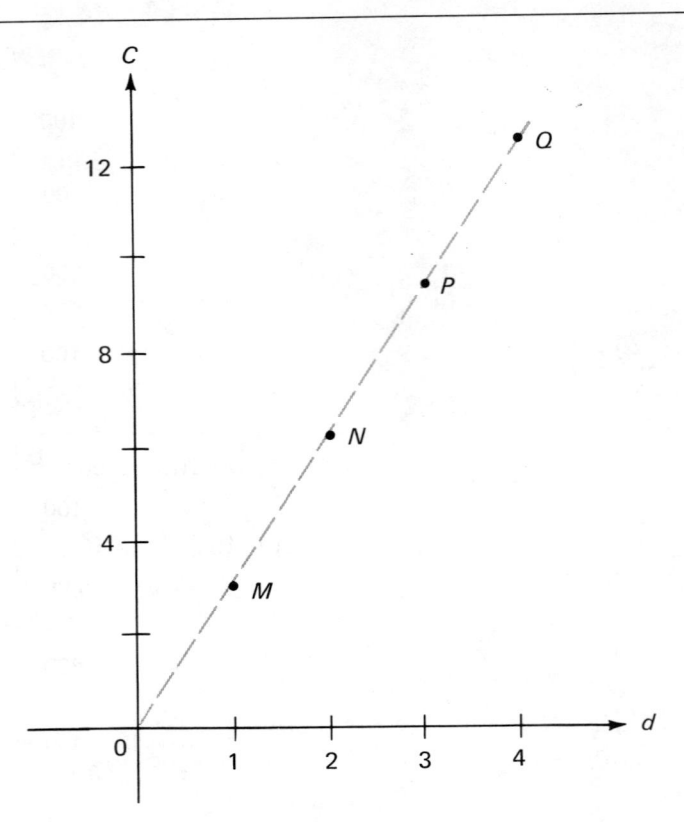

(Coordinate graph showing points M, N, P, Q along a line through the origin, with horizontal axis d and vertical axis C.)

EXERCISES

1. Draw a coordinate system and plot the ordered pairs (−5, −3), (−4, 0), (0, 6), (7, 2), (8, −3), (−4, 6), (0, 0), (0, −5).

2. Write the coordinates of points A, B, C, D, E, and F in Figure 9-5.

3. Draw a coordinate system, plot the given points, and connect the points in the order listed: A(−1, 0), B(−1, 1), C(2, 1), D(2, 5), E(5, 1.5), F(8, 1.5), G(8, −.5), H(5, −.5), I(2, −4), J(2, 0). Also draw segments CJ, EH, and AJ.

4. a. Using the formula C = .5x + 40, determine a set of ordered pairs using x = 10, 20, 30, 40, 50.

   b. Plot the ordered pairs on an appropriate coordinate system.

5. The formula for the depth of the screw thread shown in Figure 9-6 is H = P/2 + .01.

   a. Determine the set of ordered pairs for P = .083, .1, .125, .167, .25.

   b. Plot the ordered pairs on an appropriate coordinate system.

262 **Technical and industrial formulas**

Figure 9-5

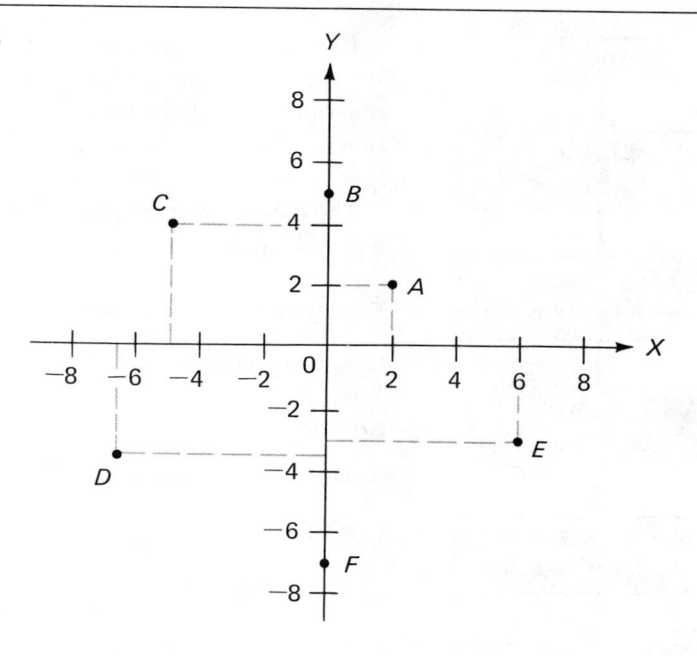

Figure 9-6

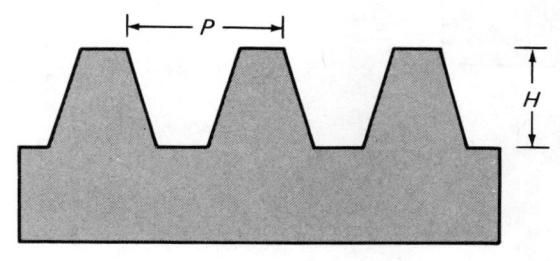

6. Assuming that the average driver has a reaction time of .75 sec, the formula $d = 1.1r$ describes the distance d in feet that an automobile will travel during the reaction time at a speed of r miles per hour.
   a. Determine a set of ordered pairs for $r = 10, 20, 30, 40, 50, 60$.
   b. Graph the ordered pairs on a coordinate system. Do you see a pattern in the graph?

9-2 GRAPHING LINEAR EQUATIONS

In Section 9-1 we learned to graph sets of ordered pairs of numbers on a coordinate system. On several of the graphs in the examples and the exercises, the ordered pairs followed a pattern, namely, the points appeared to be on a line. For example, on the graph of the equation $F = \frac{9}{5}C + 32$, which described the relationship between the Fahrenheit and Celsius temperature scales, the points fall on a line.

In general, any equation that can be written in the form $ax + by + c = 0$, where x and y are variables and a, b, and c are constants, is called a *linear equation*, and the graph of the equation is a line. The Fahrenheit-Celsius equation can be written as $F - \frac{9}{5}C - 32 = 0$, so that it fits the definition of a linear equation.

The flow chart in Figure 9-7 describes the procedure for graphing a linear equation.

Example 1

Graph the equation $3x - 2y = 6$.

Solution

a. Let $x_1 = 3$, $x_2 = 0$, $x_3 = -3$.

b. $3(3) - 2y_1 = 6$ or $y_1 = \frac{3}{2}$.

   $3(0) - 2y_2 = 6$, or $y_2 = -3$.

   $3(-3) - 2y_3 = 6$, or $y_3 = \frac{-15}{2}$

c. $\left(3, \frac{3}{2}\right)$, $(0, -3)$, $\left(-3, \frac{-15}{2}\right)$

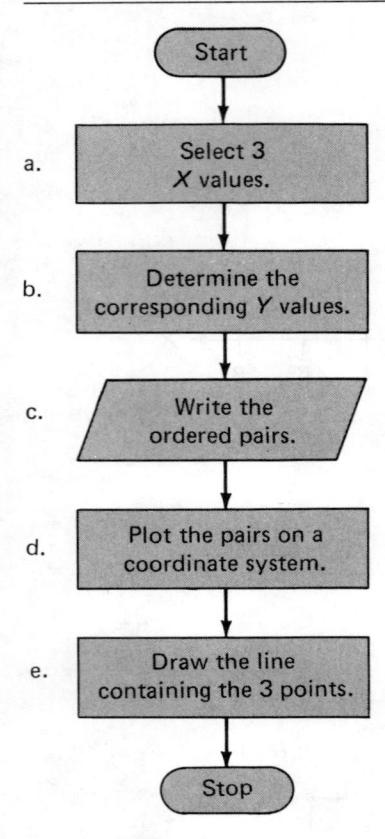

d-e.

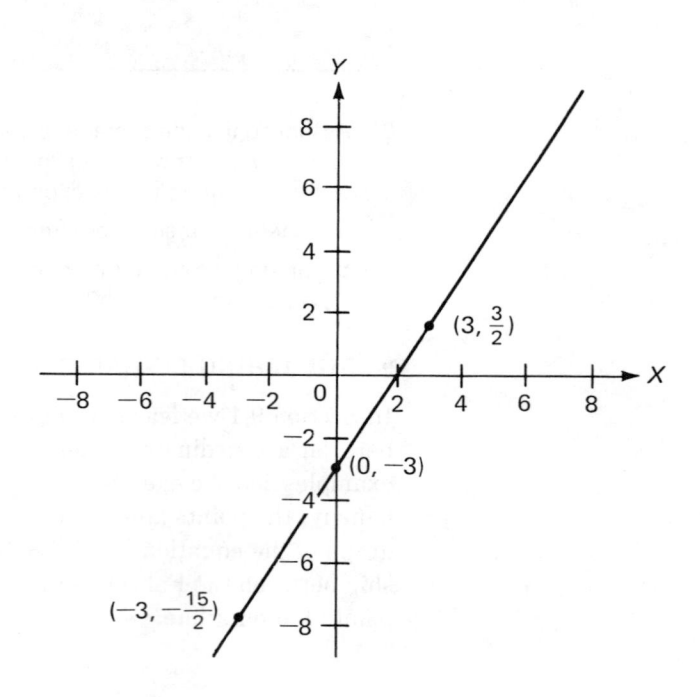

Figure 9-7

a. Select 3 X values.
b. Determine the corresponding Y values.
c. Write the ordered pairs.
d. Plot the pairs on a coordinate system.
e. Draw the line containing the 3 points.

264 Technical and industrial formulas

Note that we could have obtained the graph from only two ordered pairs, but it is a good idea to plot a third point as a check on the placement of the other two points.

Example 2

The EZ Manufacturing Company is producing widgets at a cost described by the equation $c = 800 + 1.5x$, where 800 represents the fixed costs of operating and the $1.5x$ the variable cost of producing x widgets. Graph the cost equation for x between 0 and 1000.

Solution

a. Select three x values. Let $x_1 = 0$, $x_2 = 500$, $x_3 = 1000$.
b. Determine the corresponding c values.

$c_1 = 800 + 1.5(0) = 800$

$c_2 = 800 + 1.5(500) = 1550$

$c_3 = 800 + 1.5(1000) = 2300$

c. Write the ordered pairs: (0, 800), (500, 1550), (1000, 2300).

d-e. Plot the pairs on a coordinate system and draw the line containing the three points.

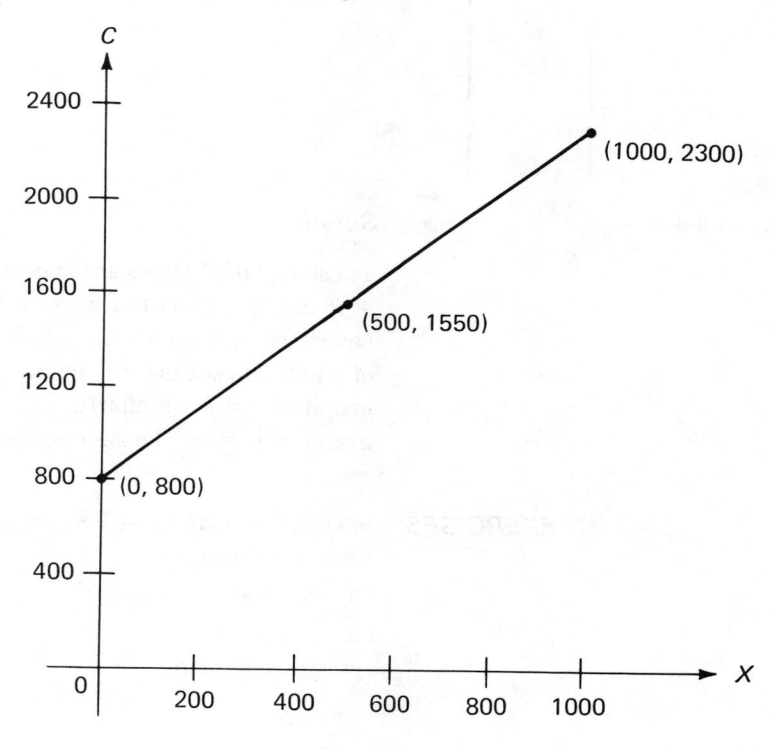

Note that the x values we selected were nonnegative, since it does not make sense to manufacture a negative number of items. Also, we selected the x values so that the points plotted were not too closely grouped. This makes drawing the graph easier. Note also that in scaling the axes we do not have to use the same units.

Example 3

Draw graphs of the equations $x = k$ and $y = m$, where k and m are fixed constants.

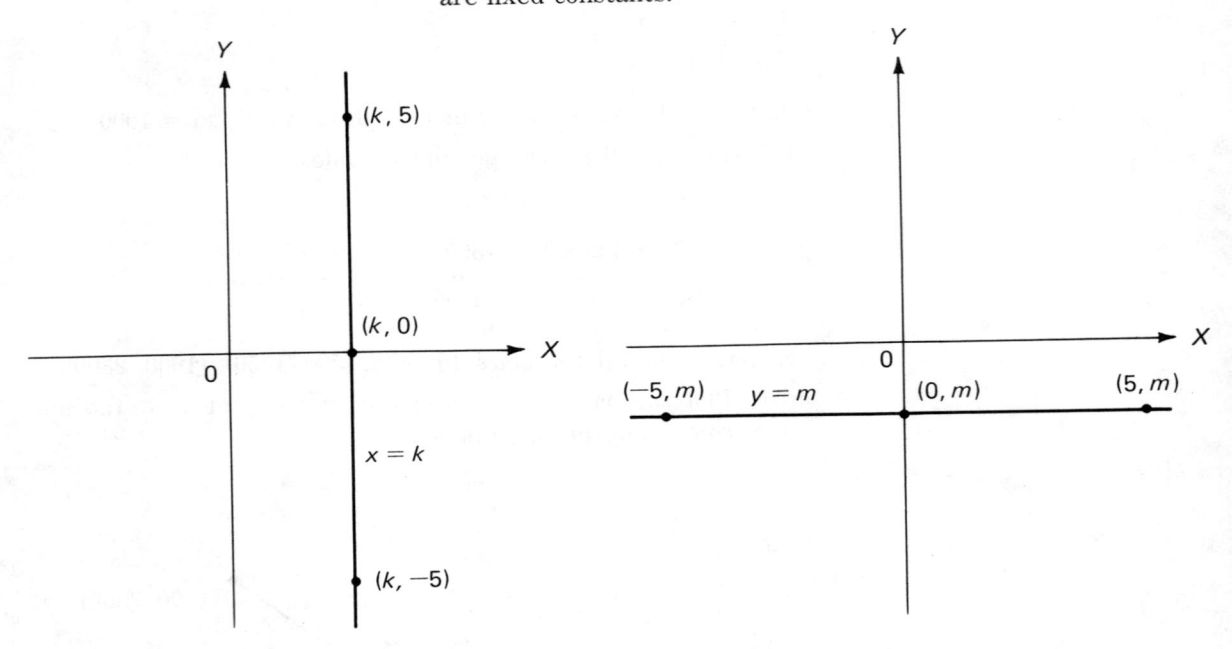

Figure 9-8

Solution

These equation types are special cases of linear equations, since $x = k$ can be rewritten as $x + 0y - k = 0$, and $y = m$ can be rewritten as $0x + y - m = 0$. Since $x = k$ regardless of our choice of y in the first case, the points $(k, -5)$, $(k, 0)$ and $(k, 5)$ are on the graph of $x = k$. Similarly, $(-5, m)$, $(0, m)$, and $(5, m)$ are on the graph of $y = m$. These results are shown in Figure 9-8.

EXERCISES In exercises 1–10, draw the graph of the linear equation on an appropriate coordinate system.

1. $2x - y - 7 = 0$
2. $x + y = 0$
3. $2x - 3y = 0$

Technical and industrial formulas

4. $y = -2$
5. $x - 2y - 10 = 0$
6. $2x + 3y = 0$
7. $3x - 3y - 2 = 0$
8. $7x + 2y + 27 = 0$
9. $x = 0$
10. $.05x + .04y = 100$
11. A manufacturer buys a machine for $20,000, and it depreciates over a period of 10 years to a value of $5000. The value v of the machine at any given time t is given by the equation $v = 20,000 - 1500t$. Graph the equation for t between 0 and 10.
12. The EZ Manufacturing Company has a fixed cost of $1500 and a variable cost of $17 per item produced. The cost equation is $C = 1500 + 17x$. Graph this equation for x between 0 and 1000.
13. The current through a 10-ohm resistor varies according to the number of volts across the resistor. This relationship is described by the equation $c = 10v$, where c represents the current in amperes and v represents the voltage in volts. Graph this equation for v between 0 and 110.
14. A particular type of copper tubing expands according to the formula $e = .000117t$, where t represents the temperature change in Fahrenheit degrees and e represents the expansion in inches per foot of length.
    a. Graph the equation for t between 0 and 250.
    b. How much will 1 ft of the tubing expand for a 50° change in temperature?
    c. How much will 25 ft of the tubing expand for a 50° change in temperature?

9-3 SOLVING PAIRS OF LINEAR EQUATIONS GRAPHICALLY

The graph of a linear equation is a line. Suppose we have a pair of linear equations graphed on the same coordinate system. There are three possibilities.

1. The lines intersect at a single point, and the equations have a single ordered pair as the common solution.
2. The lines are parallel, and the equations have no common solution.
3. The lines coincide, and the equations have an infinite number of common solutions.

Example 1

Draw the graphs of the equations $x + y = 7$ and $x - y = 5$, and estimate the coordinates of the point that is common to the lines.

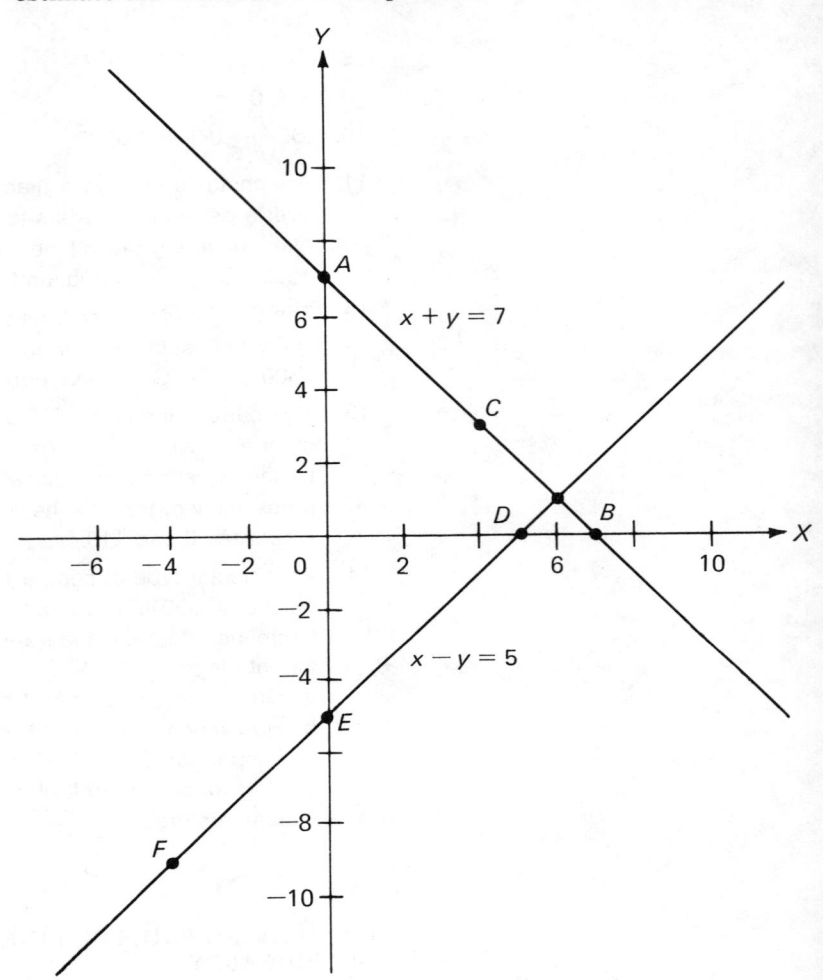

Solution

1. Graph the line $x + y = 7$. A (0, 7), B (7, 0), and C (4, 3) are points on this line. Draw the line.

2. Graph the line $x - y = 5$. D (5, 0), E (0, −5), and F (−4, −9) are points on this line. Draw the line.

3. The point of intersection P has the approximate coordinates (6, 1). We can check our estimate by substituting $x = 6$, $y = 1$ into each of the equations. Since $6 + 1 = 7$ and $6 - 1 = 5$, we see that the estimate is correct.

One of the applications of linear equations is in *break-even analysis*. The break-even point in an industrial operation is the point where the graphs of the cost equation and the revenue equation cross. Example 2 illustrates this application.

Example 2

The EZ Corporation has found that the cost y of producing x items is represented by the equation $y = 1200 + 2x$. The revenue y from the sale of x items is represented by the equation $y = 3.5x$. Find the break-even point.

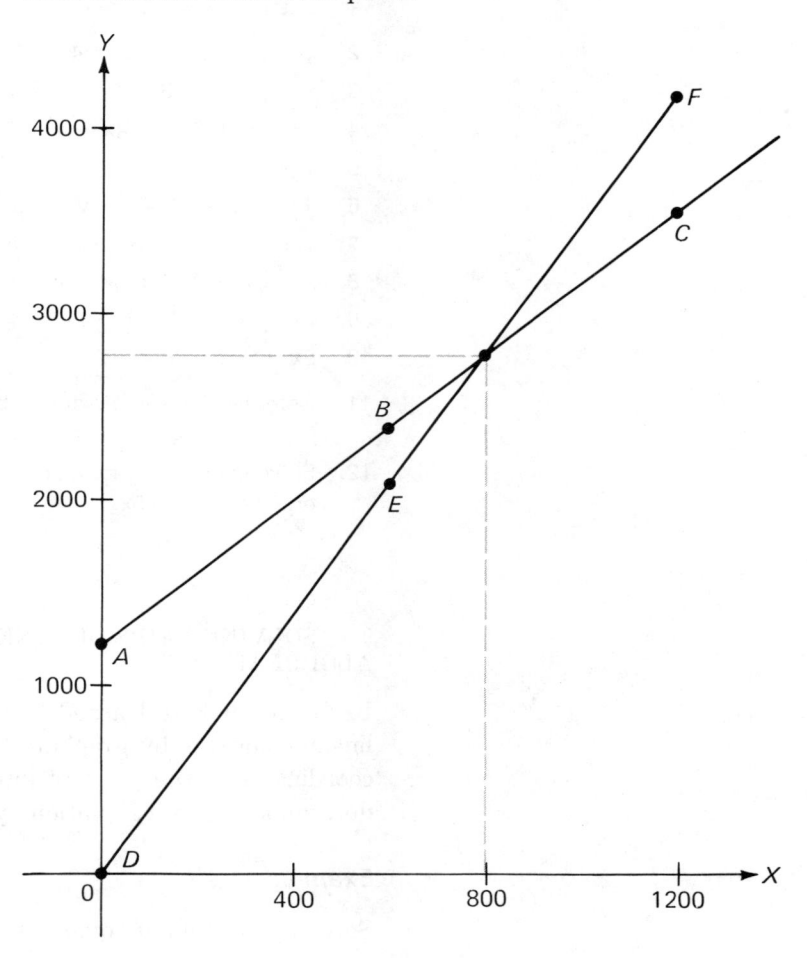

Solution

1. Graph the line $y = 1200 + 2x$. A (0, 1200), B (600, 2400), and C (1200, 3600) are on this line.

Graphing 269

2. Graph the line $y = 3.5x$. D (0, 0), E (600, 2100), and F (1200, 4200) are on this line.
3. Estimate the coordinates of the break-even point P. A reasonable estimate is (800, 2800).

The company needs to manufacture and sell 800 items to break even.

EXERCISES Graph each pair of equations on a coordinate system and estimate the coordinates of the points (if any) that are common to the graphs.

1. $x + y = 7$; $x - y = 1$
2. $x + y = 3$; $2x + y = 4$
3. $x - y = -1$; $3x - y = -5$
4. $4x - y + 3 = 0$; $4x - y + 5 = 0$
5. $x + y = 2$; $x - y = 3$
6. $4x + 3y = 26$; $3x - y = 13$
7. $x - 2y = -11$; $5x - 3y = 8$
8. $x + 3y - 8 = 0$; $5x - y + 6 = 0$
9. $2x + 3y = 13$; $5x - 2y = 4$
10. $2x + y = 0$; $x = -7$
11. Determine the approximate break-even point for a cost equation $y = 500 + 3x$ and a revenue equation $y = 4x$.
12. Solve exercise 11 for a cost equation $y = 300 + 3x$ and a revenue equation $y = 5.5x$.

9-4 SOLVING PAIRS OF LINEAR EQUATIONS ALGEBRAICALLY

In Section 9-3 we learned to estimate the solution to a pair of linear equations by graphing the equations and estimating the coordinates of the point of intersection of the graphs. We can determine the exact solution by algebraic methods.

Example 1

Solve the system of equations:

$3x - 3y = 6$

$5x + 2y = 10$

Solution

If we multiply the first equation by 2 and the second equation by 3, both equations will have a $6y$ term. The new system is:

$6x - 6y = 12$

$15x + 6y = 30$

We now add these two equations. The resulting equation does not contain the variable y.

$21x = 42$

We now solve this equation for x.

$x = 2$

To determine y, we substitute $x = 2$ into any of the previous equations involving x and y and solve the resulting equation for y. Suppose we select the equation

$5x + 2y = 10$

$5(2) + 2y = 10$

$10 + 2y = 10$

$2y = 0$

$y = 0$

The desired solution is $x = 2$, $y = 0$.

The steps in solving a pair of linear equations by the algebraic method are:

1. Multiply each equation by a constant chosen so that one of the variables can be eliminated from both equations by adding or subtracting the equations.
2. Obtain an equation in one variable by adding or subtracting the equations in step 1.
3. Solve this equation.
4. Substitute this value in one of the original equations and solve for the other variable.

Example 2

Solve the pair of equations $y = 1200 + 2x$; $y = 3.5x$ of Example 2 in Section 9-3.

Solution

Since both equations contain the same y term, we can subtract the equations.

$$y = 3.5x$$
$$\underline{y = 2x + 1200}$$
$$0 = 1.5x - 1200$$

$$1.5x = 1200$$

$$x = \frac{1200}{1.5} = 800$$

Substitute $x = 800$ into the equation $y = 3.5x$.

$$y = 3.5(800) = 2800$$

Example 3

Solve the system: $2x + 4y = 13$; $3x - 5y = 3$.

Solution

1. If we multiply the first equation by 5 and the second equation by 4, both the resulting equations will have a $20y$ term.

   $$10x + 20y = 65$$
   $$12x - 20y = 12$$

2. Add the equations.

   $$22x = 77$$

3. Solve this equation.

   $$x = \frac{77}{22} = \frac{7}{2}$$

4. Substitute $x = \frac{7}{2}$ in the equation $2x + 4y = 13$, and solve for y.

$$2(\tfrac{7}{2}) + 4y = 13$$
$$7 + 4y = 13$$
$$4y = 6$$
$$y = \tfrac{6}{4} = \tfrac{3}{2}$$

The solution is $x = \tfrac{7}{2}$, $y = \tfrac{3}{2}$.

Example 4

Solve a demand equation $y + 20p = 800$, where y represents the demand for a product at a price p, and a supply equation $y - 30p = 200$, where y represents the supply of the product that manufacturers are willing to furnish at a price p. The point of *market equilibrium* is the point $E\,(p,\,y)$ where the graphs of the supply and demand equations cross. Find the coordinates of this point.

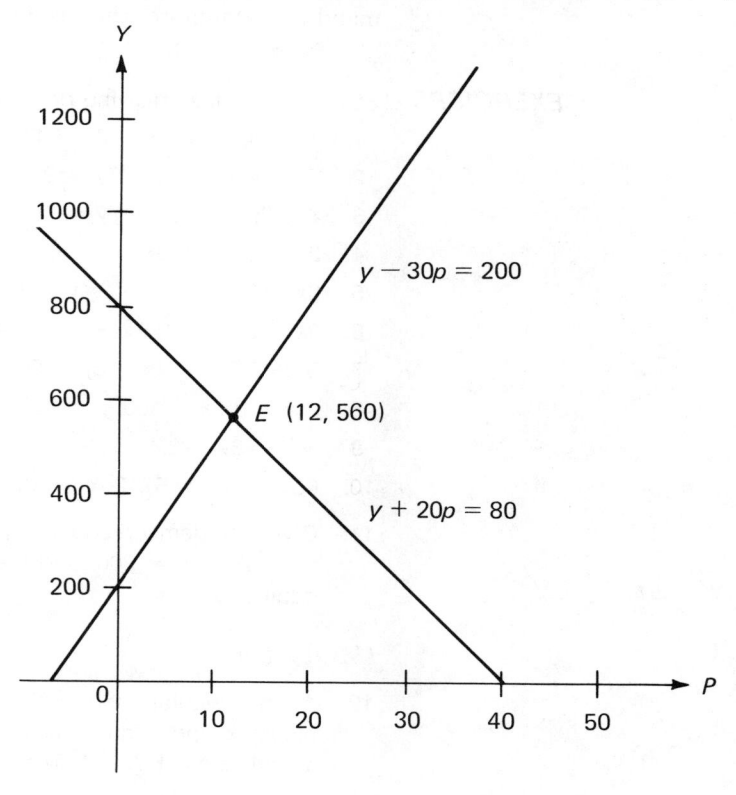

Graphing 273

Solution

$y + 20p = 800$

$y - 30p = 200$

Subtract the equations.

$50p = 600$

$p = 12$

Substitute $p = 12$ into the equation $y = 20p + 800$.

$y + 20(12) = 800$

$y + 240 = 800$

$y = 800 - 240 = 560$

The coordinates of the equilibrium point are $E\ (12, 560)$. Market equilibrium is attained with a price of 12 and a supply and demand of 560 units of the product.

EXERCISES Solve each of the following pairs of equations algebraically.

1. $3x + 4y = 24$; $x + 2y = 11$
2. $x - 4y = 3$; $2x - 7y = 9$
3. $x + 2y = 3$; $3x + 5y = 10$
4. $2x + 5y = 4$; $3x - 5y = 0$
5. $3x + 2y = 11$; $5x - 4y = 11$
6. $2x - 3y = -1$; $3x + 4y = 24$
7. $5x - 2y = 7$; $7x - 3y = 5$
8. $y = -100x + 400$; $y = 300x - 300$
9. $.1x - .3y = 2.4$; $.5x - 1.5y = 12$
10. $60x + 20y = 37$; $20x + 20y = 17$
11. Given the demand equation $y + 20p = 740$ and the supply equation $y - 30p = 200$, find the coordinates of the point of market equilibrium.

U.S. system

12. A rectangular lot with length x and width y has a perimeter of 720 ft. The

Metric system

15. A rectangular lot with length x and width y has a perimeter of 360 m. The

lot is 60 ft longer than it is wide. Find the length and the width.

13. For the three discs in Figure 9-9, $AB = 1.26$ in., $BC = .98$ in., and the radius of disc *C* is .55 in. Find *R* and *r*.

lot is 30 m longer than it is wide. Find the length and the width.

16. For the three discs in Figure 9-9, $AB = 3.2$ cm, $BC = 2.5$ cm, and the radius of disc *C* is 1.4 cm. Find *R* and *r*.

Figure 9-9

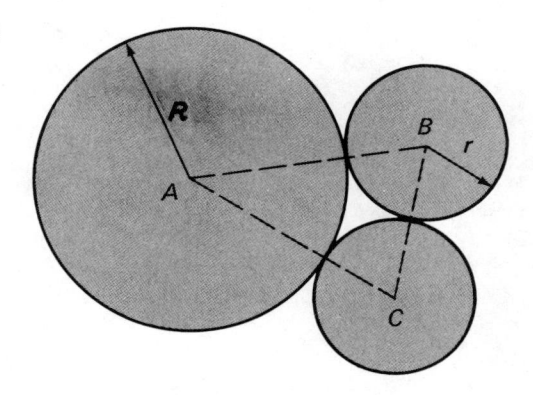

14. A pair of metal rods has a combined length of 16.64 in. The longer rod is 6.3 in. longer than the shorter rod. Find the lengths of the rods.

17. A pair of metal rods has a combined length of 42.27 cm. The longer rod is 16.01 cm longer than the shorter rod. Find the lengths of the rods.

Nonlinear equations

9-5 GRAPHING NONLINEAR EQUATIONS

Many relationships in the real world cannot be modeled by linear equations.

The graph in Figure 9-10 shows the relationship between the spark gap peak voltages obtained from 2.5-cm spherical electrodes and the length of the spark gap. From the graph we estimate that for a peak voltage of 50,000 volts, a spark gap of 2 cm is

Graphing 275

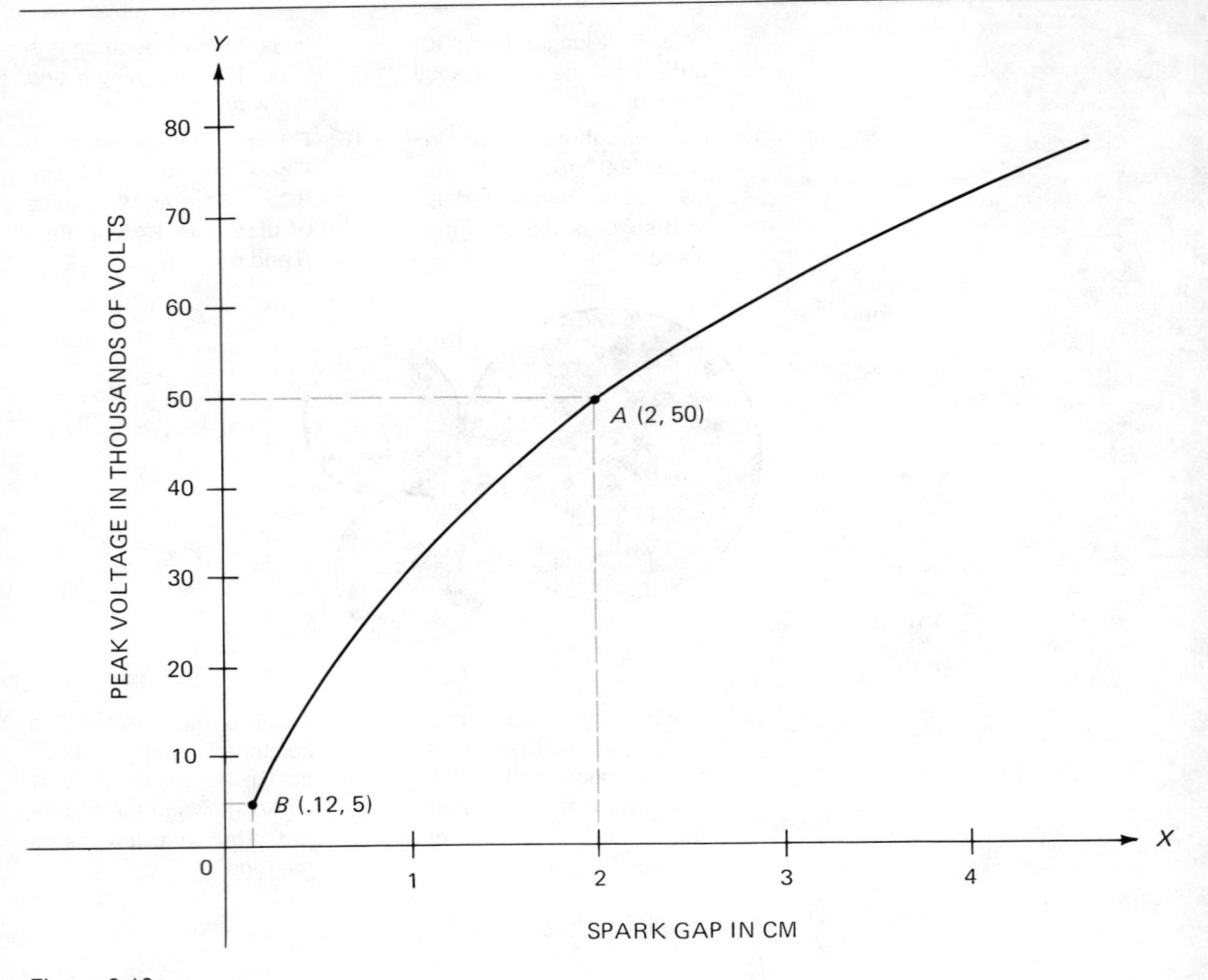

Figure 9-10

required (point A on the graph). For a .12-cm spark gap, a peak voltage of approximately 5000 volts is obtained (point B on the graph).

We construct such a graph using the steps outlined in the flow chart in Figure 9-11.

Example 1

The table shows the relationship between the diameter of standard copper wire and the resistance of the wire. Graph this relationship.

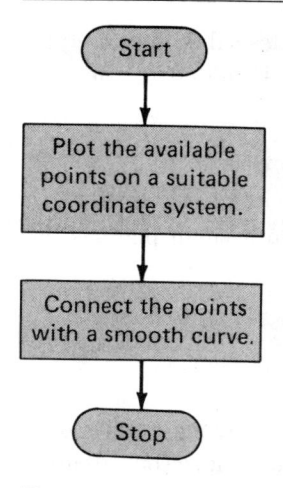

Figure 9-11

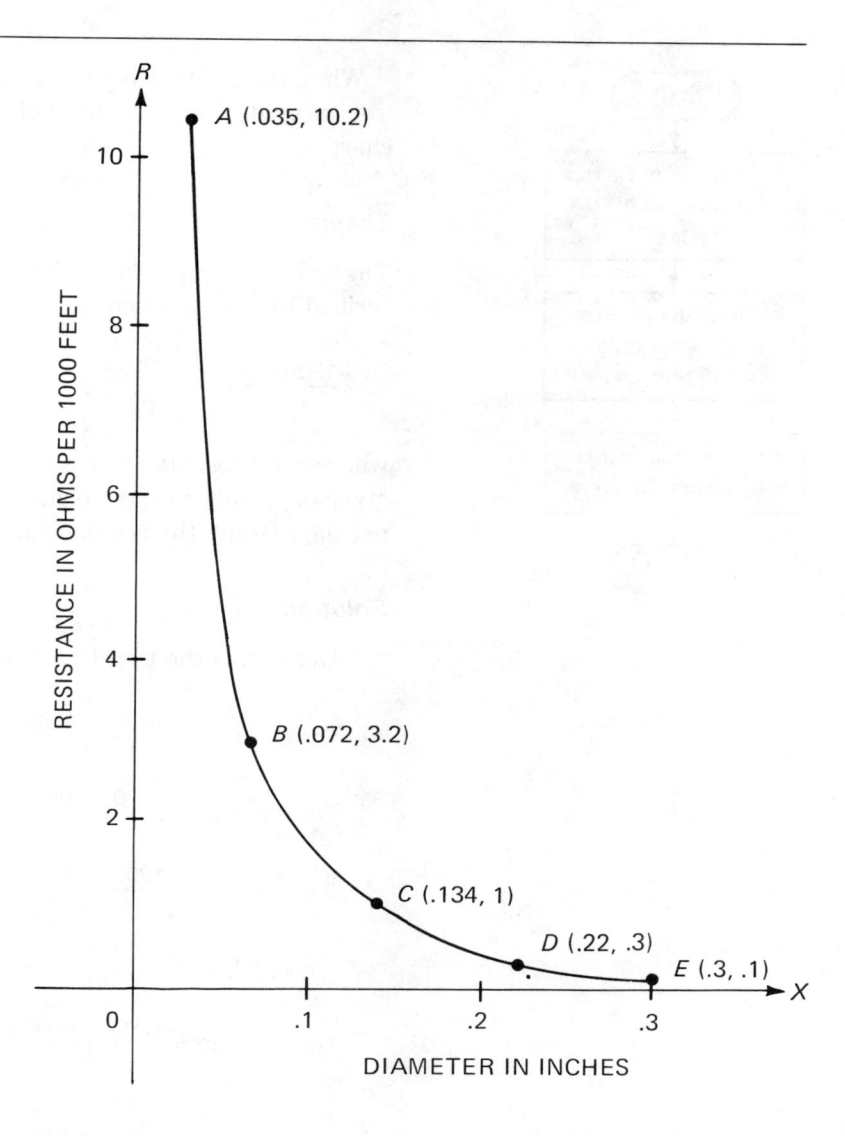

Diameter in inches	.035	.072	.134	.220	.300
Resistance in ohms per 1000 ft	10.2	3.2	1.0	.3	.1

Solution

1. Graph the points A (.035, 10.2), B (.072, 3.2), C (.134, 1.0), D (.220, .3), and E (.300, .1).
2. Connect the points with a smooth graph.

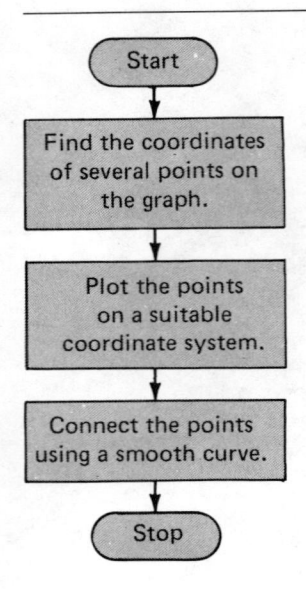

When the relationship to be graphed is described by an equation instead of a table, an additional step is added to the flow chart.

Example 2

The unit cost of producing heating units in a small plant is described by the equation

$$c = \frac{1200 + 400x}{x}$$

where c represents the cost per unit of producing x units on a given day. However, the plant cannot produce more than 10 units per day. Graph the relationship.

Solution

1. Determine the coordinates of several points on the graph.

$$\text{For } x = 1, c = \frac{1200 + 400(1)}{1} = 1600$$

$$\text{For } x = 2, c = \frac{1200 + 400(2)}{2} = 1000$$

$$\text{For } x = 3, c = \frac{1200 + 400(3)}{3} = 800$$

$$\text{For } x = 4, c = \frac{1200 + 400(4)}{4} = 700$$

$$\text{For } x = 5, c = \frac{1200 + 400(5)}{5} = 640$$

$$\text{For } x = 6, c = \frac{1200 + 400(6)}{6} = 600$$

$$\text{For } x = 7, c = \frac{1200 + 400(7)}{7} = 571$$

$$\text{For } x = 8, c = \frac{1200 + 400(8)}{8} = 550$$

$$\text{For } x = 9, c = \frac{1200 + 400(9)}{9} = 533$$

$$\text{For } x = 10, c = \frac{1200 + 400(10)}{10} = 520$$

2. Plot the points on a suitable coordinate system. The x axis must include $x = 0$ and $x = 10$. The c axis must run from $c = 0$ to $c = 1600$.

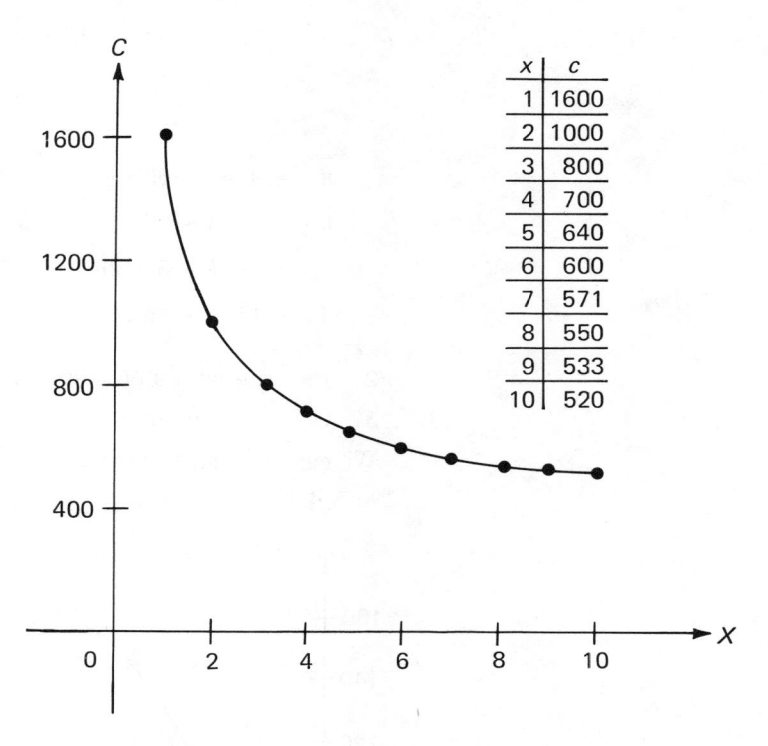

x	c
1	1600
2	1000
3	800
4	700
5	640
6	600
7	571
8	550
9	533
10	520

3. Connect the points with a smooth curve.

Sometimes we can use a graph to obtain information about a problem that is not obvious from a set of data.

Example 3

A piece of sheet metal 36 in. wide is to be bent to form a chute, as shown in Figure 9-12. Find the dimensions of the chute so that the cross-sectional area will be as large as possible.

Solution

The cross-sectional area (a rectangle) is described by the relationship $A = x(36 - 2x)$ or $A = 36x - 2x^2$. We assume that x is a positive number.

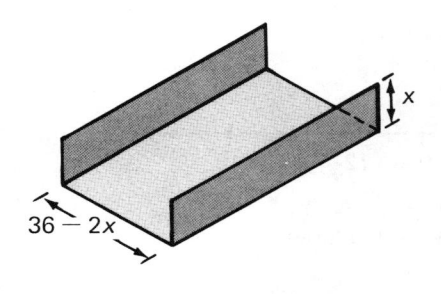

Figure 9-12

1. Determine the coordinates of several points on the graph.

   If $x = 1$, $A = 36(1) - 2(1)^2 = 34$.
   If $x = 2$, $A = 36(2) - 2(2)^2 = 64$.
   If $x = 4$, $A = 36(4) - 2(4)^2 = 112$.
   If $x = 8$, $A = 36(8) - 2(8)^2 = 160$.
   If $x = 9$, $A = 36(9) - 2(9)^2 = 162$.
   If $x = 10$, $A = 36(10) - 2(10)^2 = 160$.
   If $x = 12$, $A = 36(12) - 2(12)^2 = 144$.
   If $x = 16$, $A = 36(16) - 2(16)^2 = 64$.

2. Plot the points on a suitable coordinate system.
3. Connect the points with a smooth curve.

We can see that the maximum cross-sectional area is obtained at $x = 9$, $A = 162$.

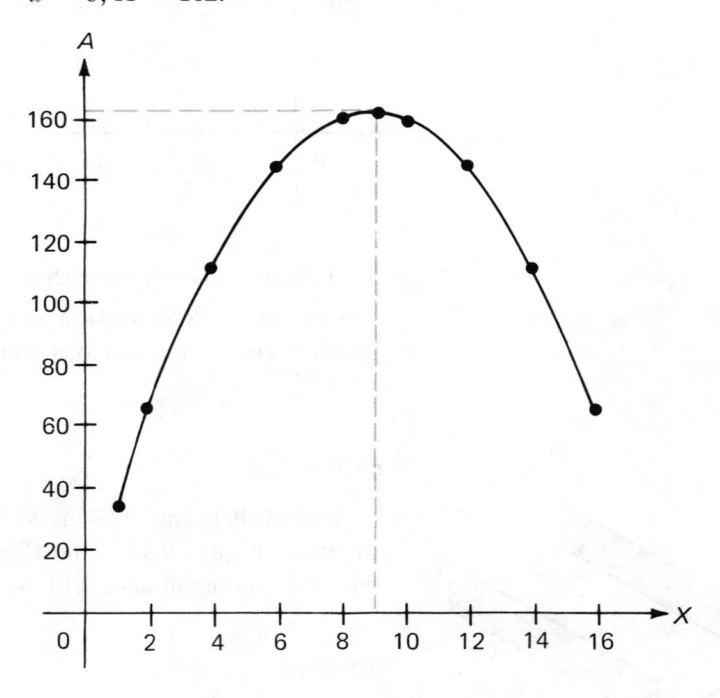

EXERCISES

1. Graph the relationship shown in the table.

p	1	2	3	4	5	6	8	10
y	15	7.5	5	3.75	3	2.5	1.875	1.5

2. Graph the relationship shown in the table.

x	0	10	20	30	40	50
y	1000	500	330	250	200	167

3. Graph the relationship determined by the equation $y = 20/(x + 1)$, for x between 0 and 9.

4. The demand for a product is described by the equation $d = 200 - 3x^2$, where d represents the demand for the product at a price x. Graph the equation for x between 1 and 8.

5. The available supply of a product is described by the equation $s = 20x^2 + 4x$, where s represents the supply available at price x. Graph the equation for x between 0 and 6.

6. The equation $p = -x^2 + 50x - 150$ describes the relationship between the profit p and the selling price per item x. Graph the equation for x between 5 and 45.

7. A trucking company has found from experience that the hourly cost of operating a truck is described by the equation $c = 800/r + r$, where r represents the speed in miles per hour that the truck is run. Graph the equation for r between 10 and 60. Estimate the most economical speed of operation.

8. Repeat exercise 7 using the equation $c = \dfrac{1550}{r} + \dfrac{r}{2}$.

9. Repeat exercise 7 using the equation $c = \dfrac{6200}{r} + 2r$.

U.S. system

10. The table shows the weight w in pounds per 1000 feet of standard copper wire with a diameter d inches. Graph the relationship.

d	w
.182	100.2
.102	31.2
.057	9.9
.032	3.1
.018	1.0

Metric system

13. The table shows the weight w in kilograms per 1000 meters of standard copper wire with a diameter d centimeters. Graph the relationship.

d	w
.462	149.10
.259	46.77
.145	14.67
.081	4.60
.045	1.44

Graphing

11. The table shows the weight w in pounds per foot for square steel bars with a width of d inches. Graph the relationship.

d	w
.5	.85
1.0	3.40
1.5	7.65
2.0	13.60
2.5	21.25
3.0	30.60

14. The table shows the weight w in kilograms per meter for square steel bars with a width of d centimeters. Graph the relationship.

d	w
1	8
2	31
3	69
4	123
5	193
6	277
7	377

12. A rectangular region is to be fenced for storage of equipment. If 160 yds of fencing is available, what is the maximum area that can be fenced?

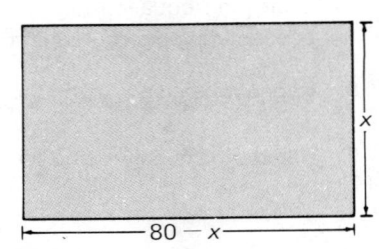

[Rectangle with height x and width 80 − x]

15. A rectangular region is to be fenced for storage of equipment. If 140 m of fencing is available, what is the maximum area that can be fenced?

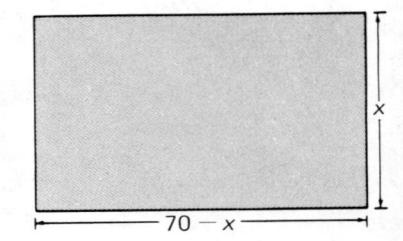

[Rectangle with height x and width 70 − x]

Taking inventory

1. A point $P(x, y)$ is plotted on a coordinate system by moving x units to the right or left of the y axis and y units above or below the x axis. (p. 260)
2. To graph the linear equation $ax + by + c = 0$:
   a. Select three x values.
   b. Determine the corresponding y values.
   c. Write the ordered pairs.
   d. Plot the pairs on a coordinate system.
   e. Draw the line containing the three points. (p. 264)

3. To solve a pair of linear equations graphically:
   a. Graph the linear equations.
   b. Estimate the coordinates of the point of intersection of the lines. (p. 268)
4. To solve a pair of linear equations algebraically:
   a. Multiply each equation by a constant chosen so that one of the variables can be eliminated from both equations by adding or subtracting the equations.
   b. Obtain an equation in one variable by adding or subtracting the equations in step a.
   c. Solve this equation.
   d. Substitute this value in one of the original equations and solve for the other variable. (p. 271)
5. To graph a nonlinear equation:
   a. Find the coordinates of several points on the graph.
   b. Plot these points on a suitable coordinate system.
   c. Connect the points, using a smooth curve. (pp. 278–279)

Measuring your skills

1. Using the formula $C = .80x + 50$ and $x = 5, 10, 15, 20, 25, 30, 35, 40$, determine a set of ordered pairs. Plot the ordered pairs on an appropriate coordinate system. (9-1)
2. Using the formula $30x + 40y = 160$ and $x = -3, -2, -1, 0, 1, 2, 3$, find a set of ordered pairs. Plot the ordered pairs on an appropriate coordinate system. (9-1)
3. Graph each of the given linear equations on a coordinate system. (9-2)
   a. $x - y = 0$
   b. $4x - 3y = 12$
   c. $x = 7$
   d. $.1x + .15y = 50$
4. Graph each pair of equations on a coordinate system and estimate the coordinates of the point of intersection of the graphs. (9-3)
   a. $2x - y + 6 = 0$; $4x + y - 12 = 0$
   b. $x + 2y = 8$; $2x + 2y = 13$
   c. $4x - 3y - 120 = 0$; $3x + 2y - 600 = 0$

Graphing 283

5. Solve each pair of equations algebraically. (9-4)
   a. $x - y + 4 = 0$; $x + y - 20 = 0$
   b. $x - 2y + 11 = 0$; $2x + y - 7 = 0$
   c. $.3x + .4y = 114$; $.5x + .6y = 178$
6. Graph the relationship shown in the table. (9-5)

x	5	10	15	20	25	30
y	21	10.5	7	5.25	4.2	3.5

7. Graph the relationship determined by the equation $y = \dfrac{5000}{x + 5}$ for x between 0 and 60. (9-5)
8. Graph the relationship given by the equation $y = -x^2 + 4x + 5$ for x between -1 and 5. Estimate the coordinates of the maximum point on the graph. (9-5)

U.S. system

9. There are 100 ft of fencing available to fence three sides of a storage area. Find the dimensions of the rectangle that fences the maximum area. (Write an equation for the area and draw the graph of the equation.) (9-5)

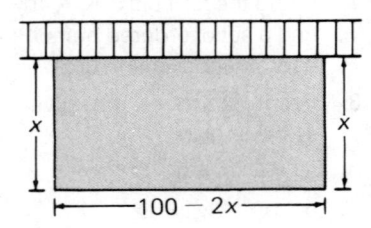

Metric system

10. A rectangular box is made by cutting out the corners of a piece of metal 60 cm by 40 cm. Find the value of **x that gives the maximum volume of the box.** (Write an equation for the volume and draw the graph of the equation.) (9-5)

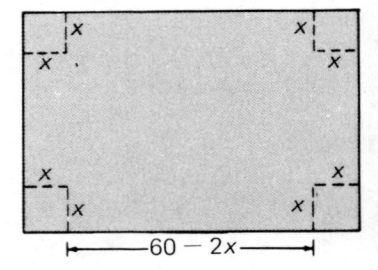

10 Geometry

After completing this chapter, you should be able to:

1. Use a protractor, a compass, and a straightedge to perform geometric constructions.
2. Apply geometric constructions to solve problems.
3. Use similar triangles and congruent triangles to solve measurement problems.
4. Use the rule of Pythagoras to solve problems involving right triangles.

Angles

10-1 MEASURING ANGLES

From earlier courses you will recall that angles such as $\angle A$ and $\angle XOY$, can be measured in *degrees*. In most work in the industry and the trades you will often find you need angles. Bolts and screws, for example, differ in the angle between the threads.

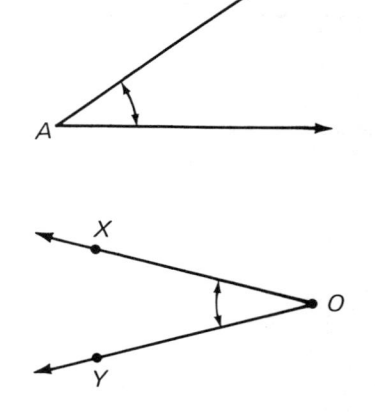

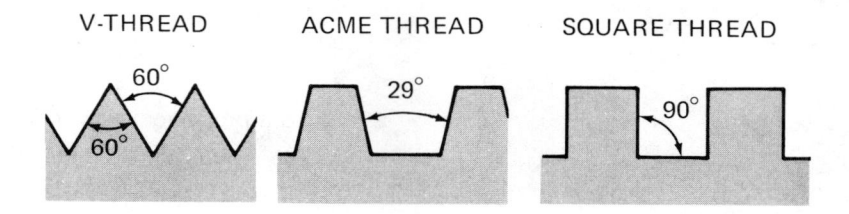

V-THREAD   ACME THREAD   SQUARE THREAD

60° / 60°   29°   90°

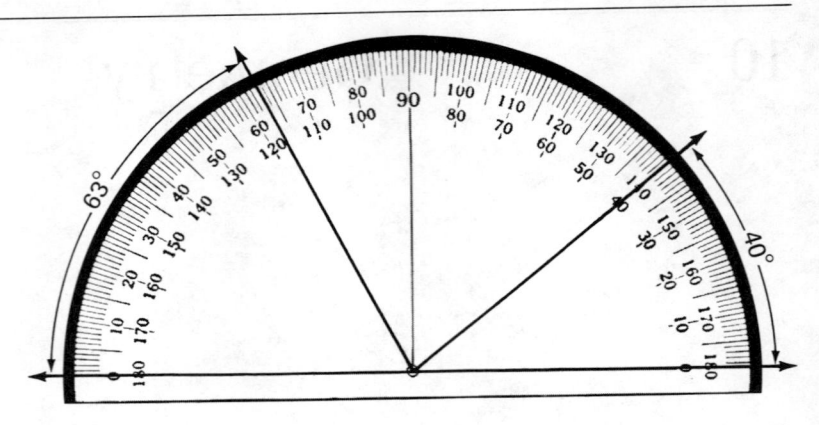

Figure 10-1

To measure an angle, you can use a protractor, as illustrated in Figure 10-1. There are two scales. Use whichever is more convenient.

As you can easily see, you can use a protractor either to measure a given angle or to draw one of a certain measure. A bevel protractor, shown in Figure 10-2, is used to measure the angle of beveled edges like those in picture frames.

Two special angles are shown here. $\angle ABC$ is a *right angle*, and $\angle XOY$ is a *straight angle*.

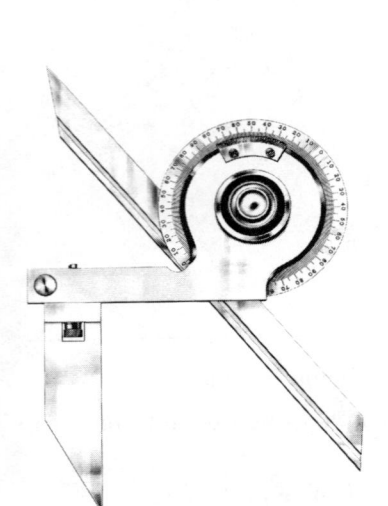

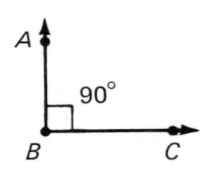

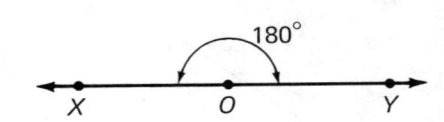

Figure 10-2

EXERCISES

1. In Figure 10-3, read the measures of:
   a. $\angle BAC$
   b. $\angle BAD$
   c. $\angle EAC$
   d. $\angle DAE$
   e. $\angle CAD$

2. Using a protractor, draw angles that have the following measures:
   a. 45°   b. 75°   c. 90°   d. 125°   e. 78°   f. 112°

286 Technical and industrial formulas

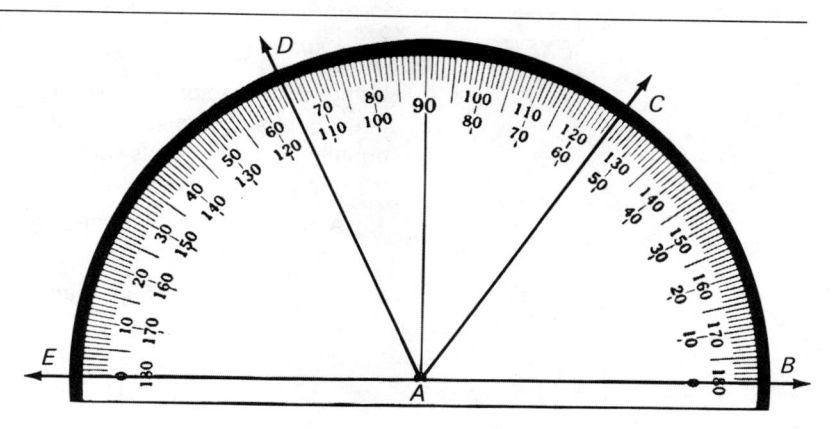

Figure 10-3

3. Draw a triangle that has a right angle, and make the two sides of the triangle forming the right angle 4 in. long. Find the measures of the two remaining angles.

4. Draw triangle ABC with the following dimensions: side AB is 10 cm long; ∠ABC is a right angle; ∠BAC is a 30° angle.

   a. What is the measure of ∠BCA?
   b. How long is side BC?
   c. How long is side AC?

5. Draw a parallelogram that has two sides 7 cm long, two sides 4.5 cm long, and two angles of 50°.

10-2 CENTRAL ANGLES

The parts, or *arcs*, of a circle have the same measure as the corresponding central angles. The number of degrees around a point is 360° (2 straight angles); the number of degrees in a circle is 360°. In the watch shown, we found equal divisions on the circle by drawing 12 equal angles at the center, or 12 equal *central angles*. How do you find the size of each angle? (360° ÷ 12) 360° ÷ 12 = 30°, so each central angle is 30°.

A car designer needs to know where to position the five lugs that hold the wheel to the car. Once the center of the wheel has been located, the designer can use central angles to locate the position of the lugs. The central angle for each lug would be 360° ÷ 5, or 72°.

Geometry 287

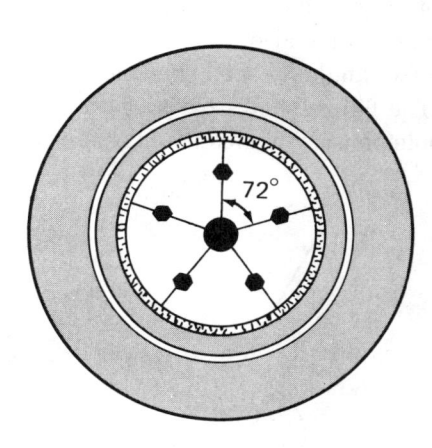

EXERCISES **U.S. system**

Using a protractor to measure the central angles, locate equally spaced points around the circle described.

1. A circle with a 5-in. diameter and 4 points.
2. A circle with a 4.5-in. diameter and 5 points.

Metric system

Using a protractor to measure the central angles, locate equally spaced points around the circle described.

3. A circle with a 15-cm diameter and 4 points.
4. A circle with a 10.3-cm diameter and 10 points.

10-3 ANGLE RELATIONSHIPS

Angles BAC and CAD are *adjacent angles*. They have the same *vertex*, A, and a *side*, AC, in common.

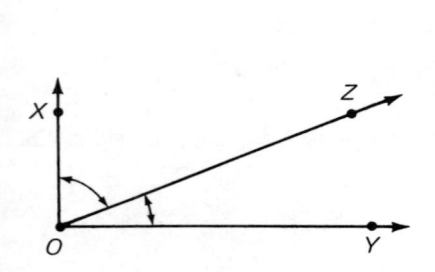

Figure 10-4

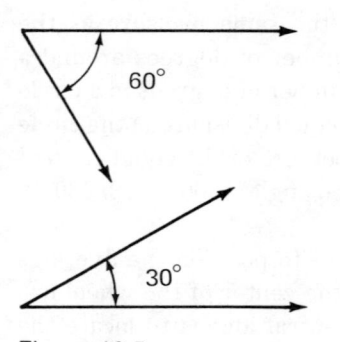

Figure 10-5

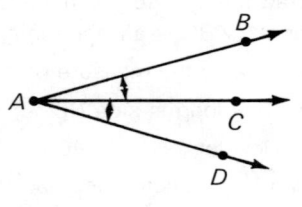

In Figure 10-4 angles XOZ and ZOY are adjacent angles. They are also *complementary angles* because their sum is 90°.

Complementary angles do not have to be adjacent angles. The angles in Figure 10-5 are complementary also.

If the sum of the measures of two angles is 180°, the angles are called *supplementary angles*. The figure shows pairs of supplementary angles. Note that supplementary angles do not have to be adjacent angles.

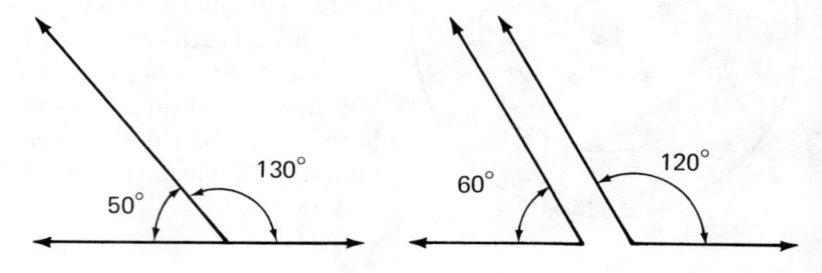

288   Technical and industrial formulas

Another very special sum also gives us 180°. This is the sum of the angles of a triangle. Measure ∠A, ∠B, and ∠C and find their sum.

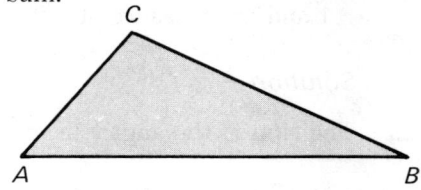

The sum of the angles of a triangle is 180°.

Example 1

Suppose we want to splice two boards to form a tight joint. If ∠ABC = 45°, what must ∠DEX measure?

Solution

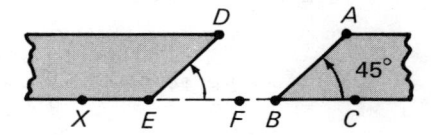

Points X, E, and C must lie in a straight line. That is, ∠ABC and ∠DEX *must* be supplementary. Thus,

∠DEX = 180° − ∠ABC
     = 180° − 45°
     = 135°

Example 2

For the right-angled mitre joint shown, assuming ∠1 = ∠2, what are the sizes of ∠1 and ∠2?

Solution

∠1 and ∠2 are complementary. So the sum of their measures is 90°. Since ∠1 = ∠2, each must be half of 90°.

90 ÷ 2 = 45

∠1 and ∠2 each measure 45°.

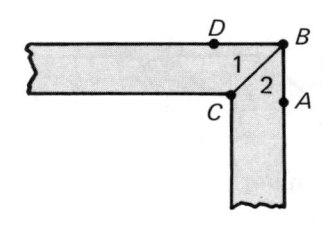

Geometry 289

Example 3

In the diagram, $\angle ACD$ is a straight angle. Find the measure of $\angle 1$ and the measure of $\angle 2$.

Solution

The sum of the angles in the triangle is 180°.

a. $\angle 1 + 30° + 40° = 180°$

$\angle 1 + 70° = 180°$

$\angle 1 = 180° - 70°$

$\angle 1 = 110°$

b. $\angle 2$ and $\angle BCA$ are supplementary angles.

$\angle 2 + 40° = 180°$

$\angle 2 = 180° - 40°$

$\angle 2 = 140°$

EXERCISES

1. Two angles of a triangle measure 35° and 55°. What is the measure of the third angle?

2. Draw a pair of adjacent supplementary angles, one of which has a measure of 72°. What is the measure of the other angle?

3. Two angles are complementary. One of them has a measure of 45°. What is the measure of the other angle?

Figure 10-6

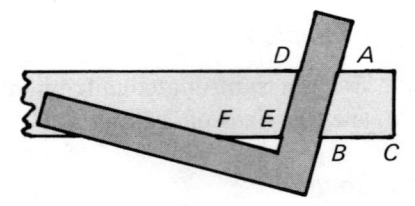

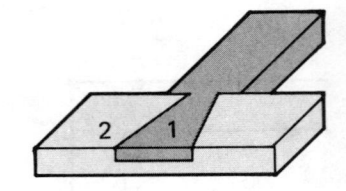

Figure 10-7

4. If $\angle ABC$ in Figure 10-6 measures 75°, how large is angle *DEF*? (Assume that there is one board under the L square.)

5. For the dovetail joint shown in Figure 10-7, if $\angle 1 = 82°$, how large is $\angle 2$?

6. The L square can be used to size the lumber for a rafter. If the plate angle in Figure 10-8 is 40°, what is the ridge angle?

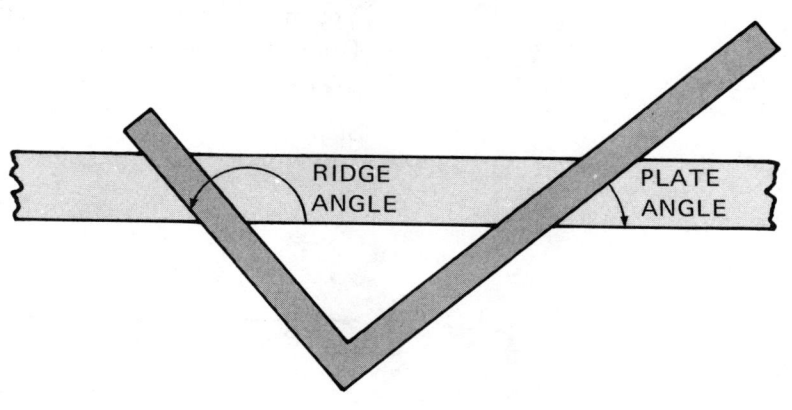

Figure 10-8

Constructions involving perpendiculars

10-4 PERPENDICULAR LINES

Lines that meet at right angles are called *perpendicular lines*, or simply *perpendiculars*. Perhaps you are familiar with the drafter's triangle, shown in Figure 10-9, or the carpenter's square, shown in Figure 10-10. Both are useful for drawing or testing perpendiculars.

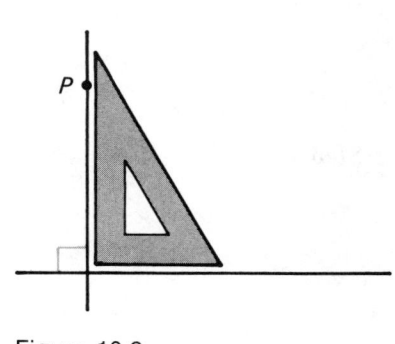

Figure 10-9

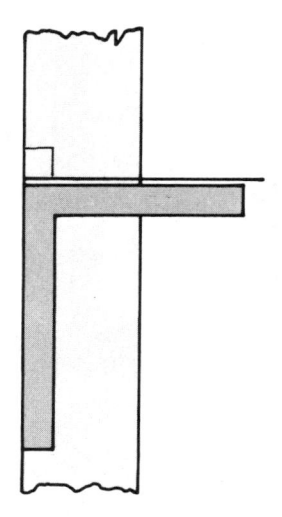

Figure 10-10

Sometimes, however, you will need or prefer to use just the compass and straightedge. On your own paper, copy the constructions shown step by step on the following pages.

Geometry 291

Constructing a perpendicular at a point on a line

You are given line ℓ and point A.
Draw DA perpendicular to BC through A.

Step 1

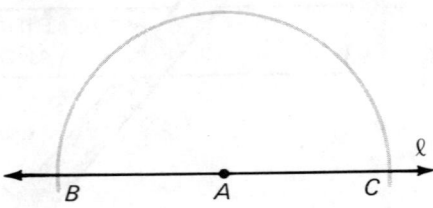

Step 2

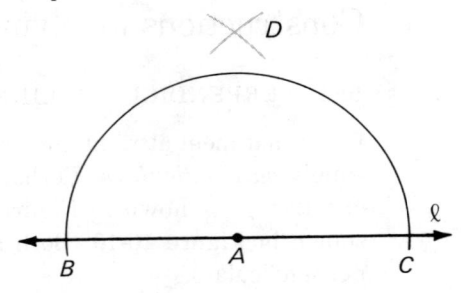

Step 3

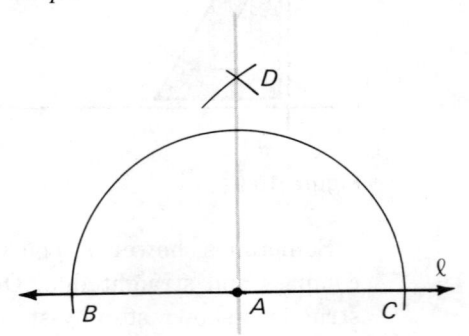

Constructing a perpendicular from a point to a line

You are given line ℓ and point P.
Draw PS perpendicular to ℓ from P.

Step 1

Step 2

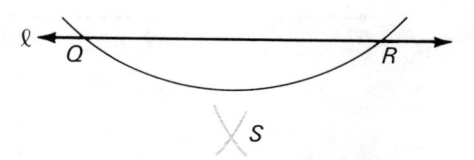

Step 3

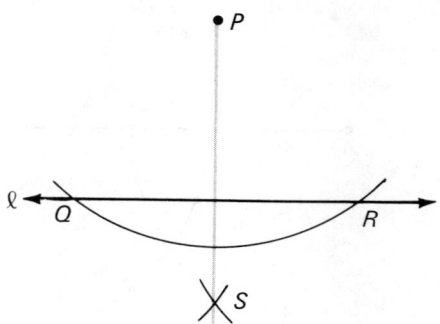

Geometry 293

Constructing the perpendicular bisector of a line segment

You are given line segment AB.
Bisect (cut in half) AB with a perpendicular CD.

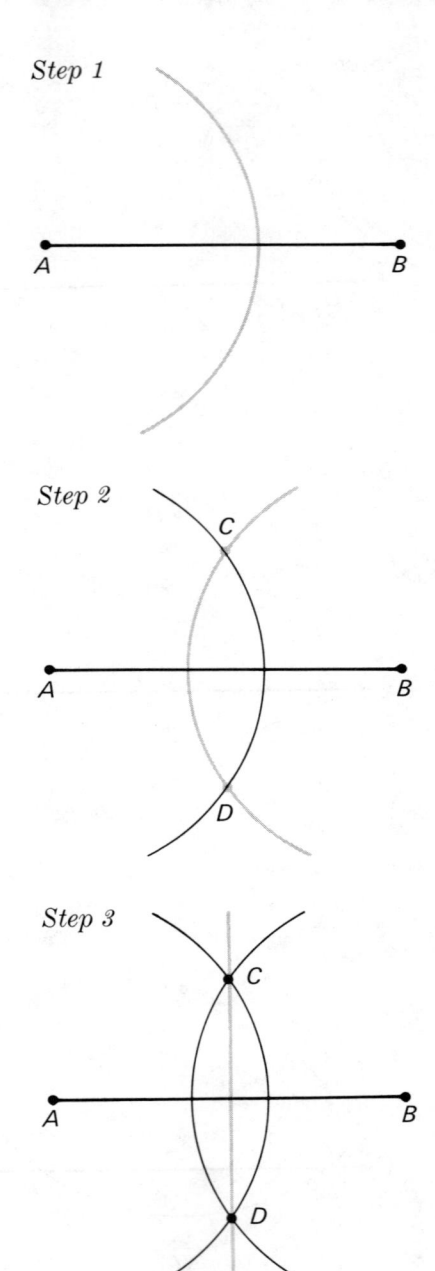

Step 1

Step 2

Step 3

EXERCISES

U.S. system

1. Draw a line and locate a point on it. Draw a perpendicular to the line through the point.
2. Draw a line 3 in. long. Construct a perpendicular at each of its endpoints.
3. Draw a line segment 5 in. long. Construct its perpendicular bisector. Divide it into 4 equal parts with 2 more perpendicular bisectors.
4. Construct a square with sides 3 in. long.
5. Draw a circle with a 2-in. radius. Divide the circular region into 4 equal parts.
6. Draw a circle with a radius of 3 in. Divide the circular region into 8 equal parts.
7. Draw a circle with a 2-in. diameter. Construct 2 perpendicular diameters. Connect the endpoints of the diameters in order around the circle. What figure have you made?

Metric system

8. Draw a line and locate a point above it. Draw a perpendicular to the line through the point.
9. Draw a line 10 cm long. Construct a perpendicular at each of its endpoints.
10. Draw a line segment 13 cm long. Construct its perpendicular bisector. Divide it into 4 equal parts with 2 more perpendicular bisectors.
11. Construct a rectangle with a length of 9 cm and a width of 4.5 cm.
12. Draw a circle with a 9-cm radius. Divide the circular region into 4 equal parts.
13. Draw a circle with a radius of 8 cm. Divide the circular region into 3 equal parts.
14. Draw a circle with a 5-cm radius. Construct 2 perpendicular diameters. Connect the endpoints of the diameters in order around the circle. What figure have you made?

10-5 CIRCLES

In Figure 10-11 lines CA and CB are tangents to circle O. *Tangents* are lines that touch a circle at just one point.

Figure 10-11

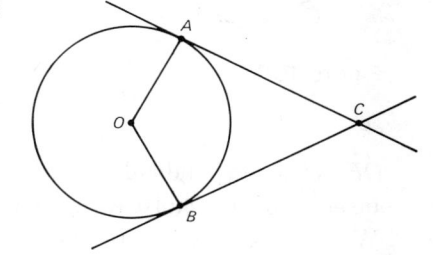

Notice that radius OA and radius OB are perpendicular to the tangents at A and B. Knowing this, we can construct a tangent at any given point on a circle. On your own paper, copy the constructions shown step by step below.

Constructing a tangent to a given point on a circle

You are given circle O and point A on the circle.
Draw a tangent ℓ through point A.

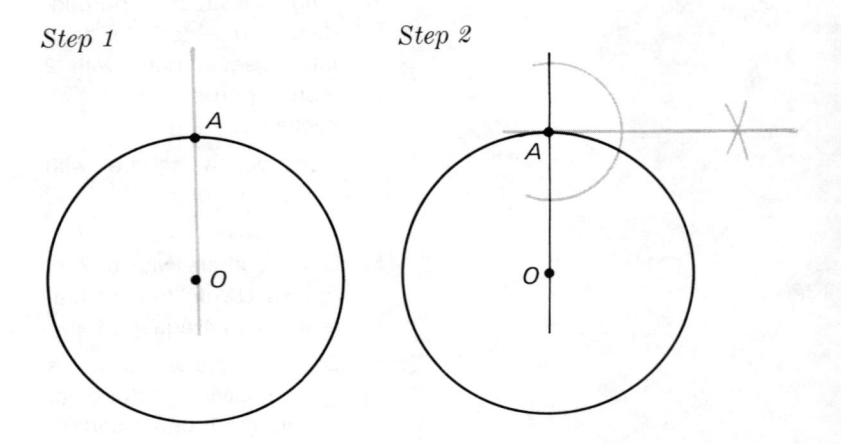

Step 1

Step 2

In Figure 10-12, AB and CD are chords of circle O. A *chord* is any line segment that has its endpoints on the circle.

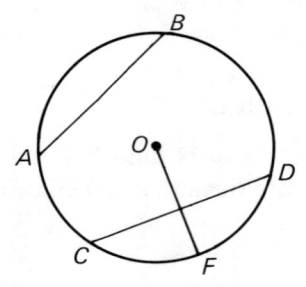

Figure 10-12

OF is the perpendicular bisector of chord CD. The perpendicular bisector of any chord always passes through the center of a circle. We can use this fact to find the center of any circle.

Locating the center of a circle

You are given a circle.
Draw any two chords.
Construct the perpendicular bisector of each to locate the center, O.

Step 1

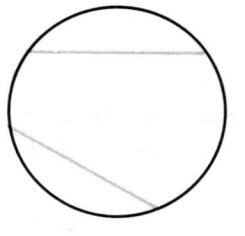

Step 2

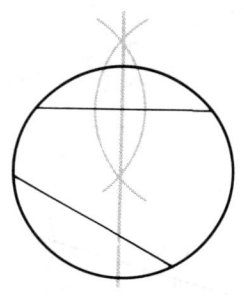

Step 3

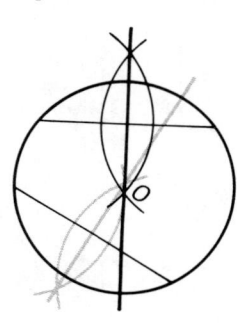

EXERCISES

1. Draw a large circle O and radius OX. Construct the tangent at X.
2. Trace a circular object such as a can. Locate the center of the circle you have drawn.
3. Draw a circle O with a diameter AB. Construct tangents at A and B. Construct the perpendicular bisector of AB. At each of its endpoints, construct a tangent. What figure have you made?
4. Draw a triangle and construct the perpendicular bisectors of all three sides. What do you notice about the three lines you have constructed? Construct a circle passing through the corners of the triangle.
5. Figure 10-13 suggests a method for "rounding off" the corner of a drawing. "Round off" the corners of a sheet of paper, using this method.

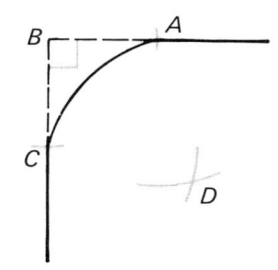

Figure 10-13

Geometry 297

Basic angle constructions

10-6 ANGLES

In Section 10-4 you learned how to bisect, or cut in half, a line segment. We can also use just a compass and straightedge to *bisect an angle*. By doing this, we will form two smaller angles, each half the measure of the original angle.

Bisecting an angle

You are given $\angle A$.
Construct bisector AD so that $\angle BAD = \angle CAD$.

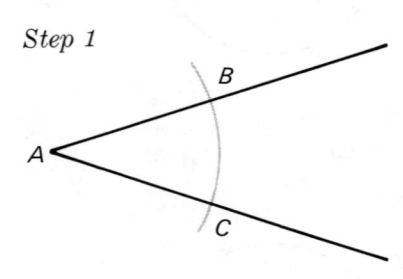

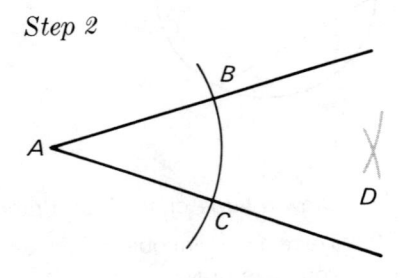

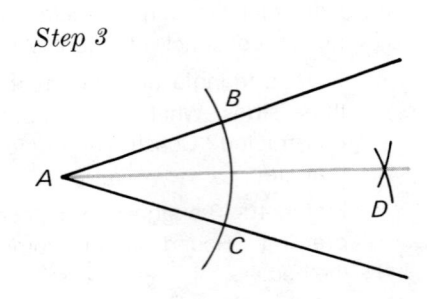

Step 1

Step 2

Step 3

A drafter was asked to extend this drawing of a cross-section of an Acme screw to include another thread cross-section. This meant constructing at point C an angle equal to $\angle ABC$, and at points D and E angles equal to $\angle BAF$.

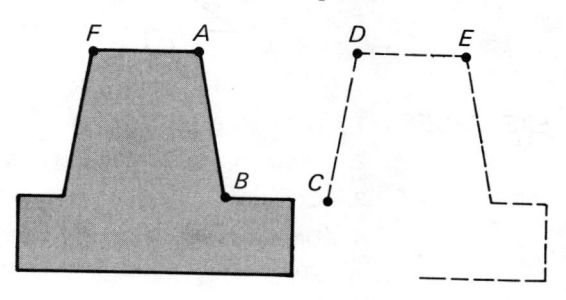

Practice the following construction until you can reproduce an angle quite precisely. A sharp lead will be essential.

Constructing an angle equal to a given angle

You are given $\angle B$ and line l.
Construct an angle at P on l equal to $\angle B$.

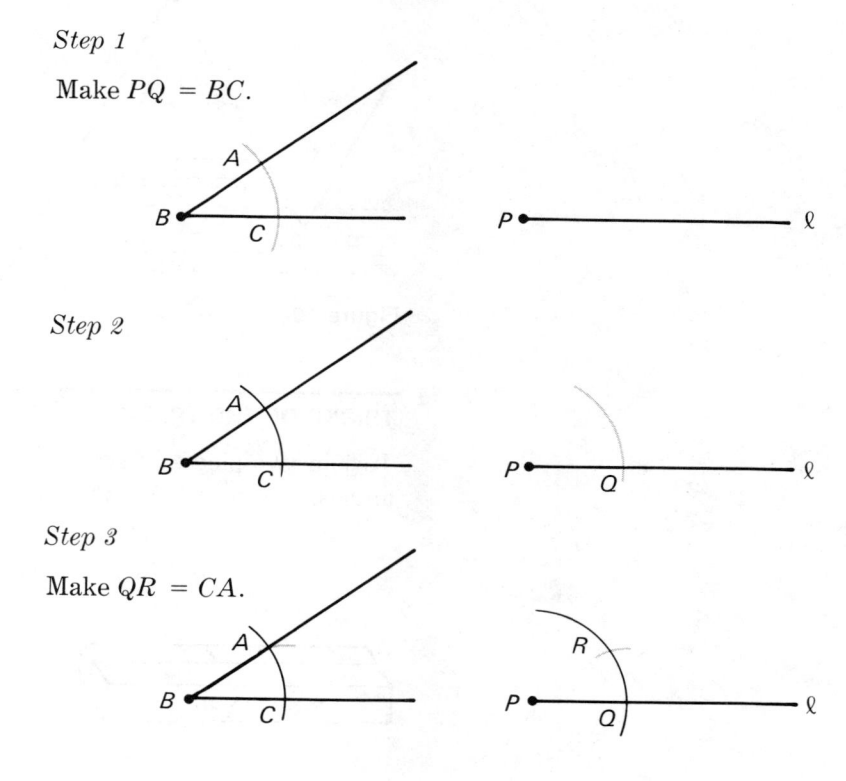

Step 1

Make $PQ = BC$.

Step 2

Step 3

Make $QR = CA$.

Geometry 299

Step 4

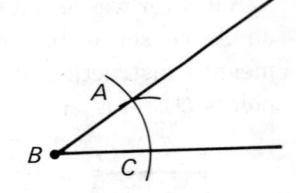

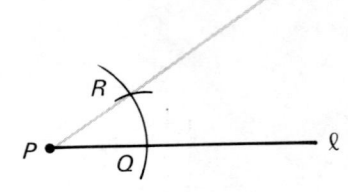

EXERCISES

1. Draw an angle less than 90°. Construct its bisector.
2. Draw an angle that measures more than 90°. Construct its bisector.
3. Construct a triangle that has two equal angles.
4. Draw a triangle. Using a ruler to copy the length of one side only, construct another triangle identical to the one you drew.
5. Draw a triangle. Bisect each of its three angles. What do you notice about the angle bisectors? Draw a circle inside the triangle to which all three sides are tangent.
6. Construct an angle that measures 45°, using only a compass and a straightedge.
7. Figure 10-14 shows a cross-section of a V thread. Trace the drawing and extend it to include two more threads.

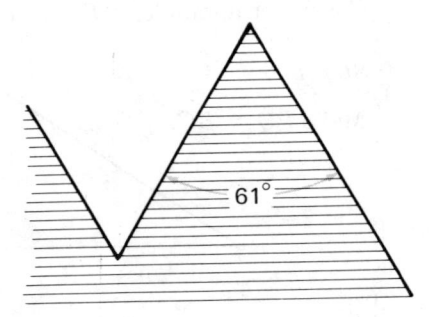

Figure 10-14

TRICKS OF THE TRADE

To join two pieces of stock at a 90° angle, cut the stock at 45° angles. The pieces fit as shown.

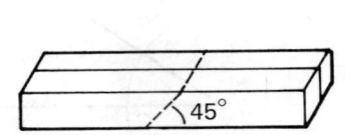

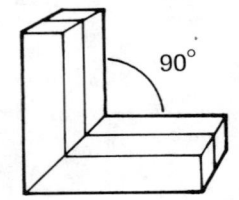

10-7 PARALLEL LINES

Parallel lines are lines on a flat surface that do not meet. The opposite edges of a table are an example of parallel lines.

A triangle or a bevel may be used to draw parallel lines, as shown in Figures 10-15 and 10-16. However, it is a good idea to learn how to use the compass for this purpose. Two methods are shown below.

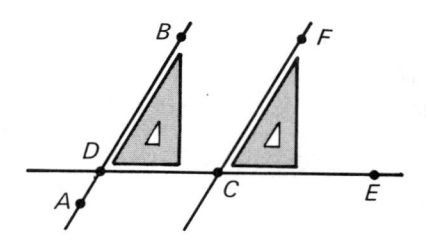

Figure 10-15

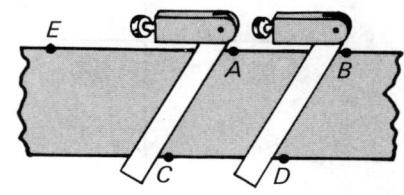

Figure 10-16

Constructing a line parallel to a given line through a given point

You are given line ℓ and point P.
Draw a line m through P parallel to ℓ.

Step 1

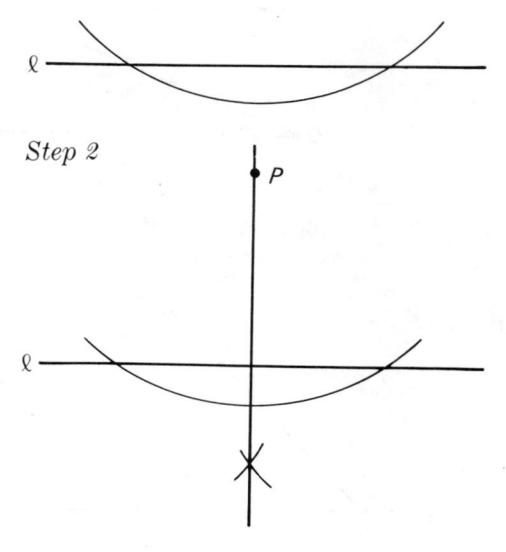

Step 2

Geometry 301

Step 3

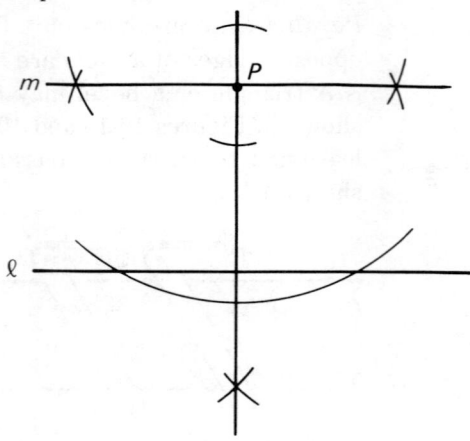

EXERCISES 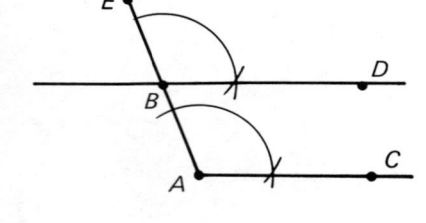
1. Draw a horizontal line and place a point above the line. Construct through the point a line that is parallel to the original line.
2. Draw a vertical line and place a point to the right of the line. Construct through the point a line parallel to the original line.
3. Draw an angle greater than 90°. Label it ∠BAC. Copy ∠BAC at point B as shown. Measure ∠DBA and ∠BAC and compare their sizes.
4. Draw an angle less than 90°. Label it ∠PQR. Copy angle PQR on the opposite side of P as shown. Measure ∠PQR and ∠SPQ to check your construction.
5. Copy Figure 10-17 and construct parallels to line AE through each of points B, C, and D. 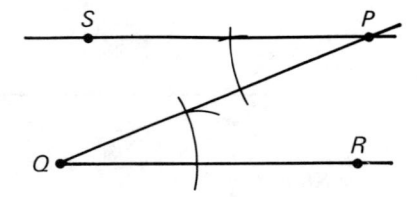

Figure 10-17

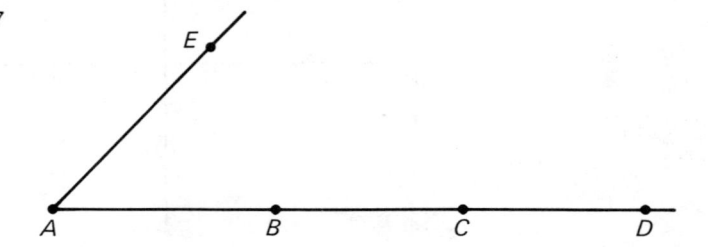

302 Technical and industrial formulas

Triangles

10-8 CONGRUENT AND SIMILAR TRIANGLES

We use the word *congruent* to describe figures having the same shape and the same size. For example, $\triangle ABC$ and $\triangle DEF$ are congruent.

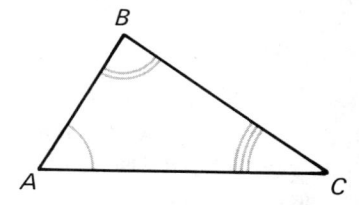

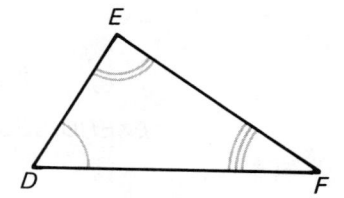

Notice that the corresponding angles are equal as well as the corresponding sides. That is,

$\angle A = \angle D$, $\angle B = \angle E$, $\angle C = \angle F$

$AB = DE$, $BC = EF$, $AC = DF$

We give a special name to triangles that have the same shape, but not necessarily the same size. They are *similar triangles*.

Similar triangles, such as $\triangle ABC$ and $\triangle XYZ$ below, have these important properties:

1. Corresponding angles are equal.
2. Corresponding sides are in proportion.

Figure 10-18

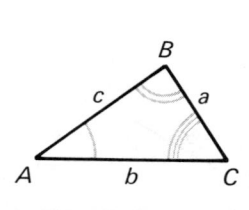

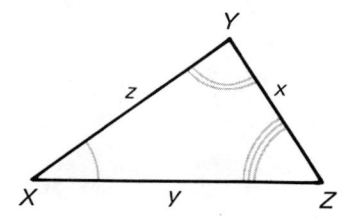

That is,

$\angle A = \angle X$, $\angle B = \angle Y$, $\angle C = \angle Z$.

$$\frac{a}{x} = \frac{b}{y}, \quad \frac{a}{x} = \frac{c}{z}, \quad \frac{b}{y} = \frac{c}{z}$$

Example

In Figure 10-18, if $a = 30$ m, $x = 10$ m, and $z = 20$ m, find the length of side c.

Geometry 303

Solution

Use the proportion containing the three given measurements.

$$\frac{a}{x} = \frac{c}{z}$$

$$\frac{30}{10} = \frac{c}{20}$$

$$c = 60 \text{ m}$$

EXERCISES

1. Draw a triangle *XYZ* with all angles less than 90°. Construct △*MNQ* congruent to it. Use three different methods.
2. Draw a triangle *ABC* with ∠*B* greater than 90°. Construct △*LMN* congruent to △*ABC*. Which method did you use?
3. Draw a large triangle *ABC*. Construct one similar to it with sides half as long. Show all construction marks.
4. △*MNO* is similar to △*PQR*. What is the measure of ∠*O*; ∠*P*; ∠*Q*; ∠*R*?

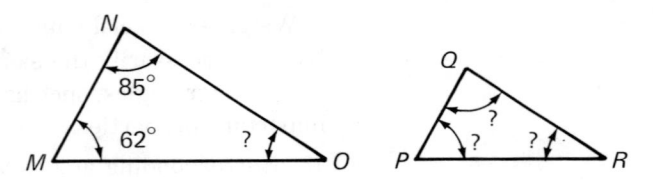

5. △*ABC* is similar to △*XYZ*. Use proportions to find the lengths of *XZ* and *YZ*.

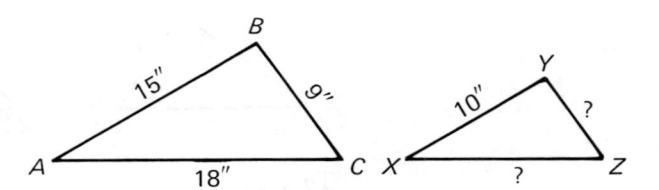

10-9 RIGHT TRIANGLES

One of the oldest and most useful formulas in all sorts of work is the one called the *rule of Pythagoras*. It can be applied to every right triangle, that is, a triangle with one right angle.

In *every* right triangle, the square of the *hypotenuse* (side opposite the right angle) equals the sum of the squares of the other two sides.

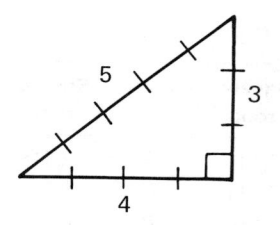

Figure 10-19

In Figure 10-19, does $5^2 = 3^2 + 4^2$?

$5 \times 5 = (3 \times 3) + (4 \times 4)$

$25 = 9 + 16$

$25 = 25$

Yes, 5^2 does equal $3^2 + 4^2$.

We can use the formula

$c^2 = a^2 + b^2$

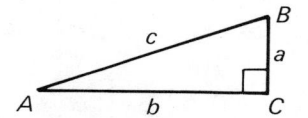

where c is the hypotenuse and a and b the other two sides.

Example 1

Find $\sqrt{62}$ to the nearest tenth.

Solution

Locate 62 in the third column headed "Number." Its square root is shown in the next column to the right.

$\sqrt{62} = 7.874$, or 7.9 to the nearest tenth

Example 2

Find the length of PR as shown.

Solution

We determine that PQ is the hypotenuse. So we can use the formula:

$c^2 = a^2 + b^2$

$13^2 = a^2 + 12^2$

$169 = a^2 + 144$

$a^2 = 169 - 144$

$a^2 = 25$

$a = \sqrt{25}$

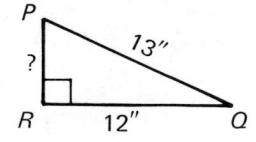

Refer to the table of square roots on page 306.

$\sqrt{25} = 5$, so $PR = 5$ in.

Geometry

Table 10-1 *Square roots of whole numbers from 1 to 100*

Number	Square root	Number	Square root	Number	Square root	Number	Square root
1	1.000	26	5.099	51	7.141	76	8.718
2	1.414	27	5.196	52	7.211	77	8.775
3	1.732	28	5.292	53	7.280	78	8.832
4	2.000	29	5.385	54	7.348	79	8.888
5	2.236	30	5.477	55	7.416	80	8.944
6	2.449	31	5.568	56	7.483	81	9.000
7	2.646	32	5.657	57	7.550	82	9.055
8	2.828	33	5.745	58	7.616	83	9.110
9	3.000	34	5.831	59	7.681	84	9.165
10	3.162	35	5.916	60	7.746	85	9.220
11	3.317	36	6.000	61	7.810	86	9.274
12	3.464	37	6.083	62	7.874	87	9.327
13	3.606	38	6.164	63	7.937	88	9.381
14	3.742	39	6.245	64	8.000	89	9.434
15	3.873	40	6.325	65	8.062	90	9.487
16	4.000	41	6.403	66	8.124	91	9.539
17	4.123	42	6.481	67	8.185	92	9.592
18	4.243	43	6.557	68	8.246	93	9.644
19	4.359	44	6.633	69	8.307	94	9.695
20	4.472	45	6.708	70	8.367	95	9.747
21	4.583	46	6.782	71	8.426	96	9.798
22	4.690	47	6.856	72	8.485	97	9.849
23	4.796	48	6.928	73	8.544	98	9.899
24	4.899	49	7.000	74	8.602	99	9.950
25	5.000	50	7.071	75	8.660	100	10.000

Example 3

In the right triangle shown, find the length of side c. Use the table of square roots above.

Solution

$c^2 = a^2 + b^2$

$c^2 = 5^2 + 4^2$

$ = 25 + 16$

$c^2 = 41$

$c = \sqrt{41}$

$c = 6.4$ (approx.)

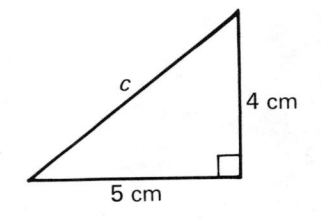

EXERCISES In the following problems, c represents the length of the hypotenuse of a right triangle, and a and b represent the lengths of the other two sides. Find the missing dimensions. Use the table of square roots.

	a	b	c
1.	3 cm	5 cm	?
2.	7 m	?	9 m
3.	8 in.	?	10 in.
4.	2 ft	4 ft	?

5. Find the length of the diagonal of a square with a side whose length is
   a. 5 cm
   b. 2 in.
   c. 3 ft

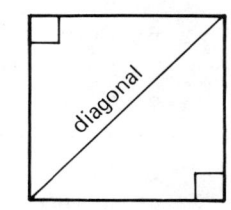

6. Find the lengths of the diagonals of rectangles with the following dimensions.
   a. $\ell = 4$ in., $w = 3$ in.
   b. $\ell = 6$ mm, $w = 8$ mm

7. For the V block in Figure 10-20, find the depth of the cut (BD). *Hint:* BD = BC

8. A square bar 1 in. on a side is to be milled from a circular rod. If we have available rods which are 1 in., $1\frac{1}{4}$ in., and $1\frac{1}{2}$ in. in diameter, which rod should we use? (Refer to Figure 10-21.)

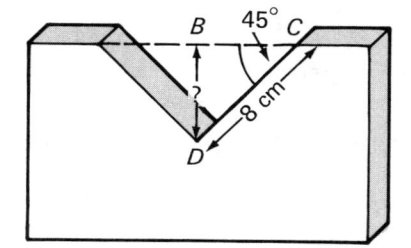

Figure 10-20

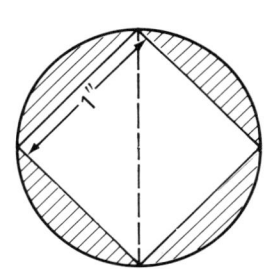

Figure 10-21

Taking inventory

You have learned to:

1. Measure angles and know the relationship between complementary angles and supplementary angles. (**p. 286 and p. 288**)
2. Construct perpendicular lines. (**p. 291**)
3. Construct tangents to a circle and locate the center of a circle. (**p. 296**)
4. Copy and bisect an angle. (**p. 298**)
5. Construct parallel lines. (**p. 301**)
6. Find the missing dimensions of similar triangles. (**p. 303**)
7. Find a missing dimension for a right triangle when two of the sides are given. (**p. 304**)

Measuring your skills

1. Using a protractor, draw an angle whose measure is 33°. **(10-1)**
2. How large must a central angle be for each spoke on a wheel if there are six spokes? **(10-2)**
3. Two angles of a triangle measure 46° and 38°. What is the measure of the third angle? **(10-3)**
4. What is the measure of ∠1 in Figure 10-22? *AB* is a straight line. **(10-3)**
5. Draw a line segment 4 in. long. Construct its perpendicular bisector. **(10-4)**
6. Draw a circle *O* and a radius *OB*. Construct a tangent to circle *O* at point *B*. **(10-5)**
7. Trace a circular object and locate the center of the circle you drew. **(10-5)**
8. Draw an angle whose measure is greater than 90° and bisect it. **(10-6)**
9. Draw a vertical line ℓ and locate point *P* to the right of it. Construct a line parallel to ℓ through *P*. **(10-7)**
10. Find the measure of *JK* and *JL* in Figure 10-23. △*ABC* is similar to △*JKL*. **(10-8)**

Figure 10-22

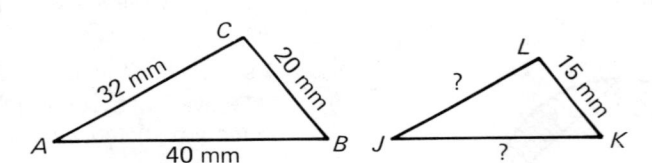

Figure 10-23

11. Find the measure of *a* in Figure 10-24. Use the table of square roots. **(10-9)**

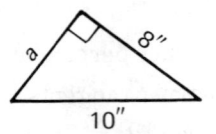

Figure 10-24

11 Ratio, proportion, scale

After completing this chapter, you should be able to:
1. Use ratios to compare quantities.
2. Find the missing term of a proportion.
3. Find lengths and areas of similar objects.
4. Read and draw scale drawings.

Ratios

11-1 MEANING OF RATIO

In Figure 11-1 gear A has 72 teeth. Gear B has 32 teeth. The ratio of the number of teeth of gear A to the number of teeth of gear B is 72:32, or $\frac{72}{32} = \frac{9}{4}$.

Concrete mixtures like those used in Figure 11-2 are also given in terms of ratios. For example, a 1:2:4 mixture for the base coat consists of 1 part cement, 2 parts sand, and 4 parts crushed rock. A 1:2 finish coat consists of 1 part cement and 2 parts sand.

As these examples show, a *ratio* is a comparison of numbers by division. Ratios are usually written as a fraction like ¾, although sometimes the form 3:4 is used.

Figure 11-1

Figure 11-2

Base coat
1:2:4

Finish coat
1:2

Ratios occur widely in industry and the trades. Consider the following examples.

Example 1

What is the ratio of the diameter of pulley A to the diameter of pulley B in Figure 11-3?

Solution

$$\frac{\text{Diameter of pulley A}}{\text{Diameter of pulley B}} = \frac{28 \text{ cm}}{12 \text{ cm}} = \frac{7}{3}$$

The ratio is 7:3.

Notice that we are comparing two dimensions, 28 cm and 12 cm, which are expressed in the same units of measure. The ratio 7:3, however, is a fraction. It is not expressed in units of measure.

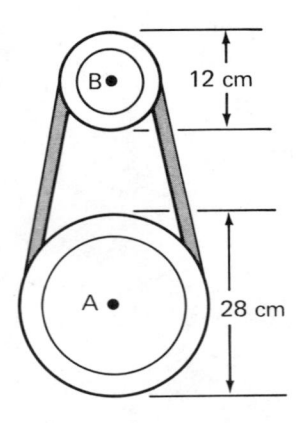

Figure 11-3

Example 2

What is the ratio of 30″ to 7′6″?

Solution

To compare these lengths, we need to express them in the same units.

Change the 7′6″ to inches.

$$7'6'' = (7 \times 12) + 6'' = 84 + 6 = 90''$$

The ratio of 30″ to 90″ is 1:3.

310   Technical and industrial formulas

EXERCISES

U.S. system

Express the following ratios in their simplest form.

1. 18″ to 24″
2. 3′ to 30′
3. 3″ to 12″
4. 2″ to $3\frac{1}{2}$″ (*Hint:* Multiply by 2.)
5. 4″ to 18″
6. 2 yd to 1.5 yd

Measure the line segments below in inches. Write the ratios in their simplest form.

7. a:c   8. e:c   9. d:a
10. a:b   11. c:d   12. e:a

Metric system

Express the following ratios in their simplest form.

14. 9 cm to 24 cm
15. 3 m to 6 m
16. 15 cm to 100 cm
17. 15 mm to 1 mm
18. 4 cm to 150 cm
19. 3 m to 1.5 m

20–25. Measure the line segments below in millimeters. Write the ratios in exercises 7–12 in their simplest form.

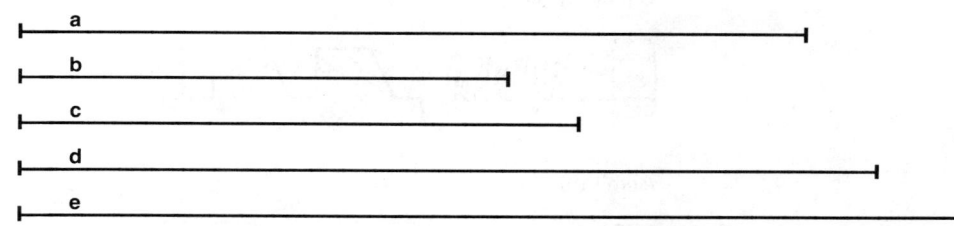

The *pitch*, or steepness, of a roof is the ratio of the rise to the span.

$$\text{Pitch} = \frac{\text{Rise}}{\text{Span}}$$

13. Find the pitch of a roof with these dimensions:
    a. rise = 8′, span = 20′
    b. rise = 6′, span = 24′
    c. rise = 6′, span = 18′
    d. rise = 5′, span = 22′6″
    e. rise = $4\frac{1}{2}$′, span = 20′3″

26. Find the pitch of a roof with these dimensions.
    a. rise = 1.5 m, span = 7.5 m
    b. rise = 1.2 m, span = 8.4 m
    c. rise = 1.6 m, span = 6.4 m
    d. rise = 1.75 m, span = 10.5 m
    e. rise = .95 m, span = 3.8 m

Ratio, proportion, scale

11-2 SCREW THREADS

Screw threads are designed to meet a variety of industrial needs. Three common screw threads are shown in Figure 11-4. The relationship between the depth of the thread D and the number of threads per inch N for each kind of thread is given by the formulas.

Screws are machined to fine tolerances. Therefore, a machinist must be familiar with these formulas in order to set the controls on the threading lathe.

Figure 11-4

American National Standard thread

$$D = \frac{0.6495}{N}$$

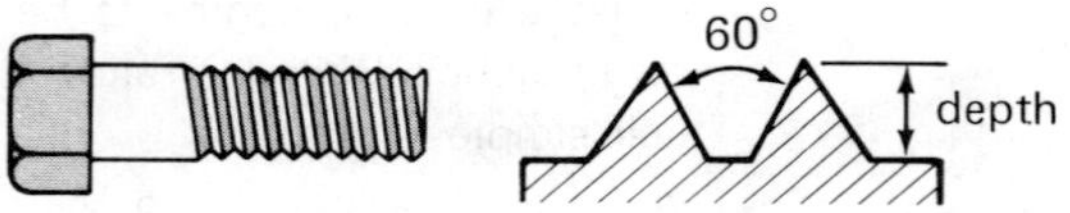

Square thread

$$D = \frac{0.5000}{N}$$

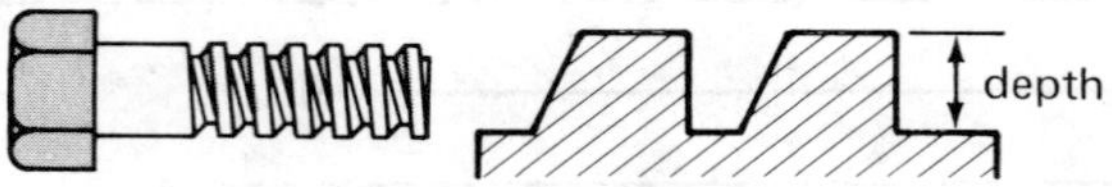

Acme thread

$$D = \frac{0.500}{N} + 0.01$$

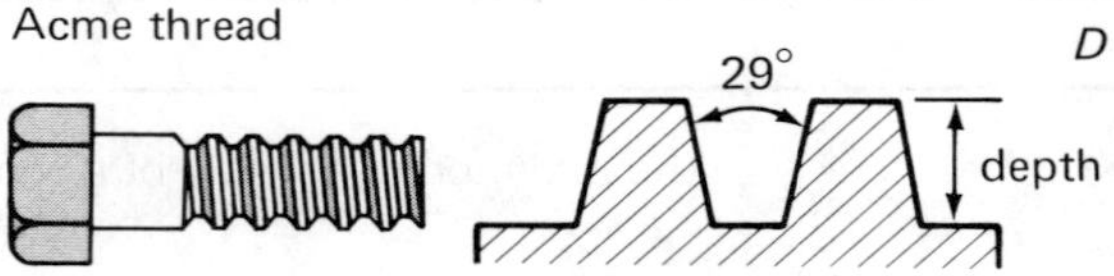

Example 1

What must be the depth setting on a lathe to machine an American National Standard screw having 12 threads per inch?

Solution

$$D = \frac{.6495}{N} = \frac{.6495}{12} = .0541 \quad \text{(approx.)}$$

The thread depth must be .0541″.

Example 2

How many threads per inch does an Acme screw have if its thread depth is .04125″?

312 Technical and industrial formulas

Solution

$$D = \frac{.5000}{N} + .01$$

$$.04125 = \frac{.5000}{N} + .01$$

$$.03125 = \frac{.5000}{N} \quad \text{(subtracting .01)}$$

$$.03125\, N = .5000 \quad \text{(multiplying by } N\text{)}$$

$$N = \frac{.5000}{.03125} \quad \text{(dividing by .03125)}$$

$$N = 16$$

There are 16 threads per inch.

EXERCISES Exercises 1 and 2 refer to the American National Standard thread.
1. Find the depth of the thread if there are
   a. 13 threads per inch.
   b. 40 threads per inch.
2. Find the number of threads per inch
   a. if the depth is .027".
   b. if the depth is .059".
   c. if the depth is .0203".
3. In an Acme screw, the number of threads per inch is 25. Find the depth.
4. In a square thread, the number of threads per inch is 36. Find the depth.

Proportions

11-3 THE MEANING OF PROPORTION

Sometimes we need to change the size of a drawing. To enlarge or reduce a drawing, we can use a photocopying machine that enlarges or reduces.

No matter which method we use, we must be sure that we change the height and the width *by the same ratio*. For example, a template for extruding a plastic grille is shown in a ½-size drawing in Figure 11-5.

Figure 11-5

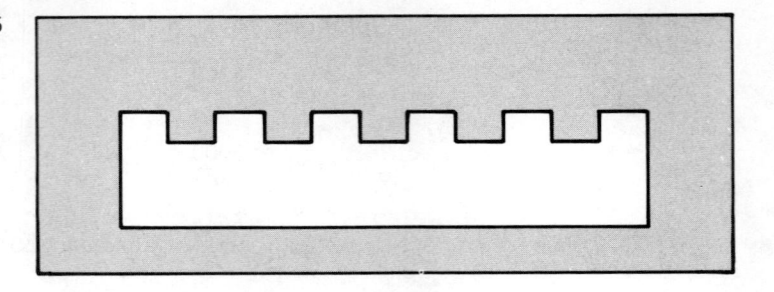

In the drawing the length of the block is $3\frac{5}{8}''$. The height is $1\frac{3}{8}''$. The actual dimensions of the block are:

$w = 2 \times 3\frac{5}{8}'' = 7\frac{1}{4}''$ $\qquad$ $h = 2 \times 1\frac{3}{8}'' = 2\frac{3}{4}''$

Would you agree with the following statement?

$$\frac{\text{Width of drawing}}{\text{Width of actual template}} = \frac{\text{Height of drawing}}{\text{Height of actual template}}$$

Or, Ratio of 1 to 2 $\quad \dfrac{3\frac{5}{8}''}{7\frac{1}{4}''} = \dfrac{1\frac{3}{8}''}{2\frac{3}{4}''} \quad$ Ratio of 1 to 2

The ratio of $3\frac{5}{8}''$ to $7\frac{1}{4}''$ is the same ratio as $1\frac{3}{8}''$ to $2\frac{3}{4}''$. An equation of two ratios is called a *proportion*. The *terms* of a proportion have special names, as shown below.

First term (extreme) — Third term (mean)
$$\frac{3\frac{5}{8}}{7\frac{1}{4}} = \frac{1\frac{3}{8}}{2\frac{3}{4}}$$
Second term (mean) — Fourth term (extreme)

Another way of writing this proportion is:

⎯Extremes⎯
$3\frac{5}{8} : 7\frac{1}{4} :: 1\frac{3}{8} : 2\frac{3}{4}$
⎯Means⎯

This statement is read as, $3\frac{5}{8}$ is to $7\frac{1}{4}$ as $1\frac{3}{8}$ is to $2\frac{3}{4}$.

Notice that the product of the means is $7\frac{1}{4} \times 1\frac{3}{8} = 9\frac{31}{32}$. The product of the extremes is $3\frac{5}{8} \times 2\frac{3}{4} = 9\frac{31}{32}$. This fact suggests the following rule.

> In a proportion the product of the means is equal to the product of the extremes.

We can solve a proportion with a missing term using the rule of means and extremes.

Example 1

Solve $\dfrac{7}{x} = \dfrac{21}{24}$.

Solution

$$\dfrac{7}{x} = \dfrac{21}{24}$$

$21x = 7 \times 24$ (means, extremes)

$21x = 168$

$\dfrac{21x}{21} = \dfrac{168}{21}$ (dividing by 21)

$x = 8$

Example 2

A pattern for a rectangular end table is printed in a book. The pattern measures 7″ by $10\frac{1}{2}$″. If the width of the finished table is to be $17\frac{1}{2}$″, what will be the finished length?

Solution

The proportion for the finished length is

$$\dfrac{7}{17\frac{1}{2}} = \dfrac{10\frac{1}{2}}{\ell}$$

$7\ell = 10\frac{1}{2} \times 17\frac{1}{2}$ (means, extremes)

$\ell = \dfrac{10\frac{1}{2} \times 17\frac{1}{2}}{7}$ (dividing by 7)

$\ell = 26\frac{1}{4}$

The table will be $26\frac{1}{4}$″ long.

Ratio, proportion, scale

EXERCISES Solve the following proportions.

1. $\dfrac{x}{2} = \dfrac{3}{4}$
2. $\dfrac{3}{y} = \dfrac{9}{21}$
3. $\dfrac{7}{8} = \dfrac{\ell}{24}$
4. $\dfrac{3}{8} = \dfrac{12}{p}$
5. $\dfrac{x}{3\frac{1}{5}} = \dfrac{6}{16}$
6. $\dfrac{5}{4\frac{1}{2}} = \dfrac{y}{27}$
7. $\dfrac{\frac{1}{2}}{400} = \dfrac{a}{100}$
8. $\dfrac{\frac{3}{4}}{150} = \dfrac{2}{s}$

9. A watch is shown here twice the actual size. Use proportions to find the dimensions of parts A, B, and C.

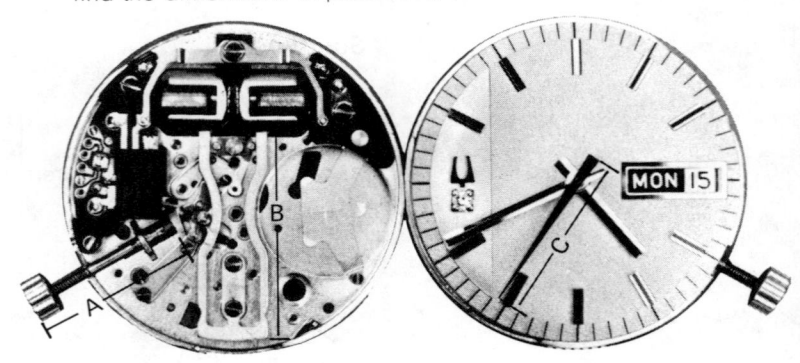

10. The plan for a 12′ by 14′ rectangular room is drawn 18″ by 21″. Use a proportion to find the dimensions of a closet shown in the drawing as $4\frac{1}{2}$″ by 9″.

11-4 INVERSE PROPORTIONS

The speeds at which a pair of gears turn are related to the number of teeth in the gears. In Figure 11-6 the ratio of the number of teeth in gear A to the number of teeth in gear B is 12:25 or $\frac{12}{25}$.

Gear A will make more revolutions per minute (rpm) than gear B. In particular, gear A will turn 25 times while gear B turns just 12 times. The ratio, $\frac{25}{12}$, is the *reciprocal* of the ratio of the numbers of teeth. It is called the gear ratio.

The following formula describes the relationship:

$$\dfrac{\text{Speed of gear A}}{\text{Speed of gear B}} = \dfrac{\text{Number of teeth in gear B}}{\text{Number of teeth in gear A}}$$

A proportion in which the ratios are based on reciprocal relationships is called an *inverse proportion*.

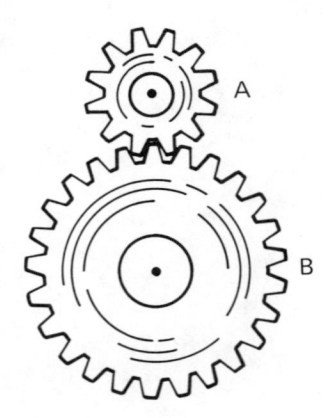

Figure 11-6

Example 1

If gear A in Figure 11-6 is turning at 300 rpm, what is the speed of gear B?

Solution

$$\frac{\text{Speed of A}}{\text{Speed of B}} = \frac{\text{Number of teeth in B}}{\text{Number of teeth in A}}$$

$$\frac{300}{s} = \frac{25}{12}$$

$$25s = 12 \times 300$$

$$25s = 3600$$

$$s = 144$$

The speed of gear B is 144 rpm.

Another example of an inverse proportion is illustrated in Figure 11-7. The speeds at which a pair of pulleys, connected by a belt, revolve are inversely proportional to the diameters of the pulleys. Thus,

$$\frac{\text{Speed of A}}{\text{Speed of B}} = \frac{\text{Diameter of B}}{\text{Diameter of A}}$$

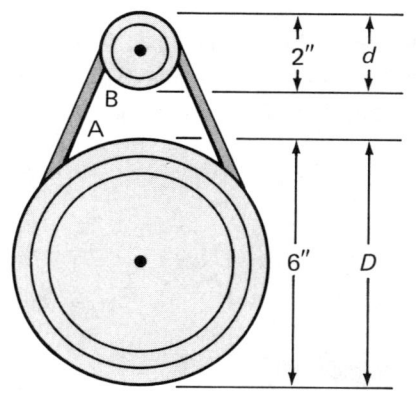

Figure 11-7

Example 2

If pulley A in Figure 11-7 is rotating at 200 rpm, what is the speed of pulley B?

Solution

$$\frac{\text{Speed of A}}{\text{Speed of B}} = \frac{\text{Diameter of B}}{\text{Diameter of A}}$$

$$\frac{200}{x} = \frac{2}{6}$$

$$2x = 1200$$

$$x = 600$$

The speed of pulley B is 600 rpm.

A third example of an inverse proportion is shown in Figure 11-8. In an automobile engine the rocker arm is used to change the upward force of the push rod to a downward force on the valve. The adjusting nut serves as a *fulcrum*, or pivot point, for the rocker arm.

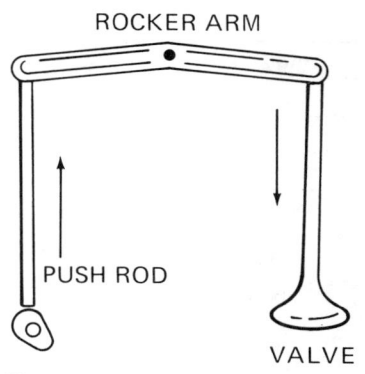

Figure 11-8

Ratio, proportion, scale 317

The proportion below describes the operations of the valve system:

$$\frac{\text{Upward force on push rod}}{\text{Downward force on valve}} = \frac{\text{Pivot-to-valve distance}}{\text{Pivot-to-push-rod distance}}$$

Can you see why this is an inverse proportion?

Example 3

Find the downward force on the valve in Figure 11-8. The upward force of the push rod is 220 lb. The valve stem is .875" from the pivot point. The push rod seat is 1.0625" from the pivot.

Solution

The proportion will be

$$\frac{220}{F} = \frac{.875}{1.0625}$$

$.875F = 220 \times 1.0625$ (means, extremes)

$$F = \frac{220 \times 1.0625}{.875}$$ (dividing by .875)

$F = 267.14$ (approx.)

The downward force on the valve is 267.14 lb.

EXERCISES Copy and complete the table. Use the formula in Example 1.

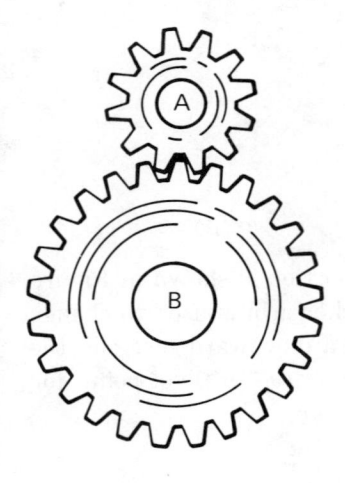

	Teeth in gear A	Teeth in gear B	Speed of gear A	Speed of gear B
1.	12	25	400 rpm	?
2.	12	24	?	1170 rpm
3.	?	15	750 rpm	350 rpm
4.	20	?	360 rpm	800 rpm
5.	15	10	320 rpm	?
6.	48	18	?	360 rpm

318 Technical and industrial formulas

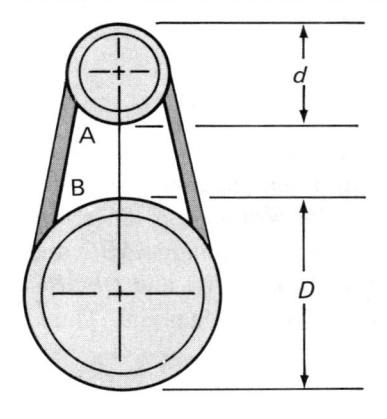

Copy and complete the table. Use the formula in Example 2.

	Diameter of pulley A	Diameter of pulley B	Speed of pulley A	Speed of pulley B
7.	4"	?	300 rpm	400 rpm
8.	4¼"	8½"	?	350 rpm
9.	15"	10"	350 rpm	?
10.	45 cm	30 cm	350 rpm	?
11.	15.3 cm	45.9 cm	?	240 rpm
12.	?	15.5 cm	600 rpm	750 rpm

U.S. system

13. The motor for a table saw runs at 1750 rpm and has a pulley 4" in diameter. The saw blade is turned by a shaft with a pulley 6" in diameter. At what speed does the saw blade revolve?

14. A 28-tooth gear is set on the shaft of a motor running at 1200 rpm. It meshes with a 64-tooth gear. What is the speed of the second gear?

15. A machinist must replace a broken pressure spring on a printing press. The spring is activated by an arm that receives 360 lb of force. The location of the pivot point on the arm is shown. What must be the strength of the replacement spring in order to "balance" the force of the arm?

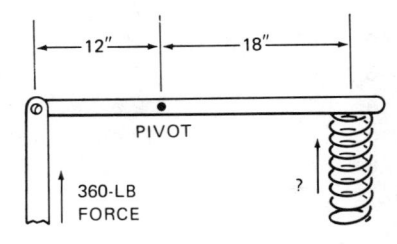

Metric system

16. The motor used to run a drill press turns at 1950 rpm. It has a pulley 10 cm in diameter. The drill is turned by a shaft with a pulley 15 cm in diameter. At what speed does the drill bit revolve?

17. A 42-tooth gear is on the shaft of an electric motor running at 3600 rpm. The gear meshes with a 70-tooth gear. What is the speed of the second gear?

18. To replace a defective valve spring, a mechanic measures the upward force of the push rod and also measures the distance from the pivot point of the rocker arm to the valve stem and to the push rod set. What should be the strength of the spring?

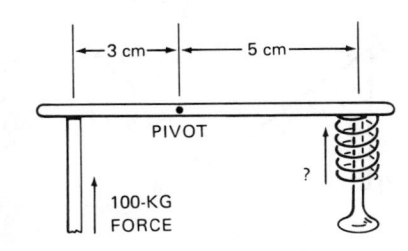

Ratio, proportion, scale 319

Similar figures

11-5 FINDING LENGTHS WITH SIMILAR FIGURES

In Chapter 10 you learned about similar triangles like those shown in Figure 11-9. Recall that in similar figures the corresponding sides have equal ratios. By corresponding sides, we mean the sides opposite the equal angles. In triangles *ABC* and *DEF* the corresponding sides are *AB* and *DE*, *AC* and *DF*, *BC* and *EF*. The ratios of the corresponding sides are equal. Thus

$$\frac{AB}{DE} = \frac{AC}{DF} = \frac{BC}{EF}$$

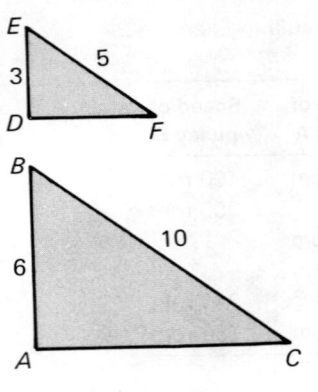

Figure 11-9

Example

A graphic arts specialist wants to make a "blow up" of an 8" by 10" photo. If the width of the blow up is to be 44", what will be the height?

Solution

The proportion for this enlargement is

$$\frac{\text{Width of original photo}}{\text{Width of enlargement}} = \frac{\text{Height of original photo}}{\text{Height of enlargement}}$$

$$\frac{8}{44} = \frac{10}{h}$$

$$8h = 44 \times 10$$

$$h = \frac{44 \times 10}{8}$$

$$h = 55''$$

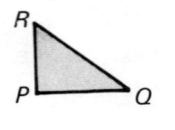

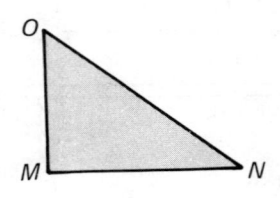

EXERCISES Find the missing lengths of the similar triangles *MNO* and *PQR*.

	PQ	PR	QR	MN	MO	ON
1.	4"	6"	7"	8"	?	?
2.	18"	15"	12"	4½"	?	?
3.	6"	7½"	5½"	9"	?	?
4.	?	?	2.5 cm	6 cm	9 cm	10 cm
5.	?	?	12 cm	8 cm	7 cm	4 cm
6.	?	?	15 m	7 cm	5 cm	3 cm

Technical and industrial formulas

11-6 FINDING AREAS WITH SIMILAR FIGURES

Figure 11-10 shows a pair of similar triangles. Would you agree that the ratio between any pair of corresponding sides is 2:3?

Using the formula for area which we studied in Chapter 7, we see that the area of triangle *ABC* is:

($\frac{1}{2}$ × 4 × 3) sq cm = 6 sq cm

The area of triangle *DEF* is:

($\frac{1}{2}$ × 6 × 4.5) sq cm = 13.5 sq cm

The ratio of the area of triangle *ABC* to the area of triangle *DEF* is $\frac{6}{13.5}$, or $\frac{4}{9}$.

Notice that if we take the squares of each pair of corresponding sides, we get the following ratios:

AB:DE = 4²:6² = 16:36 = 4:9

AC:DF = 3²:(4$\frac{1}{2}$)² = 9:$\frac{81}{4}$ = 4:9

BC:EF = 5²:(7$\frac{1}{2}$)² = 25:$\frac{225}{4}$ = 4:9

This pattern suggests the following rule.

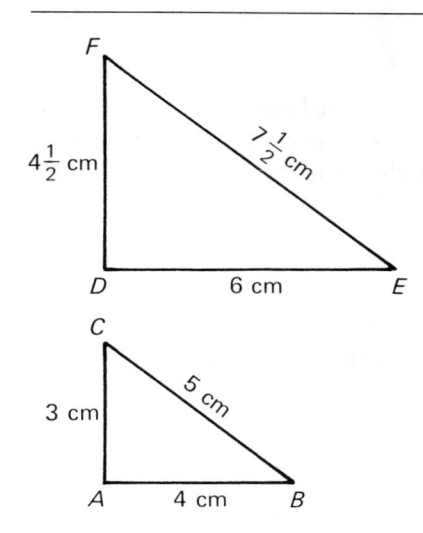

Figure 11-10

The ratio of the areas of similar figures is equal to the ratio of the squares of any pair of corresponding sides of the figures.

Example 1

What is the ratio of the areas of two circles with diameters of 2″ and 3″?

Solution

Since all circles are similar, the ratio of the areas is equal to the ratio of the squares of the diameters. The ratio is

2²:3² = 4:9

Thus, the area of the 2″ circle is $\frac{4}{9}$ of the area of the 3″ circle. Or, the area of the 3″ circle is 2$\frac{1}{4}$ times the area of the 2″ circle.

Ratio, proportion, scale

Example 2

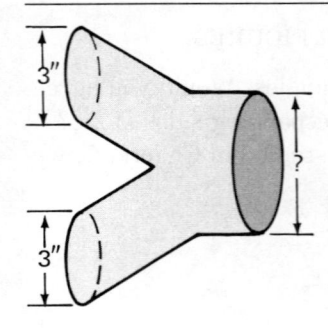

Two pipes, each 3" in diameter, are joined into a single pipe by a Y. What must be the diameter of the larger pipe in order for it to carry off the flow of the two smaller pipes?

Solution

We need to know the diameter D of a pipe whose cross-section has twice the area of a pipe 3" in diameter. Since the areas of circles are in the same ratio as the squares of their diameters, we can write the following proportion:

$$\frac{\text{(Diameter of large pipe)}^2}{\text{(Diameter of small pipe)}^2} = \frac{2}{1}$$

or

$$\frac{d^2}{3^2} = \frac{2}{1}$$

$$d^2 = 18$$

$$d = 4.25 \quad \text{(approx.)}$$

We need a pipe with a diameter of $4\frac{1}{4}''$ or more.

EXERCISES

U.S. system

1. What is the ratio between the areas of a 4" circle and a 3" circle?

2. How many times greater than the area of a 1" circle is the area of
   a. a 3" circle?
   b. a 4" circle?
   c. a $5\frac{1}{2}''$ circle?

3. A triangle has an area of 144 sq ft. One of the sides of the triangle is 9' long. Find the area of a similar triangle having a corresponding side with a length of
   a. 3'
   b. $4\frac{1}{2}'$
   c. 21'
   d. 14.5'

Metric system

8. What is the ratio between the areas of a 3-cm circle and a 5-cm circle?

9. How many times greater than the area of a 1-cm circle is the area of
   a. a 4-cm circle?
   b. a 45-cm circle?
   c. a 5.5-cm circle?

10. A triangle has an area of 72 sq cm. One of the sides of the triangle is 8 cm long. Find the area of a similar triangle having a corresponding side with a length of
    a. 2 cm
    b. 5 cm
    c. 6.5 cm
    d. 20 cm

4. What is the ratio of the areas of two squares when

   a. the ratio of the sides is $\frac{1}{2}$?

   b. the ratio of the sides is $\frac{2}{3}$?

   c. the sides are 4″ and 16″?

5. A 4″ water pipe is to be replaced by 2″ pipes. How many of the smaller pipes are needed to carry the same amount of water?

6. A circular heating duct is 12″ in diameter. How many ducts 8″ in diameter are needed to carry the same volume of air?

7. How many 1″ pipes are needed to carry as much water as one 2″ pipe?

11. What is the ratio of the areas of two squares when

    a. the ratio of the sides is 2 to 4?

    b. the ratio of the sides is 3 to 5?

    c. the sides are 8 cm and 24 cm?

12. A 10-cm water pipe is to be replaced with 4-cm pipes. How many of the smaller pipes are needed to carry the same amount of water?

13. A circular heating duct is 30 cm in diameter. How many ducts 20 cm in diameter are needed to carry the same volume of air?

14. How many 2.5-cm pipes are required to carry as much water as one 5-cm pipe?

Scales

11-7 THE ARCHITECT'S SCALE

So far in this chapter we have studied ways by which we can find dimensions if we know the size of a similar object. The methods that we studied can be used in many industrial fields, such as graphic arts, manufacturing, and so on.

A drafter who wants to find certain dimensions, as on a *scale drawing* or a blueprint, uses a special tool called an *architect's scale*. Figure 11-11 shows an architect's triangular ruler.

The architect's rule contains 10 different scales that are

Figure 11-11

frequently used in drafting. These scales, in order of increasing size, are $\frac{3}{32}$, $\frac{3}{16}$, $\frac{1}{8}$, $\frac{1}{4}$, $\frac{3}{8}$, $\frac{1}{2}$, $\frac{3}{4}$, 1, $1\frac{1}{2}$, and 3.

A section of the $\frac{1}{2}$-scale is shown in Figure 11-12.

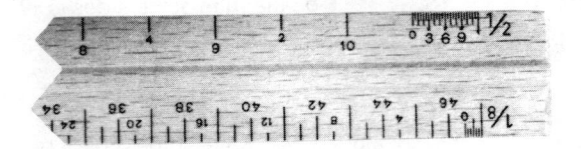

Figure 11-12

Notice that the $\frac{1}{2}$-scale runs from the right to the left across the ruler. The numbers 10, 9, 8, and 7 refer to the scale that begins at the other end of the ruler. We ignore these numbers when we are using the $\frac{1}{2}$-scale.

Example 1

Represent a length of 4′, using the $\frac{1}{2}$-scale.

Solution

To do this, we draw a line 4 units long, beginning at the 0 mark and ending at the 4 mark. Thus, in a scale drawing where all the lengths are given in the scale of $\frac{1}{2}'' = 1'$, the length of AB would represent 4′.

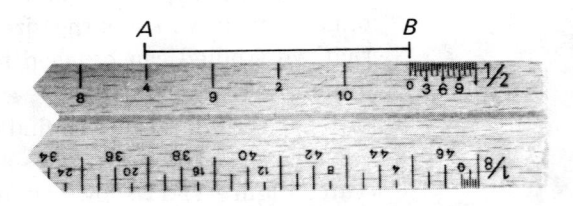

Example 2

Represent a length of 5′, using the $\frac{1}{2}$-scale.

Solution

To do this we draw a line 5 units long, beginning at the 0 mark and ending at the 8 mark. Note carefully that even though line

324 Technical and industrial formulas

CD ends at the 8 mark, CD is only 5 units long. We ignore the 8, of course, because it is part of the scale that runs from left to right.

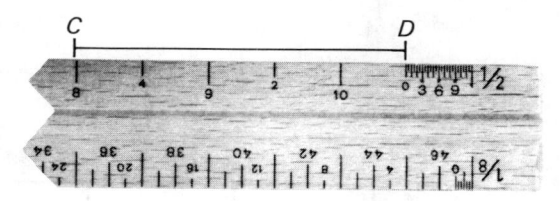

Example 3

Represent a length of 5′3″ on the ½-scale.

Solution

As in Example 2, the distance CD represents 5′. How do we represent an additional 3″ on the ½-scale? Notice the small subdivisions at the end of the scales. The numbers 3, 6, 9, 12 (not shown) represent the lengths $\frac{3}{12}$, $\frac{6}{12}$, $\frac{9}{12}$, and $\frac{12}{12}$ of the scale unit. Since 3″ is $\frac{3}{12}$ of a foot, the distance from D to E represents 3″. Thus, we show a length of 5′3″ on the ½-scale by line CE, as shown in the figure.

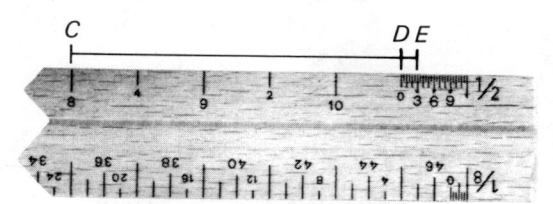

EXERCISES

1. Using the ½-scale on an architect's rule, represent the following lengths:
   a. 18′   b. 10′   c. 5′
   d. 3′   e. 7′6″   f. 9′9″

2. Using the ⅛-scale, represent the following lengths:
   a. 12′   b. 18′   c. 9′
   d. 13′   e. 6′5″   f. 2′11″

3. Using the ⅜-scale, represent the following lengths:
   a. 20'   b. 12'   c. 13'
   d. 17'   e. 9'6"   f. 19'2"
4. Assuming that Figure 11-13 was drawn using the 1-scale (1" = 1'), redraw it:
   a. Using the ¾-scale.   b. Using the ⅛-scale.
   c. Using the ½-scale.

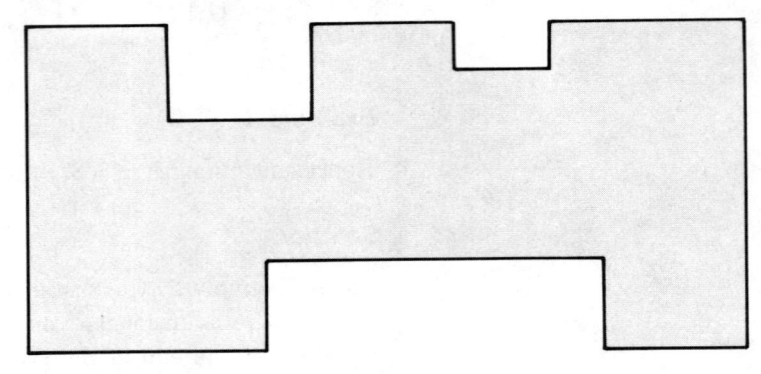

Figure 11-13

5. Using the architect's scale, make an oversize drawing of the lock plate shown, using the 1½-scale.

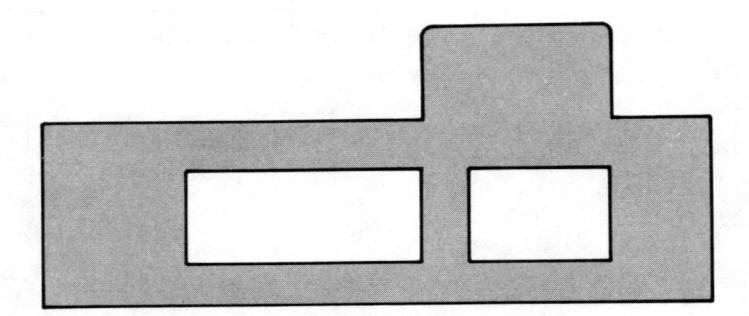

11-8 SCALE DRAWINGS

Because it is not practical to make full-size drawings or models of objects, scale drawings and scale models are used widely in industry. If the actual object is very large, it is often represented by a scale model. In other cases, a drawing must be larger than the object in order to show the important details, as with the cross-section of an American National Standard screw thread in Figure 11-14.

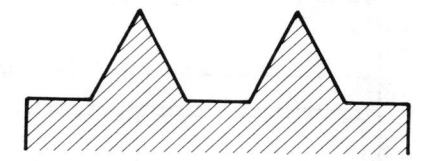

Figure 11-14

Scale drawings usually have the scale indicated on the drawing. Which scale is used depends on the space available to make the drawing and the amount of detail on the drawing.

The following rule can be helpful in preparing scale drawings.

> To find the measurements to be used in a scale drawing, multiply the dimensions of the actual object by the scale ratio.

Example 1

What lengths of lines should be used to represent dimensions A and B of the V block in a scale of $1'' = 6'''$?

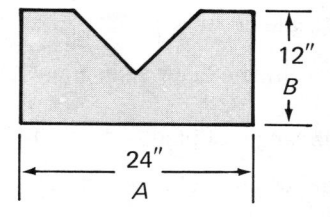

Solution

The scale ratio is $\frac{1''}{6''}$, or $\frac{1}{6}$. Therefore, each dimension in the drawing must be multiplied by $\frac{1}{6}$. Thus

Scale length of $A = \frac{1}{6} \times 24 = 4$

Scale length of $B = \frac{1}{6} \times 12 = 2$

Therefore, on the scale drawing of the V block A will be drawn $4''$ long. B will be drawn $2''$ long.

Example 2

The floor plan shown below must be drawn in a scale of $\frac{3''}{4} = 1'$. What length of lines will be used to represent the length and width of the floor plan?

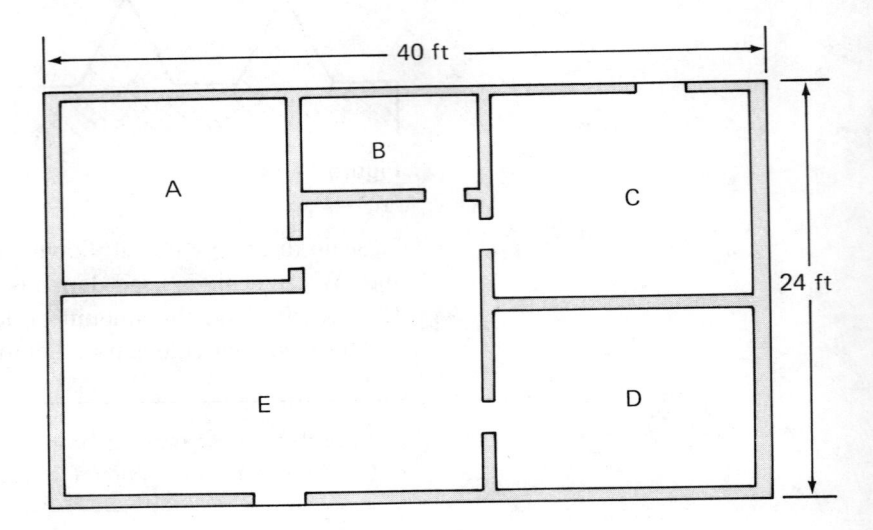

```
|←――――――― 40 ft ―――――――→|
┌─────────┬──┬──────────┐  ↑
│         │ B│          │  │
│    A    │  │    C     │  │
│         └┐ └┐         │  │
├──┐      │   │         │  │ 24 ft
│  │      │   ├─────────┤  │
│  │      │   │         │  │
│    E    │   │    D    │  │
│         │   │         │  │
└─────────┴───┴─────────┘  ↓
```

Solution

The scale ratio is $\frac{\frac{3''}{4}}{1'}$, or $\frac{\frac{3''}{4}}{12''} = \frac{1}{16}$. Therefore, each dimension shown in the floor plan must be multiplied by $\frac{1}{16}$.

Scale length of plan = $\frac{1}{16} \times 40' = 2\frac{1}{2}'$, or 30"

Scale width of plan = $\frac{1}{16} \times 24' = 1\frac{1}{2}'$ or 18"

Occasionally plans are given in a scale version only, so that they can be adapted to fit a variety of size requirements. In such cases the following rule can help us.

To find the dimensions of the actual objects from a scale drawing, multiply each length on the drawing by the *reciprocal* of the scale ratio.

Example 3

A plan for an automated assembly line is drawn in a scale of $\frac{1''}{4} = 1'$. The overall length of the assembly line on the drawing is $12\frac{1}{4}''$. How long should the actual assembly line be?

Solution

The scale ratio is $\frac{\frac{1''}{4}}{1'}$ or $\frac{\frac{1''}{4}}{12''} = \frac{1}{48}$. The reciprocal of $\frac{1}{48}$ is 48. Thus,

$$48 \times 12\frac{1}{4}'' = 48 \times \frac{49}{4} = 588'', \text{ or } 49'$$

EXERCISES Complete the following table.

	Scale	Scale ratio	Dimension being represented	Length of line on drawing
1.	$1'' = 6''$	$\frac{1}{6}$	$24''$	?
2.	$1'' = 3''$	$\frac{1}{3}$	$48''$	?
3.	$\frac{1}{2}'' = 6''$	?	$90''$	?
4.	$\frac{1}{4}'' = 6''$	?	?	$3''$
5.	$\frac{1}{8}'' = 6''$	?	?	$4\frac{1}{8}''$
6.	$\frac{1}{8}'' = 1'$	$\frac{1}{96}$	$36'$	?
7.	$\frac{1}{4}'' = 1'$	$\frac{1}{48}$	$48\frac{1}{2}'$	?
8.	$\frac{1}{16}'' = 1'$	?	$240'$	?
9.	$\frac{3}{16}'' = 1'$	?	?	$14\frac{3}{4}''$
10.	$\frac{3}{16}'' = 1'$	?	?	$22\frac{3}{8}''$

11. Measure the length of lines *A, B,* and *C.* Using the scale given for each line, find the length being represented by lines *A, B,* and *C.*

    A   Scale $\frac{1''}{4}$ to $1'$

    B   Scale $\frac{1''}{8}$ to $1''$

    C   Scale $\frac{3''}{4}$ to $1'$

12. Find the actual dimensions of rooms B, C, D, and E in the floor plan in Example 2. The scale of the plan is $\frac{3}{32}'' = 1'$.

13. The chair leg below must be drawn in a scale of $1\frac{1}{2}'' = 1'$. How long should each of the dimensions be on the scale drawing?

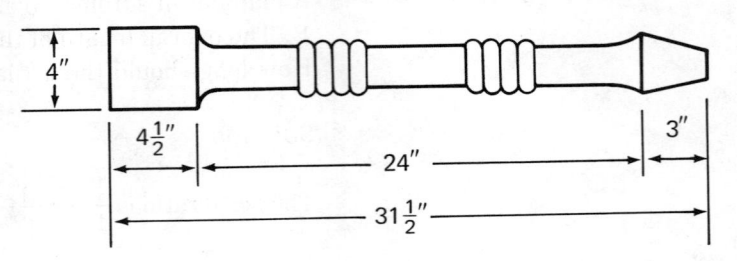

Taking inventory

1. A *ratio* is a comparison of numbers by division. (p. 309)
2. A *proportion* is a statement of two equal ratios. (p. 314)
3. In a proportion the product of the means is equal to the product of the extremes. (p. 315)
4. Corresponding sides of similar figures are proportional. (p. 320)
5. The ratio of the areas of similar figures is equal to the ratio of the squares of the corresponding sides in the figures. (p. 321)
6. The *scale* of a drawing is the ratio of a length on the drawing to the corresponding length on the actual object. (p. 327)

Measuring your skills

Express as a ratio in simplest form. (11-1)

1. 2' to 3"
2. 2.5 cm to 1.2 m
3. Find the depth of an American National Standard thread if there are 18 threads per inch. (11-2)

Solve the following proportions. (11-3)

4. $\dfrac{9}{x} = \dfrac{12}{16}$
5. $\dfrac{3}{x} = \dfrac{9}{1}$
6. $\dfrac{3}{8} = \dfrac{7}{x}$
7. $\dfrac{5}{8} = \dfrac{x}{20}$

8. Two gears have a gear ratio of 2:5. The smaller gear has 16 teeth. How many teeth does the larger gear have? (11-4)

9. For the gears in exercise 8, the larger gear revolves at a speed of 1800 rpm. What is the speed of the smaller gear? **(11-4)**

10. The sides of triangle *ABC* are 3", 4", and 5" long. The shortest side of similar triangle *DEF* is $2\frac{1}{2}$" long. What are the lengths of the other sides of *DEF*? **(11-5)**

11. The corresponding sides of a pair of similar triangles are 30 cm and 60 cm. The area of the smaller triangle is 160 sq cm. What is the area of the larger triangle? **(11-6)**

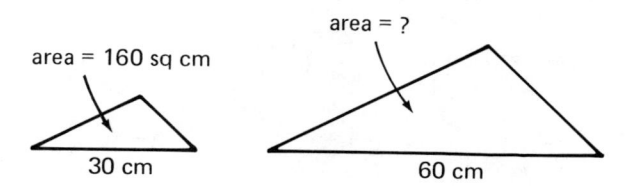

area = 160 sq cm
30 cm

area = ?
60 cm

12. Using an architect's rule, draw the following:
    a. a line representing 9'8", on the $\frac{3}{4}$-scale. **(11-7)**
    b. a line representing 6'2", on the $\frac{1}{8}$-scale. **(11-7)**

13. A drawing is made in a scale of $\frac{3}{8}$" = 1'. A line on the drawing is $7\frac{3}{8}$" long. Find the actual dimension that is being represented by the line. **(11-8)**

14. In a scale of $\frac{3}{16}$" = 1', how long a line must be drawn to represent a distance of $4'2\frac{5}{8}$"? **(11-9)**

Ratio, proportion, scale 331

12 Trigonometry

After completing this chapter, you should be able to:
1. Find the sine, cosine, and tangent of angles with measures between 0° and 180°.
2. Use the sine, cosine, and tangent ratios to solve right triangles.
3. Use the law of sines and the law of cosines to solve general triangles.

Right triangles

12-1 THE TANGENT RATIO

Figure 12-1 shows a metal plate that must be bent according to the specifications in the drawing. We want to determine the measure of $\angle A$. To aid in the solution of this problem, we introduce a function called the *tangent* of an angle.

In Figure 12-2 there is a 60° angle at A; $AB_1 = 3$, $AB_2 = 6$, and $AB_3 = 9$. We estimate $B_1C_1 = 5.2$, $B_2C_2 = 10.4$, and $B_3C_3 = 15.6$, using the blocks on the graph paper. Since triangles AB_1C_1, AB_2C_2, and AB_3C_3 are similar, we know that the ratios of corresponding sides are equal. That is,

$$\frac{B_1C_1}{AB_1} = \frac{B_2C_2}{AB_2} = \frac{B_3C_3}{AB_3}$$

Figure 12-1

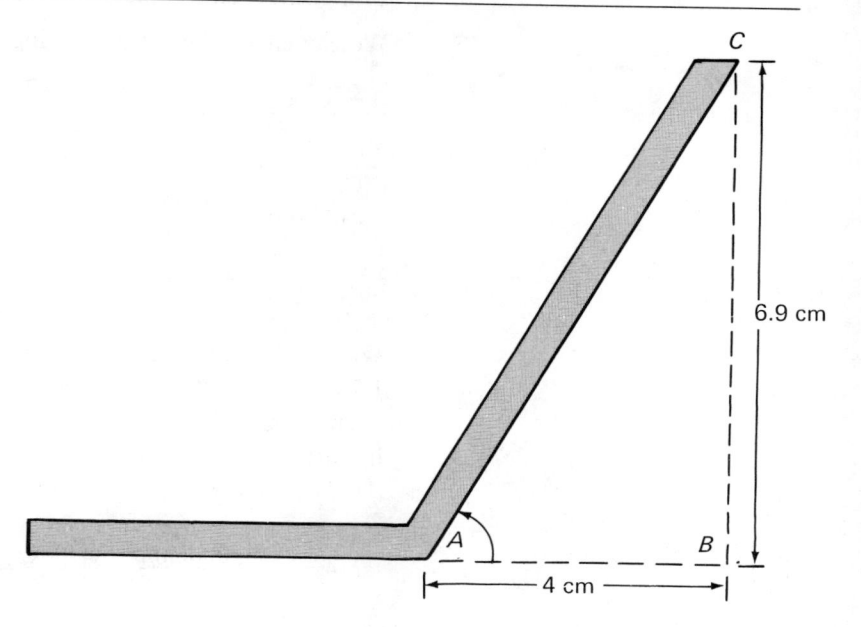

6.9 cm

4 cm

Figure 12-2

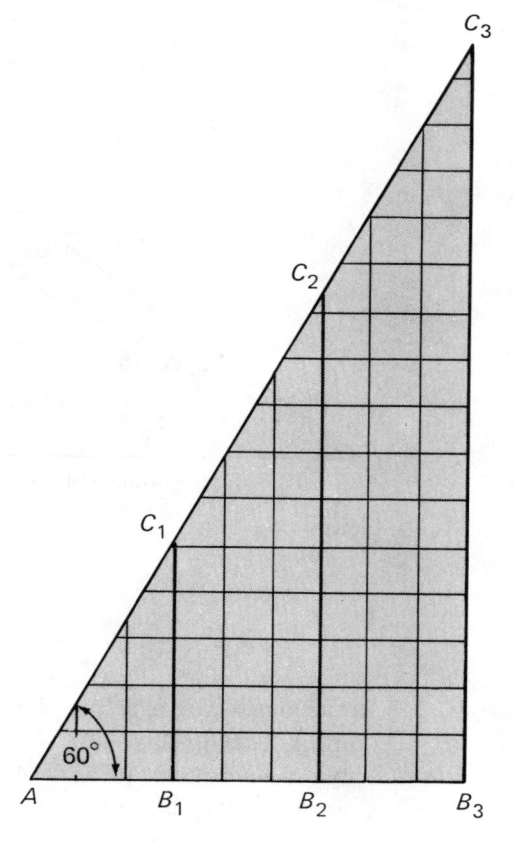

60°

Trigonometry 333

We can check these ratios using our estimates from Figure 12-2.

$$\frac{B_1C_1}{AB_1} = \frac{5.2}{3} = 1.73$$

$$\frac{B_2C_2}{AB_2} = \frac{10.4}{6} = 1.73$$

$$\frac{B_3C_3}{AB_3} = \frac{15.6}{9} = 1.73$$

We call this ratio the *tangent* of $\angle A$, and write $\tan A = \tan 60° = 1.73$.

In a right triangle, recall that the longest side is called the hypotenuse. For $\angle A$, side BC is called the *opposite side* and side AB is called the *adjacent side*. Thus, in Figure 12-3,

$$\tan A = \frac{\text{Length of opposite side}}{\text{Length of adjacent side}} = \frac{BC}{AB}$$

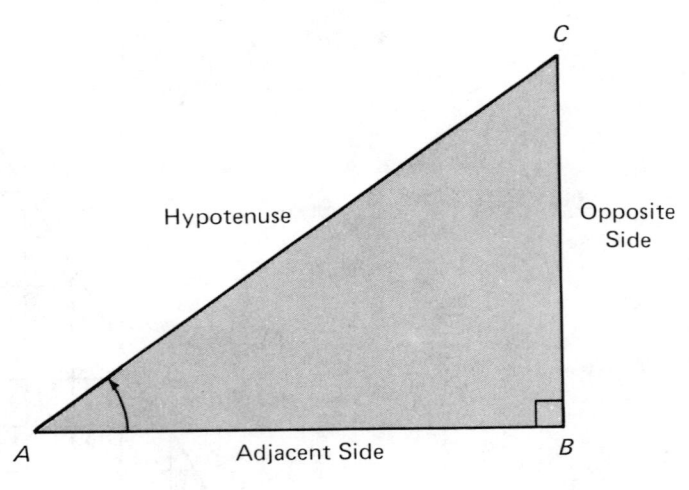

Figure 12-3

Returning to the example of Figure 12-1, $\tan A = 6.9/4 = 1.73$ (approx.), and $\angle A = 60°$.

We use a tangent table to obtain an approximation to the tangent of a particular angle.

Table 12-1 *Table of tangents*

Angle	Tangent	Angle	Tangent	Angle	Tangent
1°	.0175	31°	.6009	61°	1.8040
2°	.0349	32°	.6249	62°	1.8807
3°	.0524	33°	.6494	63°	1.9626
4°	.0699	34°	.6745	64°	2.0503
5°	.0875	35°	.7002	65°	2.1445
6°	.1051	36°	.7265	66°	2.2460
7°	.1228	37°	.7536	67°	2.3559
8°	.1405	38°	.7813	68°	2.4751
9°	.1584	39°	.8098	69°	2.6051
10°	.1763	40°	.8391	70°	2.7475
11°	.1944	41°	.8693	71°	2.9042
12°	.2126	42°	.9004	72°	3.0777
13°	.2309	43°	.9325	73°	3.2709
14°	.2493	44°	.9657	74°	3.3874
15°	.2679	45°	1.0000	75°	3.7321
16°	.2867	46°	1.0355	76°	4.0108
17°	.3057	47°	1.0724	77°	4.3315
18°	.3249	48°	1.1106	78°	4.7046
19°	.3443	49°	1.1504	79°	5.1446
20°	.3640	50°	1.1918	80°	5.6713
21°	.3839	51°	1.2349	81°	6.3138
22°	.4040	52°	1.2799	82°	7.1154
23°	.4245	53°	1.3270	83°	8.1443
24°	.4452	54°	1.3764	84°	9.5144
25°	.4663	55°	1.4281	85°	11.4301
26°	.4877	56°	1.4826	86°	14.3007
27°	.5095	57°	1.5399	87°	19.0811
28°	.5317	58°	1.6003	88°	28.6363
29°	.5543	59°	1.6643	89°	57.2900
30°	.5774	60°	1.7321	90°	undefined

Example 1

Determine the tangent of 54°.

Solution

From the table, tan 54° = 1.3764.

Example 2

Find the measure of $\angle A$ in triangle ABC in Figure 12-4.

Figure 12-4

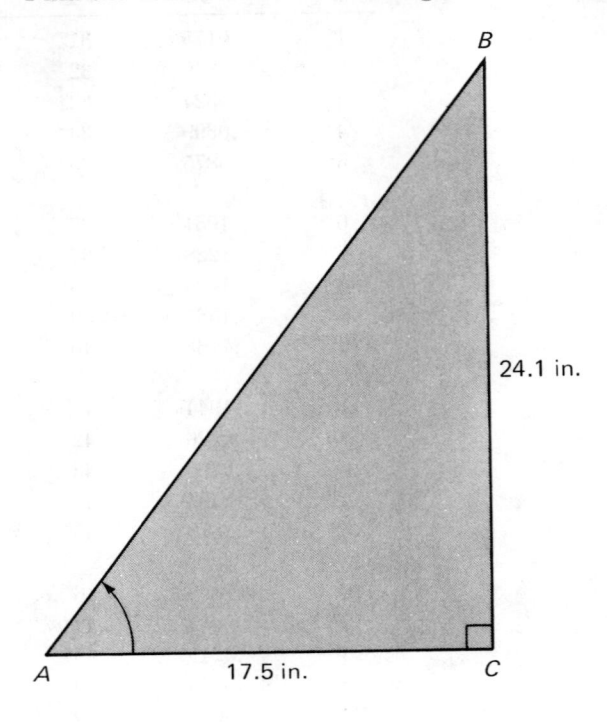

Solution

In this problem, $\tan A = \dfrac{24.1}{17.5} = 1.3771$. From the table of tangents, $\angle A = 54°$ (to the nearest degree).

Example 3

Find the length of side BC in triangle ABC in Figure 12-5.

Solution

We know that $\tan A = \dfrac{BC}{AC}$. In this problem, $\tan 36° = .7265$ and $AC = 26.1$ cm. We substitute these values in the equation and solve for BC.

$$.7265 = \dfrac{BC}{26.1}$$

$$BC = (.7265)(26.1) = 19.0$$

Figure 12-5

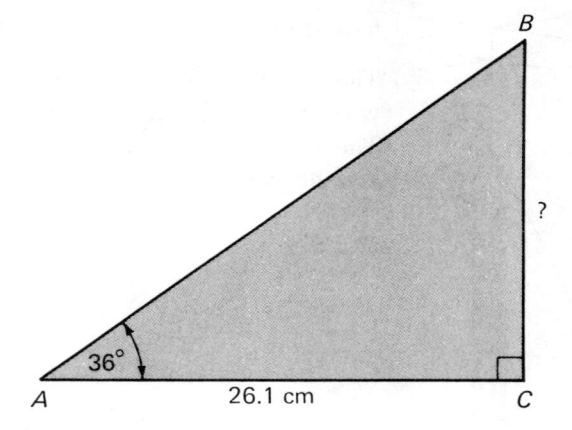

The length of BC is 19.0 cm.

Example 4

Determine the taper angle, the measure of $\angle DEA$ in Figure 12-6.

Figure 12-6

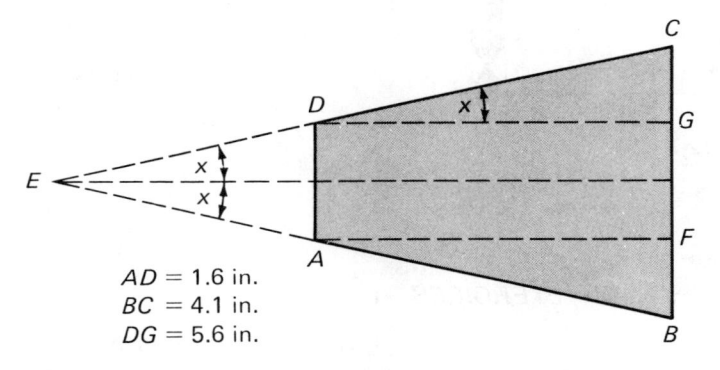

$AD = 1.6$ in.
$BC = 4.1$ in.
$DG = 5.6$ in.

Solution

The measure of $\angle DEA$ is twice the measure of $\angle CDG$. We first determine $\angle CDG$.

$$CG = \frac{1}{2}(BC - AD) = \frac{1}{2}(4.1 - 1.6) = 1.25$$

$$\tan \angle CDG = \frac{CG}{DG} = \frac{1.25}{5.6} = .2232$$

$\angle CDG = 13°$ (to the nearest degree)

$\tan 12° = .2126$ and $\tan 13° = .2309$

Since .2232 is closer to .2309 than it is to .2126,

$\angle DEA = 2(\angle CDG) = 2(13°) = 26°$

Trigonometry

Example 5

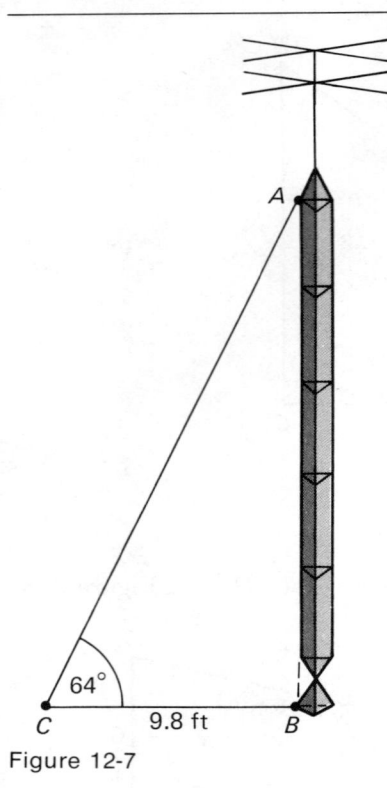

The television tower in Figure 12-7 is to be supported by brace wires anchored at a distance of 9.8 ft from the base of the tower. The angle at C measures 64°. How tall is the tower and how long are the guy wires?

Solution

a. $\tan C = \dfrac{AB}{BC}$

$AB = (BC) \tan C$

$AB = (9.8) \tan 64°$

$ = (9.8)(2.0503)$

$ = 20.1$ ft (approx.)

b. $(AC)^2 = (AB)^2 + (BC)^2$ (rule of Pythagoras)

$(AC) = \sqrt{(20.1)^2 + (9.8)^2} = \sqrt{404.01 + 96.04} = \sqrt{500.05}$

$ = 22.4$ ft (approx.)

Figure 12-7

EXERCISES

U.S. system

Using Figure 12-8 as a guide, find the missing dimensions and angles.

Metric system

Using Figure 12-8 as a guide, find the missing dimensions and angles.

Figure 12-8

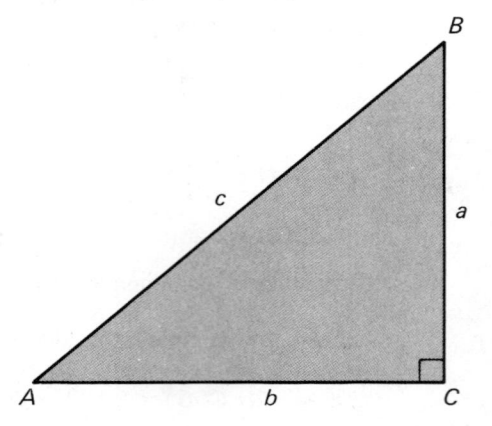

338 Technical and industrial formulas

	∠A	∠B	a	b		∠A	∠B	a	b
1.	23°	67°	10 in.	?	13.	23°	67°	25 cm	?
2.	30°	60°	14 in.	?	14.	30°	60°	36 cm	?
3.	14°	76°	7.5 ft	?	15.	14°	76°	2.3 m	?
4.	55°	35°	?	2.25 in.	16.	55°	35°	?	5.72 cm
5.	16°	74°	?	164 yd	17.	16°	74°	?	150 m
6.	70°	?	?	1.5 ft	18.	70°	?	?	45 cm
7.	?	45°	59 in.	?	19.	?	45°	150 cm	?
8.	?	?	16 in.	28 in.	20.	?	?	40 cm	71 cm

9. The roof in Figure 12-9 has a rise of 4 ft and a run of 12 ft. Find the angle that a rafter makes with the horizontal.

21. The roof in Figure 12-9 has a rise of 1.2 m and a run of 3.7 m. Find the angle that a rafter makes with the horizontal.

Figure 12-9

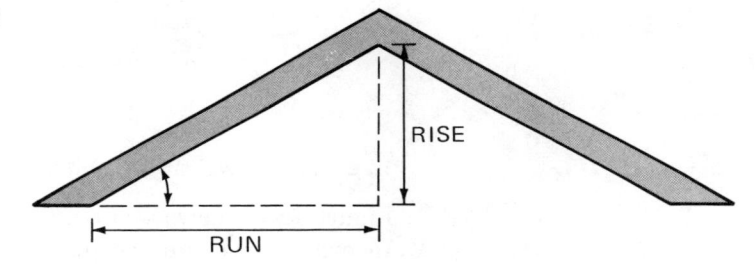

10. Find the measure of the taper angle DEA in Figure 12-6 if BC = 4 in., AD = 1¾ in., and DG = 2½ in.

11. A guy wire on a television antenna that is 20 feet tall is attached to a point 8 ft from the base of the antenna.
    a. What angle does the wire make with the ground?
    b. What angle does the wire make with the pole?
    c. How long is the wire? (Use the rule of Pythagoras.)

22. Find the measure of the taper angle DEA in Figure 12-6 if BC = 10 cm, AD = 7 cm, and DG = 5.75 cm.

23. A guy wire on a television antenna that is 6 m tall is attached to a point 2.5 m from the base of the antenna.
    a. What angle does the wire make with the ground?
    b. What angle does the wire make with the pole?
    c. How long is the wire? (Use the rule of Pythagoras.)

Trigonometry 339

12. For the hexagon in Figure 12-10, $AB = \frac{7''}{8}$ and $\angle BAC = 30°$. Determine the length of AC.

24. For the hexagon in Figure 12-10, $AB = 2.2$ cm and $\angle BAC = 30°$. Determine the length of AC.

Figure 12-10

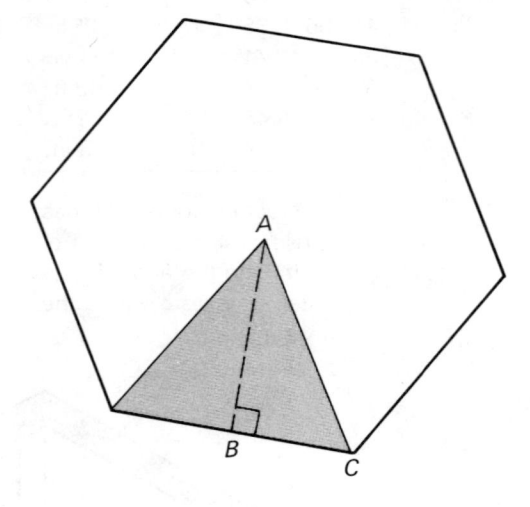

12-2 THE SINE AND COSINE RATIOS

Figure 12-11 shows a cutout to be made in a piece of sheet metal. In order to set up the punch press to make the cutout, it is necessary to determine distance AE. To solve this problem we introduce the sine and cosine ratios.

Figure 12-11

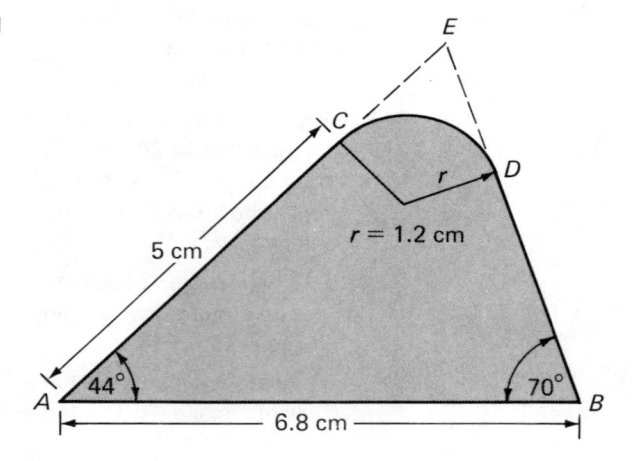

The *sine* of $\angle A$ is the ratio of the length of the side opposite $\angle A$ to the length of the hypotenuse of a right triangle. The *cosine* of $\angle A$ is the ratio of the length of the side adjacent to $\angle A$ to the

340 Technical and industrial formulas

length of the hypotenuse of a right triangle. Figure 12-12 can be used to help you to see these ratios.

Figure 12-12

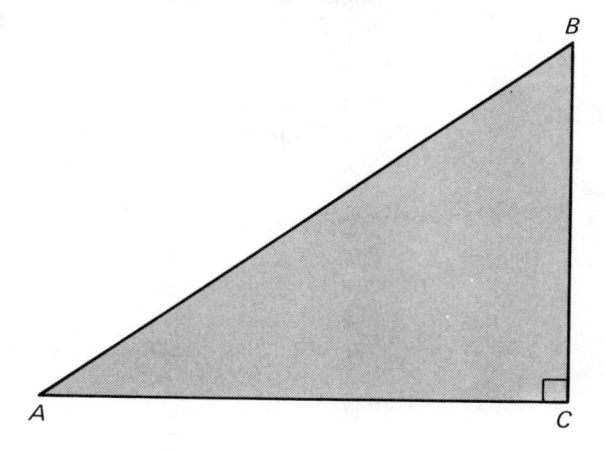

Thus,

$$\sin A = \frac{BC}{AB}, \quad \cos A = \frac{AC}{AB}$$

Similarly,

$$\sin B = \frac{AC}{AB}, \quad \cos B = \frac{BC}{AB}$$

We can use these ratios to determine unknown lengths of sides of right triangles in a way similar to the way we solved problems using the tangent ratio.

Example 1

In triangle ABC of Figure 12-12, let $AB = 13$ in., $AC = 12$ in., and $BC = 5$ in. Determine $\sin A$, and $\cos A$.

Solution

$$\sin A = \frac{BC}{AB} = \frac{5}{13} = .3846$$

$$\cos A = \frac{AC}{AB} = \frac{12}{13} = .9231$$

To determine the measure of $\angle A$, we use the table of sines and cosines.

Thus, $\angle A = 23°$ (to the nearest degree).

Table 12-2 Table of Sines, Cosines, and Tangents

Angle	Sine	Cosine	Tangent	Angle	Sine	Cosine	Tangent
1°	.0175	.9998	.0175	46°	.7193	.6947	1.0355
2°	.0349	.9994	.0349	47°	.7314	.6820	1.0724
3°	.0523	.9986	.0524	48°	.7431	.6691	1.1106
4°	.0698	.9976	.0699	49°	.7547	.6561	1.1504
5°	.0872	.9962	.0875	50°	.7660	.6428	1.1918
6°	.1045	.9945	.1051	51°	.7771	.6293	1.2349
7°	.1219	.9925	.1228	52°	.7880	.6157	1.2799
8°	.1392	.9903	.1405	53°	.7986	.6018	1.3270
9°	.1564	.9877	.1584	54°	.8090	.5878	1.3764
10°	.1736	.9848	.1763	55°	.8192	.5736	1.4281
11°	.1908	.9816	.1944	56°	.8290	.5592	1.4826
12°	.2079	.9781	.2126	57°	.8387	.5446	1.5399
13°	.2250	.9744	.2309	58°	.8480	.5299	1.6003
14°	.2419	.9703	.2493	59°	.8572	.5150	1.6643
15°	.2588	.9659	.2679	60°	.8660	.5000	1.7321
16°	.2756	.9613	.2867	61°	.8746	.4848	1.8040
17°	.2924	.9563	.3057	62°	.8829	.4695	1.8807
18°	.3090	.9511	.3249	63°	.8910	.4540	1.9626
19°	.3256	.9455	.3443	64°	.8988	.4384	2.0503
20°	.3420	.9397	.3640	65°	.9063	.4226	2.1445
21°	.3584	.9336	.3839	66°	.9135	.4067	2.2460
22°	.3746	.9272	.4040	67°	.9205	.3907	2.3559
23°	.3907	.9205	.4245	68°	.9272	.3746	2.4751
24°	.4067	.9135	.4452	69°	.9336	.3584	2.6051
25°	.4226	.9063	.4663	70°	.9397	.3420	2.7475
26°	.4384	.8988	.4877	71°	.9455	.3256	2.9042
27°	.4540	.8910	.5095	72°	.9511	.3090	3.0777
28°	.4695	.8829	.5317	73°	.9563	.2924	3.2709
29°	.4848	.8746	.5543	74°	.9613	.2756	3.4874
30°	.5000	.8660	.5774	75°	.9659	.2588	3.7321
31°	.5150	.8572	.6009	76°	.9703	.2419	4.0108
32°	.5299	.8480	.6249	77°	.9744	.2250	4.3315
33°	.5446	.8387	.6494	78°	.9781	.2079	4.7046
34°	.5592	.8290	.6745	79°	.9816	.1908	5.1446
35°	.5736	.8192	.7002	80°	.9848	.1736	5.6713

Technical and industrial formulas

Angle	Sine	Cosine	Tangent	Angle	Sine	Cosine	Tangent
36°	.5878	.8090	.7265	81°	.9877	.1564	6.3138
37°	.6018	.7986	.7536	82°	.9903	.1392	7.1154
38°	.6157	.7880	.7813	83°	.9925	.1219	8.1443
39°	.6293	.7771	.8098	84°	.9945	.1045	9.5144
40°	.6428	.7660	.8391	85°	.9962	.0872	11.4301
41°	.6561	.7547	.8693	86°	.9976	.0698	14.3007
42°	.6691	.7431	.9004	87°	.9986	.0523	19.0811
43°	.6820	.7314	.9325	88°	.9994	.0349	28.6363
44°	.6947	.7193	.9657	89°	.9998	.0175	57.2900
45°	.7071	.7071	1.0000	90°	1.0000	.0000	undefined

Example 2

In triangle ABC of Figure 12-12, let $AB = 31.3$, $BC = 13.7$, and $AC = 28.1$. Determine the measures of $\angle A$ and $\angle B$.

Solution

$$\sin A = \frac{BC}{AB} = \frac{13.7}{31.3} = .4377$$

$$\angle A = 26°$$

$$\cos B = \frac{BC}{AB} = \frac{13.7}{31.3} = .4377$$

$$\angle B = 64°$$

Example 3

For the triangle in Figure 12-13, determine the lengths of AC and BC.

Figure 12-13

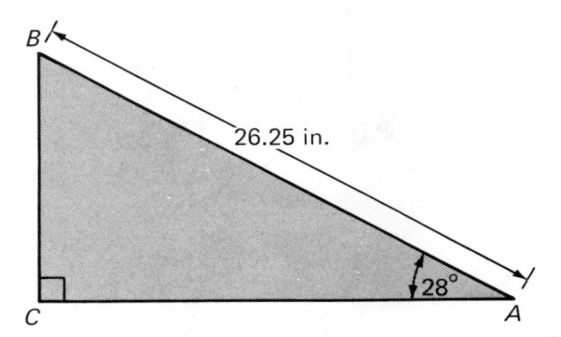

Solution

$$\sin A = \frac{BC}{AB}; \sin 28° = \frac{BC}{26.25}$$

$$BC = \sin 28° \times 26.25 = .4695 \times 26.25 = 12.32 \text{ in.}$$

$$\cos A = \frac{AC}{AB}; \cos 28° = \frac{AC}{26.25}$$

$$AC = \cos 28° \times 26.25 = .8829 \times 26.25 = 23.18 \text{ in.}$$

Example 4

Determine OE, CE, and AE for the metal cutout in Figure 12-14.

Figure 12-14

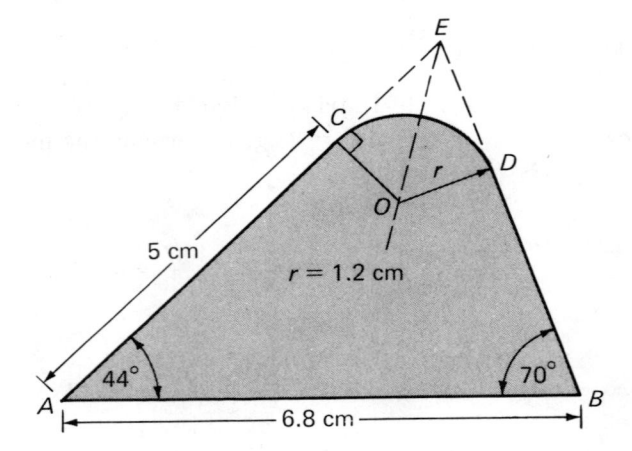

We can see that triangle OCE is a right triangle. Since point O is on the bisector of angle E,

$$\angle OEC = \frac{1}{2} \angle E$$

$$\angle E = 180° - (44° + 70°) = 66°$$

Thus,

$$\angle OEC = 33°$$

$$\sin OEC = \frac{OC}{OE}; OE = \frac{OC}{\sin OEC}$$

$$OE = \frac{1.2}{\sin 33°} = \frac{1.2}{.5446} = 2.2 \text{ cm}$$

344 **Technical and industrial formulas**

$$\cos OEC = \frac{CE}{OE}; CE = OE \times \cos OEC$$

$$CE = 2.2 \times \cos 33° = 2.2 \times .8387 = 1.85 \text{ cm}$$

$$AE = AC + CE = 5 + 1.85 = 6.85 \text{ cm}$$

Example 5

The electrical conduit shown in Figure 12-15 is bent so that $\angle A$ measures 27°. Determine the lengths of BC and AC and the measure of $\angle BCA$.

Figure 12-15

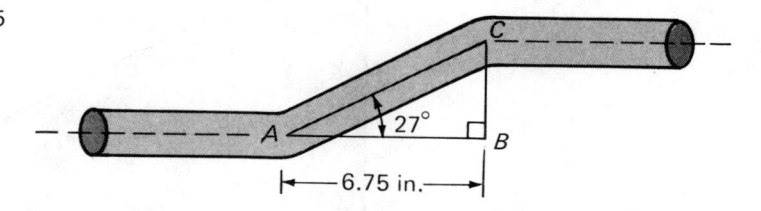

Solution

a. $\cos A = \dfrac{AB}{AC}$

$\cos 27° = \dfrac{6.75}{AC}$

$AC = \dfrac{6.75}{\cos 27°} = \dfrac{6.75}{.8910} = 7.58$ in. (approx.)

b. $\angle C = 90° - \angle A. = 90° - 27° = 63°$

c. $\cos C = \dfrac{BC}{AC}$

$\cos 63° = \dfrac{BC}{7.58}$

$BC = 7.58 \cos 63° = (7.58)(.4540) = 3.44$ in. (approx.)

EXERCISES

U.S. system

Using Figure 12-16 as a guide, find the missing dimensions and angles.

Metric system

Using Figure 12-16 as a guide, find the missing dimensions and angles.

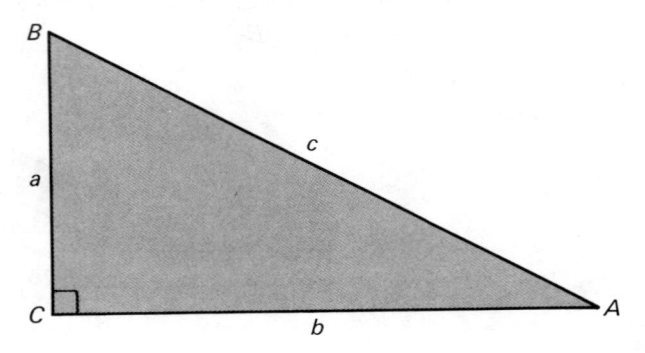

	∠A	∠B	a	b	c		∠A	∠B	a	b	c
1.	?	?	3 in.	4 in.	5 in.	11.	?	?	7.5 cm	10 cm	12.5 cm
2.	?	23°	4.8 in.	2 in.	?	12.	?	23°	12 cm	5 cm	?
3.	30°	?	?	?	3.25 in.	13.	30°	?	?	?	8.25 cm
4.	45°	?	?	4 ft	?	14.	45°	?	?	1.2 m	?
5.	?	?	28 ft	?	53 ft	15.	?	?	8.5 m	?	16.2 m
6.	?	?	19.2 in.	?	29.2 in.	16.	?	?	48 cm	?	73 cm

Figure 12-16

7. For the roof truss in Figure 12-17, calculate the measures of ∠CDE, ∠FED, and ∠ECD if DE = 12 ft, CE = 5 ft, and EF = 4.6 ft.

17. In Figure 12-17, DE = 3.7 m, CE = 1.5 m, and EF = 1.4 m. Calculate the measures of ∠CDE, ∠FED, and ∠ECD.

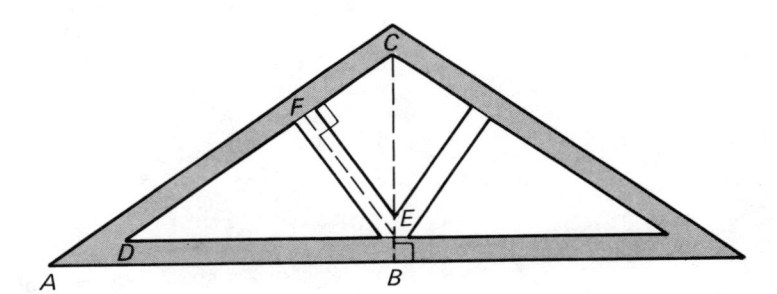

Figure 12-17

8. Calculate the measure of ∠ECD for the machine part shown in Figure 12-18.

18. Calculate the measure of ∠ECD for the machine part shown in Figure 12-19.

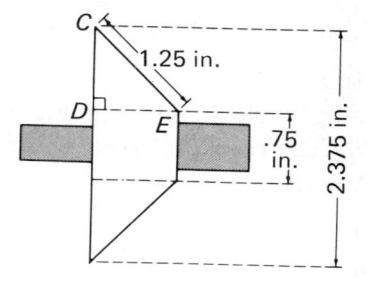

Figure 12-18

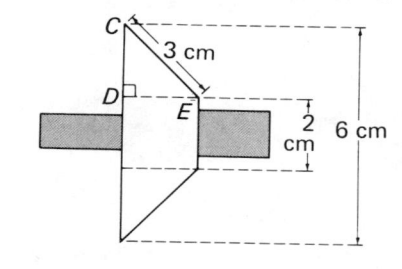

Figure 12-19

9. Calculate the groove depth *BD* for the V block in Figure 12-20.

19. Calculate the groove depth *BD* for the V block in Figure 12-21.

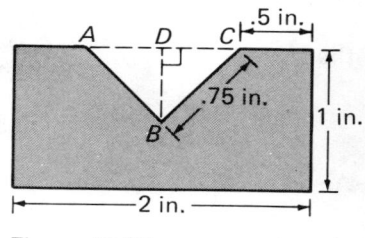

Figure 12-20

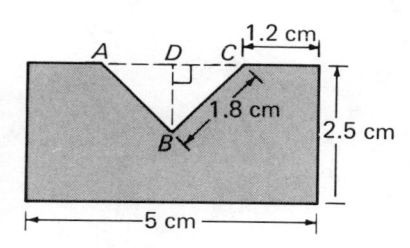

Figure 12-21

10. Calculate the length of *AB* in Figure 12-22, if *AC* = ¾ in. and ∠*BAC* = 48°.

20. Calculate the length of *AB* in Figure 12-22, if *AC* = 1.8 cm and ∠*BAC* = 48°.

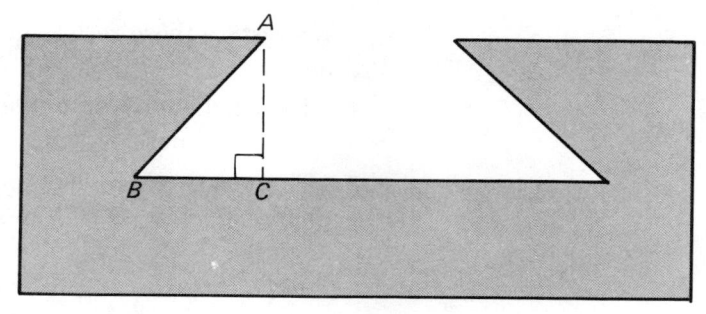

Figure 12-22

Trigonometry   347

12-3 SPECIAL TRIANGLES

In an *equilateral triangle* each of the angles measures 60°. The altitude BD of $\triangle ABC$ in Figure 12-23 is also the angle bisector of B. The resulting triangles ABD and CBD are 30°-60°-90° right triangles.

Figure 12-23

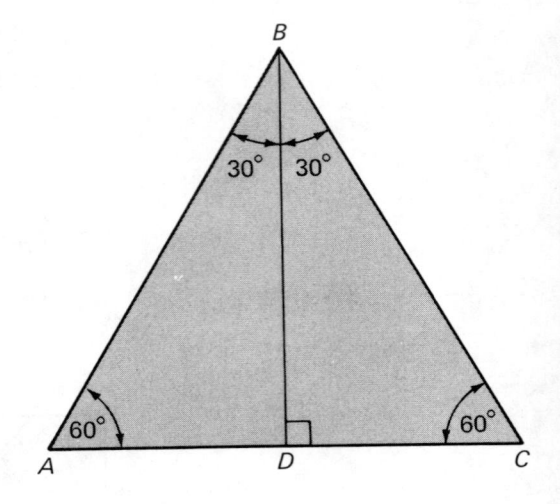

Since

$$\sin 30° = \frac{CD}{BC} = \frac{1}{2}$$

then

$$CD = \tfrac{1}{2}BC$$

Also,

$$(BC)^2 = (BD)^2 + (CD)^2$$

$$(BC)^2 = (BD)^2 + \frac{1}{4}(BC)^2$$

$$(BD)^2 = \frac{3}{4}(BC)^2$$

$$(BD) = \frac{\sqrt{3}}{2}BC$$

We restate these results in the form of rules. In a 30°-60°-90° triangle:

1. The length of the side opposite the 30° angle is $\tfrac{1}{2}$ the length of the hypotenuse.
2. The length of the side opposite the 60° angle is $\sqrt{3}/2$, or approximately .8660 times the length of the hypotenuse.

Example 1

If $BC = 14$ in. in Figure 12-23, find the lengths of DC and BD.

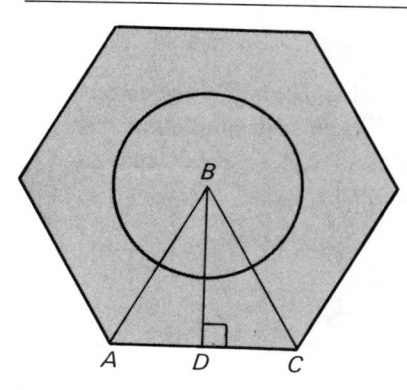

Figure 12-24

Solution

$DC = \frac{1}{2}BC = \frac{1}{2}(14) = 7$ in.

$BD = .8660(BC) = (.8660)(14) = 12.12$ in.   (approx.)

Example 2

If $BD = 37.5$ mm for the hex nut in Figure 12-24, calculate AC and BC.

Solution

$BD = .8660(BC)$ or $BC = \dfrac{BD}{.8660}$

$BC = \dfrac{37.5}{.866} = 43.3$ mm

$DC = \frac{1}{2}BC = \frac{1}{2}(43.3) = 21.65$ mm

$AC = 2DC = 2(21.65) = 43.3$ mm

Example 3

Figure 12-25 shows a layout for drilling three holes. Distance d is used to determine whether or not the holes at B and C are in the correct position, assuming that triangle ABC is isosceles.

Figure 12-25

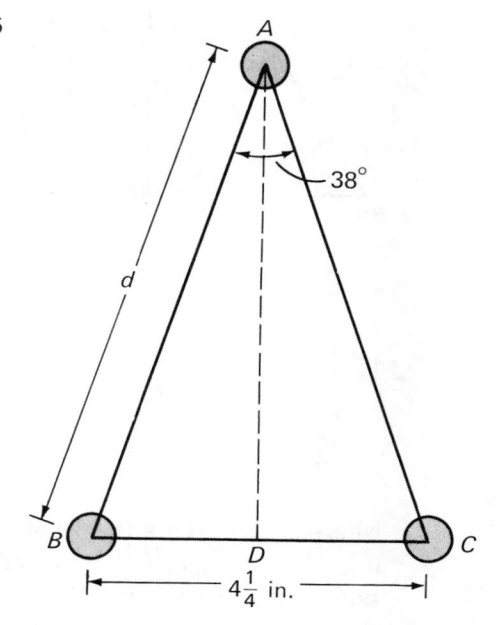

Solution

In an isosceles triangle the base angles, B and C, are congruent, and the altitude AD to the base BC is the perpendicular bisector of the base and the angle bisector of $\angle A$. $\angle BAD$ and $\angle CAD$ measure 19°, and BD and DC are $2\frac{1}{8}$ in. long.

Thus,

$$\sin BAD = \frac{BD}{AB}$$

$$\sin 19° = \frac{2\frac{1}{8}}{AB}$$

$$AB = \frac{2.125}{\sin 19°} = \frac{2.125}{.3256} = 6.526 \text{ in.}$$

For the general isosceles triangle ABC (Figure 12-26), in which $AB = AC = d$, let m represent the measure of $\angle A$.

Figure 12-26

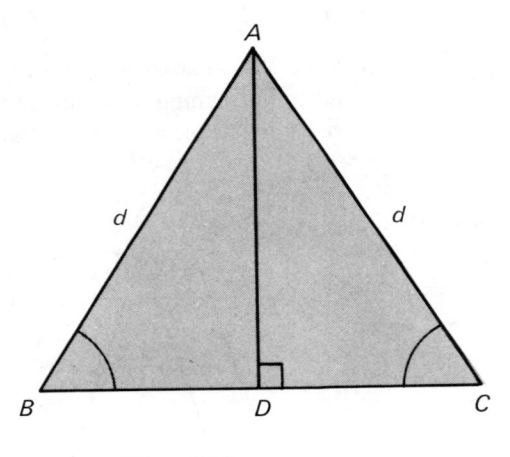

$$\sin \frac{m}{2} = \frac{BD}{AB} = \frac{\frac{1}{2}BC}{d}$$

Thus,

$$d = \frac{BC}{2 \sin (m/2)}$$

where d represents the length of one of the equal sides of an isosceles triangle, BC represents the length of the remaining side, and m represents the measure of the angle opposite the side BC.

Example 4

In triangle ABC of Figure 12-26, let $BC = 24$ cm and $A = 58°$. Find d.

Solution

$$d = \frac{BC}{2 \sin (A/2)} = \frac{24}{2 \sin 29°}$$

$$= \frac{24}{2(.4848)} = \frac{12}{.4848} = 24.75 \text{ cm}$$

An isosceles triangle that is of particular interest to us is the 45°-45° right triangle, shown in Figure 12-27.

Suppose we let x represent the length of the equal sides, BC and AC. Using the rule of Pythagoras,

$$(AB)^2 = (AC)^2 + (BC)^2$$
$$= x^2 + x^2$$
$$= 2x^2$$
$$AB = \sqrt{2x^2} = \sqrt{2}x$$

Thus,

$$\sin 45° = \sin A = \frac{BC}{AB} = \frac{x}{\sqrt{2}x} = \frac{1}{\sqrt{2}} = \frac{\sqrt{2}}{2}$$

$$\cos 45° = \frac{AC}{AB} = \frac{x}{\sqrt{2}x} = \frac{1}{\sqrt{2}} = \frac{\sqrt{2}}{2}$$

$$\tan 45° = \frac{BC}{AB} = \frac{x}{x} = 1$$

Figure 12-27

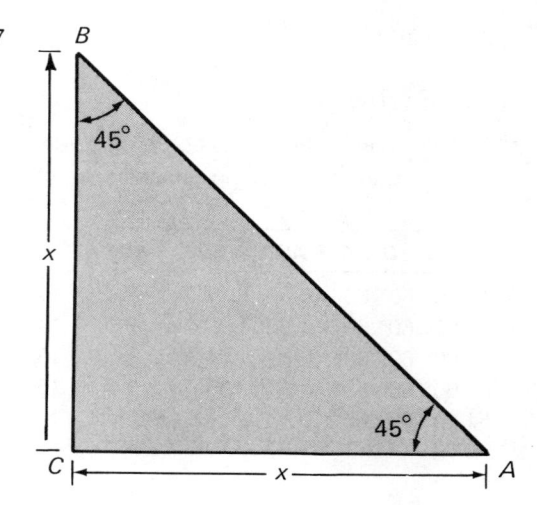

Example 5

A pipe is offset 12.5 cm with 45° bends as shown in Figure 12-28. Find the length of XY.

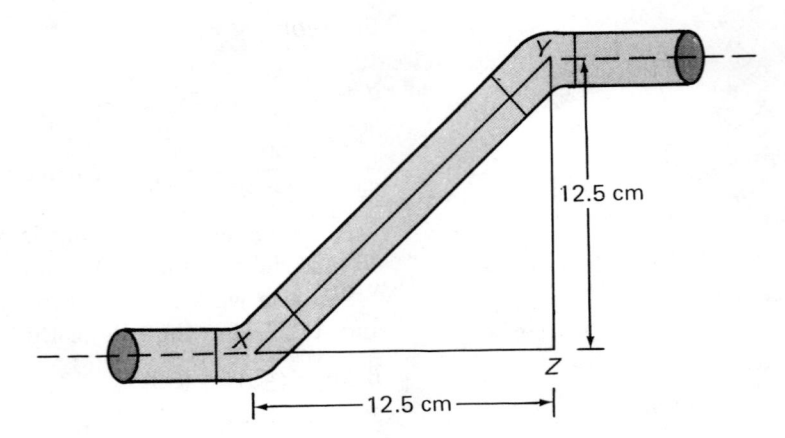

Figure 12-28

Solution

$$XY = \frac{\sqrt{2}}{2}(12.5) = .7071(12.5) = 8.84 \text{ cm} \quad (\text{approx.})$$

EXERCISES   *U.S. system*

Using Figure 12-23 as a guide, find the missing dimensions.

	∠A	∠B	∠C	AB	BD	AD
1.	60°	60°	60°	10 in.	?	?
2.	60°	60°	60°	1.5 ft	?	?
3.	60°	60°	60°	?	14 in.	?
4.	60°	60°	60°	?	100 ft	?
5.	60°	60°	60°	?	?	6.5 in.

Metric system

Using Figure 12-23 as a guide, find the missing dimensions.

	∠A, ∠B, ∠C	AB	BD	AD
13.	60°	25 cm	?	?
14.	60°	45 cm	?	?
15.	60°	?	35.6 cm	?
16.	60°	?	30 m	?
17.	60°	?	?	16.5 cm

352   Technical and industrial formulas

Using Figure 12-26 as a guide, find the missing dimensions.

	∠A	∠B	∠C	d	BC	AD
6.	?	45°	45°	20 in.	?	?
7.	?	18°	18°	1.5 ft	?	?
8.	?	28°	28°	?	10 in.	?
9.	?	70°	70°	?	?	16 in.

Using Figure 12-26 as a guide, find the missing dimensions.

	∠A	∠B	∠C	d	BC	AD
18.	?	45°	45°	50 cm	?	?
19.	?	18°	18°	45 cm	?	?
20.	?	28°	28°	?	25 cm	?
21.	?	70°	70°	?	?	41 cm

10. In Figure 12-24, if $BD = \frac{7}{8}$ in., find AC and BC.

11. In Figure 12-29, if $d = 6$ in., find the center-to-center distance between the holes.

12. In Figure 12-30, if $AB = 1.5$ in. and $BC = 1.75$ in., determine the taper angle, $\angle ACB$.

22. In Figure 12-24, if $BD = 2.2$ cm, find AC and BC.

23. In Figure 12-29, if $d = 15$ cm, find the center-to-center distance between the holes.

24. In Figure 12-30, if $AB = 12.7$ mm and $BC = 44.5$ mm, determine the taper angle, $\angle ACB$.

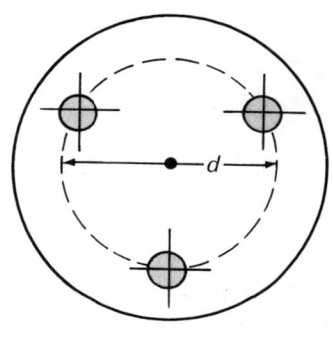

Figure 12-29

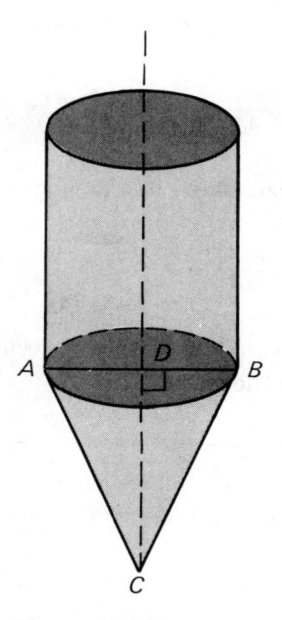

Figure 12-30

General triangles

12-4 THE LAW OF SINES

So far we have solved problems involving right triangles or special triangles like isosceles and equilateral triangles. We now consider the problem of finding the sides or the angles of a general triangle.

Figure 12-31

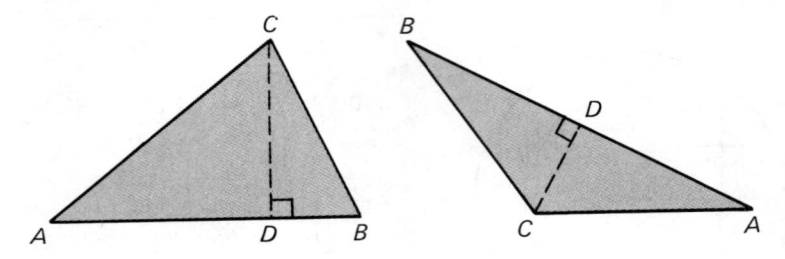

Any triangle ABC has at least one side where the point D that is the foot of the altitude to that side is on that side. For either of the right triangles, ACD and BCD, in Figure 12-31,

$$\sin A = \frac{CD}{AC}, \sin B = \frac{CD}{BC}$$

Thus, $CD = AC \sin A = BC \sin B$, or

$$\frac{\sin A}{BC} = \frac{\sin B}{AC}$$

We can show, in a similar manner, that

$$\frac{\sin B}{AC} = \frac{\sin C}{AB}$$

The equation for these results is called the *law of sines*. We restate the law as follows:

If a, b, and c represent the lengths of the sides of a triangle ABC that are opposite angles A, B, and C,

$$\frac{\sin A}{a} = \frac{\sin B}{b} = \frac{\sin C}{c}$$

For the case of the obtuse triangle in Figure 12-32, the measure of angle C is greater than 90°. In this case, the formula for the law of sines must be altered slightly.

$$\sin A = \frac{BD}{AB}, \text{ or } BD = AB \sin A$$

$$\sin (180° - \angle C) = \frac{BD}{BC}, \text{ or } BD = (BC) \sin (180° - \angle C)$$

Thus,

$$AB \sin A = BC \sin (180° - \angle C)$$

and

$$\frac{\sin A}{BC} = \frac{\sin (180° - \angle C)}{AB} = \frac{\sin B}{AC}$$

Figure 12-32

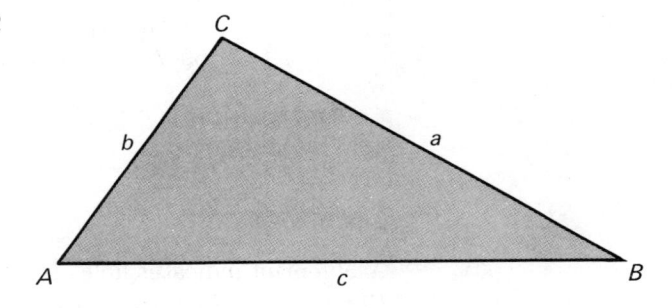

Example 1

If $\angle A = 50°$, $\angle B = 44°$, and $b = 10$ cm in Figure 12-32, find a and c.

Solution

1. We select the proportion

$$\frac{\sin A}{a} = \frac{\sin B}{b}$$

Substitute $\sin 50° = .7660$, $\sin 44° = .6947$, and $b = 10$ into the proportion and solve for a.

$$\frac{.7660}{a} = \frac{.6947}{10}$$

$$(10)(.7660) = (.6947)(a)$$

$$a = \frac{(10)(.7660)}{.6947} = 11.0 \quad \text{(approx.)}$$

The length of BC is 11.0 cm.

2. $\angle C = 180° - (44° + 50°) = 86°$

$$\frac{\sin B}{b} = \frac{\sin C}{c}$$

Substitute $\sin 44° = .6947$, $b = 10$, and $\sin 86° = .9976$ into the proportion.

$$\frac{.6947}{10} = \frac{.9976}{c}$$

$$c = \frac{(.9976)(10)}{.6947} = 14.4 \quad \text{(approx.)}$$

The length of AB is 14.4 cm.

Example 2

A blueprint indicates holes to be drilled at points A, B, and C in Figure 12-33. Determine the length of AC.

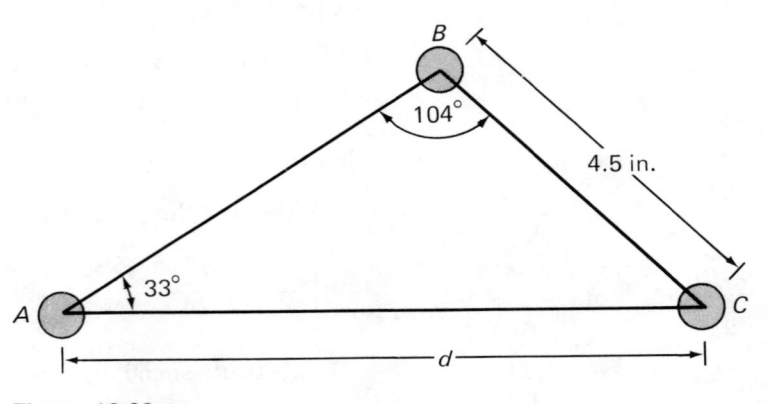

Figure 12-33

Solution

We use the proportion

$$\frac{\sin A}{a} = \frac{\sin B}{b}$$

$$\frac{\sin 33°}{4.5} = \frac{\sin 104°}{AC}$$

Thus,

$$AC = \frac{(4.5)(\sin 104°)}{\sin 33°}$$

Our table of sines does not list any angle with a measure greater than 90°. However, the sine of an angle between 90° and 180° can be found using the formula

$$\sin x = \sin (180° - x).$$

For this example, $\sin 104° = \sin (180° - 104°)$, or $\sin 104° = \sin 76° = .9703$.

$$AC = \frac{.9703 \times 4.5}{.5446} = 8.0 \text{ in.} \quad (\text{approx.})$$

We note in passing the formulas for determining the tangent and cosine of an angle between 90° and 180°.

$$\cos x = -\cos (180° - x)$$
$$\tan x = -\tan (180° - x)$$

Example 3

Determine $\cos 162°$ and $\tan 162°$.

Solution

$$\cos 162° = -\cos (180° - 72°)$$
$$= -\cos 18° = -.9511$$
$$\tan 162° = -\tan 18° = -.3249$$

Examples 1 and 2 show how to use the law of sines to determine the remaining parts of a triangle when two angles and a side are given. We can also use the law of sines when we are given the lengths of two sides and the measure of an angle opposite one of the given sides. There may, however, be more than one solution to the problem.

Example 4

In Figure 12-32, let $a = 12$, $b = 8$, and $\angle B = 30°$. Determine $\angle A$, $\angle C$, and c.

Solution

$$\frac{\sin B}{b} = \frac{\sin A}{a}$$

$$\frac{\sin 30°}{8} = \frac{\sin A}{12}$$

$$\sin A = \frac{12(.5)}{8} = .75$$

From the table of sines,

$$\angle A = 49°$$

However, we know that

$$\sin 49° = \sin (180° - 49°) = \sin 131°$$

Thus, $\angle A$ can have two possible measures, 49° and 131°. There are also two possibilities for the measure of angle C.

1. $\angle C = [180° - (30° + 49°)] = 101°$
2. $\angle C = [180° - (30° + 131°)] = 19°$

These solutions determine two possible values for c.

1. Using $\angle C = 101°$,

$$\frac{\sin 101°}{c} = \frac{\sin 30°}{8}$$

$$c = \frac{(\sin 101°)(8)}{\sin 30°} = \frac{(.9816)(8)}{.5} = 15.7$$

358   Technical and industrial formulas

2. Using $\angle C = 19°$,

$$\frac{\sin 19°}{c} = \frac{\sin 30°}{8}$$

$$c = \frac{(\sin 19°)(8)}{.5} = \frac{(.3256)(8)}{.5} = 5.2$$

EXERCISES Determine the sine, cosine, and tangent for each of the given angles.

1. $\angle A = 105°$ 2. $\angle B = 127°$ 3. $\angle C = 98°$

U.S. system

Using Figure 12-32 as a pattern, find the indicated values in the table.

Metric system

Using Figure 12-32 as a pattern, find the indicated values in the table.

	$\angle A$	$\angle B$	$\angle C$	a	b	c
4.	80°	70°	30°	10 in.	?	?
5.	62°	85°	?	?	6.2 in.	?
6.	41°	38°	?	?	?	18 in.
7.	30°	30°	?	150 ft	?	?

	$\angle A$	$\angle B$	$\angle C$	a	b	c
9.	80°	70°	30°	25 cm	?	?
10.	62°	85°	?	?	15.8 cm	?
11.	41°	38°	?	?	?	45 cm
12.	30°	30°	?	45 m	?	?

8. In Figure 12-34, holes are to be drilled at points B, C, D, E, and F. If the holes are equally spaced around a circle with a radius of 4 in., calculate BC.

13. In Figure 12-34, holes are to be drilled at points B, C, D, E, and F. If the holes are equally spaced around a circle with a radius of 10 cm, calculate BC.

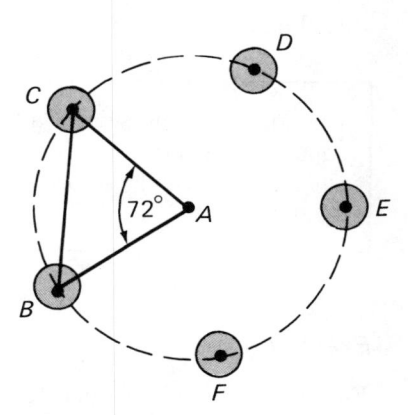

Figure 12-34

Trigonometry 359

12-5 THE LAW OF COSINES

In Section 12-4 we determined the missing parts of a triangle when two angles and a side of the triangle were given using the law of sines. We now introduce a method for determining the remaining side of a triangle when two sides and the included angle are given.

For the triangle in Figure 12-35, it can be shown that

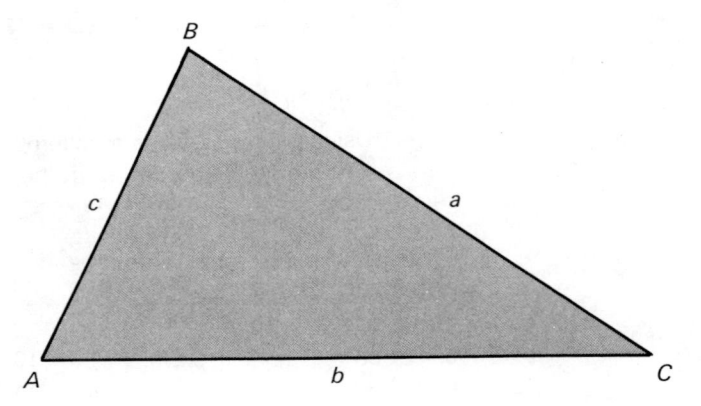

Figure 12-35

1. $c^2 = a^2 + b^2 - 2ab \cos C$
2. $b^2 = a^2 + c^2 - 2ac \cos B$
3. $a^2 = b^2 + c^2 - 2bc \cos A$

These formulas illustrate the *law of cosines*, which can be stated as follows:

> The square of any side of a triangle is equal to the sum of the squares of the other two sides minus two times the product of the other two sides and the cosine of the angle opposite the first side of the triangle.

Example 1

For the triangle in Figure 12-35, suppose $\angle A = 65°$, $b = 10$ cm, and $c = 14.5$ cm. Calculate a.

Solution

Use the formula

$$a^2 = b^2 + c^2 - 2bc \cos A$$
$$a^2 = (10)^2 + (14.5)^2 - 2(10)(14.5)(.4226)$$
$$= 100 + 210.25 - 122.55$$
$$= 187.70$$
$$a = \sqrt{187.70} = 13.7 \quad (\text{approx.})$$

The length of side BC is 13.7 cm.

Example 2

A surveyor wishes to determine the distance between two points C and D that are on the opposite side of a river without measuring the distance directly (Figure 12-36). Assume that the surveyor measures the distance between points A and B and angles 1, 2, 3, and 4 as follows: $AB = 100$ ft, $\angle 1 = 40°$, $\angle 2 = 68°$, $\angle 3 = 58°$, and $\angle 4 = 71°$.

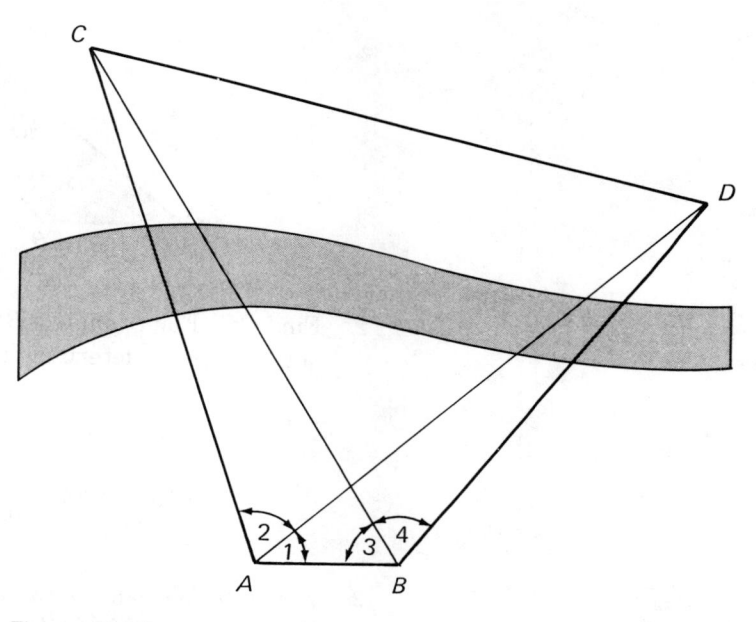

Figure 12-36

Solution

CD is a side of triangle BCD. If we can find the lengths of BC and BD, we can use the law of cosines on triangle BCD to find CD.

1. Find BD. In triangle ABD (Figure 12-37), we can use the law of sines to determine the length of BD.

$$\angle ADB = 180° - (40° + 58° + 71°) = 11°$$

$$\frac{BD}{\sin 40°} = \frac{AB}{\sin 11°} \quad \text{or} \quad BD = \frac{(AB)(\sin 40°)}{\sin 11°}$$

$$BD = \frac{(100)(.6428)}{.1908} = 337 \text{ ft} \quad (\text{approx.})$$

Figure 12-37

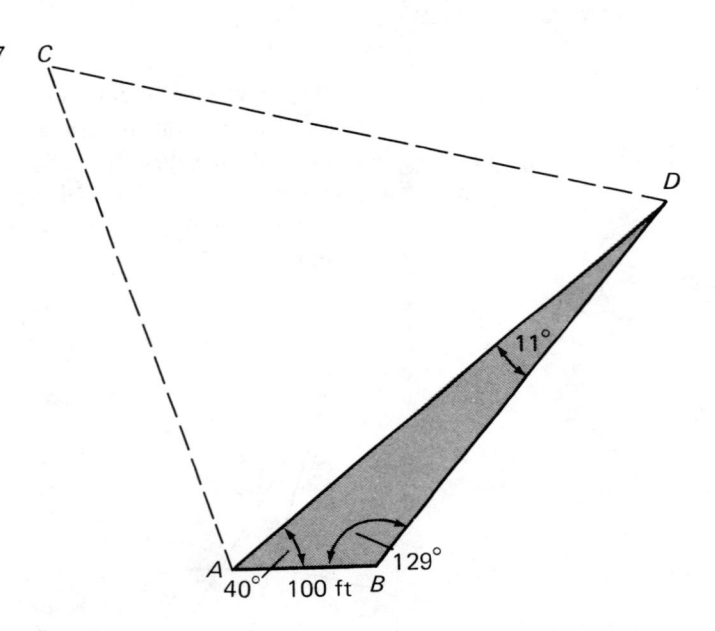

2. Find BC. For triangle ABC (Figure 12-38), we can use the law of sines to determine the length of BC.

$$\frac{BC}{\sin 108°} = \frac{AB}{\sin 14°} \quad \text{or} \quad BC = \frac{(\sin 108°)(AB)}{\sin 14°}$$

$$BC = \frac{(.9511)(100)}{.2419} = 393 \text{ ft} \quad (\text{approx.})$$

3. Find CD. We can use the law of cosines to find the length of CD, using triangle BCD (Figure 12-39).

Figure 12-38

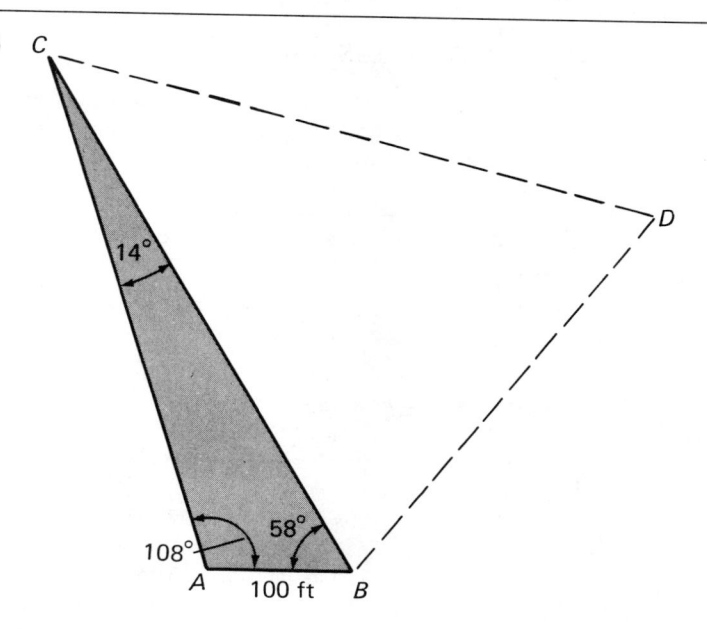

Figure 12-39

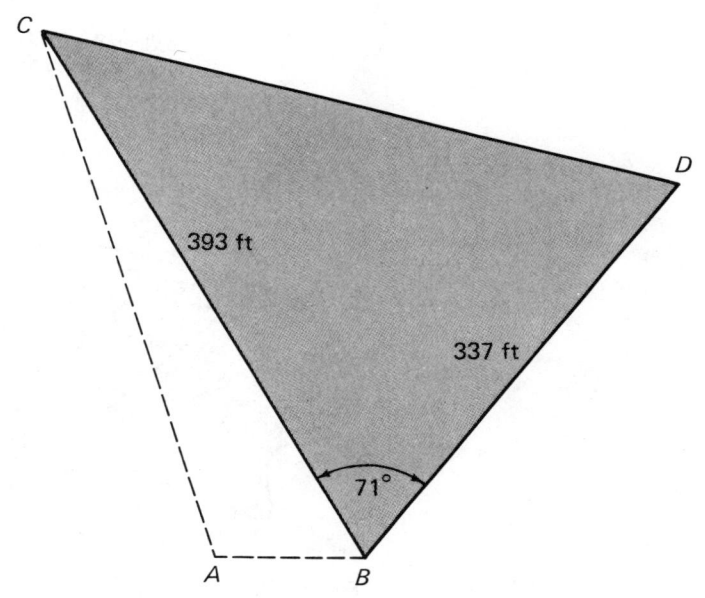

$$(CD)^2 = (BC)^2 + (BD)^2 - 2(BC)(BD)\cos 71°$$
$$= (393)^2 + (337)^2 - 2(393)(337)(.3256)$$
$$= 154449 + 113569 - 86246$$
$$= 181772$$
$$CD = \sqrt{181772} = 426 \text{ ft} \quad (\text{approx.})$$

EXERCISES

U.S. system

Using Figure 12-35 as a guide, find the indicated values in the table.

	$\angle A$	$\angle B$	$\angle C$	a	b	c
1.	80°			?	10 in.	14 in.
2.	40°			?	5 in.	6 in.
3.		60°		28 ft	?	16 ft
4.		35°		$\frac{7}{8}$ in.	?	$\frac{1}{2}$ in.
5.			67°	450 ft	180 ft	?
6.			108°	16 in.	18 in.	?

Metric system

Using Figure 12-35 as a guide, find the indicated values in the table.

	$\angle A$	$\angle B$	$\angle C$	a	b	c
9.	80°			?	25 cm	21 cm
10.	40°			?	12.5 cm	15 cm
11.		60°		9 m	?	5 m
12.		35°		22 mm	?	12 mm
13.			67°	150 m	90 m	?
14.			108°	24 cm	27 cm	?

7. A surveyor needs to determine the distance across a lake from B to C (Figure 12-40). If $\angle A$ measures 48°, AB = 480 ft, and AC = 360 ft, find BC.

15. A surveyor needs to determine distance BC across the lake in Figure 12-40. If $\angle A$ measures 48°, AB = 146.4 m, and AC = 109.8 m, find BC

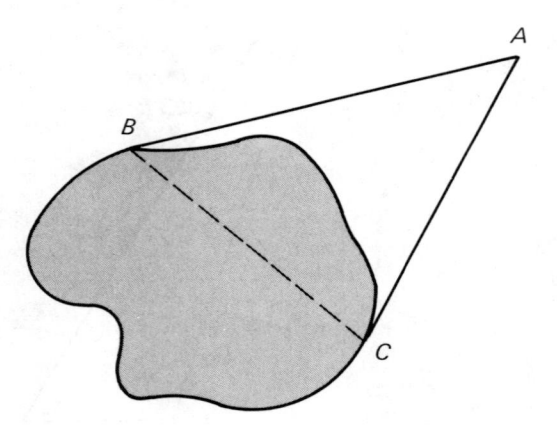

Figure 12-40

8. Using Figure 12-36, if AB = 100 ft, $\angle 1$ = 30°, $\angle 2$ = 58°, $\angle 3$ = 48°, and $\angle 4$ = 61°, find CD.

16. Using Figure 12-36, if AB = 50 m, $\angle 1$ = 30°, $\angle 2$ = 58°, $\angle 3$ = 48°, and $\angle 4$ = 61°, find CD.

Taking inventory

1. The *tangent* of an angle in a right triangle is the ratio between the lengths of the opposite side and the adjacent side of the triangle. (p. 334)
2. The *sine* of an angle in a right triangle is the ratio between the lengths of the opposite side and the hypotenuse of the triangle. (p. 340)
3. The *cosine* of an angle in a right triangle is the ratio between the lengths of the adjacent side and the hypotenuse of the triangle. (p. 340-341)
4. In a 30°-60°-90° right triangle, the length of the side opposite the 30° angle is half the length of the hypotenuse, and the length of the side opposite the 60° angle is $\sqrt{3}/2$ or approximately .8660 times the length of the hypotenuse. (p. 348)
5. If a, b, and c are the lengths of the sides of a triangle opposite angles A, B, and C, respectively, then

$$\frac{\sin A}{a} = \frac{\sin B}{b} = \frac{\sin C}{c}$$

(p. 354)

6. If a and b are the lengths of the sides forming angle C of a triangle, and c is the length of the side opposite angle C, then

$$c^2 = a^2 + b^2 - 2ab \cos \angle C$$

(p. 360)

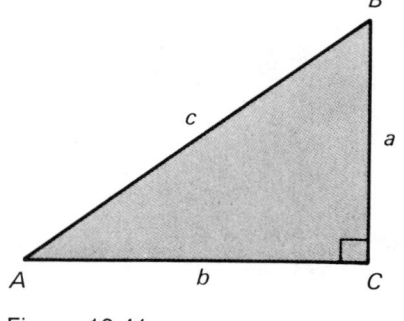

Figure 12-41

Measuring your skills

U.S. system

1. Using Figure 12-41, if a = 14 in. and b = 26 in., find tan A. (12-1)
2. Using Figure 12-41, if a = 14 in. and c = 29.5 in., find sin A. (12-2)

Metric system

11. Using Figure 12-41, if a = 3 cm and b = 14 cm, find tan A. (12-1)
12. Using Figure 12-41, if a = 3 cm and c = 14.3 cm, find sin A. (12-2)

3. Using Figure 12-41, if $b = 50$ in. and $c = 52$ in., find cos A. **(12-2)**

4. For exercises 1–3, determine the measure of $\angle A$. **(12-1, 12-2)**

5. Using Figure 12-41, if $a = 14$ in. and $\angle A = 48°$, find c and b. **(12-2)**

6. Using Figure 12-41, if $\angle A = 33°$, $\angle B = 57°$, and $c = 32$ ft, find a and b. **(12-2)**

7. If $AB = 17.5$ in. in Figure 12-42, find AC and BC. **(12-3)**

13. Using Figure 12-41, if $b = 17$ cm and $c = 36.7$ cm, find cos A. **(12-2)**

14. For exercises 11–13, determine the measure of $\angle A$. **(12-1, 12-2)**

15. Using Figure 12-41, if $a = 31$ cm and $\angle A = 48°$, find c and b. **(12-2)**

16. Using Figure 12-41, if $\angle A = 33°$, $\angle B = 57°$, and $c = 10.7$ m, find a and b. **(12-2)**

17. If $AB = 44.5$ cm in Figure 12-42, find AC and BC. **(12-3)**

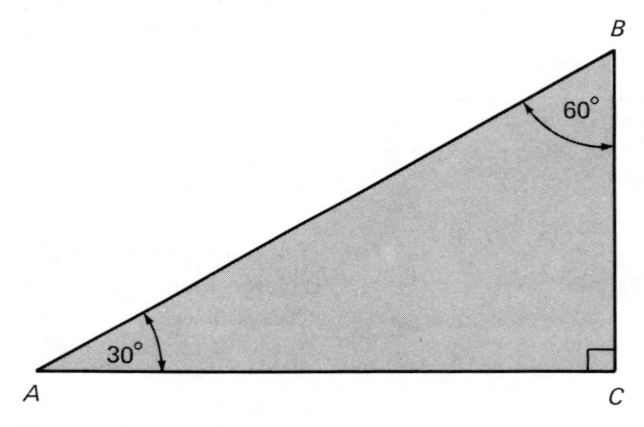

Figure 12-42

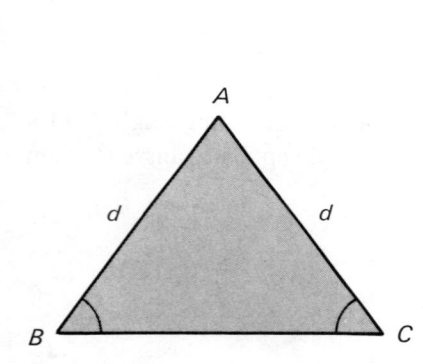

Figure 12-43

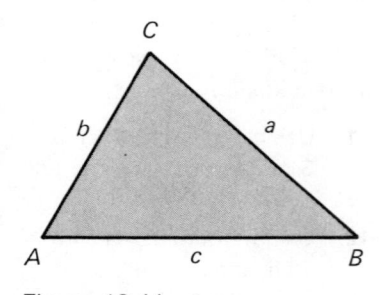

Figure 12-44

8. For the isosceles triangle in Figure 12-43, let $BC = 19.25$ in. and $\angle A = 72°$. Find d. **(12-3)**

9. For the triangle in Figure 12-44, let $b = 7.25$ in., $c = 15.5$ in., and $\angle A = 38°$. Find a. **(12-5)**

10. For the triangle in Figure 12-44, let $\angle A = 46°$, $\angle B = 72°$, and $c = 19$ ft. Find a, b, and $\angle C$. **(12-4)**

18. For the isosceles triangle in Figure 12-43, let $BC = 14.3$ cm and $\angle A = 72°$. Find d. **(12-3)**

19. For the triangle in Figure 12-44, let $b = 19.5$ mm, $c = 24.6$ mm, and $\angle A = 38°$. Find a. **(12-5)**

20. For the triangle in Figure 12-44, let $\angle A = 46°$, $\angle B = 72°$, and $c = 14$ m. Find a, b, and $\angle C$. **(12-4)**

366 Technical and industrial formulas

Measuring your progress

Simplify the expressions. (6-1, 6-2, 6-3)

1. $3x + 4x$
2. $12y - 4y$
3. 7^2
4. 4^3
5. $6 + (3 \times 9) - 8$
6. $8 \div 4 + 1 - 3$

Solve the equations. (6-4, 6-5, 6-6, 6-7, 6-8)

7. $x + 4.3 = 7.1$
8. $y - \frac{3}{8} = 1\frac{3}{16}$
9. $6s = 4.56$
10. $\frac{4}{9}x = 2\frac{1}{4}$

Substitute the values into the formulas. Then solve the equations. (6-9, 6-10)

11. $d = W/V$ where $W = 18$ and $V = 64$
12. $P = 2\ell + 2w$ where $\ell = 3.6$ and $w = 5.9$
13. Find the perimeter of a square storage area that is 300 ft on each side. (7-3)
14. Find the area of the hole made by an expansion-bit drill with a radius of 1.5 in. (7-9)
15. Find the surface area of a spherical water tank with a radius of 12.5 m. (7-10)
16. Find the volume of a cylindrical tank that has a radius of 9 ft and a height of 25 ft. (7-13)
17. Find the volume of a pyramid having a base with an area of 27 sq cm and a height of 7 cm. (7-14)

Complete the statements. (7-1, 7-2, 7-5, 7-11)

18. 2.6 m = __?__ cm
19. 10 in. = __?__ cm
20. 5 sq ft = __?__ sq m
21. 900 cu cm = __?__ cu in.
22. A truck scale has a capacity of 40,000 lb. What is its capacity in kilograms? (7-15)
23. The kindling temperature of paper is 451°F. What is this temperature in degrees Celsius? (7-17)

Perform the indicated operations. (8-1, 8-2)

24. $(3x + 2y - z) + (4x - 5y + 3z)$
25. $(4xy^2)(\frac{1}{2}x^2y^3)$
26. $(7x - 3y)(4x + 2y)$

Factor each of the polynomials. (8-3, 8-4)

27. $25x^2 - 36y^2$
28. $3x^2 - 3x - 18$

Solve the equations. (8-5, 8-6)

29. $x^2 - 2x - 80 = 0$

Trigonometry 367

30. $x^2 + 4x + 1 = 0$
31. Graph the equation $3x + 4y = 12$. (**9-1, 9-2, 9-3**)
32. Solve the pair of equations $3x + 4y = 15$, $x + 2y = 6$. (**9-4**)
33. Graph the equation $y = \dfrac{100}{x + 5}$ for x between 0 and 45. (**9-5**)
34. If two angles of a triangle measure 67° and 38°, respectively, how large is the third angle? (**10-1, 10-2, 10-3**)
35. Use a protractor to draw an angle measuring 150°. Use a compass and straightedge to divide it into two equal angles. (**10-6**)
36. A right triangle has a side 14.5 cm long and a hypotenuse 16.5 cm long. Find the length of the other side. (**10-9**)

Express as a ratio in simplest form. (**11-1, 11-2**)

37. 12.5 oz to 16 oz
38. 9 mm to 3 cm

Solve the proportions.

39. $\dfrac{x}{3.2} = \dfrac{4.6}{9.2}$ 40. $\dfrac{4}{8} = \dfrac{x}{16}$

41. A pair of gears has a 1:3 gear ratio. If the smaller gear turns at 1200 rpm, what is the speed of the larger gear? (**11-4**)
42. The sides of a triangle are 3.6 cm, 7.2 cm, and 10.8 cm long. The shortest side of a similar triangle is 5.4 cm long. What are the lengths of the other sides of the second triangle? (**11-5, 11-6**)
43. A blueprint of a factory is to be drawn to a scale of $\tfrac{3}{4}$ in. = 1 ft. How long a line should be used to represent a 28-ft hallway? (**11-7, 11-8**)

In right triangle ABC, $\angle A = 35°$ and $b = 10$ in. (**12-1, 12-2**)

44. $\angle B = \underline{\ ?\ }$ 45. $\tan A = \underline{\ ?\ }$
46. $\sin A = \underline{\ ?\ }$ 47. $a = \underline{\ ?\ }$

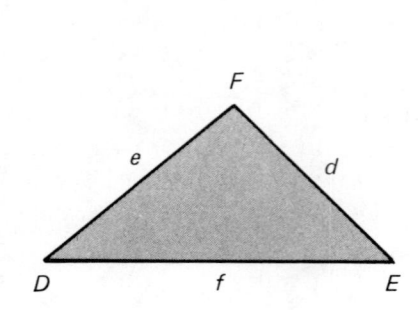

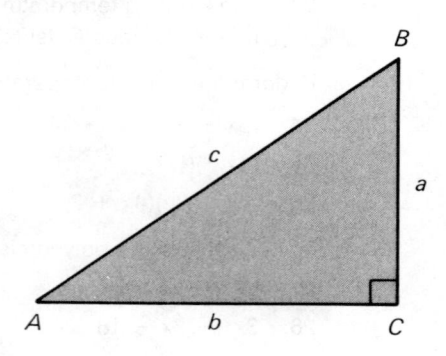

In triangle DEF, $\angle D = 40°$, $\angle E = 45°$, and $f = 15$ cm. (**12-4, 12-5**)

48. $e = \underline{\ ?\ }$ 49. $d = \underline{\ ?\ }$

368 **Technical and industrial formulas**

3 Technical and industrial applications

In Part 1 of this book you redeveloped and practiced your basic computational skills and learned how to use the hand-held calculator. The calculator made many of the computations easier and the results more reliable.

In Part 2 you expanded your knowledge of algebra, geometry, and trigonometry. With this knowledge, you could solve many additional technical and industrial problems.

Part 3 applies the mathematics that was developed in Part 1 and Part 2 to four broad technical areas. These are power and energy, construction, manufacturing, and the graphic arts. Chapter 13 develops many of the formulas that are related to measuring power and energy and applies these formulas to a variety of practical problems. Chapter 14 involves mathematical applications in the construction industry and in many of the related building trades. In Chapter 15, mathematics is applied to the field of manufacturing. The final chapter of this book, entitled "Graphic Arts," deals with applications of mathematics in the printing trades and in other aspects of the publishing business.

The hand-held calculator, which you learned to use in Chapter 5, would be a very useful device for you to have available as you work the applications that are found in these last four chapters.

13 Power and energy

After completing this chapter, you should be able to:
1. Describe power and how it is measured.
2. Describe some of the factors that affect horsepower, such as engine displacement and compression ratios.
3. Describe how power is transmitted by gears and pulleys.
4. Describe how we measure the power of hydraulic, pneumatic, and electrical systems.

Power

13-1 POWER AND ENERGY

What is it that drives our machines, heats our buildings, and runs our autos and appliances? Energy, in all its forms, is what sustains our lives and makes it possible for us to produce. *Energy* really means the ability to do work. There are many sources of energy, such as electrical, mechanical, chemical, atomic, and solar energy.

We measure energy by measuring how much work it can do. Here we are not using the everyday meaning of work, we are using a special meaning. Doing work means applying a force through a distance. So *work* is the product of the *force* times the *distance*.

Work = Force × Distance

W = f × d

A common U.S. unit of work is the *foot-pound*. You do one foot-pound of work when you move one pound a distance of one foot.

Example 1

Find the work done in lifting a 3-lb weight a distance of 2 ft.

Solution

W = f × d

   = 3 × 2

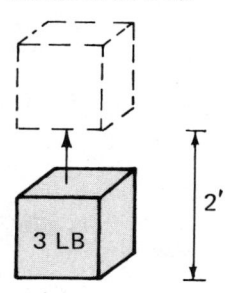

W = 6 ft-lb

The work done is 6 foot-pounds.

Sometimes the terms "power" and "work" are confused. *Power* means the rate of doing work. So power is measured in work per unit time.

$$\text{Power} = \frac{\text{Work}}{\text{Time}} \text{ or } P = \frac{f \times d}{t}$$

Sometimes power is measured in foot-pounds per second or foot-pounds per minute.

Example 2

Find the power needed to lift 100 lb of steel 12 ft in 3 sec.

Solution

$$P = \frac{f \times d}{t}$$

$$= \frac{100 \times 12}{3}$$

P = 400 ft-lb/sec

A standard U.S. unit of power is the *horsepower*. This unit was introduced by James Watt, who developed the steam engine. He wanted to compare the amount of work his engines could do with the work his horses could do. He found that an average horse could lift 550 pounds a distance of one foot in one second. This rate of doing work, 550 foot-pounds per second, he called one horsepower (hp).

1 hp = 550 ft-lb/sec

In one *minute*, Watt's average horse could do 60 times the amount of work done in one second.

550 × 60 = 33,000

1 hp = 33,000 ft-lb/min

Since 1 horsepower = 550 foot-pounds per second, we divide a power in foot-pounds per second by 550 to find the horsepower.

$$hp = \frac{ft\text{-}lb/sec}{550} \text{ or } \frac{ft \times lb}{sec \times 550}$$

If the power is given in foot-pounds per minute, we divide by 33,000 to find horsepower.

$$hp = \frac{ft\text{-}lb/min}{33,000} \text{ or } \frac{ft \times lb}{min \times 33,000}$$

Example 3

What horsepower is needed to raise 330 lb of steel 20 ft in 4 sec?

Solution

$$hp = \frac{ft \times lb}{sec \times 550}$$

$$= \frac{20 \times 330}{4 \times 550}$$

$$= \frac{6600}{2200}$$

hp = 3

We can also talk about work, force, and power in metric units. A common metric unit of work is the *newton-meter*. You do one newton-meter of work when you move one newton a distance of one meter. A newton-meter is also called a *joule*. A *dyne-centimeter* is the unit of work that results when a force of one dyne moves through a distance of one centimeter. A dyne-centimeter is also called an *erg*.

Table 13-1 shows units of length, force, work, time, and power in three different systems: the U.S. system, the MKS system, and the CGS system. The U.S. system is often called the English engineering system. The MKS system is a metric system with base units of meters, kilograms, and seconds. The CGS system is also a metric system. Its base units are centimeters, grams, and seconds.

Table 13-1 Units of length, force, work, time, and power

	English engineering	Metric MKS	Metric CGS
Length	foot	meter	centimeter
Force	pound	newton	dyne
Work	foot-pound	newton-meter (joule)	dyne-centimeter (erg)
Time	second	second	second
Power	horsepower	watt	erg per second

Example 1

Find the work done by a force of 5 newtons acting through a distance of 4 meters.

Solution

$W = f \times d$

$ = 5 \times 4$

$W = 20$ newton-meters

Example 2

How many watts of power are used when a force of 6 newtons lifts an object a distance of 3 meters in 2 seconds?

Solution

$$P = \frac{f \times d}{t}$$

$$= \frac{6 \times 3}{2}$$

$$= 9 \text{ newton-meters per second}$$

$$= 9 \text{ joules per second}$$

$$P = 9 \text{ watts}$$

EXERCISES

U.S. system

1. How much work is done when a force of 60 lb moves an object 15 ft?

2. Find the work done in lifting a 200-lb object a distance of 35 ft.

3. A force of 30 lb moves an object 8 ft in 2 sec. How much power is used?

4. Find the horsepower needed to lift a 5500-lb object 20 ft in 5 sec.

5. A gasoline engine lifts 84 lb a distance of 60 ft in 12 sec. How much power is used?

6. What horsepower is needed to raise a 2100-lb object 50 ft in 3 min? Give the answer to the nearest horsepower.

7. It takes a 50-lb force to pull a 200-lb crate 3 ft up a platform. How much work is done?

Metric system

10. How much work is done when a force of 6 newtons moves an object 5 meters?

11. Find the work done when a force of 30 newtons moves an object a distance of 10 meters.

12. A force of 5 newtons moves an object 2 meters in 2 seconds. How much power is used?

13. How many watts of power are used when a 35-newton force moves an object 6 meters in 5 seconds?

14. An electric motor moves an object 20 meters in 5 seconds by exerting a force of 32 newtons. How many watts of power are used?

15. A force of 2100 newtons moves an object 28 meters in 3 minutes. How much power is used? Give the answer to the nearest watt.

16. It takes a 70-newton force to pull a 2-kilogram crate 2 meters up a ramp. How much work is done?

8. An object was moved by a 5-lb force. How far was it moved if the work done was 350 ft-lb?

9. It takes a 5-hp engine 1 min to raise a 1650-lb car to the top of a junk pile. How high is the pile?

17. An object was moved by a 30-newton force. How far was it moved if the work done was 810 newton-meters?

18. It takes 500 watts of power to raise an object to the top of a junk pile in 1 minute. How much work was done?

13-2 MEASURING HORSEPOWER

How can the horsepower of an engine or electric motor be measured? One device that measures the power output of a machine is called a dynamometer. One way a dynamometer can measure the horsepower output of an engine is by measuring the force required to brake the engine. This method measures *brake horsepower (bhp)*, the most widely used measure of engine horsepower. One device used to measure horsepower output is a prony brake. Figure 13-1 shows a diagram of a prony brake.

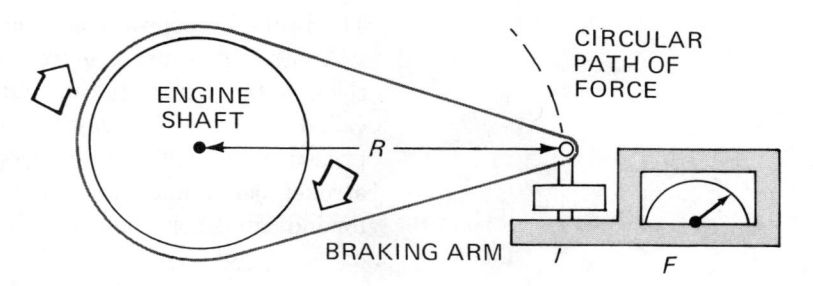

Figure 13-1

This method of finding brake horsepower uses the force needed to brake a running engine to a specific number of revolutions per minute (rpm). The engine crankshaft moves in a circular path. So the distance the crankshaft moves in one minute is the circumference ($2\pi R$) times the number of revolutions per minute (N).

Distance = $2\pi R \times N$

$$\text{bhp} = \frac{2\pi RN}{33,000} \times F$$

Power and energy 375

Since $\dfrac{2\pi}{33{,}000} = \dfrac{1}{5252}$, the formula can be written more simply as:

$$\text{bhp} = \dfrac{RNF}{5252}$$

where R = length of the arm in feet

N = rpm of engine

F = force in pounds

Example 1

If the arm is 3 ft long, the force needed to brake is 100 lb, and the engine speed is 1000 rpm, find the brake horsepower.

Solution

$$\text{bhp} = \dfrac{RNF}{5252}$$

$$\text{bhp} = \dfrac{3 \times 1000 \times 100}{5252}$$

$$\text{bhp} = 57.12$$

The brake horsepower is about 57.

Another measure of horsepower is based on the power input through the pistons to the engine. This measurement of power is called *indicated horsepower* (*ihp*). It is based on the average pressure to each square inch of piston, the number of power strokes per minute, and the length of each stroke. The formula for indicated horsepower is

$$\text{ihp} = \dfrac{PLAN}{33{,}000}$$

where P = average pressure in pounds per square inch

L = length of the stroke in feet

A = area of piston in square inches

N = number of power strokes per minute

Example 2

Find the indicated horsepower of an engine with a bore of 6″, a stroke of $2\tfrac{3}{4}''$, average pressure of 125 lb per sq in., and 6000 power strokes per minute. ($\pi = 3.14$)

Solution

First we must change to the proper dimensions:
The stroke of $2\frac{3}{4}''$ must be changed to feet. To convert inches to feet, we divide by 12.

$$2\frac{3}{4} \div 12 = \frac{11}{4} \times \frac{1}{12}$$

$$= \frac{11}{48} = .229$$

The stroke is about .23'.

The area of the piston is found by using the formula for the area of a circle. Since the bore is 6", the radius of the piston is 3".

$$A = \pi r^2$$

$$= 3.14 \times (3)^2$$

$$= 3.14 \times 9$$

$$= 28.26$$

The area of the piston is about 28.3 sq in.

So

$P = 125 \qquad A = 28.3$

$L = .23 \qquad N = 6000$

$$\text{ihp} = \frac{PLAN}{33,000}$$

$$= \frac{125 \times .23 \times 28.3 \times 6000}{33,000}$$

$$= \frac{4881.75}{33}$$

ihp = 148   (approx.)

The indicated horsepower is about 148.

EXERCISES   Find the answer to the nearest horsepower.

1. What is the brake horsepower of a gasoline engine if the arm is 3' long, the force is 300 lb, and $N = 3000$ rpm?

2. Find the brake horsepower of an engine that registers 135 lb at the end of a prony brake arm $2\frac{1}{2}$ ft long. The engine is running at 4600 rpm.

3. Find the indicated horsepower for an engine that has an average pressure of 100 lb per sq in., a stroke of .5 ft, and pistons each having an area of 20 sq in., and that makes 3000 strokes per minute.

4. Find the ihp of an engine that has an average pressure of 125 lb per sq in., a stroke of .75′, a piston with an area of 22.5 sq in., and 3500 strokes per minute.

5. Find the brake horsepower of an engine where the load on the scale is 125 lb at the end of a 2′6″ arm when the engine is running at 5500 rpm.

6. Find the indicated horsepower of an engine whose piston has an area of 33.2 sq in., a stroke of 8″, and an average pressure of 120 lb per sq in., and that makes 4000 strokes per minute.

7. Find the indicated horsepower of an engine with a bore of 6″, a stroke of $7\frac{1}{2}$″, an average pressure of 130 lb per sq in., and 3000 power strokes per minute. ($\pi = 3.14$)

Factors that affect horsepower

13-3 ENGINE DISPLACEMENT

Why are some engines more powerful than others? One factor that affects the horsepower rating of an engine is its displacement. The *displacement* of one cylinder is the amount of space through which the piston travels in one stroke. Figure 13-2 shows the relationship of the bore and the stroke to the displacement.

Figure 13-2

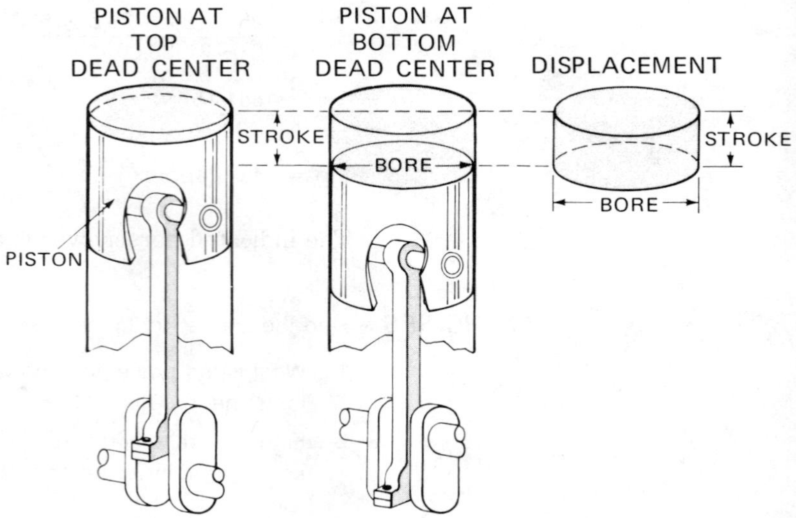

378 Technical and industrial applications

Do you see that the displacement of a cylinder is actually a cylinder of space? Recall the formula for the volume of a cylinder:

$$V = \pi r^2 h$$

Since h = the stroke of the engine and $r = \left(\dfrac{\text{Bore}}{2}\right)$, we can rewrite the formula for the displacement of a cylinder as

$$D = \pi \times \left(\dfrac{\text{Bore}}{2}\right)^2 \times \text{Stroke}$$

$$= \pi \times \dfrac{(\text{Bore})^2}{4} \times \text{Stroke}$$

$$= \dfrac{1}{4} \times \pi \times (\text{Bore})^2 \times \text{Stroke}$$

The total displacement of an engine is the sum of the displacement of all the cylinders. We may use the following formula to find the total displacement of an engine.

$$D = \tfrac{1}{4} \times \pi \times (\text{Bore})^2 \times \text{Stroke} \times \text{No. of cylinders}$$

Example 1

Mr. Porter's V-8 engine has a $3\tfrac{3}{4}''$ bore and a $3\tfrac{1}{4}''$ stroke. What is its displacement? ($\pi = \tfrac{22}{7}$)

Solution

$$D = \tfrac{1}{4} \times \tfrac{22}{7} \times \left(\tfrac{15}{4}\right)^2 \times \tfrac{13}{4} \times 8$$

$$= \tfrac{1}{4} \times \tfrac{22}{7} \times \tfrac{225}{16} \times \tfrac{13}{4} \times 8$$

$$= \tfrac{32175}{112} = 287 \text{ cu in.} \quad (\text{approx.})$$

Frequently, decimals are used to state engine dimensions, especially dimensions given in metric units. In such cases, we can use a simpler form of the displacement formula.

$$D = \tfrac{1}{4} \times \pi \times (\text{Bore})^2 \times \text{Stroke} \times \text{No. of cylinders}$$

$$= .25 \times 3.1416 \times (\text{Bore})^2 \times \text{Stroke} \times \text{No. of cylinders}$$

$$D = .7854 \times (\text{Bore})^2 \times \text{Stroke} \times \text{No. of cylinders}$$

Example 2

Find the displacement of a 4-cylinder engine with a 9.0-cm bore and a 6.7-cm stroke.

Solution

$$D = .7854 \times (9.0)^2 \times 6.7 \times 4$$
$$= .7854 \times 81 \times 6.7 \times 4$$
$$= 1705 \text{ cu cm} \quad (\text{approx.})$$

Displacement figures involving metric units are often stated in liters as well as cubic centimeters. Recall that 1 liter = 1000 cubic centimeters. Therefore, the displacement of the engine in Example 2 can be stated as

$$D = 1705 \text{ cu cm or } D = 1.705 \text{ l}$$

Sometimes we can use a flow chart to help us solve a problem. Study the flow chart for finding engine displacement below.

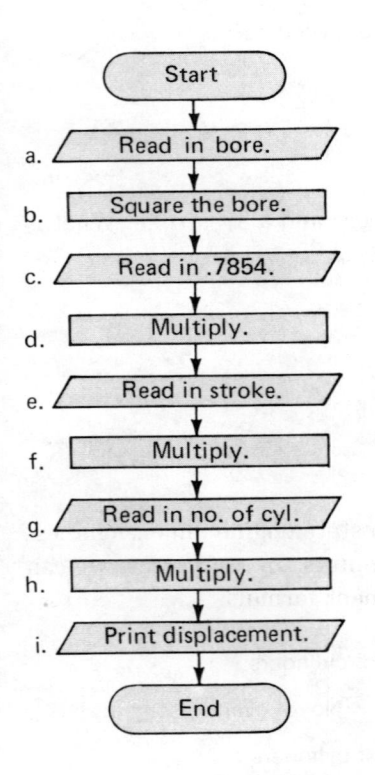

a. Read in bore.
b. Square the bore.
c. Read in .7854.
d. Multiply.
e. Read in stroke.
f. Multiply.
g. Read in no. of cyl.
h. Multiply.
i. Print displacement.

Start → End

Example 3

What is the displacement of a V-8 engine with a 3.5″ bore and a 3.25″ stroke?

Solution

a. 3.5
b. 3.5 × 3.5 = 12.25
c. .7854
d. .7854 × 12.25 = 9.62115
e. 3.25
f. 3.25 × 9.62115 = 31.2687375
g. 8
h. 8 × 31.2687375 = 250.1499
i. 250.1499 = 250 (approx.)

EXERCISES

U.S. system

Find the displacement of the following engines in cubic inches.

1. A V-8, with a 3″ bore and a 3″ stroke.
2. A 6, with a 3½″ bore and a 3″ stroke.
3. A 6, with a 4″ bore and a 3⅜″ stroke.
4. A 6-cylinder pick-up-truck engine has a displacement of 265 cu in. If the bore is 3¾″, what is the stroke?
5. Mrs. Brown's 6-cylinder engine has a bore of 3.5″ and a displacement of 255 cu in. What is the stroke?
6. An economy 6-cylinder engine has a displacement of 196 cu in. If the stroke is 2.75″, what is the bore?

Metric system

Find the displacement of the following engines (a) in cubic centimeters and (b) in liters.

7. A V-8, with a 9.0-cm bore and a 7.0-cm stroke.
8. A V-8, with a 8.3-cm bore and a 8.0-cm stroke.
9. A 6, with a 8.75-cm bore and a 7.7-cm stroke.
10. A standard 6-cylinder engine has a displacement of 3780 cu cm. If the bore is 9.8 cm, what is the stroke?
11. An optional V-8 engine has a displacement of 4770 cu cm. If the bore is 9.75 cm, what is the stroke?
12. Find the bore of a 4-cylinder engine if the displacement is 1340 cu cm and the stroke is 7.25 cm.

13-4 COMPRESSION RATIOS

Mr. Peters and Ms. Salvador each have a car with a 270 cu in. V-8 engine. Mr. Peters buys regular gas for his car, but Ms. Salvador must buy premium, higher octane gas for her car. The difference is that Mr. Peters' car has a higher compression ratio.

Recall that engine displacement is the difference between the amount of space when the piston is at the top of its stroke and the amount of space when it is at the bottom of its stroke.

Figure 13-3

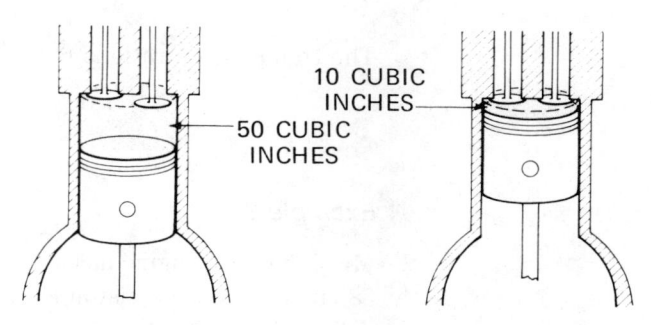

Power and energy

Figure 13-3 shows a cylinder that has a maximum of 50 cu in. at the bottom of the piston's stroke and a minimum of 10 cu in. at the top of the piston's stroke. The ratio of the maximum space in the cylinder to the minimum space in the cylinder is called the *compression ratio* of an engine. So the compression ratio of the engine in Figure 13-3 is

$$\frac{50}{10} \text{ or } \frac{5}{1}$$

A ratio of $\frac{5}{1}$ is usually written as 5 to 1 or 5:1. Compression ratios are always written so that the second number in the ratio is 1. Many car engines today have compression ratios of $8\frac{1}{2}$:1 to $9\frac{1}{2}$:1.

Note the relationship between the compression ratio and the displacement of the cylinder in Figure 13-3. The displacement in the cylinder is the difference between the two numbers used to find the compression ratio.

50 cu in. − 10 cu in. = 40 cu in.

So the displacement of the cylinder is 40 cu in.

Example 1

What is the compression ratio if there are 42 cu in. of space when the piston is at the bottom of its stroke and only 7 cu in. when the piston is at the top of its stroke?

Solution

$$\frac{42}{7} = \frac{6}{1}$$

The compression ratio is 6:1.

Example 2

Mr. Jones' car engine has a compression ratio of 9:1. If there are 38 cu cm when the piston is at the top of its stroke, what is the cylinder's displacement?

Solution

First we must find the number of cubic centimeters when the piston is at the bottom of its stroke

$$\text{Compression ratio} = \frac{\text{Maximum space}}{\text{Minimum space}}$$

$$\frac{9}{1} = \frac{\text{Maximum space}}{38}$$

$$\text{Maximum space} = 9 \times 38$$

$$= 342 \text{ cu cm}$$

$$\text{Displacement} = \text{Maximum space} - \text{Minimum space}$$

$$= 342 - 38$$

$$= 304$$

The displacement of the cylinder is 304 cu cm.

EXERCISES

U.S. system

1. A cylinder contains 40 cu in. of space when the piston is at bottom dead center and 5 cu in. when the piston is at top dead center. What is the compression ratio?

2. The compression ratio of an engine is 9:1. When the piston is at the top of its stroke, there are 6 cu in. of space in the cylinder. How much space is there when the piston is at the bottom of its stroke?

3. A cylinder has a maximum amount of 45 cu in. of space when the piston is at the bottom of its stroke. The compression ratio is 8:1. How much space is there when the piston is at the top of its stroke?

Metric system

8. A cylinder has 220 cu cm of space when the piston is at the bottom of its stroke and 22 cu cm of space when the piston is at the top of its stroke. What is the compression ratio?

9. The compression ratio is 8:1 and there are 328 cu cm of space in the cylinder when the piston is at the bottom of its stroke. How much space is there when the piston is at the top of its stroke?

10. There are 35 cu cm of space in a cylinder when the piston is at the top of its stroke. The compression ratio is 9:1. How much space is there in the cylinder when the piston is at the bottom of its stroke?

4. An engine has a compression ratio of $8\frac{1}{2}:1$. There are 4 cu in. of space in the cylinder when the piston is at the top of its stroke. How much space is there when the piston is at the bottom of its stroke?

5. There are 50 cu in. of space in the cylinder when the piston is at bottom dead center and the displacement of the cylinder is 40 cu in. How many cubic inches of space are there when the piston is at top dead center?

6. Mrs. Matthews' car engine has a compression ratio of 8:1. If one of its cylinders has 4 cu in. of space when the piston is at the top of its stroke, what is the cylinder's displacement?

7. Mr. Peron has a 6-cylinder car with a compression ratio of 9:1. In each cylinder, there are 36 cu in. of space when the piston is at the bottom of its stroke. What is the engine's displacement?

11. There are 42 cu cm of space in a cylinder when the piston is at the top of its stroke. There are 399 cu cm of space when the piston is at the bottom of its stroke. What is the compression ratio?

12. Mrs. Perella's car engine has a displacement of 280 cu cm in each cylinder. There are 40 cu cm of space in each cylinder when the piston is at the top of its stroke. How much space is there in each cylinder when the piston is at the bottom of its stroke?

13. A car engine has a compression ratio of 9:1. Each of its cylinders has a maximum of 450 cu cm of space when the piston is at the bottom of its stroke. What is the displacement in each of its cylinders?

14. Mr. Tamura's car has a displacement of 2025 cu cm in 6 cylinders. There are 37.5 cu cm of space in each cylinder when the piston is at the top of its stroke. What is the engine's compression ratio?

Gears and pulleys

13-5 GEARS

Machines must be able to transfer their energy output to the proper moving parts. One method of transferring power is to use gears.

Gears can be used to change the speed, force, and direction of motion. In *Figure 13-4*, two spur gears are shown.

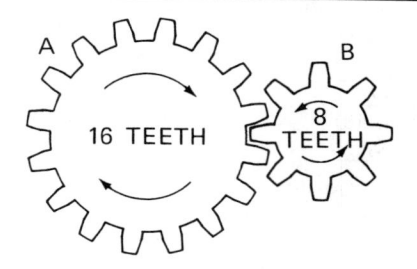

Figure 13-4

In this case, the gears produce a change in both direction and speed of the motion produced. You can see that when gear A moves clockwise, gear B moves counterclockwise. Notice that gear A has twice as many teeth as gear B. So whenever gear A makes one turn, gear B makes two turns. This means that gear B is turning twice as fast as gear A. We say that the *gear ratio* of A to B is 1 to 2. The gear ratio of B to A is 2:1. The gear ratio of two gears is the ratio of their speeds. Notice how the gear ratio is related to the number of teeth in each gear.

$$\text{Gear ratio of A to B} = \frac{\text{Speed of A}}{\text{Speed of B}} = \frac{\text{Teeth in B}}{\text{Teeth in A}}$$

So you see that the gear ratio is always the *reciprocal* of the ratio of the number of teeth.

Example 1

If gear C has 60 teeth and gear D has 20 teeth, what is the gear ratio of C to D?

Solution

The teeth ratio of C to D is $\frac{60}{20}$.
The gear ratio is the reciprocal of $\frac{60}{20}$, so it is $\frac{20}{60}$ or $\frac{1}{3}$.
The gear ratio of C to D is 1:3.

Power is transmitted from the engine to the wheels of an auto by a series of gears. There are many common gear ratios that describe how the engine speed is changed to the rear axle speed. The transmission in the automobile is a gear train, or series of connected gears, that changes the speed of the engine to the speed of the drive shaft.

The *transmission ratio* is the ratio of the engine speed to the drive shaft speed.

$$\text{Transmission ratio} = \frac{\text{Engine speed}}{\text{Drive shaft speed}}$$

A transmission ratio of 4:1 means the rpm (revolutions per minute) of the engine is 4 times the rpm of the drive shaft.

Example 2

If the engine is running at 2000 rpm and the transmission ratio is 5:1, what is the speed of the drive shaft?

Solution

$$\text{Transmission ratio} = \frac{\text{Engine speed}}{\text{Drive shaft speed(s)}}$$

$$\frac{5}{1} = \frac{2000}{s}$$

$$5 \times s = 2000 \times 1$$

$$s = \frac{2000}{5}$$

$$s = 400 \text{ rpm}$$

The *rear axle ratio* describes how the speed of the drive shaft is changed by gears to the speed of the rear axle.

$$\text{Rear axle ratio} = \frac{\text{Drive shaft speed}}{\text{Rear axle speed}}$$

The two gears involved in the rear axle are shown in Figure 13-5.

Figure 13-5

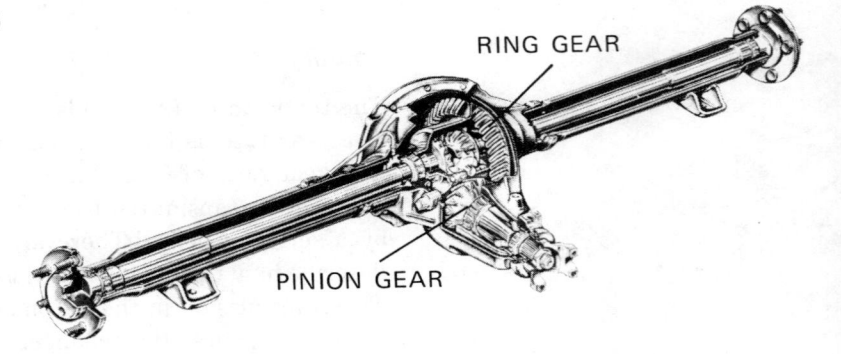

Connected to the drive shaft is the *pinion gear*, which drives the *ring gear*. The ring gear is connected to the rear axle. Recall that the gear speed ratio is the reciprocal of the ratio of the number of teeth. So another way to find the rear axle ratio is from the number of teeth in the ring gear and in the pinion gear.

$$\text{Rear axle ratio} = \frac{\text{Teeth in ring gear}}{\text{Teeth in pinion gear}}$$

Example 3

If the pinion gear has 20 teeth and the ring gear has 60 teeth, what is the rear axle ratio?

Solution

$$\text{Rear axle ratio} = \frac{\text{Teeth in ring gear}}{\text{Teeth in pinion gear}}$$

$$= \frac{60}{20}$$

$$= \frac{3}{1}$$

The rear axle ratio is 3:1.

To determine the gear ratio of the engine speed to the rear axle speed, we can multiply the transmission ratio by the rear axle ratio.

Example 4

If the transmission ratio is 4:1 and the rear axle ratio is 5:1, what is the gear ratio from the engine to the rear axle?

Solution

$$\frac{4}{1} \times \frac{5}{1} = \frac{20}{1}$$

The gear ratio from the engine to the rear axle is 20:1.

EXERCISES

1. If gear A has 20 teeth and gear B has 60 teeth, what is the gear ratio of A to B?
2. If gear A has 30 teeth and gear B has 70 teeth, what is the gear ratio of A to B? What is the gear ratio of B to A?
3. The engine speed is 2400 rpm while the drive shaft speed is 800 rpm. What is the transmission ratio?
4. The rear wheels are turning at 240 rpm while the drive shaft is turning at 1200 rpm. What is the rear axle ratio?
5. If the gear ratio of A to B is 3:1 and gear A has 9 teeth, how many teeth does gear B have?
6. The transmission ratio is 3:1. The engine speed is 1500 rpm. Find the drive shaft speed.
7. The rear axle ratio is 4:1. The rear axle is turning at 600 rpm. What is the speed of the drive shaft?
8. If the rear axle ratio is 4.2:1 and the pinion gear has 15 teeth, find the number of teeth in the ring gear.

9. Find the engine-to-rear-axle ratio if the engine speed is 4800 rpm, the drive shaft speed is 1200 rpm, and the rear axle speed is 400 rpm.

10. The engine is running at 2960 rpm. The transmission ratio is 2:1 and the rear axle ratio is 4:1. What is the speed of the rear wheels?

13-6 PULLEYS

Pulleys are another device used to transfer power from one place to another. A common use for pulleys is in lifting heavy objects. Pulleys transfer power much as gears do. Figure 13-6 shows two pulleys connected by a belt.

Figure 13-6

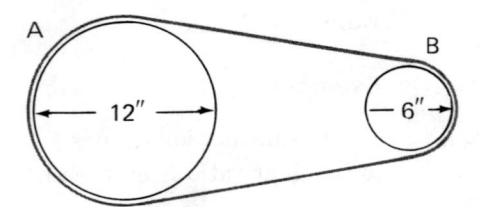

Pulley A has a diameter twice that of pulley B. Therefore the belt has to travel twice as far around pulley A as around pulley B. When pulley A makes one turn, pulley B makes two turns. *Pulley speed ratios* are like gear ratios. Recall that a gear ratio is the reciprocal of the ratio of the number of teeth. In the same way, a pulley speed ratio is the reciprocal of the ratio of the diameters of the pulleys,

$$\frac{\text{Speed of pulley A}}{\text{Speed of pulley B}} = \frac{\text{Diameter of B}}{\text{Diameter of A}}$$

Example

Pulley C is 35" in diameter and pulley D is 7" in diameter. What is the ratio of the speed of C to the speed of D?

Solution

$$\frac{\text{Speed of C}}{\text{Speed of D}} = \frac{\text{Diameter of D}}{\text{Diameter of C}}$$

$$= \frac{7}{35}$$

$$= \frac{1}{5}$$

The ratio of the speeds of C to D is 1:5.

EXERCISES

1. The diameter of one pulley is 13" and the diameter of a second pulley is 4". Find the ratio of the speed of the first pulley to the speed of the second pulley.

2. Pulley A has a 15" diameter and pulley B has a 3" diameter. Find the ratio of the speed of B to the speed of A.

3. Pulley C has a 7" radius and pulley D has a 2" radius. Find the ratio of the speed of C to that of D.

4. Pulley R has a diameter of 12" and pulley S has a radius of 3". If pulley R is turning at 500 rpm, how fast is pulley S turning?

5. One pulley turns at 100 rpm and a connected pulley turns at 25 rpm. If the diameter of the first pulley is 2", what is the diameter of the second pulley?

6. Find the rpm of the alternator pulley in Figure 13-7 if the crankshaft turns at 1200 rpm.

7. Find the rpm of the fan pulley in Figure 13-7 if the crankshaft turns at 1200 rpm.

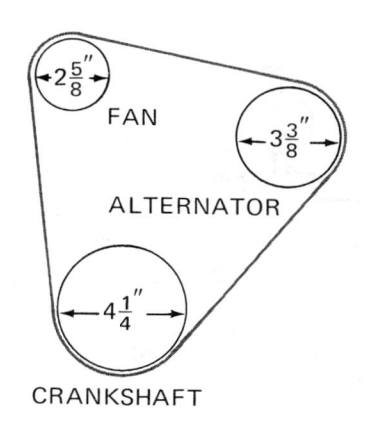

Figure 13-7

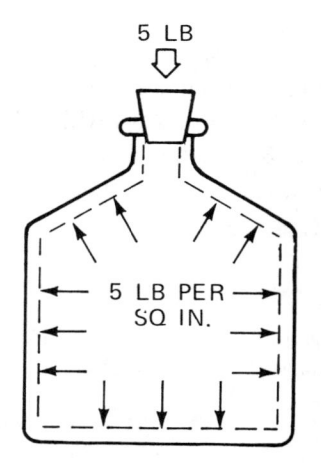

Figure 13-8

Fluid and electrical power

13-7 HYDRAULIC POWER

Fluid power systems use the fluid of liquids and gases to control and carry power. Fluid power is based on pressure applied to a confined liquid or gas. When a liquid, usually oil, transmits the power, we call it *hydraulic power*. When the power is transmitted through air or gas, it is called *pneumatic power*.

Fluid power is a fast-growing type of power system. It can produce a great force in a small space, and it is very flexible. It can carry power wherever tubes or pipes can be placed. Fluid power is used in many industrial machines.

Fluid power works because of this principle: Pressure on a confined fluid acts equally in all directions. *Pressure* is the measure of fluid force. To see how pressure is measured, look at Figure 13-8. The stopper on the bottle has an area of 1 square inch. A force of 5 pounds is pushing down on it. This pressure acts equally on all the air in the bottle. So every square inch of bottle has a force of 5 pounds pushing it out. We call this a pressure of 5 pounds per square inch, or 5 psi.

Hydraulic lifts are used in service stations to make it easier for mechanics to work on cars. To understand how a hydraulic device

Power and energy 389

works, look at Figure 13-9. An outside force of 50 pounds is pushing on the left piston. The piston has an area of 5 square inches. Each square inch receives its equal share of the 50-pound force.

Figure 13-9

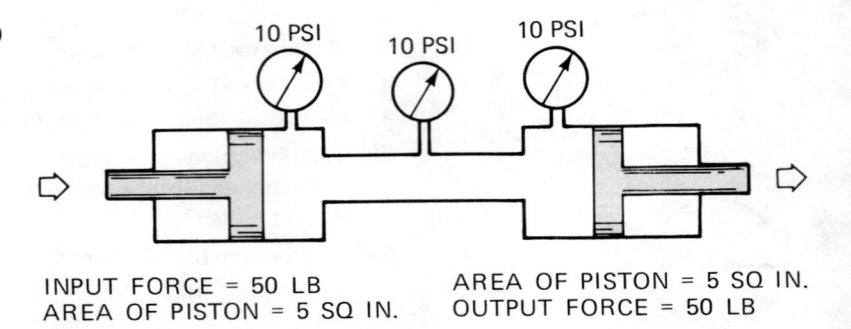

INPUT FORCE = 50 LB
AREA OF PISTON = 5 SQ IN.

AREA OF PISTON = 5 SQ IN.
OUTPUT FORCE = 50 LB

To find the pressure on the liquid, remember that pressure can be measured in pounds per square inch (psi).

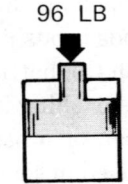

$$\text{Pressure} = \frac{\text{Force}}{\text{Area}}$$

$$= \frac{50 \text{ pounds}}{5 \text{ square inches}}$$

$$= 10 \text{ psi}$$

This pressure of 10 psi acts on all the liquid in the cylinder. So the piston on the right is getting its share, too. The force with which it is pushed depends on its area. Each of its 5 square inches gets 10 pounds of force. So the entire piston receives 50 pounds of force.

Example 1

The piston in Figure 13-10 receives a force of 96 lb. The area of the piston is 8 sq in. What is the pressure?

Figure 13-10   96 LB

Solution

$$\text{Pressure} = \frac{\text{Force}}{\text{Area}}$$

$$= \frac{96}{8}$$

$$= 12 \text{ psi}$$

Sometimes we want more output force than input force. To do this, we can make the area of the output piston larger. Figure 13-11 shows what happens.

Figure 13-11

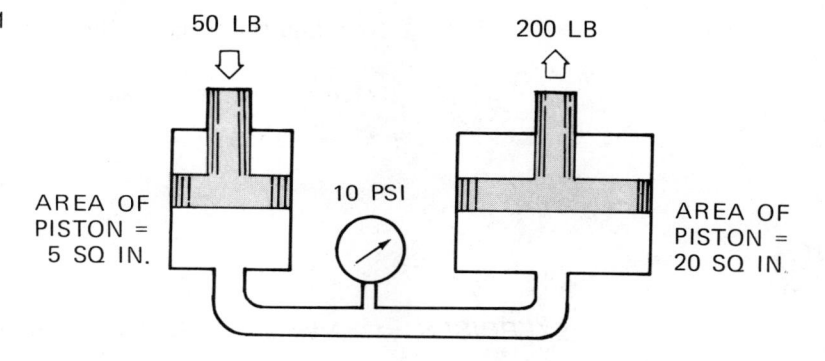

When a 50-pound force acts on 5 square inches we have a pressure of 10 pounds per square inch. The larger piston has an area of 20 square inches. So it receives a force of 10 pounds for each square inch.

Force = Pressure × Area

= 10 × 20

= 200 pounds

Thus the output piston receives a force of 200 pounds. However, the output piston can do no more *work* than the input piston. Remember that work is the product of force times distance. For the two cylinders to do equal work, the output piston moves ¼ the distance of the input piston. This is because it has 4 times the force of the input piston. So to have more output force in a hydraulic system we always sacrifice distance.

Example 2

In a hydraulic system the input piston has an area of 8 sq in. The output piston has an area of 24 sq in. If the force on the input piston is 25 lb, what is the force on the output piston?

Solution

$$\text{Pressure} = \frac{\text{pounds}}{\text{square inches}}$$

$$= \frac{25}{8}$$

$$= 3.125 \text{ psi}$$

Now find the force on the output piston.

$$\text{Force} = \text{Pressure} \times \text{Area}$$

$$= 3.125 \times 24$$

$$= 75 \text{ lb}$$

EXERCISES

1. A force of 20 lb acts on 2 sq in. What is the pressure?
2. The liquid in a cylinder has a pressure of 15 psi. The force on the input piston is 45 lb. What is the area of the piston?
3. A piston with an area of 10 sq in. produces a pressure of 45 psi. How much outside force is it receiving?
4. The pressure in a hydraulic system is 20 psi. The output piston has an area of 4 sq in. What is the output force?
5. The smaller piston in a hydraulic system has an area of 5 sq in. It receives a force of 190 lb. What is the pressure in the system?
6. The input piston in a hydraulic system has an area of 8 sq in. The area of the output piston is 25 sq in. If the input force is 40 lb, what is the output force?
7. The small piston in a hydraulic system has an area of 6 sq in. It receives a force of 30 lb. The large piston puts out a force of 120 lb. Find the area of the large piston.
8. In a hydraulic system an input force of 50 lb produces an output of 125 lb. The pressure in the cylinder is 25 psi. Find the area of both the input piston and the output piston.
9. Find the output force in Figure 13-12. How far up does the output piston move when the input piston moves down 10"?

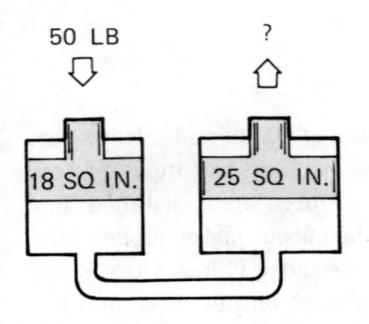

Figure 13-12 (50 LB on 18 SQ IN.; ? on 25 SQ IN.)

392 Technical and industrial applications

13-8 PNEUMATIC POWER

Hydraulic power is based on the principle that liquids under pressure do not compress. This means that when we apply a force to a liquid, the pressure goes up, but the volume stays the same. However, if you put pressure on a gas, it compresses to a smaller volume with more pressure. Look at Figure 13-13 to see how this works.

Figure 13-13

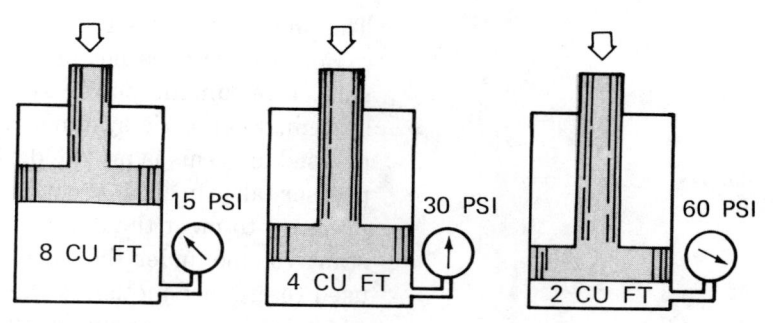

When 4 cubic feet of air is compressed to half its original volume, the pressure is doubled. When the air space is one-fourth the original volume (2 cubic feet), the pressure is multiplied four times. Notice that the product of volume and pressure remains the same.

8 × 15 = 4 × 30 = 2 × 60

120 = 120 = 120

We can use this fact to find the pressure of a gas after it has been compressed.

Old pressure × Old volume = New pressure × New volume

$$P_1 \times V_1 = P_2 \times V_2$$

Example 1

The piston in Figure 13-14 compresses 10 cu ft of gas to 2 cu ft. What is the new pressure?

Figure 13-14

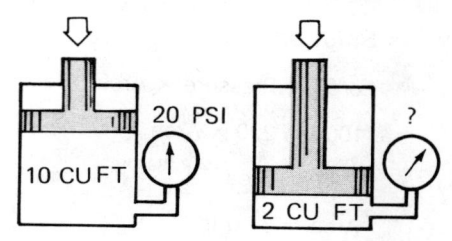

Power and energy 393

Solution

$$P_1V_1 = P_2V_2$$
$$20 \times 10 = P_2 \times 2$$
$$200 = P_2 \times 2$$
$$100 = P_2$$

The new pressure is 100 psi.

Because air does not maintain the same volume under pressure, a pneumatic power system cannot work like a hydraulic system. Pneumatic systems require a constant volume of compressed air. This is provided by an air compressor. The air compressor takes in air at the normal pressure of 15 psi and raises its pressure to meet the demands of the system. It stays at a constant volume under the same pressure. So its fluid force can be used to produce motion in the output piston.

A simplified pneumatic system is shown in Figure 13-15. Notice that the air compressor supplies a large volume of pressurized air to the air storage tank, to be used by the output piston.

Example 2

What is the area of the piston in Figure 13-15?

Figure 13-15

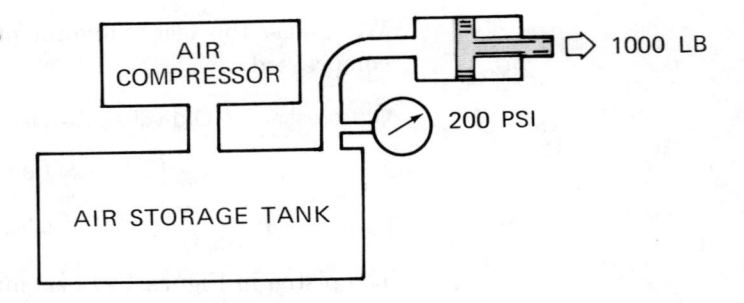

Solution

$$\text{Force} = \text{Pressure} \times \text{Area}$$
$$1000 = 200 \times \text{Area}$$
$$\frac{1000}{200} = \text{Area}$$
$$\text{Area} = 5 \text{ sq in.}$$

394 Technical and industrial applications

EXERCISES

1. A piston compresses 20 cu ft of air at 15 psi to 5 cu ft. What is the new pressure?
2. Air at 20 psi is compressed so that its new pressure is 35 psi. If the original volume was $3\frac{1}{2}$ cu in., what is the new volume?
3. How much air pressure results when 10 cu ft of air at 15 psi is compressed to 1 cu ft?
4. Air at 15 psi is compressed so that its volume is 4 cu in. and its pressure is 60 psi. What was the original volume?
5. An air compressor supplies a constant pressure of 150 psi to a piston. If the area of the piston is 5 sq in., what is the resulting force on the piston?
6. How much air pressure is needed to lift a 300-lb object if the area of the piston is 4 sq in.?
7. If 2 cu ft of air at 15 psi is compressed to 40 cu in., what is the new pressure?
8. A volume of air at 15 psi is compressed to 3 cu ft, with a resulting pressure of 100 psi. What was the original volume?

AMPERES
(ELECTRONS PER SECOND)

CURRENT FLOW

RATE
(GALLONS PER SECOND)

WATER FLOW

Figure 13-16

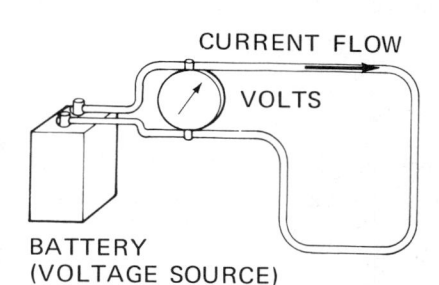

CURRENT FLOW
VOLTS
BATTERY
(VOLTAGE SOURCE)

PSI
PUMP
(PRESSURE SOURCE)
WATER FLOW

Figure 13-17

13-9 ELECTRICAL POWER

We use electricity every day for light, heat, and appliances, as well as for radio, television, and other communication. Industries depend on electricity to run tools and machines, to provide heat for processes such as steelmaking, and for many other purposes. The source of electrical power is sometimes water power from dams. This water power is changed to electrical power by huge generators. Electricity for smaller systems, like a car's ignition system, is available from the chemical energy produced by and stored in batteries.

The energy of electricity is controlled and transmitted by conductors, usually metal wires. The flow of electricity, or electric current, is measured in three ways.

1. The *rate of flow* of a current is measured in *amperes*. This can be compared to the rate at which water flows through a pipe, as shown in Figure 13-16.
2. The *pressure* under which the current flows is the measure of its *voltage*. Figure 13-17 compares the voltage of a current to pressure in a water pipe.
3. The *resistance* that the wire offers to the current's flow is measured in *ohms*. The resistance of the wire depends on the type of metal used and the length and diameter of the wire.

Power and energy 395

Volts, amperes, and ohms are all related. It takes one volt to force one ampere of current through a resistance of one ohm. This can be shown by the following formula.

Volts = Amperes × Ohms

Example 1

How large a current will 120 volts send through a resistance of 2 ohms?

Solution

$$\text{Volts} = \text{Amperes} \times \text{Ohms}$$
$$120 = \text{Amperes} \times 2$$
$$\text{Amperes} = \frac{120}{2}$$
$$= 60 \text{ amperes}$$

How do we measure the rate of work done by an electrical system? The unit of power used for electricity is the *watt*. One watt of power is used when one volt produces a current of one ampere.

Watts = Volts × Amperes

A watt is a small unit of power compared to one horsepower, used earlier in this chapter. It takes 746 watts to make one horsepower.

1 hp = 746 watts

Since the watt is such a small unit, a larger unit, the kilowatt (1000 watts), is sometimes used.

Example 2

What is the wattage of a circuit using 10 amperes when the resistance is 12 ohms?

Solution

To find the wattage we must know the voltage.

Volts = Amperes × Ohms

= 10 × 12

= 120 volts

Watts = Volts × Amperes

= 120 × 10

= 1200 watts

EXERCISES

1. How many volts are needed to send a current of 2 amperes through a resistance of 60 ohms?
2. It takes 110 volts to send a current of 5 amperes through an electric toaster. Find the resistance of the toaster.
3. How many amperes will 120 volts send through a resistance of 3 ohms?
4. How many watts of power are used by an electric range that draws 5 amperes of current on a 240-volt line?
5. In a 12-volt auto electrical system, it takes 8 amperes to light the headlights. How much power is used?
6. A 12-volt battery has an output of 75 amperes. How much power is available from this battery?
7. An industrial machine uses 50 amperes of current on a 240-volt line. How many kilowatts does it use?
8. What is the wattage produced by 12 amperes when the resistance is 15 ohms?
9. The resistance of an iron is 11 ohms when the iron is using 110 volts. How many watts are being used?
10. Find the resistance of a 60-watt light bulb on a 120-volt line.
11. A 1-horsepower electric motor is connected to a 120-volt line. How many amperes of current does it require?
12. A 3-horsepower motor is connected to a 220-volt line. How much resistance does this motor have?

Taking inventory

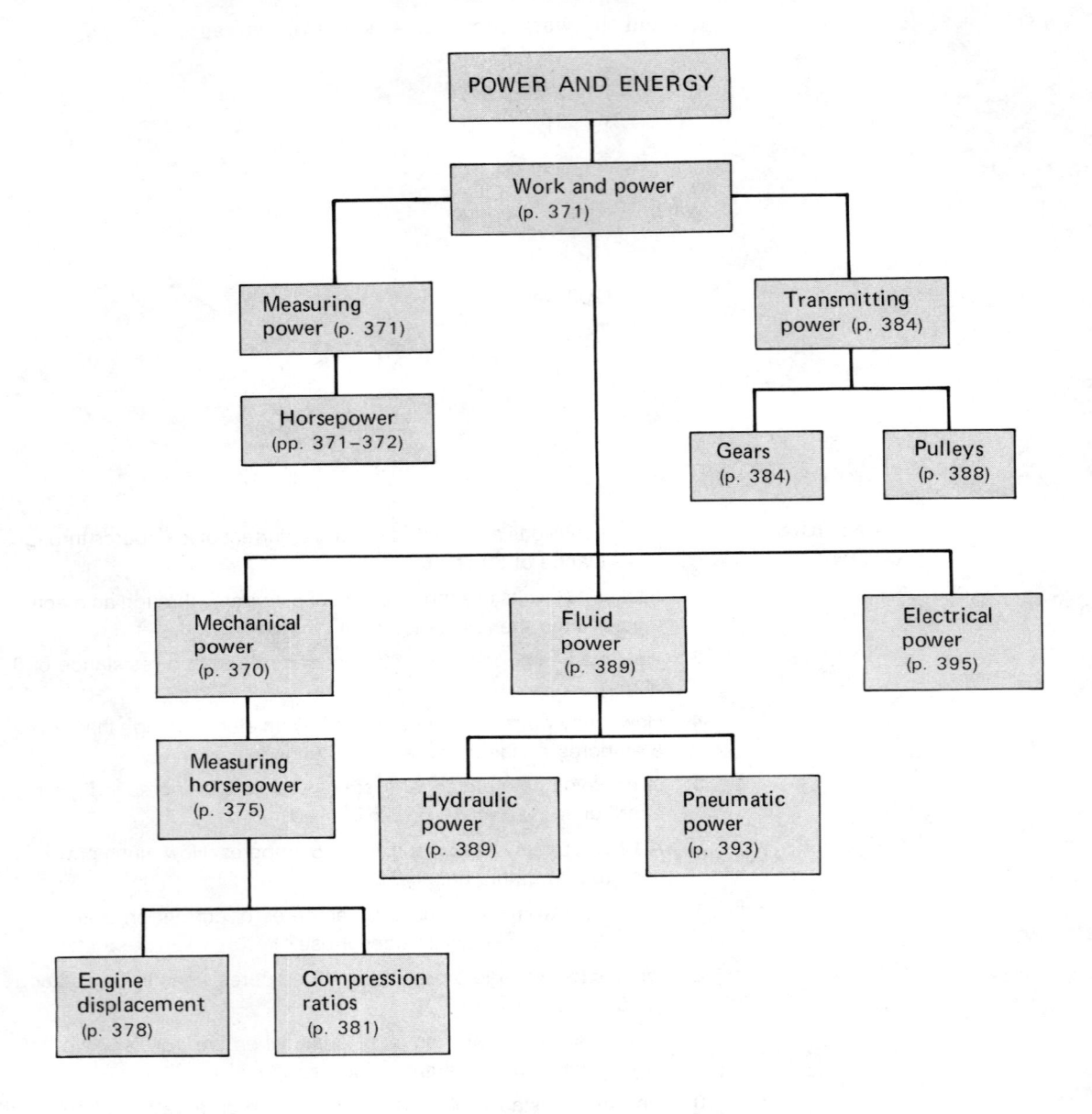

```
POWER AND ENERGY
        │
   Work and power
      (p. 371)
```

- Measuring power (p. 371)
  - Horsepower (pp. 371–372)
- Transmitting power (p. 384)
  - Gears (p. 384)
  - Pulleys (p. 388)
- Mechanical power (p. 370)
  - Measuring horsepower (p. 375)
    - Engine displacement (p. 378)
    - Compression ratios (p. 381)
- Fluid power (p. 389)
  - Hydraulic power (p. 389)
  - Pneumatic power (p. 393)
- Electrical power (p. 395)

Measuring your skills

1. Find the amount of work done in lifting 500 lb 3 ft. **(13-1)**
2. Find the horsepower needed to raise a 2750-lb car 3 ft in 5 sec. **(13-1)**

398 Technical and industrial applications

3. What is the brake horsepower of an engine that registers 200 lb on the end of an arm 2 ft long when the engine speed is 2800 rpm? Find the answer to the nearest horsepower. **(13-2)**
4. Find, to the nearest horsepower, the ihp of an engine whose pistons have an area of 15 sq in., a stroke of 5 ft, an average pressure of 80 lb per sq in., and 2000 piston strokes per minute. **(13-2)**

U.S. system

5. Find the displacement of a V-8 engine with a 4" bore and a 3½" stroke. ($\pi = \frac{22}{7}$) **(13-3)**

6. There are 3 cu in. of space in a cylinder when the piston is at the top of its stroke. If the compression ratio is 9:1, how much space is there when the piston is at the bottom of its stroke? **(13-4)**

Metric system

7. What is the displacement of a 6-cylinder engine whose bore is 9 cm and whose stroke is 9 cm? **(13-3)**

8. A cylinder contains 200 cu cm of space when the piston is at the bottom of its stroke. There are 25 cu cm of space when it is at the top of its stroke. What is the compression ratio? **(13-4)**

9. What is the gear ratio of A to B in Figure 13-18? **(13-5)**
10. The engine speed of a car is 3600 rpm, while the drive shaft speed is 1200 rpm. What is the transmission ratio? **(13-5)**
11. The rear axle ratio is 5:1. If the drive shaft is turning at 2500 rpm, at what speed is the rear axle turning? **(13-5)**
12. What is the ratio of the speed of pulley A to that of pulley B in Figure 13-19? **(13-6)**
13. The liquid in a cylinder has a pressure of 20 psi. The output piston has an area of 8 sq in. What is the output force? **(13-7)**
14. If 30 cu ft of air is compressed to 6 cu ft and the air pressure was 15 psi before compression, what is the pressure after compression? **(13-8)**
15. A 5-ampere electric drill requires 110 volts. What is its resistance? **(13-9)**
16. How much electrical power is used for a 6-ampere electric lawn mower connected to a 240-volt line? **(13-9)**

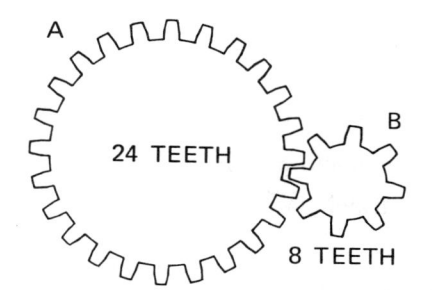

Figure 13-18

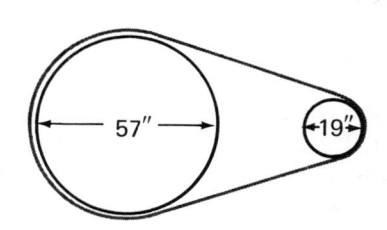

Figure 13-19

14 Construction

After completing this chapter, you should be able to:
1. Read blueprints and working drawings.
2. Estimate costs of projects.
3. Determine quantities of materials needed for various construction projects.

From drawing to foundations

14-1 PLANS AND ESTIMATING COSTS

Before you start building anything, you need a plan. It might be a simple sketch or a detailed blueprint. Figure 14-1 shows a *working drawing* for adding a patio to a house.

To build a complete house, more detailed drawings are necessary. Figure 14-2 is a *floor plan* for a two-bedroom house.

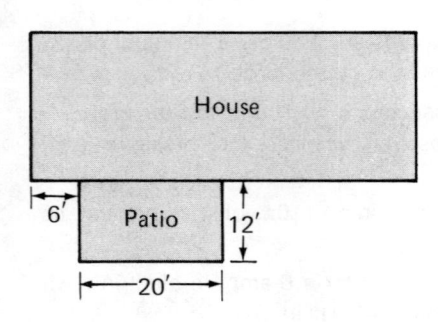

Figure 14-1

Example 1

How many square feet of floor space is in the house shown in Figure 14-2?

Solution

$54'6'' = 54\frac{1}{2}$ ft

$54.5 \times 24 = 1308$ sq ft

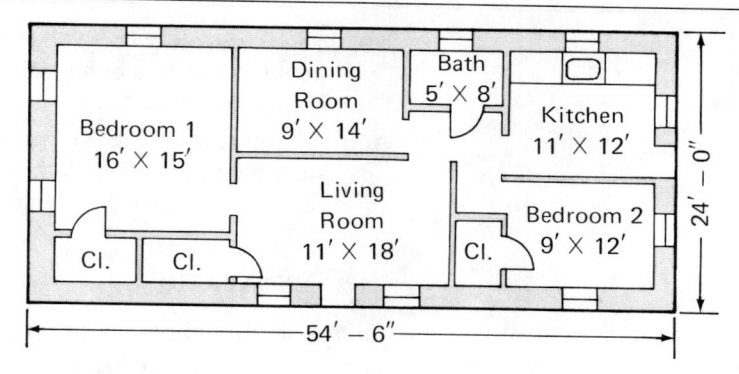

Figure 14-2

The architect's blueprints help him or her make an estimate of the cost of construction. Then the contractor uses the architect's blueprints to make another estimate. The *estimated cost* of a building is usually given in dollars per square foot.

Example 2

An architect estimates that a house can be built for $15 per square foot. The house has 1800 square feet of floor space. What is the estimated cost?

Solution

15 × 1800 = 27,000

The estimated cost is $27,000.

EXERCISES Use the floor plan in Figure 14-2 for exercises 1–10.

Find the number of square feet in each of the following rooms.

1. Living room
2. Kitchen
3. Bathroom
4. Bedroom 1
5. Bedroom 2
6. The estimated cost of the house in Figure 14-2 is $16 per square foot. What is the estimated cost of construction?
7. How many square yards of carpet are needed to carpet the two bedrooms?
8. If the price of carpeting is $12.50 per square yard, what is the cost of carpeting the living room?

Construction 401

9. If 1 square foot of floor tiling for the kitchen costs $1.20, how much would it cost to tile the kitchen floor? (The 3' × 12' countertop should be subtracted.)

10. A contractor wishes to carpet all the rooms shown except the kitchen and bathroom. If the carpet costs $8.95 per square yard, what would be the cost of the carpet?

14-2 EXCAVATIONS

Before any building can be started, *excavation* is usually necessary. Excavation for foundations, footings, or basements is usually computed in cubic yards.

The cost of excavating depends on the type of material that must be dug up. Another cost factor is the distance the material must be hauled for disposal.

Example

An underground parking lot requires an excavation 200 ft long, 150 ft wide, and 33 ft deep. The cost will be $6.00 per cu yd. What will the excavation cost?

Solution

200 ft = $\frac{200}{3}$ yd; 150 ft = 50 yd; 33 ft = 11 yd

$\frac{200}{3} \times 50 \times 11 = \frac{200}{3} \times \frac{50}{1} \times \frac{11}{1}$

$= \frac{110000}{3}$

$= 36,667$ (approx.)

There are 36,667 cubic yards to be excavated.

$36,667 \times 6 = 220,002$

The cost will be $220,002.

EXERCISES

1. A warehouse basement requires an excavation 10' deep, 78' wide, and 96' long. How many cubic yards of earth must be removed for this basement?

2. **How many hours will it take a contractor to dig the basement in Exercise 1 if the contractor can remove 60 cu yd per hr?**

3. A trucker can haul 10 cu yd of earth per trip from an excavation to a landfill. Making two trips per hour, how many cubic yards can the trucker haul in an 8-hr day?

4. An excavation of 750 cu yd is to be done by an earthmover who will charge $3.00 per cu yd. What will it cost to get the job done?
5. If the basement of the house in Figure 14-2 requires an excavation 9 ft deep, how many cubic yards of earth need to be removed?
6. The cost of excavating a 9-ft-deep basement in Figure 14-2 is $7.00 per cubic yard. What is the cost of excavating?
7. A basement requires an excavation 9 ft deep, 54 ft long, and 30 ft wide. An earthmover can dig this basement at a rate of 50 cu yd per hour. If the charge is $150.00 per hour, what is the cost of the excavation?

14-3 FOOTINGS

Before a building can be planned, the architect must know the *bearing capacity* of the soil. This is the number of pounds per square foot that the soil will support. The total weight of a building divided by the bearing capacity of the soil determines the area of *footing* on which the building should rest.

$$\text{Area of footing} = \frac{\text{Weight}}{\text{Bearing capacity}}$$

Figure 14-3 illustrates part of a foundation and the footing on which it rests.

Figure 14-3

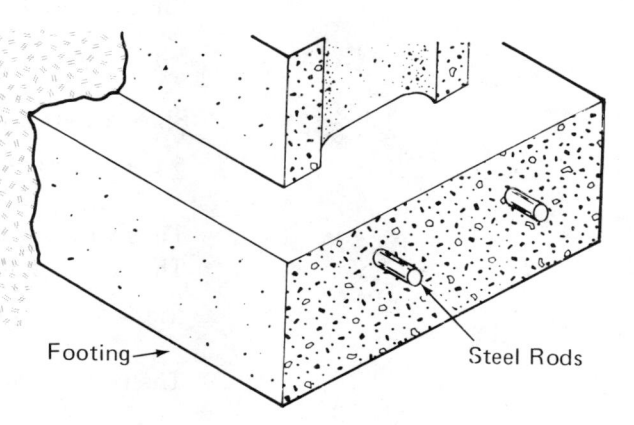

Example 1

The bearing capacity of a soil is 2500 lb per sq ft. If the footing must sustain a load of 60,000 lb, what should be the area of the footing?

Solution

$$\text{Area of footing} = \frac{\text{Weight}}{\text{Bearing capacity}}$$

$$= \frac{60{,}000}{2500}$$

$$= 24$$

24 sq ft of footing is needed.

The materials used for footings vary with the type of building constructed. Reinforced concrete is a popular material for footings. The reinforcement consists of metal rods placed in the concrete, as shown in Figure 14-3.

Reinforcing rod is usually sold by the foot and is available in different diameters. The cost of various widths of reinforcing rods is shown in Table 14-1.

Example 2

The footing for the house in Figure 14-2 has #3 reinforcing rods in it. Find the number of feet required if 2 parallel rods are to be used. What is the cost of these rods?

Solution

First we must find the perimeter.

$54\frac{1}{2} + 54\frac{1}{2} + 24 + 24 = 157$ ft

Since 2 rods are required, we get

$2 \times 157 = 314$ ft

Thus we need 314 feet of reinforcing rod.
The cost of #3 rod is $.066 per foot.

$314 \times .066 = 20.724$

The cost is $20.72.

Table 14-1 *Costs of reinforcing rods*

Number	Diameter	Price per ft
#3	$\frac{3}{8}''$	6.6¢
#4	$\frac{1}{2}''$	9.3¢
#5	$\frac{5}{8}''$	13.5¢
#6	$\frac{3}{4}''$	22.2¢

EXERCISES
1. If the bearing capacity of the soil is 2100 lb per sq ft and the area of the footing is 30 sq ft, what is the maximum weight the footing will support?
2. A building will weigh 150,000 lb. If the bearing capacity of the soil is 2000 lb per sq ft, what area of footing is required?
3. A structure weighing 50,000 lb is to be built on soil that has a bearing capacity of 1800 lb per sq ft. What is the area of footing needed?
4. How many cubic yards of concrete are needed for a footing 18" wide, 9" deep, and 120 feet long?
5. Use Table 14-1 to find the cost of 75 ft of ½" reinforcing rod.
6. A concrete footing 24 in. wide, 9 in. thick, 240 ft long, and reinforced with #4 rod is to be constructed. Three sets of reinforcing rods are placed lengthwise in the footing. Determine the cost of the reinforcing rods.
7. The concrete for the footing in Exercise 6 weighs 935 lb per cu yd. What is the weight of the concrete used?

14-4 FOUNDATIONS

The type of foundation used for a building depends on its structure and location. In areas with a high risk of flooding or earthquakes, houses may be built on a concrete slab foundation. Otherwise they are usually constructed with basements. Foundations under houses having basements are commonly built with poured concrete or concrete blocks. Figure 14-4 shows a top view of the foundation wall of a house and a cross-section of the foundation and its footing.

Example 1

How many cubic yards of concrete are needed for the foundation shown in Figure 14-4?

Figure 14-4

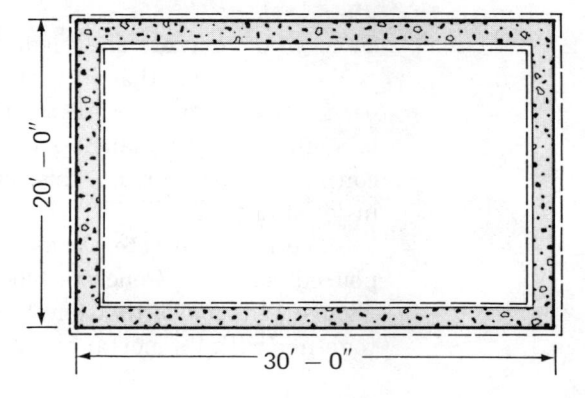

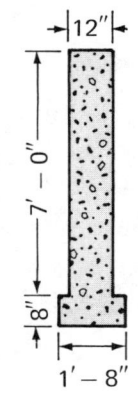

Solution 1

Two walls are 30' × 7' × 1':

30 × 7 × 1 = 210 cu ft
210 × 2 = 420 cu ft

Two walls are 20' × 7' × 1':

20 × 7 × 1 = 140 cu ft
2 × 140 = 280 cu ft
420 + 280 = 700 cu ft

There are 27 cu ft in 1 cu yd.

700 ÷ 27 = 25.9   (approx.)

26 cu yd of concrete are needed.

Solution 2

We could solve this problem by multiplying the perimeter of the foundation times the cross-sectional area of the foundation.

Perimeter = 30 + 20 + 30 + 20 = 100

Cross-sectional area = 7 × 1 = 7

100 × 7 = 700 cu ft

There are 27 cu ft in 1 cu yd.

700 ÷ 27 = 25.9   (approx.)

26 cu yd of concrete are needed.

You will notice that in both of these solutions we have used each corner two times. Although this gives us an answer that is slightly larger than it should be, the additional amount of concrete called for by this method of calculation is usually insignificant.

Sometimes concrete blocks are used for foundations instead of poured concrete. Concrete blocks are held together by mortar. Figure 14-5 shows that one *course* of blocks is the block itself, together with its mortar joint.

One Course

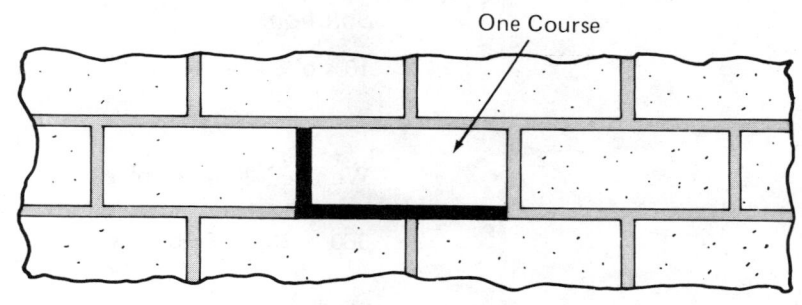

Figure 14-5

One course of concrete blocks measures 8" high and 16" long.

Example 2

How many concrete blocks are needed to build a foundation wall 8 ft high and 40 ft long?

Solution

First we find the area of the wall.

8 × 40 = 320 sq ft

Since one course of blocks is 16" × 8", or 128 sq in., we need to convert 320 sq ft to sq in.
There are 144 sq in. in 1 sq ft.

320 × 144 = 46,080

So there are 46,080 sq in. of wall. To find the number of blocks needed, divide the area of one course.

46,080 ÷ 128 = 360

360 blocks are needed.

Concrete blocks can sometimes be broken. So contractors frequently add on a percentage for breakage when ordering.

Example 3

If 10% is allowed for breakage in Example 2, how many concrete blocks should be ordered?

Construction

Solution

10% of 360 = .10 × 360

= 36

We need 36 more blocks.

360 + 36 = 396 blocks

EXERCISES

1. A concrete slab foundation 6″ thick, 20′ wide, and 25′ long is to be poured. How many cubic yards of concrete must be ordered?
2. A poured concrete foundation wall is 10″ thick, 8′ high, and 40′ long. How many cubic yards of concrete are required?
3. How many concrete blocks are needed to build a wall 10 ft high and 25 ft long?
4. A contractor determines that 450 concrete blocks will be needed for a foundation. Allowing 8% for breakage, how many blocks should the contractor order?
5. A warehouse requires a concrete foundation wall 9″ thick and 3′ high. The building measures 38 ft by 82 ft. How many cubic yards of concrete are needed for the foundation?
6. It costs Mr. Rivera $.45 each for concrete blocks. He wishes to build an 8-ft-high foundation for a 25′ × 45′ house. If he allows 5% for breakage, how much will the concrete blocks for the foundation cost?
7. Ready-mixed concrete will be delivered to the construction site for $21.00 per cu yd. What will concrete cost for the foundation and footing of a 60′ × 100′ building? The cross-section of the foundation and footing is shown in Figure 14-6.

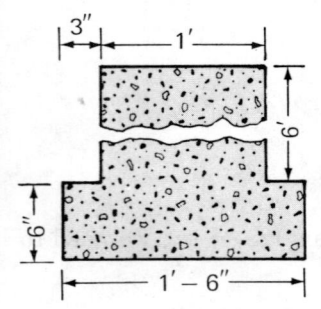

Figure 14-6

Framing and exterior

14-5 FRAMING

The framework of most houses built today is made of wood. Figure 14-7 shows the framework of a house.

The contractor orders wood for the frame in *board feet*, the standard measuring unit for lumber. One board foot is a piece of lumber one foot long, one foot wide, and one inch thick. Sometimes a larger unit, M, is used for 1000 board feet. Boards come in varying widths of 2″, 3″, 4″, 5″, 6″, 8″, 10″, and 12″. The thickness may vary, too, but standard units are 1″, $1\frac{1}{4}$″, $1\frac{1}{2}$″, 2″, 3″, 4″, 6″, 8″,

Figure 14-7

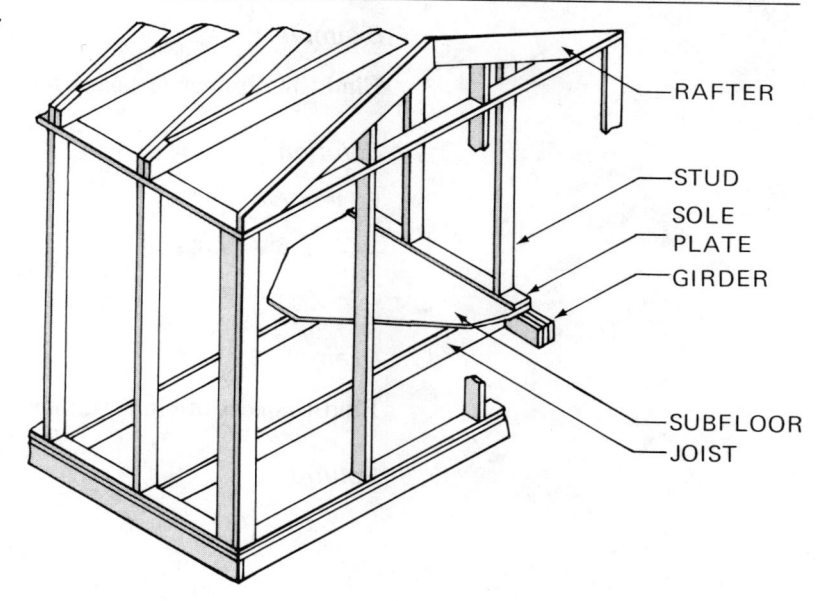

and so forth. In calculating board feet, lumber less than 1" thick is counted as 1" thick. One board foot may be any piece with the same volume as $1' \times 1' \times 1''$. Figure 14-8 shows three pieces of lumber that measure 1 board foot.

Figure 14-8

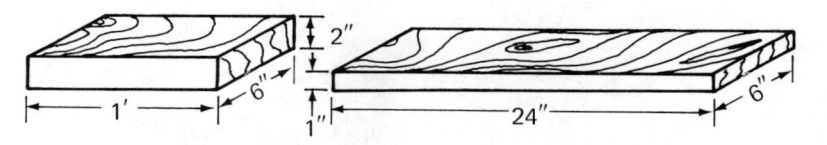

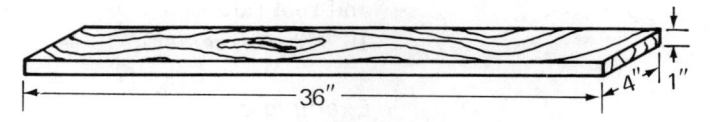

To find the number of board feet in a piece of lumber we can use the formula

bd ft = ℓwt

where ℓ = length in feet

w = width in feet

t = thickness in inches

Construction 409

Example 1

Find the number of board feet in a 2 by 4 that is 6 ft long.

Solution

Bd ft $= \ell wt$

$\qquad = 6 \times \frac{4}{12} \times 2$

$\qquad = 4$ bd ft

Example 2

Find the cost of eight timbers $6'' \times 10'' \times 20'$ at \$200 per M.

Solution

bd ft $= \ell wt$

$\qquad = 20 \times \dfrac{10}{12} \times 6$

$\qquad = 100$ bd ft

$8 \times 100 = 800$ bd ft

$\qquad \text{Cost} = \dfrac{\text{bd ft} \times \text{Price per M}}{1000}$

$\qquad\qquad = \dfrac{800 \times 200}{1000}$

$\qquad\qquad = 160$

The cost is \$160.

There are many places in construction where pieces are spaced the same distance apart. For example, floor joists, wall studs, and roof rafters are usually placed 16″ apart. We indicate this as 16″ o.c. (*on center*).

Example 3

How many 16″ o.c. floor joists are needed for an 8-ft-long room?

Solution

We must divide the length by 16″.

$\quad 8 \times 12 = 96''$ long

$96 \div 16 = 6$

Figure 14-9 illustrates the joists.

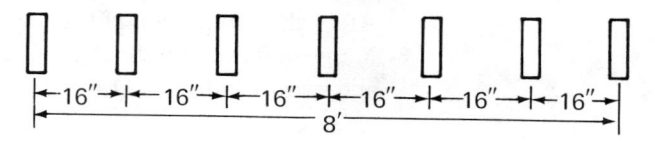

Figure 14-9

Then 6 + 1, or 7, joists are needed, because we must add one for a starter.

EXERCISES Find the number of board feet.

1. 6 boards, each 2" × 8" × 12'
2. Twenty 2 × 6's, each 14 ft long
3. 15 pieces, $\frac{3}{4}$" × 10" × 8', and 5 pieces, $\frac{3}{4}$" × 10" × 2'
4. How many wall studs are needed for a 20-ft wall if the studs are 16" o.c.?
5. How many roof rafters are needed for a 42-ft-long roof if the rafters are 16" o.c.?
6. The cost of a certain kind of lumber is $190 per M. Find the cost of 150 pieces, each 2" × 8" × 12', at this price.
7. A 12-ft wall needs 2 × 4 studs placed 16" o.c. If the height of the wall is 8 ft, find the number of board feet of studs to be ordered.
8. Good quality 2 × 6's used for rafters sell for $238 per M. Find the cost of rafters for the roof shown in Figure 14-10 if they are placed 16" o.c.

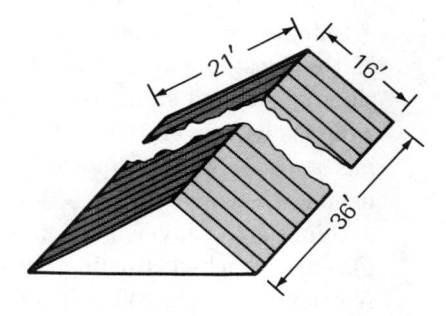

Figure 14-10

9. If the rise of a rafter is 3 ft and the run is 12 ft, find the angle that the rafter makes with the horizontal.

14-6 ROOFING

After the completion of the frame, the house is ready for roofing. Many different kinds of roofs are available. Some varieties are shown in Figure 14-11.

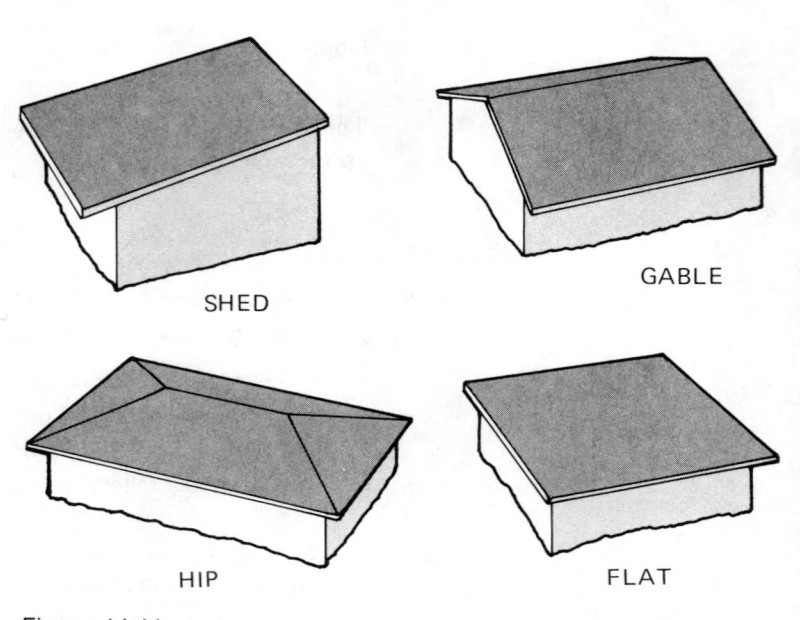

SHED

GABLE

HIP

FLAT

Figure 14-11

The materials chosen for roofing vary with the appearance desired, the climate, and the cost of materials and labor.

Shingles are commonly used for most gable or shed roofs. One of the most popular kind of shingle used is the asphalt strip shingle. Asphalt shingles are measured in *squares*. One square of shingles will cover 100 square feet of roof. For example, if a roof has an area of 900 square feet, 9 squares of shingles will be needed to cover it.

Generally, a layer of roofing felt is laid down before the shingles are attached. Roofing felt comes in rolls 3 feet wide and 144 feet long. So, one roll will cover 3 × 144, or 432 square feet.

Another material needed for the roofing process is *nails*, for attaching the shingles. About 2 pounds of nails should be allowed for one square of shingles.

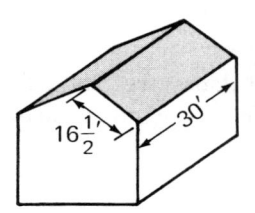

Figure 14-12

Example

The gable roof pictured in Figure 14-12 is 30 ft long and has rafters 16½ ft long. How many squares of shingles are needed to cover it?

Solution

Half of roof = 16½ × 30

= 495 sq ft

Area of roof = 2 × 495 = 990 sq ft

1 square of roofing covers 100 sq ft.

990 ÷ 100 = 9.9

10 squares of roofing are needed.

EXERCISES Find the number of squares needed for the following roofs.

1. A gable roof 42 ft long with 16-ft rafters (see Figure 14-13.)
2. A gable roof 27 ft long with 12-ft rafters
3. A shed roof 20 ft long with 14-ft rafters (see Figure 14-14.)
4. A shed roof 30 ft long with 15-ft rafters
5. Find the cost of the shingles for a gable roof 35 ft long with 15-ft rafters if shingles cost $7.50 per square.
6. Find the cost of the roofing nails needed for the roof in exercise 5 if nails cost $0.20 per lb.
7. How many rolls of roofing felt are needed for a shed roof which is 24 ft long and has rafters 16 ft long?
8. What is the cost of roofing materials for a 1680-sq-ft roof? Assume that shingles cost $7.90 per square, roofing felt costs $2.75 per roll, and nails are $.25 per lb.
9. Find the cost of roofing the building in Figure 14-15 if shingles cost $7.88 per square, roofing felt, $2.80 per roll, and nails, $.19 per lb.

Figure 14-13

Figure 14-14

Figure 14-15

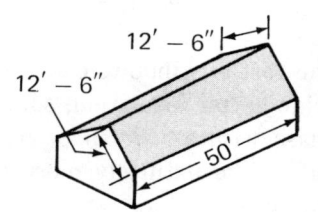

Construction

14-7 WINDOWS AND DOORS

The size, type, and number of windows and doors are planned along with the rest of the house. Various kinds of windows are available, as shown in Figure 14-16.

Figure 14-16

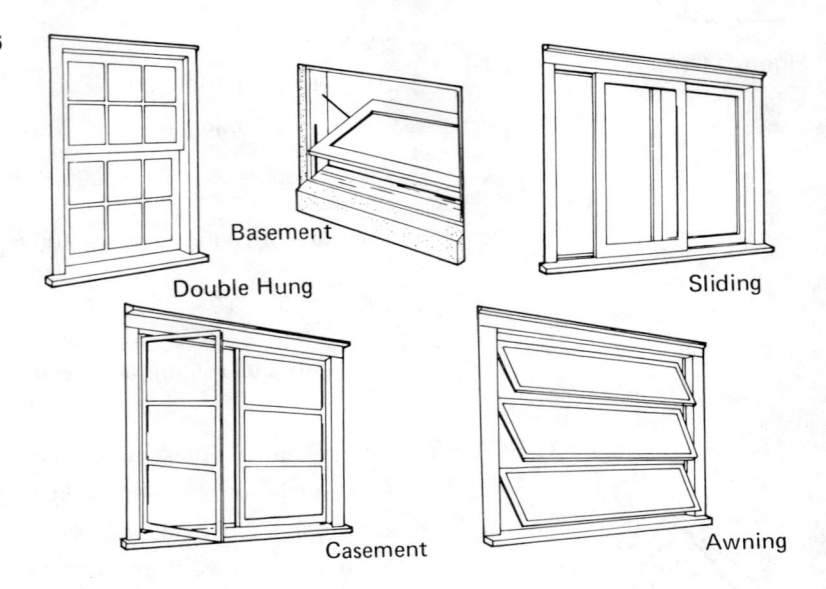

Double Hung Basement Sliding

Casement Awning

When the windows are planned, FHA (Federal Housing Administration) requirements must be followed. The FHA requires that the window area of a house be at least 10% of the floor space.

Example

What should be the minimum window area for a house whose floor area is 1800 sq ft?

Solution

10% of 1800 = .10 × 1800

= 180 sq ft

The cost of windows can vary with the type of glass used. To provide better insulation, some windows are made of two pieces of glass with an air space between them. These welded glass windows are naturally more expensive, as they have twice as much glass.

The price of doors also varies with the type of wood or metal.

414 Technical and industrial applications

EXERCISES

1. What is the price of 7 double-hung windows with 28" × 20" welded glass, if the price of each is $70.84?

2. A double-hung 28" × 20" window with regular glass costs $14.70. What will five of these windows cost?

3. Welded glass windows cost $70.84 each, and regular glass windows the same size cost $46.86 each. How much more will welded glass windows cost, if 8 windows are ordered?

4. A contractor needs nine room and closet doors. What will be the total cost if each door costs $12.90?

5. How many windows with an area of 9 sq ft are needed to meet FHA requirements for a 25' × 40' house?

6. A builder needs nine double-hung 28" × 20" windows, 2 basement windows, 1 bay window, 9 plywood doors, and 2 mahogany doors. Find the cost if the prices are as follows:

28" × 20" double-hung window	$53.90
Basement window	35.40
Bay window	85.75
Plywood door	13.85
Mahogany door	59.40

14-8 SIDING

To be finished on the outside, a house needs siding or some other material to cover the exterior walls. Siding is sometimes made of wood. More recently, aluminum, steel, and vinyl siding have become popular. These prefinished sidings require little or no maintenance and are easy to apply.

To calculate the amount of siding needed for a house, it is necessary to subtract the space for windows and doors. A common rule of thumb is to subtract 5 square feet for each window and 10 square feet for each door. So we must calculate the total number of square feet for the walls of the house and then use the rule of thumb to subtract the window and door area.

Example

The house shown in Figure 14-17 has 2 doors and 8 windows. Use the rule of thumb to find the area to be sided.

Solution

Perimeter = (2 × 30) + (2 × 48)

= 156 ft

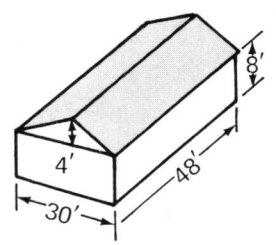

Figure 14-17

The area without the two triangular roof sections is the perimeter × height.

156 × 8 = 1248 sq ft

Triangular area = ½ × 30 × 4

= 60 sq ft

2 triangular areas = 2 × 60 = 120 sq ft

1248 + 120 = 1368 sq ft

Now we subtract 5 sq ft for each window and 10 sq ft for each door.

(8 × 5) + (2 × 10) = 60

1368 − 60 = 1308

The area to be sided is 1308 sq ft.

EXERCISES

1. A wall 30 ft long and 8 ft high is to be sided. The wall contains two windows and one door. Estimate the total area that is to be sided.
2. Estimate the number of square feet that are to be sided in the wall pictured in Figure 14-18.
3. A wall 15 ft high and 25 ft long contains four windows and a door. Estimate the area of the wall to be sided.
4. A flat roof house has exterior walls 10 ft high. The dimensions of the house are 40′ × 25′. There are 8 windows and 2 doors. Estimate the area to be sided.
5. Estimate the total area that is to be sided on the end of the house shown in Figure 14-19.
6. The house sketched in Figure 14-20 has 10 windows and 3 doors. Estimate the area to be sided.

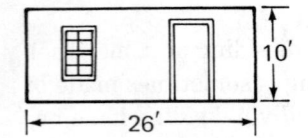

Figure 14-18

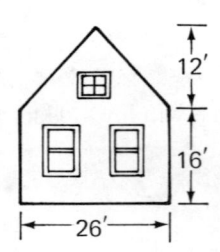

Figure 14-19

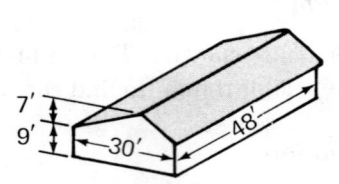

Figure 14-20

14-9 BRICKWORK

Another method of finishing the exterior of a house is with brickwork. Brick facing is commonly put on all or part of the exterior of a house.

The standard size of a brick is $2\frac{1}{4}'' \times 3\frac{3}{4}'' \times 8''$. Other sizes are available, and many different textures and colors are available in *face brick*. Where the brick will be seen, face brick is used, and it is considerably more expensive than *common brick*.

Mortar is used to bond the brick together. Several varieties of bonding are available, but the most common is the *running bond*. The thickness of the mortar joints may vary from $\frac{1}{4}''$ to $\frac{1}{2}''$. If thicker mortar joints are used, there are fewer bricks per square foot. Table 14-2 shows how the number of bricks laid in running bond varies with the thickness of the mortar.

Table 14-2 *Running bond*

Thickness of mortar joint	$\frac{1}{4}''$	$\frac{3}{8}''$	$\frac{1}{2}''$
Number of bricks per 100 sq ft	698	655	616

This information can be used to determine how many bricks should be ordered. As with concrete block, a percentage is usually allowed for breakage. Bricks are usually sold in *flats*, with 300 bricks per flat.

Example

A $25' \times 10'$ wall of a house is to have bricks laid in running bond with $\frac{1}{4}''$ mortar joints. If the bricks sell for $40.00 per flat, how much will the bricks cost?

Solution

The area to be bricked is 25×10, or 250 sq ft.
The number of bricks is 698 per 100 sq ft.

$$\frac{250 \times 698}{100}$$

$2.5 \times 698 = 1745.0$

1745 bricks are needed.

To find the number of flats, divide by 300.

1745 ÷ 300 = 5.8

6 flats are needed.

6 × $40 = $240

The cost of the bricks is $240.

EXERCISES

1. How many standard bricks laid in running bond are needed for a 200-sq-ft wall if the mortar joints are $\frac{3}{8}''$ wide?
2. The cost of a flat of face brick is $45.00. The front of a house, which measures 45' × 12', is to be faced with bricks laid in running bond with $\frac{1}{4}''$ mortar joints. How much will the bricks cost?
3. A contractor needs to face an entire house with bricks laid in running bond with $\frac{1}{4}''$ joints. The dimensions of the house are 30' × 45' with 10' high walls all around. Allowing 3% for breakage, how many bricks should the contractor order?
4. A house is to be faced with running bond brick two layers deep. The mortar joints will be $\frac{3}{8}''$ thick. The house has 12'-high walls with outer dimensions 25' × 40'. The cost of a flat of bricks is $55.00. How much will the bricks cost if 4% is added for breakage?

Finishing the job

14-10 ELECTRICAL WIRING

While carpenters are busy shingling and siding a building, the electricians and plumbers are busy installing their systems.

Recall from Chapter 13 that we measure electrical power in *watts* or *kilowatts*. Remember also that

Watts = Volts × Amperes

The voltage entering a house is about 120 volts or 240 volts. So the factor that determines how many watts are available is amperage. The amperage is restricted at the service entrance, the part of the electrical system linked to the outside power source. The service entrance is connected by a meter to the distribution panel. *Fuses* or *circuit breakers* are contained in the distribution panel to protect electrical equipment. Out of the distribution panel, branch circuits carry current to various locations in the house. There is a fuse or circuit breaker to control the amount of current for each circuit.

Example

A 20-ampere circuit breaker is placed on a 110-volt circuit. How many watts of power can be used on this circuit?

Solution

Watts = Volts × Amperes

 = 110 × 20

 = 2200

2200 watts or 2.2 kilowatts can be used.

EXERCISES

1. A 15-ampere fuse is placed in a 120-volt circuit. How many watts of power are available through this circuit?
2. A 30-ampere circuit breaker is located in a 220-volt circuit. How many kilowatts of power are available through this circuit?
3. An electric motor requires 1.7 amperes on a 115-volt circuit. How many watts of power does it use?
4. An electrical code requires an outlet every 8 ft or less. How many outlets are necessary for a room that is 14 ft wide and 20 ft long?

14-11 PLUMBING

A plumbing system uses pipes of various sizes and materials. Before the foundation is completed, a line from the water main and a water meter are positioned. The drain pipe leading to the sewer is also put in place early in the construction process. Figure 14-21 illustrates some of the basic plumbing in a house.

The pipes of a plumbing system used to be made of cast iron or galvanized iron. Copper pipes were an improvement, but now plastic pipes are used extensively. Plastic pipes are less expensive, less easily corroded, and easy to use.

PLUMBING IN A 2-STORY HOUSE

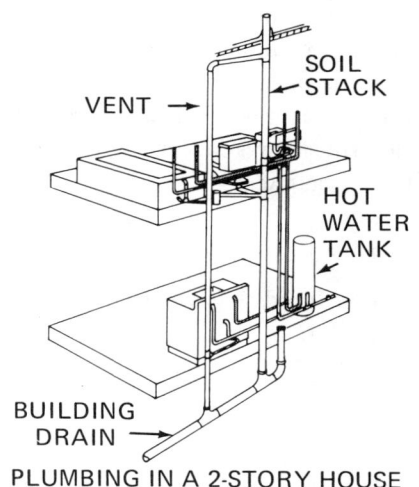

Figure 14-21

Example

The price of $\frac{1}{2}''$ copper pipe is $.50 per foot, and the price of $\frac{1}{2}''$ plastic pipe is $.26 per foot. A plumbing job requires 340 feet of $\frac{1}{2}''$ pipe. How much less will the pipe cost if plastic is used instead of metal?

Solution

Cost of copper pipe: 340 × .50 = 170.00

Cost of plastic pipe: 340 × .26 = 88.40

170.00 − 88.40 = $81.60 less

EXERCISES
1. The price of 4" cast iron soil pipe is $1.57 per foot, and plastic soil pipe of the same diameter costs $.044 per foot. What is the difference in price per foot?
2. If ½" plastic pipe costs $0.25 per ft, find the cost of 180 ft of this pipe.
3. If a plumber works for 37 hr in one week and is paid $6.75 per hr, how much does the plumber earn?
4. If 4" plastic soil pipe costs $43.90 per hundred feet, find the cost of 2000 ft of this pipe.
5. A repair project required 40 ft of ½" plastic pipe at $0.26 per ft, 12 ft of 6" soil pipe at $1.10 per ft, and 5 hr of labor at $5.25 per hr. What did the repair project cost?

14-12 INTERIOR FINISHING

In seasonal climates, insulation is one of the first steps in interior finishing. The insulation must be placed between the wall frame and the interior wall finish. A variety of materials can be used to finish the interior walls. Gypsum lath, metal lath, or plaster board (sheet rock) is commonly used as wall board. This can then be plastered, painted, wallpapered, or panelled.

To determine the amount of wall covering needed, we can use a simple rule of thumb: Four times the floor space will give the approximate wall and ceiling area to be covered.

Example 1

How much plaster board should be ordered to cover the interior walls and ceilings for a 23′ × 46′ house?

Solution

23 × 46 = 1058 sq ft of floor

1058 × 4 = 4232 sq ft

4232 sq ft of plaster board should be ordered.

Wallpaper is very popular as a wall covering. It is usually sold in 18″-wide rolls either 24 feet or 48 feet long.

Example 2

How many 24-ft rolls of wallpaper 18″ wide should be ordered for a 8′ × 10′ room with an 8-ft ceiling?

Solution

The area of the 4 walls is:

(8 × 8) + (8 × 8) + (8 × 10) + (8 × 10)

64 + 64 + 80 + 80 = 288 sq ft

Area of one roll of paper:

1½′ × 24′ = 36 sq ft

288 ÷ 36 = 8

8 rolls of paper should be ordered.

Paint is another popular wall and ceiling covering. The area that a volume of paint will cover varies with the type of paint used and whether a primer coat has been applied first.

Example 3

If 1 gal of paint will cover 460 sq ft, how many gallons and quarts should be ordered for a 15′ × 20′ room with an 8-ft ceiling?

Solution

The area of the 4 walls is:

(15 × 8) + (15 × 8) + (20 × 8) + (20 × 8)

120 + 120 + 160 + 160 = 560 sq ft

560 ÷ 460 = 1.2   (approx.)

1 gallon and 1 quart should be ordered.

Various types of floor covering are in use today. Hardwood, tile, linoleum, and carpet are available in a variety of colors, sizes, patterns, and textures. Hardwood is sold by the board foot, but the others are sold by the square yard or by the piece.

Example 4

A room 12′ × 15′ is to be carpeted with carpet costing $13.00 per square yard. What will the carpet cost?

Solution

Changing the feet to yards gives

$$\tfrac{12}{3} \times \tfrac{15}{3} = 4 \times 5$$
$$= 20 \text{ sq yd}$$

20 × 13.00 = 260.00

The carpet will cost $260.00.

TRICKS OF THE TRADE

To divide a board into strips of the same width without calculating, use this method.

a. Lay a rule at an angle across the width of the board so that the distance on the rule is easily divisible by the number of strips you want. Mark these points.
b. Repeat the procedure at another part of the board.
c. Join the marks and extend the lines to the ends of the board.

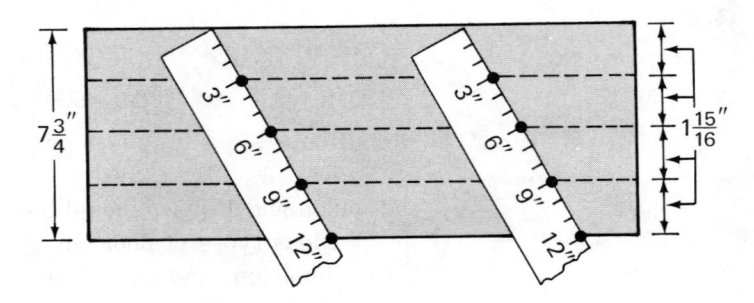

EXERCISES

1. If the floor area of a house is 1800 sq ft, estimate the number of square feet of gypsum lath needed for the interior of the house.

2. About how many square feet of wallboard are needed for a 30' × 45' house?

3. How many rolls of wallpaper 18" wide and 24' long are needed to paper a wall 8 ft high and 27 ft long?

4. A 15' × 25' living room with an 8-ft ceiling is to be papered with 18"-wide wallpaper. If the rolls are 24 ft long, how many rolls should be ordered?

5. If 1 gal of paint will cover 500 sq ft, how many gallons are needed for 2700 sq ft?

6. How many square yards of carpeting are needed for a 15' × 18' room?

7. If one piece of gypsum lath is 16" wide and 48" long, estimate the number of pieces of lath needed for a house with 2500 sq ft of floor space.

8. If indoor-outdoor carpet costs $6.00 per square yard, find the cost of carpeting a patio that is 12 ft × 18 ft.

9. A 12' × 15' master bedroom is to be wallpapered with 18" × 24' rolls costing $5.65 per roll. If the bedroom ceiling is 8' high, what will be the cost of the wallpaper?

10. A 20' × 24' living room has an 8-ft ceiling. Wall-to-wall carpeting is ordered at $9.00 per sq yd. Wallpaper at $5.75 per roll is also picked out. The rolls of wallpaper are 18" wide and 24' long. What is the cost of carpet and wallpaper for the room?

14-13 LANDSCAPING

One of the last steps in completing the construction of a house or commercial building is the landscaping. *Landscaping* means grading the earth and planting trees, shrubs, and grass.

Grading or sloping the earth may require more excavation or fill to raise the level of the ground. The fill is usually covered with topsoil, which can support grass and shrubs. Both fill and topsoil are sold by the cubic yard. Sometimes sod is applied, as a short cut to planting grass seed. Sod is sold by the square yard.

Example 1

A lot 100 ft wide and 165 ft long is to be covered with topsoil to a depth of 4″. Topsoil will be delivered to the site for $4.00 per cu yd. What will the topsoil cost?

Solution

Changing the feet and inches to yards, we multiply to get the volume.

$$V = \tfrac{100}{3} \times \tfrac{165}{3} \times \tfrac{4}{36}$$

$$= \tfrac{100}{3} \times \tfrac{55}{1} \times \tfrac{1}{9}$$

$$= \tfrac{5500}{27}$$

$$= 203.7 \quad \text{(approx.)}$$

204 cubic yards are needed.

$204 \times 4.00 = 816.00$

The topsoil will cost $816.00

Example 2

The lot in Example 1 is to be sodded. The sod costs $.75 per sq yd. How much will the sod cost?

Solution

We convert to yards to find the area.

$$A = \tfrac{100}{3} \times \tfrac{165}{3}$$

$$= \tfrac{100}{3} \times \tfrac{55}{1}$$

$$= \tfrac{5500}{3}$$

$$= 1833.3 \quad \text{(approx.)}$$

1834 sq yd of sod is needed.

$1834 \times \$.75 = \$1375.50$

The sod will cost $1375.50.

EXERCISES

1. Fill can be bought for $3.50 per cubic yard. If 35 cu yd are needed for one job and 45 cu yd for another job, how much will the fill cost for both jobs?

2. If topsoil, delivered to the site, costs $4.50 per cu yd, what will 21 cu yd cost?

3. A 60' × 75' plot needs 4" of topsoil. How many cubic yards of topsoil should be ordered?

4. Sod is being ordered for a 40' × 60' yard. How many square yards should be ordered?

5. What is the cost of 375 sq yd of sod at $.65 per square yard?

6. What is the total cost of the following list of trees and shrubs?

    1 mountain ash tree at $14.00
    5 junipers at $12.00 each
    6 rose bushes at $2.75 each
    3 seedless ash trees at $5.00 each

7. For a 60' × 90' lot, 6 in. of topsoil are needed. At $4.50 per cu yd, what will the topsoil cost?

8. A contractor needs to lay sod for a 50' × 60' lot. If the sod costs $.75 per sq yd, how much will the contractor pay for the sod?

9. Topsoil can be delivered to a lot for $4.75 per cu yd. How much will it cost to lay topsoil for a 45' × 75' lot to a depth of 3"?

10. Figure 14-22 is a sketch of a house and lot. Find the cost of sodding this lawn if sod costs $0.60 per sq yd.

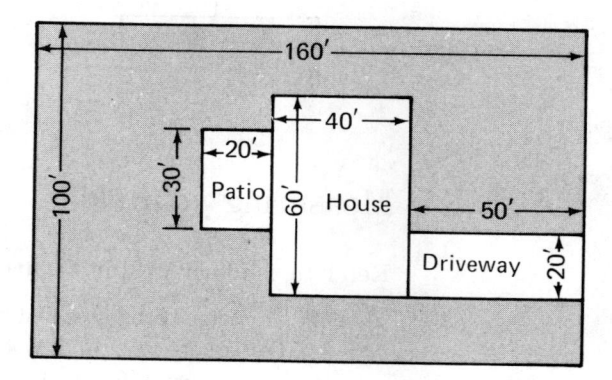

Figure 14-22

Taking inventory

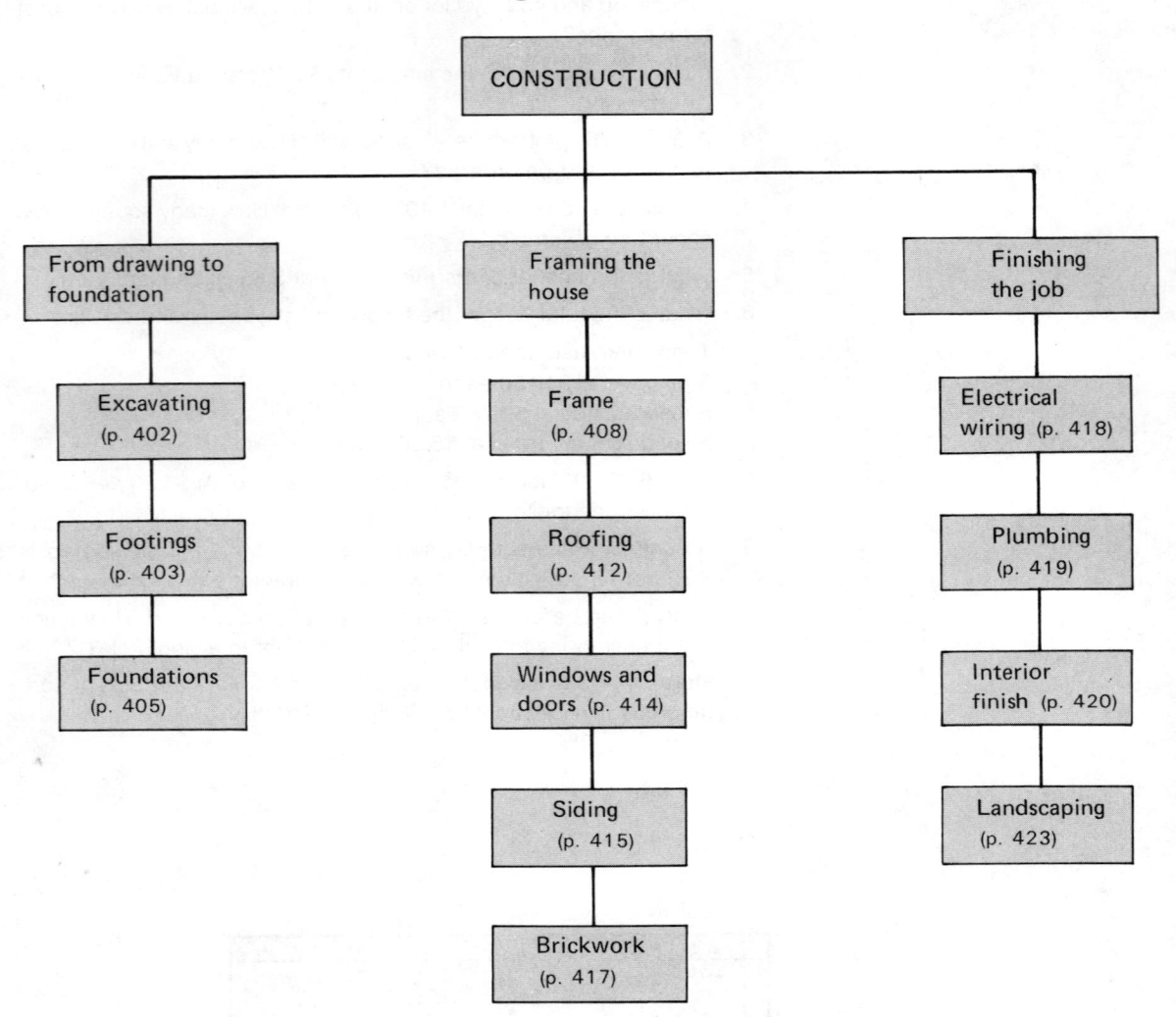

```
                    CONSTRUCTION
                         |
      ┌──────────────────┼──────────────────┐
From drawing to    Framing the         Finishing
  foundation          house             the job
      |                  |                  |
  Excavating           Frame          Electrical
  (p. 402)           (p. 408)        wiring (p. 418)
      |                  |                  |
   Footings           Roofing           Plumbing
  (p. 403)           (p. 412)           (p. 419)
      |                  |                  |
 Foundations       Windows and          Interior
  (p. 405)        doors (p. 414)      finish (p. 420)
                         |                  |
                      Siding           Landscaping
                     (p. 415)           (p. 423)
                         |
                     Brickwork
                     (p. 417)
```

Measuring your skills

Refer to the floor plan in Figure 14-23 for exercises 1–8.

1. The contractor estimates that the cost of building a house will be $16 per square foot. What is the estimated cost of construction of the floor shown in Figure 14-23? **(14-1)**

2. The excavation for the basement of the house is to be 8 ft deep. How many cubic yards of dirt must be removed? **(14-2)**

Figure 14-23

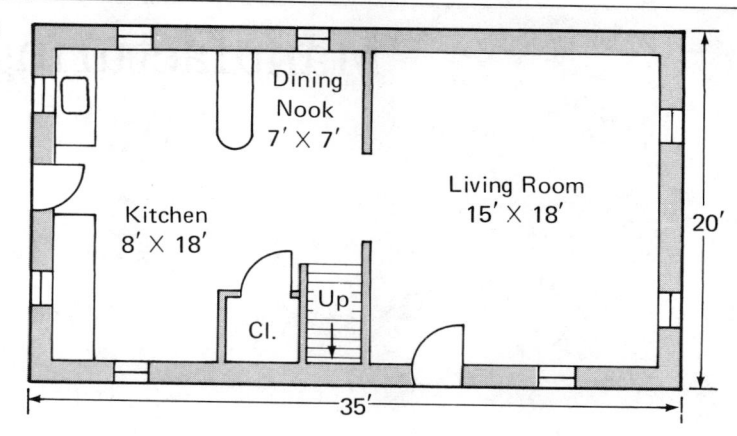

Floor plan: overall 35' × 20'. Rooms: Dining Nook 7' × 7', Kitchen 8' × 18', Living Room 15' × 18', with closet (Cl.) and stairs (Up).

3. The bearing capacity of the soil is 1600 lb per sq ft. If the house is to weigh 27,000 lb, what area of footing should the house have? **(14-3)**

4. Twenty 2 × 4 wall studs, each 8' long, are ordered for the living room. How many board feet is this? **(14-5)**

5. How many squares of shingles should be ordered for a gable roof on the house if the rafters are 12 ft long? **(14-6)**

6. What is the cost of 8 double-hung windows if the price of each is $60.55? **(14-7)**

7. How much siding is needed for the 35' × 20' wall having three windows? **(14-8)**

8. Bricks are laid in bonding that gives 660 bricks per 100 sq ft. How many bricks should be ordered for a one-layer-deep facing of a 35' × 10' wall? **(14-9)**

9. A 30-ampere fuse is placed in a 120-volt circuit. How many watts of power does this circuit provide? **(14-10)**

10. Find the cost of 38 ft of plastic piping at $.23 per ft. **(14-11)**

11. What is the cost of $9-per-sq-yd carpeting for a 15' × 18' room? **(14-12)**

15 Manufacturing

After completing this chapter, you should be able to:
1. Find the amounts of materials needed to produce various items.
2. Work problems relating to forming, joining, and separating materials.
3. Work problems relating to packaging and marketing.

Forming and conditioning

15-1 CASTING

A manufacturer must produce a product that people need or will use. The first step in making a new product is testing to see if people will buy it. Then the manufacturer must find the best way to make the product with the greatest profit.

Many different processes are involved in making products. One step is to take a material and make it into a desired shape. One method of shaping involves *casting*. This means pouring molten or fluid material into a mold. The mold forms the material, so that when it hardens, it will be the desired shape. Many kinds of metals, plastics, rubber, glass, and concrete are cast in molds. Bottles and tires are two examples of products made by casting.

Example

A new washing machine is to have a water pump made of cast aluminum. Each pump requires 2 lb of aluminum. How many tons of aluminum are needed to make 175,000 washing machines?

Solution

2 × 175,000 = 350,000

There are 2000 lb in 1 ton.

350,000 ÷ 2000 = 175

175 tons of aluminum are needed.

EXERCISES

1. A certain type of casting requires 2.54 lb of cast iron.
   a. How much cast iron is needed to make 175 castings?
   b. How many castings can be made from 1270 lb of molten iron?
2. One mold can be used to produce 5 castings per hour. How many castings can be produced by 2 molds in 3 hr?
3. After being cast in the mold, the casting cools at a rate of 15° every 2 min. How many minutes are needed for the casting to cool 75°?
4. A machine can produce cast soft drink bottles at a rate of 6 bottles per minute. How many bottles can be made by one machine in an 8-hr shift?
5. If molten iron is to be used for making castings, about 10% of it is lost during the casting process and is recycled for future use. How much iron is recycled if 1 ton of molten iron is used?
6. A machine will produce model cars at a rate of one car every 12 sec. Each car requires 1.5 ounces of plastic. How much plastic is needed when the machine runs for 3 hr?
7. After casting a block for an engine, 3% of the material is removed by drilling, tapping, grinding, and other processes that are necessary to produce a finished block. If the finished block weighs 175 lb, what was its original weight?
8. It costs a soft drink company 1.5 cents to produce one new bottle. The company uses 3 million bottles per year, of which 80% have been recycled and the rest are new. What is the cost of producing new bottles each year?

15-2 COMPRESSING AND STRETCHING

Another way of forming parts is by compressing or stretching. Sometimes this is done by large presses that shape the material.

Forging is a type of press work usually done with a heated metal that can be shaped as desired more easily. Figure 15-1 shows molten metal used in the forging process.

Figure 15-1

Example

It takes a pressure of 7 tons per square inch to press a design into a piece of metal. How much force will be needed to press a piece with a surface of 5 sq in.?

Solution

We multiply the pressure per square inch by the area.

$7 \times 5 = 35$

35 tons of force are needed.

Press work can also involve any *stamping* device that shapes the material. Parts for appliances and aluminum siding are commonly made by a stamping process.

When a material is needed in sheets, a *rolling* process is used to stretch the material. Rolling reduces the thickness of the material and increases the length or width. Some materials, like aluminum foil, are rolled at room temperature. In other cases, the piece of material is heated to make the rolling easier. This process is called *hot-rolling*.

EXERCISES

1. A pressure of 4 tons per square inch is needed to forge a metal part. How much force is required to forge a part that has an area of $9\frac{1}{2}$ sq in.?

2. A forging machine can produce 35 tons of force. An object to be forged is 17.5 sq in. How many pounds of pressure is this piece getting on each square inch?

U.S. system

3. A piece of aluminum 2 in. × 4 in. × 6 in. is rolled to a thickness of .01 in. It stays 4 in. wide. How long is it? (*Hint:* Volume is constant.)

4. Each time an iron bar passes through the roller of a rolling machine, its thickness is decreased by $\frac{3}{8}''$. How many times must a bar 4″ thick be rolled to produce a bar $2\frac{1}{8}''$ thick?

5. A press stamps metal parts. Each part contains .25 lb of metal. How many tons of metal are needed to produce 1 million parts?

Metric system

6. A piece of aluminum 10 mm × 25 mm × 100 mm is rolled to a thickness of .1 mm. If it stays 25 mm wide, how long is it?

7. Each time a metal bar is passed through a rolling machine, its thickness is decreased by 2 mm. How many times must a bar 15 mm thick be rolled to produce a bar 5 mm thick?

8. A press stamps metal parts. Each part contains 25 grams of metal. How many kilograms of metal are needed to produce 1 million parts?

15-3 CONDITIONING

Changing the properties of a material is sometimes needed. *Thermal conditioning* means changing a material by heat treatments. The exposure to extreme heat can add hardness and strength to some materials.

Materials can also be hardened by *chemical conditioning*. Many of the plastic parts in some of our machines today have been hardened by chemical treatments.

The properties of a material can also be changed by *physical conditioning*, such as magnetizing. For example, in upholstery shops, tack hammers have their heads magnetized to make it easier to apply tacks.

EXERCISES

1. A 2" round iron bar is heated and then cooled with water. The rate of cooling is 550° per sec. How many degrees will it cool in 3 sec?
2. The center of the iron rod in exercise 1 above will cool at the rate of 46° per sec. How long will it take the center to cool 230°?
3. If the same 2" round bar is cooled with oil instead of water, the rate of cooling at the surface is 81° per sec. How long will it take for the surface to cool 405°?
4. How long will it take for the surface of the bar in exercise 3 above to cool 1620°?

Separating materials

15-4 SHEARING AND CHIP REMOVAL

Materials can be altered by *separation* as well as by forming and conditioning. Here separating means to remove excess material. This can be done by shearing, by chip removal, or with chemicals, heat, and electricity.

Cutting paper with scissors is an example of *shearing*. Large industrial presses can produce enough pressure to shear metals, plastics, paper, fabrics, and other materials.

Separating materials by removing small bits and pieces is called *chip removal*. This is done by saws, drills, grinding wheels, milling cutters, and lathes.

Chip removal is usually done with a circular sawing motion. The cutting speed of any tool used for chip removal is important. The distance that the cutting edge of the tool moves in one revolution is the circumference of the tool. So the *cutting speed*, in feet per minute (fpm), is the circumference times the number of revolutions per minute (rpm).

S = Circumference in feet × rpm

$S = \pi \times$ Diameter in feet × rpm

However, the diameter of drills and grinding wheels is usually given in inches. So we can write the formula like this:

$$S = \frac{3.1416 \times \text{Diameter in inches} \times \text{rpm}}{12}$$

Since 3.1416/12 = .2618, the formula can be simplified as

$$S = .2618 \times d \text{ in inches} \times \text{rpm}$$

Example

A grinding wheel with a 9-in. diameter turns at 1850 rpm. What is its cutting speed?

Solution

S = .2618 × d × rpm

= .2618 × 9 × 1850

= 4358.97

The cutting speed is about 4359 ft per min.

EXERCISES   What is the cutting speed in each of the following?

1. A circular saw blade, 12 in. in diameter, rotating at 3450 rpm.
2. A grinding wheel, 6 in. in diameter, rotating at 1750 rpm.
3. A surface planer with blades $4\frac{1}{2}$ in. in diameter, rotating at 3600 rpm.
4. A router with a $\frac{3}{8}$-in.-diameter cutting bit, rotating at 25,000 rpm.
5. A $\frac{5}{8}$-in. drill bit rotating at 400 rpm.
6. A grinding wheel with an 8-in. diameter, rotating at 2500 rpm.
7. The recommended cutting speed for brass is 120 fpm. If a piece of stock 3 in. in diameter is being turned in a lathe, what should be the rpm of the lathe?
8. A high-speed steel drill bit has a cutting speed of 70 fpm. If the drill bit has a diameter of $\frac{3}{8}$ in., what is the rpm of the bit?
9. A piece of wood with a diameter of $\frac{1}{3}$ ft is being turned in a wood lathe at 550 rpm. What is the cutting speed?

15-5 SEPARATING BY HEAT, CHEMICALS, AND ELECTRICITY

Styrofoam can be cut by using a thin hot wire, and metals can be cut with the aid of a cutting torch. Separating with the aid of heat is called *thermal erosion.*

When acids are used to "eat away" material to create a certain pattern, design, or texture, the material is being *etched.* Etching is an example of *chemical separating.*

We can cause some materials to separate by passing electric currents through them. This is *electrochemical separating.*

If we can cause a material to break along a certain line we are using *induced-fracture separating.* Gem and glass cutters use this form of separating.

EXERCISES

1. A gas cutting torch is capable of cutting a certain type of steel at a rate of 18 in./min. How many inches of this steel can be cut in 1 hr?
2. It takes 40 sec to etch a printed circuit for a television set. A company wishes to produce 10,000 sets. How much time must be allowed for the etching of the circuits?
3. Electrochemical machining removes 2 milligrams of material per second from a fragile material. How many milligrams will be removed in 3 periods, each 10 seconds in length?
4. A glass-cutting device cuts pieces of glass for 8" × 11" picture frames. How many pieces will be cut from a sheet of glass that measures 48" × 60"?
5. A laser beam is capable of cutting the pieces for 150 suits in 15 minutes. Older techniques could cut only 5 suits in 20 minutes. How many times faster is the method of cutting with a laser beam?
6. A gas cutting torch will burn for 4 hr on one tank of gas. If the torch is capable of cutting 6 pieces per minute, how many pieces can be cut with one tank of gas?

Combining materials

15-6 MIXING AND COATING

When some items are manufactured it is necessary to combine materials or to coat them with some other material. One way of combining materials is with a *mixing process*. This means stirring ingredients together until they are spread evenly throughout the mixture. The mixing process is used in making many plastic parts. To make a plastic part, various chemical elements are mixed to form the fluid plastic, which later hardens as a result of a chemical reaction.

Materials can also be made more durable and attractive by *coating* them. Various kinds of paints are available and can now be applied easily.

Porcelain and ceramic coatings are used on large appliances, bathroom fixtures, and even rocket nose cones.

Protective coatings made of such metals as zinc, nickel, and chromium prevent rust and corrosion.

Sometimes metal coatings are applied by *electroplating*. In this process a direct current of electricity is used. The current causes particles of metal in a solution to be deposited on another metal.

EXERCISES

1. A building with 1500 sq ft of wall area is to be painted. A gallon of paint will cover about 400 sq ft. How much paint is needed to paint the building?
2. Paint sells for $6.00 per gallon or $2.00 when bought by the quart. Find the cost of paint for the building in exercise 1.
3. If a painter paints at a rate of 200 sq ft per hr, how many hours will it take to paint 1500 sq ft?
4. A current of 20 amperes per square foot is needed to nickel-plate a surface. How many amperes of current are needed to plate a surface of 3.5 sq ft?
5. A current of 20 amperes per square foot will nickel-plate a surface to a thickness of .001" per hour. How much time is needed to nickel-plate a surface to a thickness of .015" with a 20-ampere current?
6. A painter applies a primer coat at the rate of 150 sq ft per hr and is paid $5.50 per hr. How much will it cost to have the painter paint 1200 sq ft?
7. With a voltage of 115 volts, a current of 20 amperes per square foot will give a nickel plating .001" thick. How many watts of power are needed to nickel-plate an area of 1 sq ft to a thickness of .001"?

15-7 BONDING

A popular method of joining solid parts is by *bonding*. We call it *adhesive bonding* when the two solid parts are held together with an adhesive such as glue, paste, or cement. When the parts are heated until they flow together, then cooled to make a rigid bond, we call it *fusion bonding*.

Many types of adhesive materials are available for adhesive bonding. Those used in woodworking form a long-lasting rigid bond and are applied when needed. Others can be applied at a factory and are ready to use when needed, like the glue on envelopes.

Solder, which is a combination of tin and lead, is used as an adhesive for certain types of metals. Solder melts at a relatively low temperature. A plumber uses solder for joining copper pipes.

Welding is an example of fusion bonding. *Arc welding* uses electricity to fuse the materials. Electric welding equipment is used to weld many modern auto bodies.

EXERCISES

1. If 1 quart of contact cement will cover about 100 sq ft of surface, how much cement is needed for a counter top that measures 3 ft by 8 ft? Both the surfaces need to be covered with cement.

2. If 1 gal of white glue weighs 11 lb and sells for $5.44, what is the price per pound of this glue?
3. Rosin-core solder for use in printed circuits contains 60% tin and 40% lead. How much tin is contained in a 1-lb spool of this solder?
4. The glue sticks for an electric glue gun cost $2.39 for a box of 60 sticks. What is the cost per stick?
5. If one welding rod will weld a joint 8 in. long, how many rods are needed to weld a joint that is 3 ft 4 in. in length?
6. A package of hard-surface welding rods contains 23 pieces and costs $1.97. What is the price of each piece?

15-8 MECHANICAL FASTENERS

How many kinds of fasteners can you find in Figure 15-2?

Nails, hooks, snaps, staples, zippers, screws, bolts, and nuts are a few of the mechanical fasteners in use today. Bolts and screws are called *threaded fasteners*. The size, type, and nature of the thread is determined by the use that will be made of the bolts and screws. The bolts and screws used in a fine Swiss watch will differ from those used in a piece of heavy industrial equipment.

The majority of threaded fastening devices in the United States use the American National and Unified Thread Series.

Figure 15-2

Two of the most commonly used series are the National Coarse (NC) and the National Fine (NF). The NC series is used on most machine tools and the NF series is used in automobile work.

Screws and bolts are identified by their *diameter* and the *number of threads per inch*. A $\frac{5}{16}''$-18 NC bolt means a diameter of $\frac{5}{16}''$, 18 threads per inch, and National Coarse series.

EXERCISES

1. A package of 1000 staples for a stapler costs $2.41. What is the cost of 5 packages of staples?
2. How many threads per inch are there on a $\frac{3}{8}''$-16 NC threaded bolt?
3. A package of 100 assorted bolts weighing 5 lb, 8 oz sells for $3.39. What is the price per pound?
4. What is the maximum diameter bolt that will fit in a $\frac{1}{2}''$ diameter hole if $\frac{1}{64}''$ clearance is needed on all sides of the bolt?
5. If one bolt is capable of holding 2 tons, how many bolts are needed for a 60-ton load if we add 20% more bolts for safety?
6. How far will an 18 NC bolt advance in 4 revolutions?
7. How many threads are there on an 18 NF bolt $3\frac{1}{2}''$ long?

Testing and marketing the product

15-9 QUALITY CONTROL

In order to be able to sell a product, the manufacturer must be sure that it is free from major defects. It also should meet the requirements for which it was designed. The testing of a product before, during, and after its development is called *quality control*.

Inspection is one way of testing the quality of a product. Measuring various parts and even X-raying internal parts can determine the quality without destroying the product.

Another kind of testing destroys a product. An example of *destructive testing* is setting off a flash bulb to see if it will flash. Some products are tested to see how long it will take to wear out the product. Sometimes this is done by using the product the way it was designed to be used. For example, a car might be driven 50,000 miles, but in a period of only a few weeks. Other types of wear-testing use machines to wear out the product.

During the production of some items it may not be practical to test every item. Of course, this is true if destructive testing is involved. If it is impractical to test every item, a *representative sample* is tested.

Manufacturing 437

EXERCISES

1. At a bottling company the bottles pass by a visual inspection point at the rate of 5 bottles per second. How many bottles must be inspected each hour?

2. In a visual inspection on an assembly line, 18 defective items were found in an hour. If the items move through the line at the rate of 500 items per minute, what percent were defective?

3. A certain type of tire has $\frac{1}{4}''$ of usable tread. Testing indicates that under normal conditions $\frac{1}{32}''$ of tread is used for each 6000 miles of travel. For how many miles should this tire last?

4. During testing of transistors it was found that 5 out of 600 tested were defective. What percent were defective?

5. Standards set by a certain company require that the maximum percent of defective parts be .15%. How many defective parts are allowed for each 10,000 parts?

6. A factory produces 5000 light bulbs per 8-hr shift. One bulb of each 150 is tested. How many bulbs must be tested during each shift?

7. The thickness tolerance of cold rolled metal .250″ thick and 20″ wide is .0055″. What is the maximum and minimum thickness of this type of metal sheet?

15-10 PACKAGING

After a product has been completed, it must be prepared for distribution. Most finished products must be put in some type of container. This is called *packaging*.

The containers that are used for packaging are designed with a number of purposes in mind. A container should make it easier to handle the product. It should also protect the product, identify it, and look attractive.

The size, shape, and nature of the product will determine the kind of container that is needed. Cans, bottles, jars, bags, crates, boxes, and sacks are some of the types of containers being made today.

The shape of the container, as well as the type of material used, will determine its cost. For example, a cylindrical can with a radius of 2″ and a height of 9″ will contain the same volume as a can with a radius of 3″ and a height of 4″. However, about 132 sq in. of material are needed to make the can with a 3″ radius and a 4″ height, while about 138 sq in. of material are required to make the other can.

EXERCISES

U.S. system

1. A rectangular box has a length of 8″, a width of 6″, and a height of 5″. How many cubic inches does the box contain?

2. A rectangular box has a length of 10″, a width of 8″, and a height of 5″. What is the surface area of this box?

3. A container in the shape of a rectangular solid is being designed. It must have an 8″ square base and a volume of 1280 cu in. How high is the container?

4. A container must hold 450 cu in. If it is to be a box, find 2 sets of dimensions.

5. The rectangular-shaped bottom of a container has an area of 35 sq in. and a perimeter of 24 in. Find the dimensions of the bottom.

6. A container in the shape of a rectangular solid is to be designed to contain 512 cu in. Try to find the dimensions of the container using the least amount of material.

Metric system

7. A rectangular box has a length of 10 cm, a width of 8 cm, and a height of 4 cm. How many cubic centimeters does it contain?

8. A rectangular container has a length of 8 cm, a width of 6 cm, and a height of 12 cm. What is the surface area of the box?

9. A rectangular box has a 5 cm × 8 cm base. If it is to contain 480 cu cm, how high should it be?

10. A container must hold 360 cu cm. If it is to be a box, find 2 sets of dimensions.

11. A rectangular box has a volume of 224 cu cm. If the height of the box is 7 cm, find the area of the base.

12. A rectangular container 9 cm × 14 cm × 25 cm costs $.015 per sq cm to produce. How much will it cost to produce 100 containers?

15-11 MARKETING THE PRODUCT

The *marketing* of a product involves pricing, advertising, and selling. The price of the product is determined by many factors. The cost of making the product plus all the expenses of packaging, shipping, storing, advertising, and selling it determine the manufacturer's cost. In determining the *selling price*, the manufacturer adds an amount for profit. The profit cannot be too large, or the product will not sell. Also, many industries have government regulations about the amount of profit they can show. The following formula relates the selling price to the cost, expenses, and profit.

Selling price = Cost + Expenses + Profit

Example

What is the selling price of a washing machine if the cost to the dealer is $192.50, the expenses are $66.00 and the profit is $16.50?

Solution

Selling price = 192.50 + 66.00 + 16.50

= $275.00

EXERCISES

1. What is the selling price of an item if the merchant's cost is $2.98, expenses are $0.23, and profit is $0.29?
2. If an item sells for $10.95 and the cost and expenses to the dealer are $8.75, what is his profit?
3. An item sells for $12.98. The dealer knows that 20% of the selling price must be used to pay expenses. How much were the expenses on this item?
4. A company has gross sales of $1.5 million. It spends $120,000 for advertising. What percent of its gross is spent on advertising?
5. What is the percent of profits on an item that sells for $49.95 if the expenses are $5.00 and the cost is $39.95?
6. A dealer wishes to make a 15% profit on an item whose costs and expenses are $150.00. What should the selling price be? (*Hint*: The profit is a percentage of the selling price.)

Taking inventory

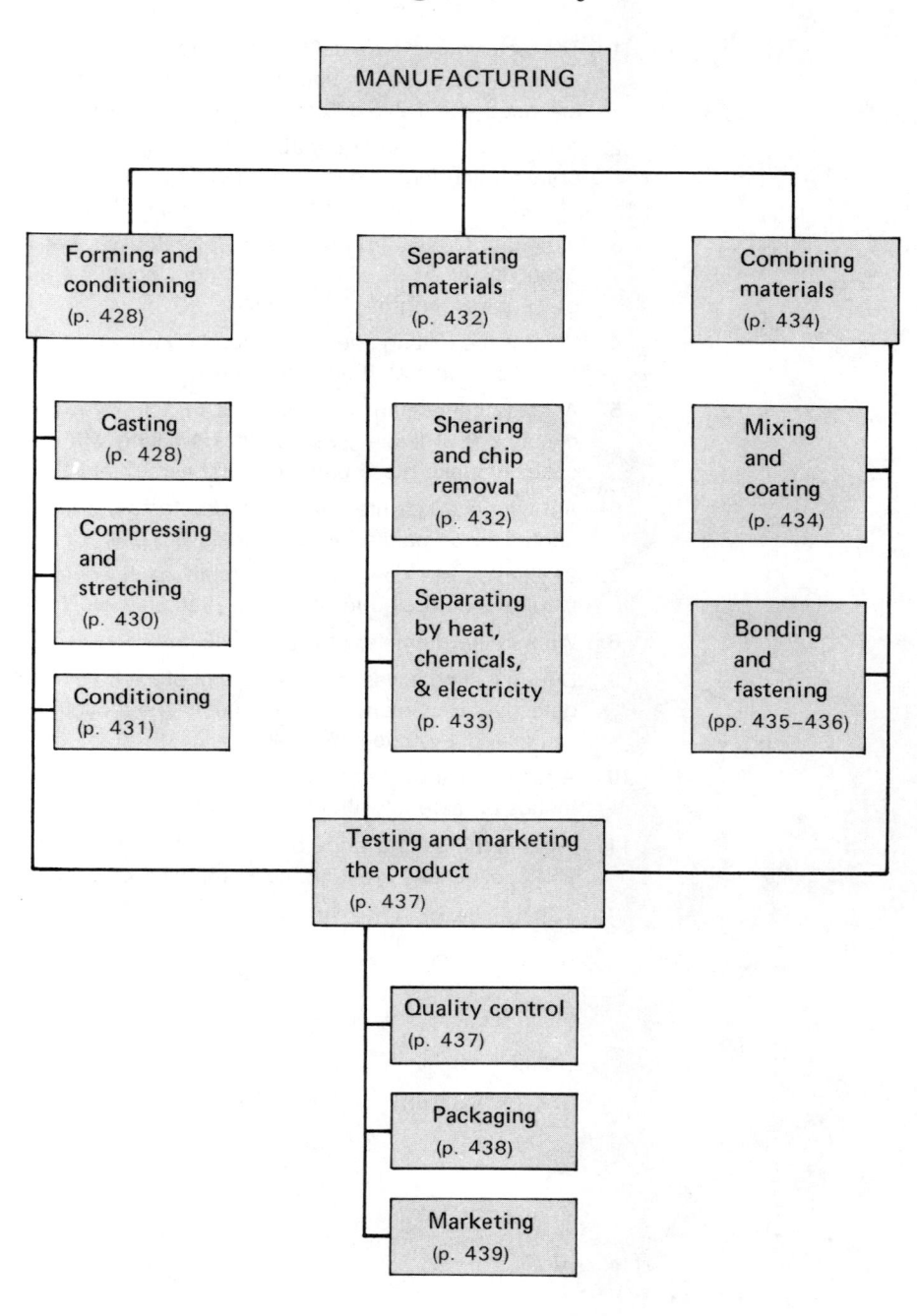

- **MANUFACTURING**
  - **Forming and conditioning** (p. 428)
    - Casting (p. 428)
    - Compressing and stretching (p. 430)
    - Conditioning (p. 431)
  - **Separating materials** (p. 432)
    - Shearing and chip removal (p. 432)
    - Separating by heat, chemicals, & electricity (p. 433)
  - **Combining materials** (p. 434)
    - Mixing and coating (p. 434)
    - Bonding and fastening (pp. 435–436)
  - **Testing and marketing the product** (p. 437)
    - Quality control (p. 437)
    - Packaging (p. 438)
    - Marketing (p. 439)

Measuring your skills

1. The water pump for a new automobile is made from cast iron. If it takes 2.74 lb of iron for each water pump, how many pounds of iron are needed if 1,275,300 cars are built? **(15-1)**

2. A forging machine is capable of exerting 80 tons of force. If the object being forged has an area of 27.5 sq in., how many pounds of pressure can be exerted on each sq in.? **(15-2)**

3. A certain type of iron part is heat-treated so that it can support 5000 lb per sq in. How much weight could be supported by a piece measuring 8" × 12"? **(15-3)**

4. What is the cutting speed of a grinding wheel with a 6" diameter if the wheel turns at 1650 rpm? **(15-4)**

5. A glass cutter can cut 500 parts in 1 hr. A manufactured item contains 8 of these glass parts. How long will it take the glass cutter to make glass parts for 5000 items? **(15-5)**

6. An automated painter can paint 135 items per minute. How many items can be painted in an 8-hr shift? **(15-6)**

7. 21 staples are needed to assemble each container. How many staples are needed for 25,000 containers? **(15-7)**

8. What is the diameter of a $\frac{9}{16}$"-12 NC bolt? **(15-8)**

9. During testing it was found that 1 out of each 150 items tested was defective. How many defective items should you expect to find if you tested 1200 items? **(15-9)**

10. A rectangular container measures 5" × 3" × $8\frac{1}{2}$". How many cubic inches of material will it hold? **(15-10)**

11. What is the selling price of an item if the cost to the merchant is $5.23, expenses are $.52, and profit is $.48? **(15-11)**

16 Graphic arts

After completing this chapter, you should be able to:
1. Calculate space with points, picas, ems, and column inches.
2. Describe the differences between letterpress, lithography, and gravure printing.
3. Calculate materials, cost, and time involved in the steps in printing.

Measuring type

16-1 THE POINT SYSTEM

The *point* and the *pica* are the units used in printing in all English-speaking countries. The *point system* is used to measure type sizes, to mark copy, and to measure distances in printed matter.

A *point* measures .01384″, or approximately $\frac{1}{72}$ inch. So there are 72 points to an inch. One *pica* is approximately $\frac{1}{6}$ inch, or 12 points. So we have these approximate measures:

6 picas = 1 inch

12 points = 1 pica

72 points = 1 inch

Figure 16-1

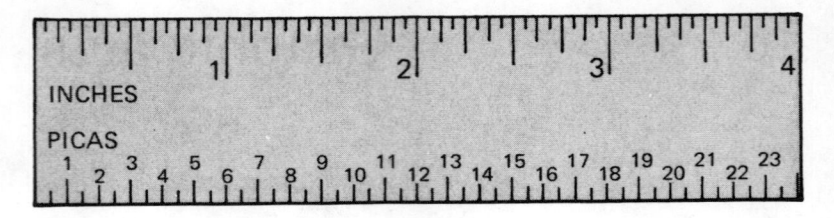

Figure 16-2 ─────────────────────────────

The pica ruler shown in Figure 16-1 shows units of ½ pica, or 6 points. To see how big 1 point is, look at the width of the line in Figure 16-2. You are looking at a 1-point line.

Many different sizes and styles of typefaces are available today. Figure 16-3 shows a variety of styles as well as sizes.

Figure 16-3

This is 24 point Lucia Script

This is 24 point *Palatino Italic*

This is 24 point Bodoni Book

This is 18 point Imperial Bold

This is 18 point Optima Semibold

THIS IS 14 POINT NEWS GOTHIC BOLD

This is 12 point Baskerville

This is 10 point Futura Bold Condensed

This is 12 point Futura Bold Condensed

For most purposes it is important for type to be readable. Typefaces that resemble handwriting or elaborate script are sometimes used to give a decorative appearance.

It is sometimes necessary to convert measurements given in picas and points to measurements in inches.

Example

How many inches long is a 39-pica line?

Solution

Since there are 6 picas to an inch, we divide 39 by 6:

$39 \div 6 = 6\frac{1}{2}$

The line is $6\frac{1}{2}$ in. long.

EXERCISES Change the following measures to inches.

1. 18 picas
2. 23 picas
3. 42 picas
4. 24 picas
5. 36 picas
6. 41 picas
7. 20 picas
8. 48 picas
9. Give the size of a card in picas if it is $4\frac{1}{2}''$ wide and $6\frac{1}{2}''$ long.
10. A form has a width of 54 picas and a depth of 78 picas. Find the area, in square inches, of this form.
11. Copy that measures 39 picas wide is centered on a page that is $8\frac{1}{2}''$ wide. How wide are the side margins in picas? in inches?

16-2 THE EM

In the printing trade, picas and points are used to measure length. A printer measures area in terms of a unit called an em. An *em* is the area occupied by the letter M in a given type size. The printed capital letter M always takes up a perfectly square space. Thus an em is as long as it is wide. As the point size of the type varies, so does the size of an em of that size. Figure 16-4 shows the size of an em in four different sizes of type.

Figure 16-4

■ Em of Six Point Type.
■ Em of Eight Point Type.
■ Em of Ten Point Type.
■ Em of Twelve Point Type.

Because the em is a measure of area, it can be used to determine how much type will fit in a given area.

Example

How many ems of 8-point type are there in a page with type set 288 points wide and 504 points deep?

Solution

To find the number of ems in one line, we divide the length in points by 8:

288 ÷ 8 = 36 ems per line

To find the number of lines on the page, divide the depth in points by 8:

504 points ÷ 8 = 63 lines

63 lines × 36 ems per line = 2268 ems

EXERCISES

1. How many ems of 7-point type are there in a line 217 points long?
2. How many ems of 5-point type are there in a line 24 picas long?
3. How many points of length are used by 27 ems of 4-point type?
4. How many picas of length are used by 15 ems of 12-point type?
5. How many lines of 6-point type are there in a page 546 points deep?
6. How many lines of 8-point type are there in a page 44 picas deep?

How many ems are there in pages of these sizes set with the given type size?

7. A page 291 points wide, 504 points deep, set with 7-point type.
8. A page $4\frac{1}{4}$ inches wide, 7 inches deep, set with 6-point type.
9. A page 24 picas wide, 36 picas deep, set with 8-point type.
10. A page $4\frac{1}{3}$ inches wide, $7\frac{1}{2}$ inches deep, set with 12-point type.

16-3 COLUMN INCH

In newspaper publishing a different standard measure is used, the *column inch*. One column inch is an area one column wide and one inch deep. The American Newspaper Publishers Association has set the standard width of a newspaper column at $11\frac{1}{2}$ picas. Advertising in newspapers is usually priced by the column inch. The price per column inch depends on the newspaper.

Example

Find the cost of an advertisement 2 columns wide and 3 inches deep at $3.60 per column inch.

Solution

First we find the number of column inches.

2 columns × 3 inches = 6 column inches

6 × $3.60 = $21.60

The cost is $21.60.

EXERCISES Find the number of column inches in an advertisement that is
1. 1 column wide and 3 inches deep.
2. 2 columns wide and 1 inch deep.
3. 3 columns wide and 2 inches deep.
4. 5 columns wide and 5 inches deep.
5. 8 columns wide and $20\frac{1}{2}$ inches deep.

Find the cost of an advertisement that is:
6. 6 column inches at $2.40 per column inch.
7. 3 columns wide and 2 inches deep at $5.60 per column inch.
8. 8 columns wide and $10\frac{1}{4}$ inches deep at $4.80 per column inch.
9. 4 inches deep and 3 columns wide at $7.50 per column inch.
10. What is the cost of a full-page advertisement if a full page is 8 columns wide and $20\frac{1}{2}$ inches long and the advertising rate is $18.50 per column inch?
11. If a full page is 8 columns wide and $20\frac{1}{2}$ inches deep, find the cost of a half-page advertisement at $9.50 per column inch.

Printing processes

16-4 LETTERPRESS

The method by which a printed image is transferred to paper depends on the kind of printing process used. The oldest method of printing is the *letterpress*. In this process, ink is applied to a raised surface and transferred directly to the paper by means of pressure.

There are three kinds of presses used in this process: the rotary press, the cylinder press, and the platen press. All kinds of printed material, including newspapers, pamphlets, catalogues, books, and packaging materials, are printed on these presses. The most common type of press used for books is the *rotary press*. Two- and four-color work is usually printed on rotary presses. A continuous roll, or *web*, of paper is used in web-fed rotary presses.

EXERCISES

1. If a cylinder press can print 5000 sheets per hour, what is the cost of printing 3000 sheets at $25.75 per hour?

2. A folded-sheet pamphlet is printed on a sheet-fed rotary press which runs at 35 sheets per minute. Two passes through the press are necessary to print each pamphlet. How many pamphlets can be printed in 1 hour?

3. A carton can be printed and cut on a flat-bed cylinder press at the rate of 800 cartons per hour. How long will it take to print and cut 22,000 cartons?

16-5 LITHOGRAPHY

One of the newest and fastest-growing printing processes is *lithography*, sometimes called *offset printing*. Lithography is printing from a *flat* surface, rather than a raised surface, as in letterpress. The basic principle behind lithography is that greasy surfaces will attract greasy ink and repel water, and wet surfaces will attract water and repel greasy ink.

Figure 16-5

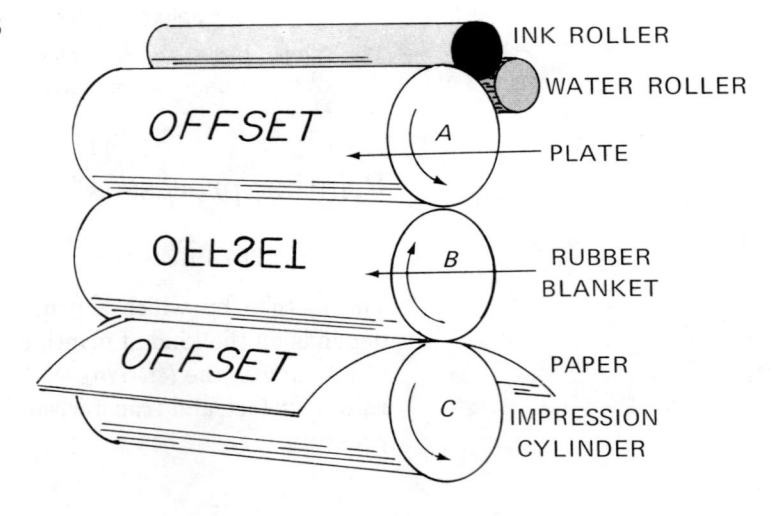

Photo-offset lithography is a type of lithographic process that is becoming very popular today. *Offset* refers to the process that uses three cylinders, instead of the two used in letterpress. Figure 16-5 illustrates the three cylinders used in an offset press. The plate is inked and watered on A, and then prints onto the rubber blanket on B. The rubber blanket *offsets* this ink impression to the paper held in the impression cylinder C. The rubber blanket prints a clearer impression on a wide variety of papers and other materials than the metallic plates used in letterpress.

Another way that photo-offset printing differs from letterpress is that anything that can be photographed can be used as original copy. So illustrations and photographs can be used with less preparation, and therefore with less expense.

The image carrier placed on the plate can be made of a variety of materials. There are plastic carriers that cost as little as 10 cents for a $8\frac{1}{2}'' \times 11''$ page. For long press runs, the plates are made of combinations of copper and aluminum. Such plates can produce up to 500,000 copies.

Small printing jobs are frequently run on *offset duplicators*. They are ideal for quick printing of office and plant forms, inventory sheets, price lists, bulletins, sales letters, programs, and schedules. Plates for offset duplicators can be prepared easily.

EXERCISES

1. If it costs $4.50 to make a copper-aluminum plate for printing 8 pages, what will the plates cost for printing 128 pages?
2. A printer will do small printing jobs for 1¢ per page plus 10¢ for the lithographic plate. What is the cost of a job that involves 250 pages?
3. What will it cost to print 250 copies of the program for a high school band concert if the printing office charges 10 cents for the lithographic plate and two cents for each copy printed?
4. If an offset press can print 50,000 pages per hour, how long will it take to print 1,000,000 pages?
5. An offset duplicator can print 300 copies per minute from an aluminum plate. How long will it take to produce 1800 copies?

16-6 GRAVURE

While letterpress uses a *raised* surface, and offset a *flat* surface, *gravure* printing uses a *sunken* or *depressed* surface for transferring the image. The lines to be printed are cut into the plate, and the plate is coated with ink and then wiped off. The ink remaining in the depressed surface is then deposited on the paper. This makes the ink stand out from the paper or printing surface. Figure 16-6 illustrates these three types of surfaces.

Figure 16-6

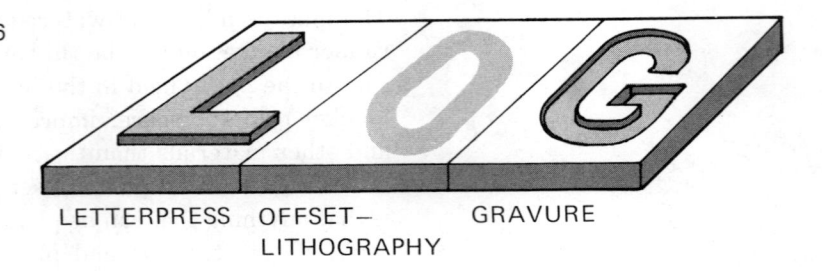

LETTERPRESS   OFFSET—   GRAVURE
              LITHOGRAPHY

Gravure printing is considered to be the finest method of reproducing pictures, but the higher cost of plate-making limits its use to long runs. As with the rotary letterpress, gravure presses are made for both sheets (sheet-fed gravure) and rolls (rotogravure) of paper. Some publications commonly printed on rotogravure presses are Sunday newspaper magazine sections, premium stamp catalogues, and large mail-order catalogues. Some interesting examples of gravure printing are vinyl floor coverings, vinyl upholstery fabrics, and pre-pasted wallpaper.

EXERCISES

1. A rotogravure press can print 10,000 copies per hour, and an offset press can print 50,000 copies per hour. If both presses run for 3 hours, how many more copies will be printed with the offset than with the rotogravure?.

2. If a rotogravure press can print 10,000 copies per hour, how long will it take to print 25,000 copies?

3. If a rotogravure plate costs $5.00 and the paper costs $.003 per copy, what will the plate and paper cost for 20,000 copies of one page?

4. A printer charges 3 cents per page for the first 5000 copies, 2 cents per page for the next 5000 copies and 1 cent for each page over 10,000. What will it cost to have 12,000 copies printed?

5. A printer charges $6.00 for making a rotogravure plate, 2 cents per copy for the first 5000 copies, and 1 cent per copy for each copy over 5000. What will it cost for a plate and 12,000 copies?

6. Each rotogravure plate costs $3.50, and each printed page costs 2 cents. What will it cost to make 10,000 copies of a 4-page bulletin? Assume one plate is needed for each page.

Steps in printing

16-7 PREPARING COPY

There are many steps in making a finished book, magazine, or advertising pamphlet. The first step is always preparing the original copy. Usually it is carefully typewritten, so that it is easy to read. The typed copy is sometimes prepared by a technical typist, who may be paid by the page or by the hour.

For some offset lithography presses, the copy is prepared on special typewriters. These machines produce the desired typeface with the proper spacing.

Another part of preparing the copy is planning space for illustrations or photographs that will appear in the finished copy.

EXERCISES

1. If a typist can type 80 words per minute, how long will it take to type a 3200-word manuscript?

2. If it takes 2 hours to type a 6000-word manuscript, what is the average number of words being typed per minute?

3. If a typist can type a page in 5 minutes and is paid 60 cents per page, what is the hourly rate?

4. If a typist can type 45 words per minute, how long will it take to type an average page containing 280 words?

5. If it takes 20 seconds to insert the paper in the typewriter, 5 minutes to type the page, $1\frac{1}{2}$ minutes to proofread the page, and 5 seconds to remove the page, how much time is needed to complete 1 page?

6. A technical manuscript is being typed for $1.20 per page. If there are 200 words per page, what is the cost of typing a 15,000-word manuscript?

7. A typist can average 40 words per minute on a certain 18,000-word manuscript. There will be a total of 90 typed pages. Should the typist take the job for 70 cents per page or $8.00 per hour?

8. A typist is hired at $5.00 per hour to type a manuscript that can be done at a rate of 1 page every 6 minutes. What is the cost per page of typing this manuscript?

9. A manuscript involves 30 pages of regular typing at 60 cents per page, 10 pages of technical typing at 80 cents per page, and 5 pages of charts and tables at $1.00 per page. What is the cost of typing this manuscript?

16-8 COMPOSITION

The next step in the printing process, after the preparation of the original copy, is the composition. *Composition* involves determining the amount of copy, calculating the area needed, selecting or producing the proper style and size of type, and setting the type.

There are many different kinds of machines for setting type. The earliest kind of typesetting, dating back to the 15th century, was hand setting. Today most type is set by machines, with increasing computer assistance. Type is cast from hot metal on *linotype* and *monotype* machines. In linotype work, a whole line is molded in one piece, called a slug. Monotype uses individual molded letters, usually set at the direction of a punched tape.

A newer method of setting type is *photocomposition*. The various kinds of photocomposing machines produce type directly on film. This film will later be used to make the plates used in the various printing processes. The machines used in photocomposition include *linofilm* and *monophoto*, similar to linotype and monotype.

Computers are being used increasingly to set up the line breaks and word breaks from a punched tape containing the characters of the words in a manuscript.

Composition is usually measured in *ems* and the rate of setting type is given in ems per hour. Recall from section 16-2 that an em is a square of space in a given type size.

Example

A linotype operator can set 3500 ems per hour. How many pages of 12-point type can the operator set in 7 hours if each page measures 42 picas × 50 picas?

Solution

Determine the number of ems on one page:
12 points = 1 pica, so the em of 12-point type measures 1 pica × 1 pica.

42 × 50 = 2100

There are 2100 ems on one page.
In 7 hours, the operator can set 7 × 3500, or 24,500 ems.

24,500 ÷ 2100 = $11\frac{2}{3}$

The operator can set $11\frac{2}{3}$ pages in 7 hours.

EXERCISES

1. One page of copy contains 1280 ems of type. A monotype operator can set 2000 ems per hour. How long will it take to set 50 pages of copy?
2. The cost of setting type in linofilm is $15.00 per hour. What is the cost of linofilm composition for a 250-page book if the operator can set 2 pages per hour?
3. Determine the number of ems of 9-point type on a page if the page measures 30 picas × 45 picas.
4. A linotype operator can set 1 page of 12-point type that is 39 picas × 55 picas in 1 hour. What is this rate, in ems per hour?
5. In 1 hour, a linotype operator can set 1 page of 10-point type that is 30 picas × 40 picas. How many ems per hour is this?
6. The operator of a computer-assisted linofilm machine can set 5000 ems per hour. How long will it take to set 50 pages of 10-point type if each page measures 27 picas × 36 picas?

16-9 PHOTOENGRAVING

Once the type has been set, in metal or on film, the next step is making the plates to be used on the printing press. At this point there may be photographs or line drawings to be added. *Camera copy* is compiled using the printed type, if hot metal has been used, along with the photos or line art to be included. If the type was set in film, then the plates are made directly from the film. *Photoengraving* is the process of etching the metal plates to be used on the press.

A *line plate*, which consists of lines and type characters, but no grey tones, is engraved using a simple photographic and chemical process. For letterpress, the plates are etched in a metal such as zinc or copper. In the lithographic process, the etching is done on aluminum or plastic bases.

To reproduce a photograph, a *halftone* engraving is made. Making a halftone is similar to making a line engraving. A halftone *screen* is placed between the camera lens and the film. This screen, or grid, breaks up the blacks and whites in a photograph into tiny dots that are etched into the printing plates. The reproduction of color photographs is similar, except that four separate plates are usually made. Red, blue, yellow, and black printed on top of each other will combine to give the image of the colors in the original photograph.

Photographs are frequently reduced or enlarged from their original size to fit the space allowed for in the copy. This reduction process may be done in the final plate-making stage, or it may be done earlier if the type is set in film. To determine the reduced size of a photograph, we can use a *proportion*. The ratio of the width and depth of the original photo is equal to the ratio of the width and depth of the reduction or enlargement desired.

$$\frac{\text{Original width}}{\text{Original depth}} = \frac{\text{Final width}}{\text{Final depth}}$$

Example

A 5" × 7" photograph must be reduced so that it is 3½" wide. What is the depth of the reduced picture?

Solution

$$\frac{\text{Original width}}{\text{Original depth}} = \frac{\text{Final width}}{\text{Final depth}}$$

$$\frac{5}{7} = \frac{3\frac{1}{2}}{d}$$

$$5 \times d = 7 \times 3.5$$

$$d = \frac{7 \times 3.5}{5}$$

$$d = 7 \times .7$$

$$d = 4.9$$

The depth of the reduced picture is 4.9".

EXERCISES

1. The cost of engraving plates for a book is $20 per page, plus $5 for each halftone. If the book has 412 pages and 42 halftones, what is the cost of the plates?

2. The cost of preparing plates for a book is $21,000. If the book has 360 pages, what is the cost per page?

U.S. system

3. An 8" × 10" photograph must be reduced so that its width is 2⅝". What is the resulting depth?

4. A 4" × 5" photograph must be reduced to fill a column that is 20 picas wide. What will be the resulting depth?

Metric system

8. A 60 mm × 70 mm photograph must be enlarged so that it is 75 mm wide. What is the resulting depth?

9. A 200 mm × 250 mm photograph must be reduced to a width of 85 mm. What will be its depth?

5. An 8″ × 10″ photograph is reduced to 4″ × 5″. Find the percent of reduction of the width and depth.

6. An 8″ × 10″ photograph is reduced to 4″ × 5″. The area of the reduced photo is what percent of the area of the original?

7. A 4″ × 5″ photograph is reduced to 80% of its original dimensions. What percent of the area of the original is the area of the reduction?

10. A 200 mm × 250 mm photograph is reduced to 50% of its original size. What are the resulting dimensions?

11. A 150 mm × 200 mm photograph is reduced to 75 mm × 100 mm. The area of the reduced photo is what percent of the area of the original?

12. A 240 mm × 320 mm photograph is reduced to 75% of its original dimensions. What percent of the area of the original is the area of the reduction?

16-10 PAPER

A variety of papers is available for printing. Paper may be selected for its appearance, for its lasting quality, for the texture of its surface, or for its cost and weight. The four basic classifications of paper and sizes presently in use are given in Table 16-1.

A *ream* of paper consists of 500 sheets of a basic size. The *substance* of paper is the weight of a ream, or 500 sheets. For example, paper is labeled "20-pound bond" if 500 sheets (one ream) of size 17″ × 22″ weigh 20 pounds. A heavier paper means each sheet is thicker.

Most countries in the world use the metric system. Like many industries, the graphic arts industry is adopting metric standards to compete in foreign markets. To standardize paper sizes while

Table 16-1 *Classification of paper and sizes*

Classification	Use	Basic sheet size
Bond	Standard office use, letterheads, typing, etc.	17″ × 22″
Book	Books	25″ × 38″
Cover	Magazine covers	20″ × 26″
Card	Posters, advertising displays	$22\frac{1}{2}$″ × $28\frac{1}{2}$″

converting to metric standards, the International Standards Organization (ISO) has planned new paper sizes. Metric units are used, and each size in a series is half the size of the one which comes before it. Figure 16-7 shows how seven different sizes of paper can be cut from one sheet. The large sheet in the A series has an area of 1 sq m, so A1 is ½ sq m, A2 is ¼ sq m, and so on. When paper sizes are changed, printing equipment and packaging materials will also have to be changed. With education and planning we can make these changes with a minimum of problems.

Occasionally paper is ordered in special sizes, that is, sizes other than the basic sizes given in Table 16-1. The weight of a special size is proportional to its increased area. So to find the weight of a special size, we can use the ratios of the area and weight of the paper.

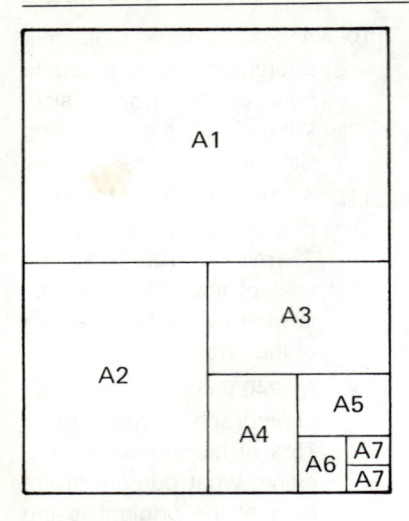

Figure 16-7

Example

Find the 500-sheet weight of 18-pound bond in a 20″ × 28″ size.

Solution

$$\frac{\text{Weight of special size}}{\text{Area of special size}} = \frac{\text{Weight of standard size}}{\text{Area of standard size}}$$

$$\frac{w}{20 \times 28} = \frac{18}{17 \times 22}$$

$$\frac{w}{560} = \frac{18}{374}$$

$$w \times 374 = 18 \times 560$$

$$w = \frac{18 \times 560}{374}$$

$$= \frac{10080}{374}$$

$$= 27 \text{ pounds} \quad \text{(approx.)}$$

500 sheets weigh about 27 pounds.

Although the basic size of bond paper is 17 × 22 inches, bond is often used for letterheads measuring 8½ × 11 inches. Figure 16-8 indicates how a 17″ × 22″ sheet can be cut to give 4 sheets each 8½″ × 11″.

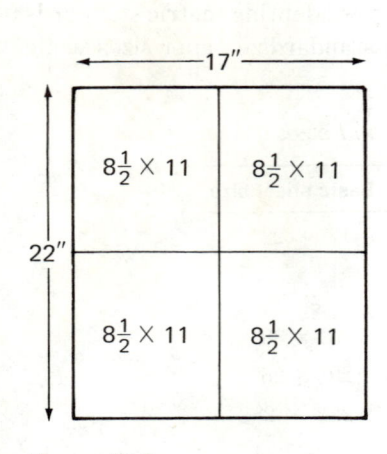

Figure 16-8

456   Technical and industrial applications

EXERCISES

1. If 500 sheets of 17″ × 22″ bond weigh 20 pounds, how much will 2000 sheets weigh?
2. What is the cost of 2500 sheets of 25″ × 38″ book paper, substance 20, if the paper costs 38 cents per pound?
3. Find the 500-sheet weight of 80-pound book paper in 35″ × 46″ sheets.
4. What is the maximum number of 8″ × 9″ sheets that can be cut from one 25″ × 38″ sheet?
5. If 1000 sheets of 17″ × 22″ bond weigh 56 pounds, what is the substance number of this paper?
6. Find the weight of 1500 sheets of substance 120 book paper in a special 38″ × 52″ size.
7. How many 8½″ × 11″ letterheads can be cut from two 500-sheet reams of 17″ × 22″ bond paper?
8. A sheet of 28″ × 42″ book paper is cut into 7″ × 9″ pieces. What is the maximum number of pieces that can be cut from one sheet?
9. A printing job requires 3 reams of substance 24 paper at 48 cents per pound and 2½ hours of labor for typesetting and printing at $6.50 per hour. What was the total cost of the job?
10. What is the area, in square meters, of A3? A4? A5? A6? A7?
11. What is the area, in square centimeters, of A2? A3? A4?

16-11 FINISHING

After the printing is done, there may be other steps necessary to produce the final product. *Finishing* involves folding, binding, trimming, gluing, or any other activity that is needed to get the product ready for use.

In producing a book, magazine, or other publication, a printer ordinarily prints several pages on each side of a single sheet of paper. This sheet is then folded and trimmed to form a *signature* and makes up one section of the publication.

The simplest signature involves two pages. This would be one sheet printed on both sides with no folding. Usually signatures will have 4, 8, 16, 32, or 64 pages. Take a sheet of paper and fold it to see if you can figure out why signatures have 4, 8, 16, 32, or 64 pages.

When printing a 16-page signature, it is necessary to arrange the printing of the pages on each side of the sheet so that after folding and trimming, the pages will be in order. The arrangement of type pages is called the *imposition*. Figure 16-9 shows an imposition for each side of a sheet for a 16-page signature.

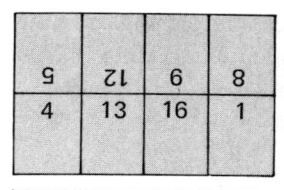

8	9	12	5
1	16	13	4

2	15	14	3
7	10	11	6

Figure 16-9

A variety of paper-folding machines is available. Some machines can fold a single sheet 5 times to produce a 64-page signature.

Paper for newspapers comes in large rolls and is fed into the presses in a continuous sheet. Rotary folders attached to the delivery end of these presses can fold paper at a rate of 1600 feet per minute. This is equivalent to about 20,000 sheets per hour. If we know the speed of the folding machine and the number of folds required, we can determine how many newspapers can be produced per hour.

Example

How many 32-page newspapers can be produced per hour if the folding machine can fold 20,000 sheets per hour?

Solution

If a sheet of newspaper is cut and folded once to produce a 4-page signature, we must divide the number of pages by 4 to get the number of folds per newspaper.

$32 \div 4 = 8$

Thus we have 8 folds per newspaper.

$20,000 \div 8 = 2500$

Thus our folding machine can produce 2500 32-page newspapers per hour.

Most of the other finishing activities, such as binding, punching, trimming, or stapling, are also done by machines. If we know the rate at which each machine works, we can determine the time that is needed to complete a job. Once we know the time that is needed, we can calculate the costs involved.

EXERCISES

1. How many times must a single sheet be folded to produce an 8-page signature?
2. If a paper-folding machine can fold 20,000 sheets per hour, how long will it take to fold 300 sheets?
3. How many folds are necessary if each sheet of a printing job is folded twice and the job involves 1350 sheets?
4. A machine can trim and bind 5 books per minute. How long will it take to trim and bind 10,000 copies?

5. It costs $3.50 per hour to operate the machine in Exercise 4. What is the cost of trimming and binding the job?
6. How many folds are necessary to produce a book that contains ten 32-page signatures?
7. If a machine can make 20,000 folds per hour, how long will it take to do the folding for 5000 copies of the book in exercise 6?
8. It takes 17 trees to make a ton of newsprint. How many trees can be saved by making 85,000 tons of newsprint with 100% recycled paper?
9. A company uses stationery made of 65% recycled paper. If the company uses 586,000 kilograms of paper in one month, how much of the paper has been recycled?

Taking inventory

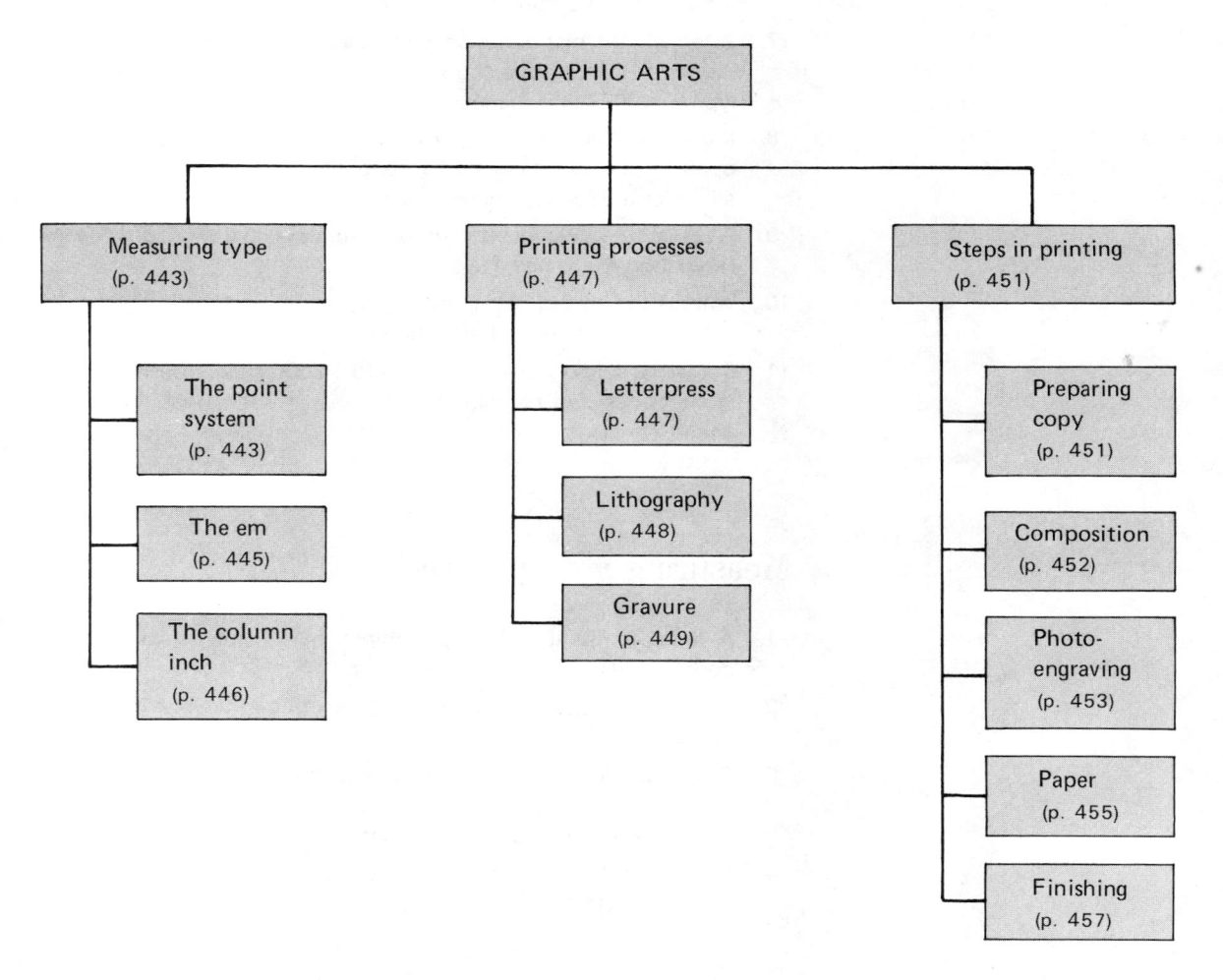

- **GRAPHIC ARTS**
  - **Measuring type** (p. 443)
    - The point system (p. 443)
    - The em (p. 445)
    - The column inch (p. 446)
  - **Printing processes** (p. 447)
    - Letterpress (p. 447)
    - Lithography (p. 448)
    - Gravure (p. 449)
  - **Steps in printing** (p. 451)
    - Preparing copy (p. 451)
    - Composition (p. 452)
    - Photo-engraving (p. 453)
    - Paper (p. 455)
    - Finishing (p. 457)

Measuring your skills

1. A column of type is 4½". How many picas is this? **(16-1)**
2. How many ems of 8-point type are there in a line 368 points long? **(16-2)**
3. Find the cost of an advertisement that is 3 columns wide and 4 inches long if the rate is $12.00 per column inch. **(16-3)**
4. A sheet-fed rotary press can print 4000 sheets per hour. How long will it take to print 5000 copies of a pamphlet if 3 sheets are used for each pamphlet? **(16-4)**
5. It costs $3.50 to make a lithographic plate and $.015 per page for paper. What is the cost of plates and paper for 100 copies of a 4-page bulletin if one plate per page is required? **(16-5)**
6. If a rotogravure press can print 8000 copies per hour, how long will it take to print 22,000 copies? **(16-6)**
7. A technical typist can type 4 manuscript pages per hour. At the rate of 50¢ per page, how much should the typist be paid for typing a 300-page manuscript? **(16-7)**
8. It costs $14.00 per hour to run the linofilm machine and pay the operator. If 3 pages can be set per hour, what is the cost of composition for a 320-page book? **(16-8)**
9. A 200 mm × 250 mm photograph is reduced to a width of 120 mm. How deep will it be? **(16-9)**
10. What is the weight of paper for a book that contains 10 sheets of substance 85 book paper? **(16-10)**
11. If a paper-folding machine can fold 10,000 sheets per hour, how many books can be folded in an hour if each book requires 8 folds? **(16-11)**

Measuring your progress

1. A force of 50 lb raises an object 14 ft. How much work is done? **(13-1)**
2. Find the displacement of a V-8 engine with a stroke of 4.5 in. and a bore of 3.25 in. **(13-3)**
3. A cylinder contains 400 cu cm of space when the piston is at the bottom of its stroke and 50 cu cm when it is at the top of its stroke. What is the compression ratio? **(13-4)**
4. Gear A has 15 teeth and gear B has 45 teeth. What is the gear ratio of A to B? **(13-5)**

5. In a hydraulic device a force of 60 lb is applied to a piston whose area is 1 sq in. What is the force on the larger piston whose area is 90 sq in.? **(13-7)**
6. What is the resistance of an 8-ampere air-conditioner using 115 volts? **(13-9)**
7. How many hours will it take a contractor to dig a basement that measures 9 ft deep, 28 ft wide, and 54 ft long if the contractor can remove 60 cu yd per hr? **(14-2)**
8. How many cubic yards of concrete are needed for a footing 16 in. wide, 8 in. deep, and 120 ft long? **(14-3)**
9. What is the cost of 10 lb of roofing nails at $.20 per lb? **(14-6)**
10. A wall 30 ft long and 8 ft high is to be sided. The wall contains two windows. Estimate the total area that is to be sided. **(14-8)**
11. A 20-ampere fuse is placed in a 115-volt circuit. How many watts of power are available through this circuit? **(14-10)**
12. If $\frac{1}{2}$-in. plastic pipe costs $0.25 per ft, find the cost of 230 ft of this pipe. **(14-11)**
13. If a room is 18 ft × 20 ft, find the cost of carpeting at $12.00 per sq yd. **(14-12)**
14. If topsoil, delivered to the site, costs $5.00 per cu yd, what will 21 cu yd cost? **(14-13)**
15. A cast aluminum head for an engine weighs about 24 lb. If a company makes 1.5 million engines, how many tons of aluminum must be available? **(15-1)**
16. A forging machine is capable of exerting 25 tons of force. If the object being forged has an area of 16 sq in., how many pounds of pressure are being exerted on each square inch? **(15-2)**
17. A grinding wheel with a diameter of 8 in. turns at 1850 rpm. What is its cutting speed? **(15-4)**
18. If 1 gal of paint will cover 500 sq ft of surface, how many gallons of paint are needed for a surface that measures 150 ft by 200 ft? **(15-6)**
19. Standards set by a certain company require that the maximum percentage of defective parts is .025%. How many defective parts are allowed for each 10,000 parts? **(15-9)**
20. A box with a base 8 cm square is to have a volume of 2560 cu cm. What is the height of the box? **(15-10)**
21. What is the selling price of an item if the cost to the merchant is $2.85, expenses are $.27, and profit is $.31? **(15-11)**
22. How many picas are there in $3\frac{1}{2}$ inches? **(16-1)**
23. Find the number of column inches in an advertisement that is 3 columns wide and 2 inches deep. **(16-3)**

24. A sheet-fed rotary press will print 3500 sheets per hour. How long will it take to print 10,000 copies of a 320-page book if there are 32 pages per sheet? **(16-4)**

25. A linotype operator who can compose 3 pages per hour is paid $8.00 per hour. What will the operator earn for composing a 256-page book? **(16-8)**

26. A 200 mm × 250 mm photograph is reduced so that its width is 160 mm. What is its reduced depth? **(16-9)**

27. Paper for a book weighs 80 pounds for 500 sheets. How many pounds of paper are used to make 25,000 copies if each copy requires 5 sheets? **(16-10)**

Answers to selected exercises

CHAPTER 1 READING AND USING MEASURING DEVICES

Page 3 1. 9824 kWh 3. 6645 kWh 5. 590 kWh

Pages 4–5 1. 7634 CCF or 763,400 cu ft 3. 8446 CCF or 844,600 cu ft 5. 9605 CCF or 960,500 cu ft

Page 7 1. 2,324,617 automobiles 3. 627,229,367 rivets 5. 47,287,539 fasteners 7. 432,627,456 bottles 9. 329,614 employees

Page 8 1. 500 3. 6300 5. 3000 7. 30,000 9. 30 ft 11. 3290 ft 13. 3300 mi

Pages 8–9 1. 3500 rpm 3. 5300 rpm 5. 4000 rpm 7. 5000 rpm

Pages 11–12 1. 1,900,000; 3,300,000 3. 1970, by 200,000; 1972, by 100,000 5. $700,000,000 7. February 9. March and April; 360,000 11. March

Pages 12–13 1. $133,000,000 3. $5,000,000 5. $113,000,000 7. 1976; 1966 9. Supply tends to adjust to demand.

Pages 14–15 1. 12 mph; 11 mph; 7 mph; 5 mph; 4 mph; 3 mph; 3 mph; 2 mph 3. 20 mph; 37 mph; 46 mph 7. 23.5 mpg; 16.5 mpg 9. 40 mph and 50 mph

Pages 17–18 1. $\frac{1}{8}''$; $\frac{1}{16}''$ 3. $\frac{5}{16}''$; $\frac{7}{16}''$; $\frac{13}{16}''$ 5. $\frac{1}{32}''$; $\frac{1}{64}''$ 7. $\frac{27}{64}''$; $\frac{43}{64}''$ 9. $\frac{1}{100}''$ 11. $\frac{17}{100}''$; $\frac{34}{100}''$; $\frac{25}{100}''$; $\frac{17}{100}''$; $\frac{76}{100}''$; $\frac{97}{100}''$

Page 21 1. Yes 3. Yes 5. Yes 7. $\frac{2}{16}''$ 9. $\frac{8}{16}''$ 11. $\frac{12}{16}''$ 13. $\frac{6}{32}''$ 15. $\frac{14}{32}''$ 17. $\frac{16}{32}''$ 19. $\frac{3}{4}$ 21. $\frac{1}{4}$ 23. $\frac{5}{8}$ 25. $\frac{2}{5}$ 27. $\frac{1}{4}$ 29. $\frac{3}{4}''$ 31. $\frac{1}{16}''$ 33. $\frac{20}{32}''$ 35. $\frac{1}{8}''$ 37. Yes

Pages 22–23 1. $2\frac{4}{16}''$ 3. $1\frac{13}{16}''$ 5. $3\frac{12}{16}''$ 7. $1\frac{6}{32}''$ 9. $2\frac{24}{32}''$ 23. $A: \frac{31}{32}''$; $B: 1''$; $C: 1''$; $D: \frac{23}{32}''$; $E: 2\frac{8}{32}''$; $F: \frac{16}{32}''$; $G: \frac{15}{32}''$; $H: 1''$; $J: 2''$; $K: \frac{11}{32}''$; $L: 2\frac{8}{32}''$; $M: \frac{12}{32}''$; $N: 2\frac{31}{32}''$

Pages 24–25 1. 3 mm; 10 mm; 17 mm; 29 mm 3. 41 mm; 29 mm; 68 mm; 18 mm; 57 mm 5. AB: 5 mm; BC: 22 mm; CD: 53 mm 7. AB: 77 mm; CD: 6 mm; EF: 10 mm; GH: 9 mm; JK: 40 mm

Pages 32–34 1. 8787 kWh 3. 600,416; 5,500,500 5. 7600 rpm 7. $100,000,000 9. $700,000,000 11. $1\frac{19}{32}''$; $2\frac{24}{32}''$ 13. $\frac{28}{32}''$ 15. $\frac{4}{5}$; $\frac{1}{4}$; $\frac{13}{16}$ 17. 86 mm; 36 mm

CHAPTER 2 WORKING WITH INTEGERS

Page 37 1. 136 in. 3. 275 m 5. 803 ft 7. 1701 days 9. 435 m 11. 57 mm, 36 mm

Page 39 1. 138 3. 1092 5. 14 7. 41 9. 68,832 11. 2,575,287 13. 60

463

Page 41   11. $^+2$ 13. $^+2$ 15. $^+4$ 17. $^-1$ 19. $^-1000$ 21. Town A

Pages 43–44   1. $^+10$ 3. $^-13$ 5. $^-101$ 7. $^+14$ 9. $^-2$ 11. $^-1$ 13. $^-4$ 15. 0   17. $^-3$ 19. 0 volts   21. a. 210   c. 232   e. 209

Pages 45–46   1. $^+15$ 3. $^+35$ 5. $^+1$ 7. $^-15$ 9. $^+30$ 11. $^-98$ 13. $^-8$ 15. $^-17$ 17. 370 cu ft   19. $^+82 - {}^-25 = \$107$ 21. $^+16$ ft/sec

Pages 47–48   1. $^-56$ 3. $^+81$ 5. 0   7. $^-112$ 9. $^-36$ 11. $^+63{,}388$ 13. $^+125$ 15. $^+720$ 17. $^+32°$ 19. $7.68

Page 49   1. $^+5$ 3. $^-3$ 5. $^-1$ 7. $^+12$ 9. $^-144$ 11. $^-11$ 13. $^-264$ 15. $^-35$ 17. $^+5$ amperes

Pages 50–51   1. 1243   3. 21,828   5. $^+12$ 7. $^-24$ 9. $^+60$ 11. $^+13$ 13. $^+1$ 15. $^-45$ 17. $^+96$ 19. $^-360$ 21. $^+7$ 23. $^+10$ 25. $^+15$ 27. 50 miles south   29. $^-\$300$

CHAPTER 3. USING FRACTIONS

Page 54   1. $\frac{5}{8}$ 3. $\frac{1}{4}$ 5. $\frac{2}{5}$ 7. $\frac{19}{25}$ 9. $\frac{1}{4}$ 11. $\frac{57}{64}$ 13. $\frac{19}{32}''$ 15. $\frac{7}{32}''$ 17. $A: \frac{1}{2}''; B: \frac{3}{8}''$

Pages 56–57   1. $\frac{7}{8}$ 3. $\frac{15}{16}$ 5. $\frac{11}{16}$ 7. $\frac{5}{8}$ 9. $\frac{19}{20}$ 11. $\frac{13}{20}$ 13. $\frac{13}{16}''$ 15. $\frac{7}{8}''$ 17. $\frac{5}{16}''$ 19. $A: \frac{5}{8}''; B: \frac{11}{16}''$ 21. $E: \frac{9}{32}''; F: \frac{3}{32}''; G: \frac{5}{16}''$

Pages 59–60   1. $\frac{7}{8}$ 3. $\frac{25}{32}$ 5. $\frac{11}{16}$ 7. $\frac{51}{64}$ 9. $\frac{93}{100}$ 11. $\frac{41}{64}$ 13. $\frac{93}{100}''$ 15. $\frac{27}{100}''$ 17. $\frac{13}{64}''$ 19. $\frac{29}{32}''$ 21. $X: \frac{5}{32}''; Y: \frac{5}{32}''; Z: \frac{5}{8}''$

Page 63   1. $\frac{6}{1}$ 3. $\frac{15}{4}$ 5. $\frac{39}{8}''$ 7. $\frac{67}{32}''$ 9. 3   11. $1\frac{13}{16}$ 13. $3\frac{3}{4}''$ 15. $2\frac{1}{32}''$

Pages 65–66   1. $9\frac{7}{8}$ 3. $19\frac{17}{100}$ 5. $11\frac{31}{40}$ 7. $15\frac{1}{64}''$ 9. $17\frac{9}{16}''$ 11. $8\frac{3}{16}''$ 13. $13\frac{1}{2}''$ 15. $A: 2\frac{3}{4}''; B: 4\frac{1}{2}''$

Page 68   1. $5\frac{1}{8}$ 3. $6\frac{5}{16}$ 5. $1\frac{3}{32}$ 7. $2\frac{3}{16}''$ 9. $3\frac{1}{16}''$ 11. $1\frac{11}{16}$ 13. $\frac{3}{8}''$ 15. $1\frac{13}{16}''$ 17. $A: \frac{11}{16}''; B: \frac{9}{16}''$

Pages 70–71   1. $23\frac{3}{4}''$ 3. $7'1\frac{1}{4}''$ 5. $7'8\frac{1}{4}''$ 7. $19'9''$ 9. $61\frac{1}{2}''$ 11. $37'9''$ 13. $154'4''$ 15. $36'6''$ 17. $17'11\frac{9}{16}''$

Pages 72–73   1. a. $2\frac{3}{8}''$ b. $2\frac{3}{4}''$ 3. b. $6'4\frac{7}{16}''$ c. $11'7\frac{9}{16}''$

Page 74   1. $\frac{7}{8}$ 3. $1\frac{1}{5}$ 5. $\frac{9}{32}$ 7. $1\frac{11}{4}$ 9. $62\frac{1}{2}''$ 11. $7''$ 13. $5\frac{1}{4}$ lb   15. $3\frac{3}{4}$ hr   17. $13'5''$

Page 76   1. $7\frac{1}{2}$ 3. $33\frac{1}{8}$ 5. $247\frac{1}{2}$ 7. $284''$ 9. $39\frac{11}{16}$ 11. a. $105''$ b. $8'9''$ 13. a. $6\frac{9}{16}''$ b. Yes

Pages 78–79   1. $\frac{1}{8}$ 3. $\frac{7}{24}$ 5. $1\frac{5}{28}$ 7. $9\frac{3}{16}''$ 9. $6\frac{51}{100}$ 11. $\frac{9}{20}$ 13. $9\frac{4}{5}$ 15. $18''$ 17. $2\frac{7}{16}$ 19. $\frac{11}{192}''$ 21. $11\frac{5}{8}$ gal   23. $14\frac{7}{8}¢$ 25. a. $71\frac{61}{100}$ sq in.   b. $2.51

Page 80   1. $\frac{3}{8}$ 3. $\frac{7}{25}$ 5. $\frac{3}{8}$ 7. $16\frac{2}{3}$ 9. 18   11. 72   13. $29''$ 15. $1''$

Page 81   1. $\times \frac{4}{3} = 1$ 3. $\times \frac{2}{7} = 1$ 5. $\times \frac{7}{22} = 1$ 7. $\times \frac{32}{31} = 1$ 9. $\times \frac{1}{53} = 1$ 11. $\times \frac{3}{26} = 1$

Pages 82–83   1. $\frac{3}{10}$ 3. $\frac{4}{35}$ 5. $\frac{5}{16}''$ 7. $\frac{9}{100}''$ 9. $\frac{2}{3}$ 11. $\frac{3}{16}''$ 13. $1\frac{5}{8}'$ 15. $5\frac{1}{4}''$ 17. a. $4\frac{7}{8}''$ b. $2\frac{17}{24}''$

Page 85   1. 14   3. $15\frac{3}{4}$ 5. $\frac{1}{2}$ 7. $2\frac{5}{14}$ 9. $1\frac{2}{3}$ 11. $\frac{3}{10}''$ 13. 9 strips   15. 16 turns   17. 32 boards   19. a. 3   b. $\frac{1}{3}$ c. $\frac{8}{7}$ d. $\frac{4}{9}$ e. $\frac{8}{19}$

464   Answers to selected exercises

Page 87   1. 28 binding posts   3. 4 strips   5. 15 bars   7. 135 blocks

Pages 90–91   1. a. $\frac{3}{4}''$ b. $\frac{1}{2}''$ 3. $8\frac{1}{4}$ ft   5. 21 cork sleeves   7. $2\frac{5}{8}''$ 9. a. 17 bricks   b. 28 layers

Pages 92–94   1. $\frac{15}{32}$ 3. $\frac{1}{2}''$ 5. $\frac{5}{8}$ 7. $\frac{5}{8}''$ 9. 60   11. $\frac{141}{8}$ 13. 7   15. $6\frac{3}{8}$ 17. $17\frac{15}{16}$ 19. $1\frac{9}{16}''$ 21. $23\frac{13}{16}''$
23. $1\frac{7}{9}$ 25. $\frac{15}{28}$ 27. $\frac{51}{64}$ 29. $2\frac{11}{12}$ 31. $\frac{3}{20}$ 33. $\frac{1}{5}$ 35. 5   37. $\frac{5}{32}$ 39. $\frac{2}{25}$ 41. $1\frac{11}{21}''$ 43. $75'10''$
45. $801'10''$ 47. $19'$ 49. $18'4\frac{1}{2}''$

CHAPTER 4   DECIMALS, PERCENTS

Pages 96–97   1. $\frac{4}{10}$; .4   3. $\frac{7}{100}$; .07   5. $\frac{93}{100}$; .93   7. $\frac{155}{1000}$; .155
9. .2   11. .006   13. .900   15. .018   17. .5   19. $\frac{12}{100}$ 21. $\frac{27}{100}$ 23. $\frac{2737}{10000}$ 25. $\frac{725}{10000}$
27. The top missing length is 1.14″; the missing width is 1.34″; the bottom missing length is 2.09″

Pages 100–101   1. .75   3. .875   5. .5625   7. .9063   9. $\frac{5}{8}$ 11. $\frac{59}{64}$ 13. $\frac{9}{32}$ 15. $\frac{25}{32}$ 17. .2   19. .875
21. .3125   23. .09375   25. .28125   27. 2.75″   29. $\frac{1}{4}$ 31. $\frac{3}{5}$ 33. $\frac{3}{4}$ 35. $\frac{7}{8}$ 37. $\frac{7}{16}$ 39. $3\frac{3}{20}''$ 41. 3.875″
43. $1.5'' = 1\frac{1}{2}''$; $.375'' = \frac{3}{8}''$; $.625'' = \frac{5}{8}''$; $.1875'' = \frac{3}{16}''$

Pages 102–104   1. 6.94   3. 21.55   5. 17.52 cm   7. 14.739   9. 71.4632   11. 1.2   13. 3.32 cm
15. 3.121   17. 2.29   19. 111.13   21. 3.275″   23. .229 m   25. $A = 7.084$ cm; $B = 3.76$ cm

Pages 106–107   1. .42   3. .147   5. .6062   7. .0126   9. .000012   11. .910   13. .0225   15. .0002961
17. 388.500   19. 102.624   21. 651.7812   23. 1.237″   25. 13.5″   27. 3.8104 cm   29. 2.5452 cm

Pages 109–110   1. .283″   3. .477″   5. .076″   7. 12.13 mm   9. 4.90 mm   11. 22.92 mm

Pages 113–114   1. 2.1   3. 8.76   5. 36.85   7. 1.37   9. .017   11. .035   13. 9   15. 3.3   17. 153.0   19. .5
21. 230   23. .85″   25. .125″   27. 7.518 cm   29. 8.2 sq m   31. $856.82

Pages 116–117   1. .03   3. .78   5. .253   7. .002   9. $\frac{1}{4}$ 11. $\frac{3}{4}$ 13. $\frac{3}{5}$ 15. $\frac{1}{8}$ 17. $\frac{5}{8}$ 19. $\frac{1}{200}$ 21. 14%
23. 145%   25. 3.2%   27. 40%   29. 50%   31. 40%   33. $37\frac{1}{2}$%   35. $87\frac{1}{2}$%   37. 65%   39. 14%   41. $\frac{1}{2}$
43. 1.00

Page 118   1. $168   3. $21   5. $24.12   7. $2.80, $2.10; $2.30, $1.61; $9.40, $7.99; $11.70

Pages 120–121   1. 2.375″, 2.625″; 5%   3. 2.443″, 2.447″; .08%   5. 1.871″, 1.879″, .21%
7. 44.5 mm, 45.5 mm; 1.1%   9. 82.8 mm, 83.2 mm; .24%
11. Figure 4-17: A: 3.497″, 3.503″; .09% B: .280″, .290″; 1.75% C: 1.247″, 1.253″; .24% D: 3.996″, 4.004″; .1%
E: .260″, .264″; .76% Figure 4-18: A: 89.93 mm, 90.07 mm; .07% B: 7.02 mm, 7.22 mm; 1.4% C: 31.43 mm,
31.57 mm; .2% D: 99.92 mm, 100.08 mm; .08% E: 2.70 mm, 2.80 mm; 1.8%   13. 855.5 mm, 856.5 mm; 1 mm;
.06%

Page 123   1. 7600 ohms ± 5%   3. 98 ohms ± 10%   5. 700,000 ohms ± 2%   7. 690,000 ohms ± 2%

Page 124   1. 80%   3. 129%   5. 9450 hp   7. 12 copies/min; $33\frac{1}{3}$%

Pages 125–127   1. .9   3. .375   5. $\frac{2}{5}$ 7. 9.41   9. The missing width is .5625″; the missing length is .25″
11. 398.382″   13. .498″   15. 2.338″   17. 7.910 cm
19. The missing width is 12.3 mm; the missing length is 9.0 mm   21. 232.685 cm   23. 3.86 mm
25. 163.375 mm   27. $\frac{4}{25}$ 29. $\frac{113}{100}$ or $1\frac{13}{100}$ 31. .25   33. 60%   35. 80%   37. $7025.25
39. 54,000,000 ohms, ±10%   41. 90%

Answers to selected exercises   465

CHAPTER 5   USING HAND-HELD CALCULATORS AND OTHER COMPUTING DEVICES

Page 130   1. correct   3. correct   5. correct   7. correct   9. incorrect (2178)   11. correct

Page 132   1. 136 in.   3. 68,832   5. $^+$6   7. $^-$98   9. $^+$720   11. 916 ft   13. 25,088   15. $^+$201   17. $^-$51,912   19. $^+$15

Pages 134–135   1. $14.7 billion   3. $5 billion   5. A is 18 mm, B is 73 mm   7. $^+$156°   9. $250.00

Page 136   1. 14   3. 41   5. 661.50   7. 2.82   9. 0.32   11. 2.59   13. 9.75   15. 0.0175

Pages 138–139   1. 0.8750   3. 0.5313   5. 1.8750   7. 1.3750   9. 56.00   11. 0.32   13. 157.12   15. 80.208   17. 145.350   19. A is 0.625, B is 0.688   21. 42.995 ft   23. 74   25. 4.525 cm

Page 140   1. 0.78   3. 0.253   5. 1.082   7. $168   9. $21   11. $24.12   13. 403,000 hp   15. 214.88 hp

Pages 142–143   1. correct   3. correct   5. correct   7. 2909   9. 34   11. $14,285   13. 20.76   15. 4.52   17. 67.2   19. 12.6

Pages 143–146   1. 7431 kWh   7. 14   9. $^-$17   11. 5850   13. $^-$36   15. $^-$5   17. $\frac{21}{32}''$ 19. $4''$ and $2\frac{1}{4}''$ 21. $2\frac{1}{2}$ 23. $3\frac{21}{32}$ 25. $1\frac{3}{4}$ 27. .375   29. .3125   31. $\frac{3}{4}$ 33. $6.145''$ 35. 1.15371   37. $7.96   39. 7   41. $^-$10,752   43. A: 10 mm; B: 56 mm   45. 18.646   47. 142.396   49. $190

CHAPTER 6   EQUATIONS, FORMULAS

Pages 151–152   1. $5a$ 3. $4x$ 5. $2x$ 7. $10m$ 9. $3s$ 11. $7r$ 13. $7x + 1$ 15. $7t + 8$ 17. $7x + 4y$ 19. a 21. $P = 2s + 6$ 23. $L = 40x$

Pages 155–156   1. 9   3. 49   5. 512   7. 125   9. 256   11. 5.29   13. 59.319   15. 157.464   17. 9.2416   19. 10^6 21. 6.7×10^6 23. 4.5×10^{-5} 25. 2.07×10^{10} 27. 400   29. $68\frac{1}{16}$ sq in.   31. 153.86 sq ft   33. 1815.8 cu cm

Page 157   1. 27   3. 27   5. 19   7. 9   9. 38   11. 16   13. 35   15. 26   17. 7   19. 24   21. 13

Pages 159–160   1. $x = 4$ 3. $a = 9$ 5. $c = 17$ 7. $b = 56$ 9. $y = 242$ 11. $a = 4\frac{1}{4}$ 13. $c = 7\frac{1}{4}$ 15. $y = 46.7$ 17. $y = 8\frac{2}{5}$ 19. $y = 116.43$ 21. $A + 2\frac{7}{8} = 4\frac{1}{8}, A = 1\frac{1}{4}''$ 23. $x + 15\frac{7}{8} + 3 = 22\frac{1}{4}, x = 3\frac{3}{8}''$ 25. $A + 7.3 = 10.4, A = 3.1$ cm   27. $x + 36.7 + 7.7 = 52.2, x = 7.8$ cm

Page 162   1. $x = 26$ 3. $a = 27$ 5. $c = 26$ 7. $y = 28$ 9. $a = 53$ 11. $y = 1$ 13. $x = 31.0$ 15. $z = 11.4$ 17. $h = 13\frac{1}{8}$ 19. $d = 8.248$ 21. $s - .023 = 1.375, s = 1.398''$ 23. $L - 1\frac{3}{4} = 22\frac{1}{2}, L = 24\frac{1}{4}''$ 25. $D - 2(.51) = 4.31, D = 5.33$ cm   27. $D - 2(4.6) = 23.4, D = 32.6$ cm

Pages 164–165   1. $a = 4$ 3. $x = 4$ 5. $t = 11$ 7. $d = 3$ 9. $c = 7\frac{7}{9}$ 11. $y = .9$ 13. $p = 90$ 15. $y = .004$ 17. $b = .03$ 19. $80L = 9600, L = 120$ ft   21. $17L = 374, L = 22$ in.   23. $60L = 7200, L = 120$ cm   25. $3w = 43.5, w = 14.5$ cm

Pages 166–167   1. $x = 30$ 3. $z = 36$ 5. $c = 153$ 7. $d = 70$ 9. $w = 4\frac{2}{3}$ 11. $z = 1\frac{7}{8}$ 13. $x = 3.38$ 15. $d = 26.6$ 17. $x = 2.15$ 19. $.34 = \dfrac{F}{.866}, F = .294$ in.   21. $r = 21.36$ ohms

23. $\dfrac{9.2}{12.4} = \dfrac{1850}{s}$, $s = 2493.5$ rpm

Pages 169–171 1. $x = 11$ 3. $x = 17$ 5. $x = 16$ 7. $t = 6$ 9. $p = 10$ 11. $a = 25$ 13. $y = 39.0$
15. $y = \frac{13}{42}$ 17. $x = 2.86$ 19. $a = \frac{1}{24}$ 21. $26 = 2s + 8\frac{3}{4}$, $s = 8\frac{5}{8}$ 23. $2x + 2\frac{3}{8} = 4\frac{7}{16}$, $x = 1\frac{1}{32}$ in.
25. $46.6 = 3x + 5x + 16.2$, $x = 3.8$; $AB = 3x = 11.4$ mm; $BC = 5x = 19.0$ mm 27. $2x + 64 = 113$, $x = 24.5$ mm

Page 173 Letters used in the formulas may vary. 1. $I = 12F$ 3. $K = .454P$ 5. $A = LW$
7. $D = 1.414S$ 9. $D = HS$

Pages 176–177 1. a. $C = (3.14)(.6) = 1.884$ b. $10.048 = 3.14d$, $d = 3.2$
3. a. $d = (1.41)(8.2) = 11.562$ b. $12.831 = 1.41s$, $s = 9.1$
5. a. $P = 2(16) + 2(12) = 56$ b. $48 = 2\ell + 2(9\frac{1}{2})$, $\ell = 14\frac{1}{2}$ c. $36.5 = 2(12.75) + 2w$, $w = 5.5$ 7. 1120
9. $5413\frac{1}{3}$ 11. 12.726 ft 13. 17.26 cm 15. 43.98 m

Pages 178–179 1. $20a$ 3. 81 5. 4.913 7. 28 9. $t = 2$ 11. $y = 1\frac{1}{2}$ 13. $b = 5.8$ 15. $m = 54$
17. $a = 2.43$ 19. $A = \dfrac{bh}{2}$ 21. $H = .5725$ in. 23. $s = 1500$ rpm

CHAPTER 7 LENGTH, AREA, VOLUME

Pages 183–184 1. A: 2.5 cm; B: 4.3 cm; C: 6.7 cm; D: 8.1 cm 3. 9 cm 5. 1.52 7. 14,000 9. 300,000
11. 1.784 13. 10; .1; .01 15. .1; .01; .001 17. 12,500 micrometers

Page 186 1. 4 cm 3. $\frac{1}{2}$ in. 5. $\frac{1}{2}$ m; $\frac{1}{2}$ yd; 1 ft 7. 6.1 9. 10.2 11. 91.4 13. 1610
15. .236 in.; .315 in.; .394 in.; .473 in.; .552 in. 17. 30 mm; 1.1811 in.

Pages 187–188 1. 14 in. 3. 16 ft 6 in. 5. 13 ft 6 in. 7. 344 yd 9. 99 mm 11. 380 mm 13. 16 m

Pages 190–191 1. 16 in. 3. 113 yd 5. 22.67 in. 7. 16 in. 9. 25 in. 11. 4.1 m 13. 139.1 m
15. 400 cm 17. 16 cm

Page 193 1. $\frac{1}{144}$ or .0069 3. .01 5. $\frac{1}{9}$ 7. 350 9. 360 11. 2232 13. $117 15. $65.61

Page 195 1. 144 sq ft 3. 21 sq ft 5. 256 sq ft 7. 27 sq ft 9. 95.14 sq ft 11. 144 sq m 13. 7.0 sq. cm
15. 225 sq cm 17. 126 sq cm 19. 50.676 sq m

Pages 197–198 1. 70 sq in. 3. 4800 sq ft 5. 140 sq in. 7. 9200 sq ft 9. a. 1147.5 sq in. b. 652.5 sq in.
11. 70 sq cm 13. .3 sq m 15. 70 sq cm 17. .2125 sq m 19. a. 7593.75 sq cm b. 3656.25 sq cm

Page 201 1. 320 sq ft 3. 220.5 sq ft 5. 1450 sq ft 7. 1.6 sq m 9. 13,750 sq mm

Pages 203–204 1. 79 sq in. 3. 707 sq yd 5. 63.6 sq in. 7. 113 sq ft 9. a. 38.465 sq in. b. 10.205 sq in.
11. 31,400 sq mm 13. 43.0 sq cm 15. 3.27 sq m 17. 26.06 sq cm

Pages 206–207 1. 471 sq in. 3. 42.4 sq in. 5. 201 sq in. 7. 44 sq in. 9. 57,776 sq yd 11. 31.4 sq in.
13. 603 sq cm 15. 11.1 sq m 17. 314 sq cm 19. 20,410 sq mm 21. 104.4 sq m 23. 28,260 sq mm

Page 210 1. 1728 3. .001 5. .028 7. 2100 9. 249.75 11. 468 13. $509.33 15. 27.8 cu yd
17. 96 cu yd 19. about 499 cu m

Answers to selected exercises

Pages 211–212 1. 240 cu in. 3. 128.25 cu in. 5. 237 cu yd 7. 1350 gal 9. 6.696 lb 11. 180 cu cm
13. 49 cu m 15. 223.2 cu m 17. 4050 l 19. 29,025 g or 29.025 kg

Page 214 1. 2512 cu in. 3. 98.1 cu in. 5. .331 cu in. 7. 5.5 cu in. 9. 3815 cu cm 11. 219.3 cu m
13. 763,020 l

Pages 216–217 1. 628 cu in. 3. 11.78 cu in. 5. 165 cu in. 7. 900 cu in. 9. 4187 cu in.
11. 113,040,000 cu yd 13. 8.9 gal 15. 942,000 cu mm 17. 79.8 cu m 19. 330 cu mm
21. 42,500 cu mm 23. 4,186,667 cu mm 25. 150.5 cu m 27. 32 l

Pages 219–220 1. $\frac{1}{2000}$ 3. 1000 5. $\frac{1}{2204}$ 7. 5500 9. 2.7 11. 454,000 13. 2.27 15. .5 17. 5.94
19. 336 lb; 152.544 kg 21. 11.33 t 23. $9922.50

Pages 221–222 1. 2.25 3. 298 5. 10,500 7. .02533 9. 85.1625 11. 7.5768 13. 904,320 l 15. 7.4 l
17. 111 ml

Page 223 1. 89.6 3. 71.1 5. 93.3 7. 113 9. 36° 11. −38.2°F

Pages 225–226 1. 430; 4300 3. 30.5 5. 5 7. 11.96 9. 1300 11. 1.6 13. 3.6 15. $.36 per liter
17. 29.4 19. 15.7 in. 21. 3.5 sq in. 23. 50 sq ft 25. 270 cu in. 27. 1696 cu in. 29. 25 cm
31. 144 sq cm 33. 78.5 sq m 35. 2400 cu mm 37. 4187 cu mm

CHAPTER 8 POLYNOMIALS

Pages 230–231 1. 5 3. 4 5. 3 7. 4 9. $5x^2 + 6x + 5$ 11. $6x^3y - 9x^2y^2 + xy^3$ 13. $-x^3 + 2x^2 + 5x - 7$
15. $3x^3 + 7x^2 - 13x + 2$ 17. $x^3 - x^2 + x - 2$ 19. $2x^2 - 2x + 14$

Pages 236–237 1. x^7 3. y^5 5. x^4y^4 7. $15x^2 - 10x$ 9. $-2x^4 + 10x^3 - 8x^2 + 6x$ 11. $x^2 + x - 6$
13. $x^3 - 1$ 15. $x^3 - 2x^2y + y^2x - 2y^3$ 17. $8x^4 - 2x^2y^2 - 3y^4$ 19. $20x^2 + 20x - 15$

Pages 242–243 1. $9(x + 2)$ 3. $5a(x + y)$ 5. $2x^2y(x^2 + 2xy + 4y^2)$ 7. $(x + 3)(x - 2)$ 9. $(x + 5)(x - 4)$
11. $(6y + 7)(2y - 2)$ 13. $(x - 9)(x + 4)$ 15. $(x + 12y)(x + 2y)$ 17. $2(x + 5)(x + 5)$
19. $2(2x + 1)(2x + 1)$ 21. $A = r^2(4 - \pi)$

Pages 245–246 1. $(x + 3)(x - 3)$ 3. $(2x + 5)(2x - 5)$ 5. $(4x + 5y)(4x - 5y)$ 7. $5(x + 5)(x - 5)$
9. $2(3x + 2)(3x - 2)$ 11. $(y + x)(y - x)$ 13. $\pi\left(\frac{y}{2} + x\right)\left(\frac{y}{2} - x\right)$

Pages 250–251 1. $x = 3, x = 2$ 3. $x = -4, x = 3$ 5. $x = 15, x = 3$ 7. $x = 6, x = -6$ 9. $x = \frac{1}{5}, x = \frac{3}{2}$
11. 30 by 40 13. $x = .4, 5x = 2, x + 8 = 8.4$

Pages 254–256 1. $x = 3.41, x = .59$ 3. $x = 2.41, x = -.41$ 5. $x = 3, x = -1$ 7. $x = 4.45, x = -.45$
9. $x = 7.23, x = -9.23$ 11. $w = 7.31, \ell = 12.31$ 13. $x = 100$

Page 257 1. $5x^2 + 2x - 5$ 3. $-x^2 + 6x - 7$ 5. x^{12} 7. $x^2 + 2x - 35$ 9. $x^3 + 1$ 11. $4x(x - 4)$
13. $(x - 3)(x + 2)$ 15. $2x(x - 9)(x + 4)$ 17. $5(2x - 5)(2x + 5)$ 19. $x = 3, x = -\frac{2}{3}$
21. $x = -2 \pm \sqrt{2}$ or $x = -.59, x = -3.41$ 23. $x = 3 \pm \sqrt{10}$ or $x = 6.16, x = -.16$

CHAPTER 9 GRAPHING
Pages 262–263

1.

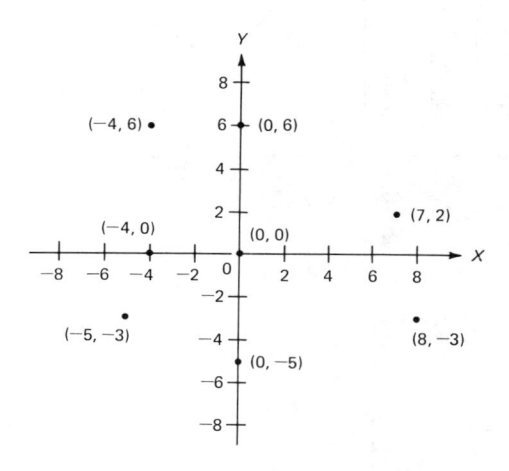

Points plotted: (−4, 6), (0, 6), (−4, 0), (0, 0), (7, 2), (−5, −3), (8, −3), (0, −5)

3.

Figure with labeled points A, B, C, D, E, F, G, H, I, J on coordinate plane.

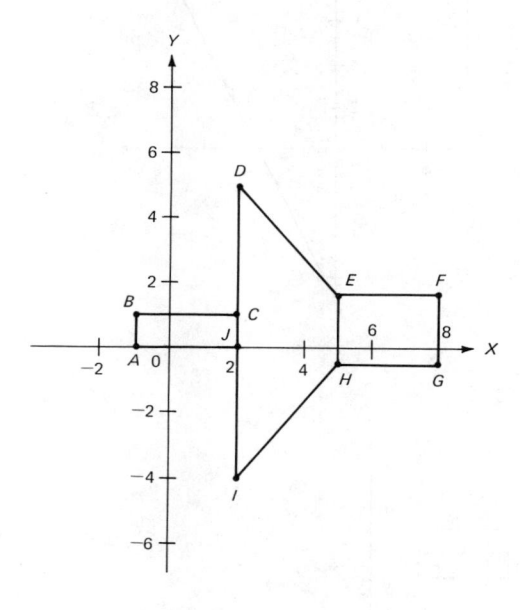

5.

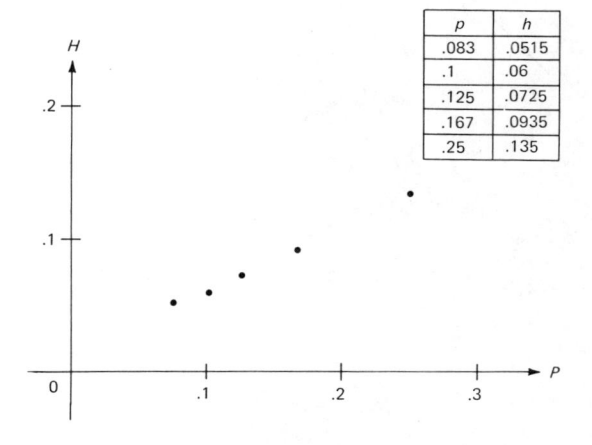

p	h
.083	.0515
.1	.06
.125	.0725
.167	.0935
.25	.135

Answers to selected exercises 469

Pages 266-267

1.

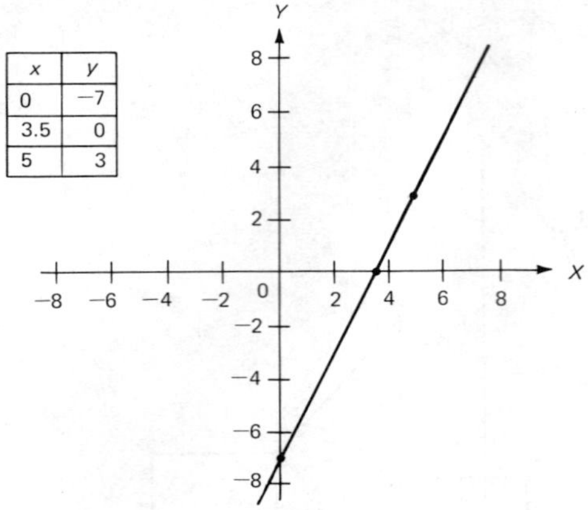

x	y
0	−7
3.5	0
5	3

3.

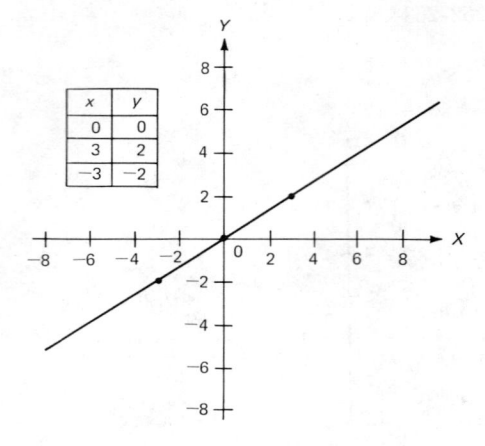

x	y
0	0
3	2
−3	−2

5.

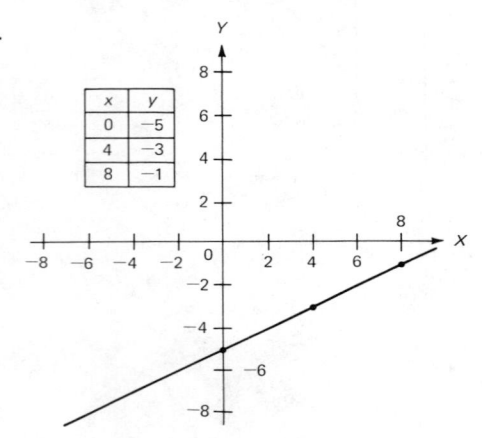

x	y
0	−5
4	−3
8	−1

7.

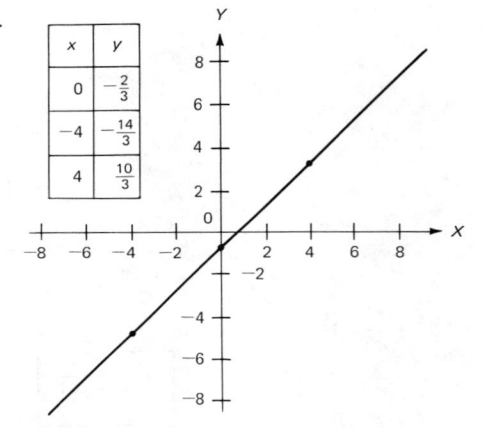

x	y
0	$-\frac{2}{3}$
−4	$-\frac{14}{3}$
4	$\frac{10}{3}$

9. A (0, 5); $x = 0$; B (0, 1); C (0, −3)

470 Answers to selected exercises

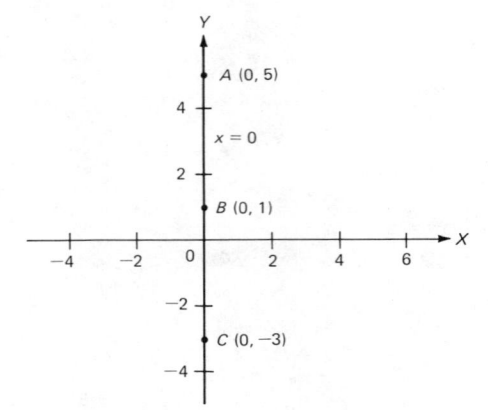

11.

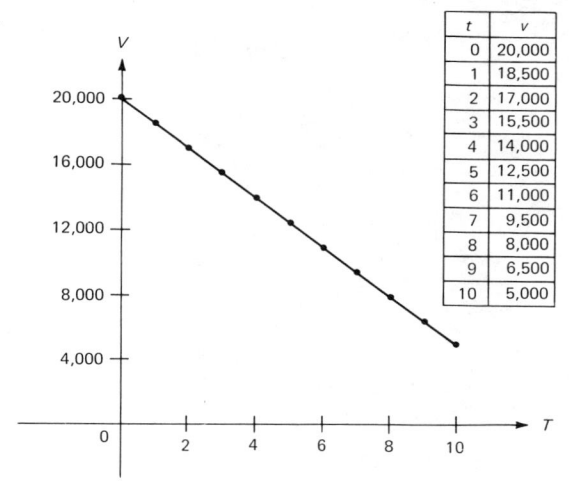

t	v
0	20,000
1	18,500
2	17,000
3	15,500
4	14,000
5	12,500
6	11,000
7	9,500
8	8,000
9	6,500
10	5,000

13.

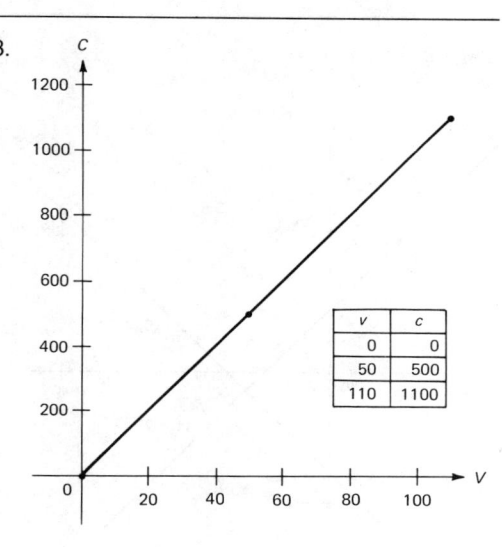

v	c
0	0
50	500
110	1100

Page 270

1.

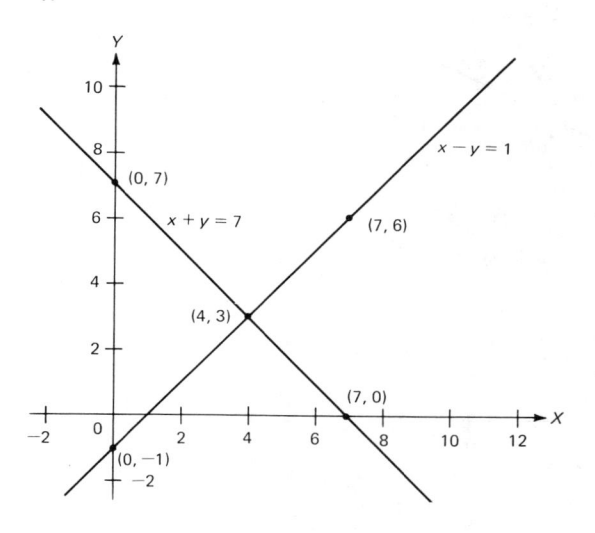

3.

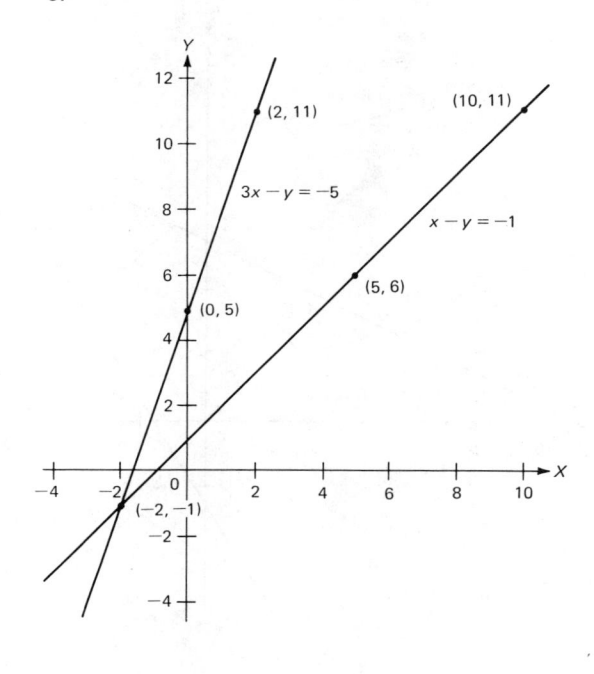

Answers to selected exercises 471

Page 270 (*cont.*)
5.

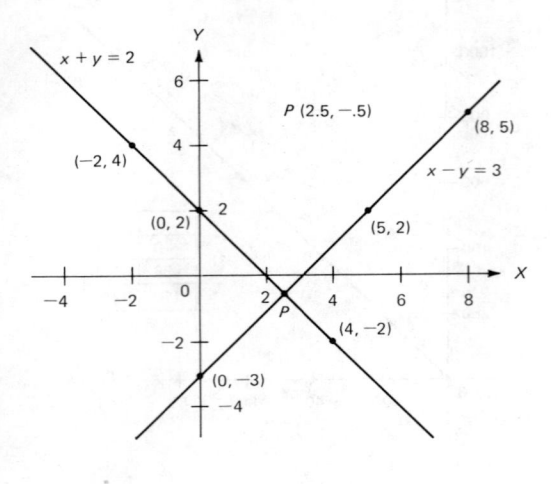

7.

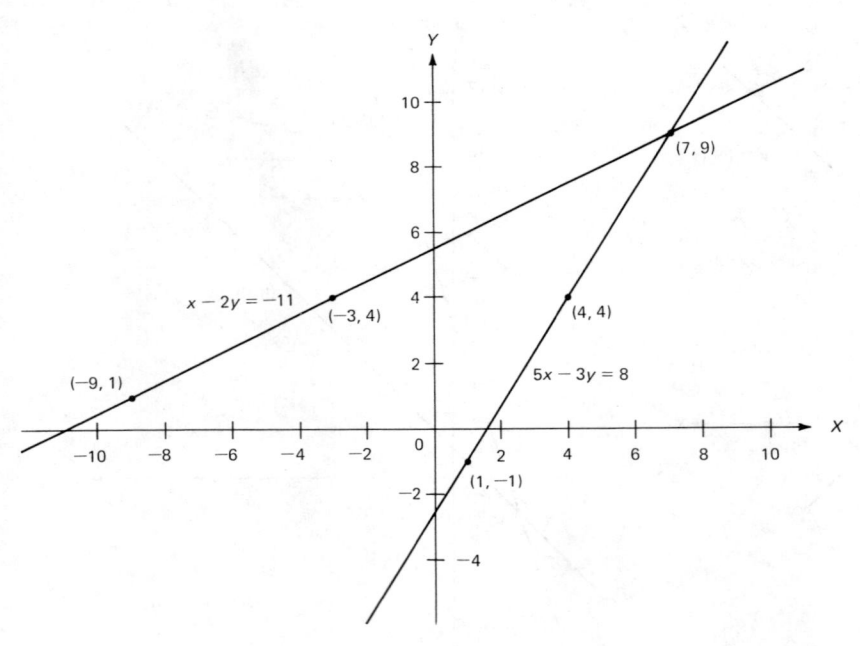

472 Answers to selected exercises

9.

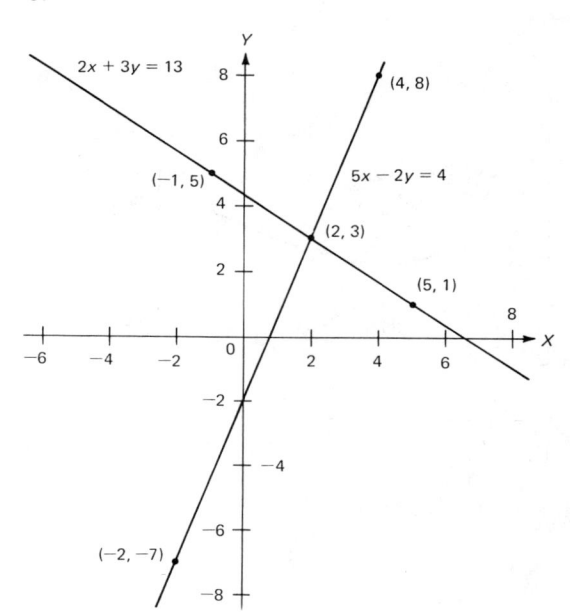

11.

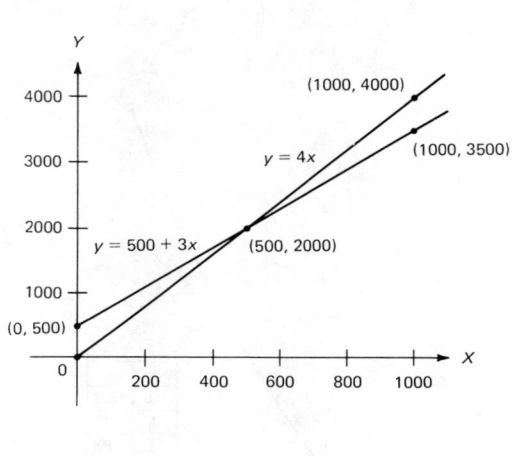

Pages 274–275 1. $x = 2, y = 4.5$ 3. $x = 5, y = -1$ 5. $x = 3, y = 1$ 7. $x = 11, y = 24$ 9. no solution   11. $p = 10.8, y = 524$ 13. $R = .83$ in., $r = .43$ in.   15. $x = 105$ m, $y = 75$ m   17. 13.13 cm and 29.14 cm

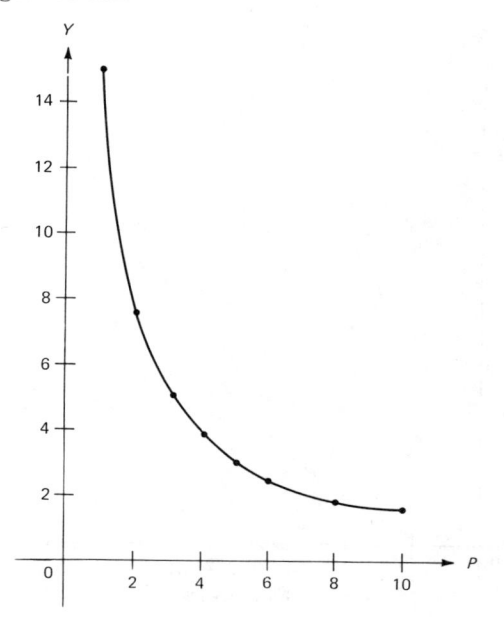

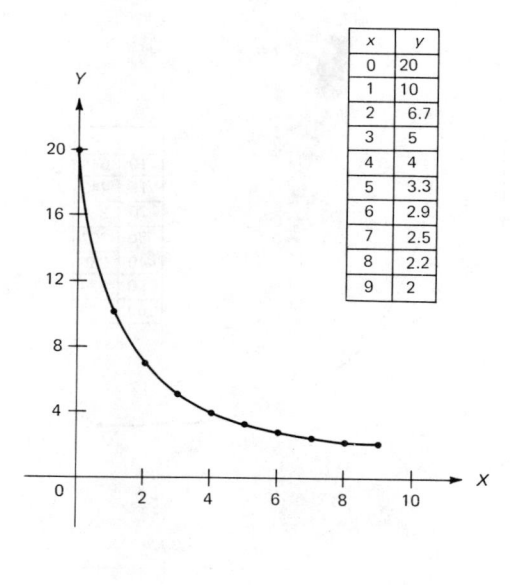

Pages 280–282

1.

3.

x	y
0	20
1	10
2	6.7
3	5
4	4
5	3.3
6	2.9
7	2.5
8	2.2
9	2

Pages 280–282 (cont.)

5.

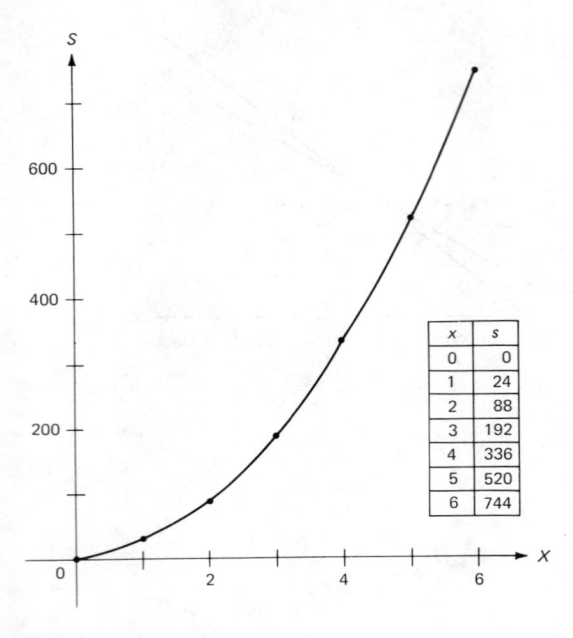

x	s
0	0
1	24
2	88
3	192
4	336
5	520
6	744

7.

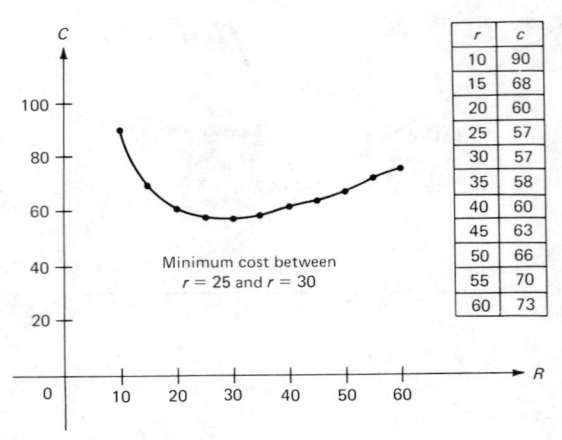

Minimum cost between $r = 25$ and $r = 30$

r	c
10	90
15	68
20	60
25	57
30	57
35	58
40	60
45	63
50	66
55	70
60	73

9.

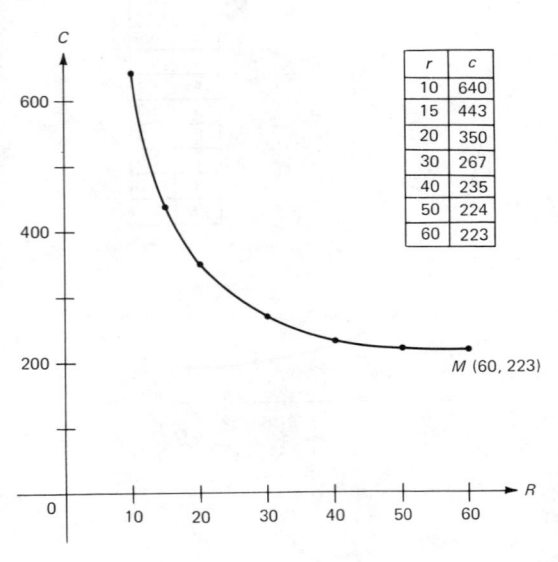

r	c
10	640
15	443
20	350
30	267
40	235
50	224
60	223

$M\ (60, 223)$

11.

WEIGHT IN POUNDS PER FOOT

DIAMETER IN INCHES

474 Answers to selected exercises

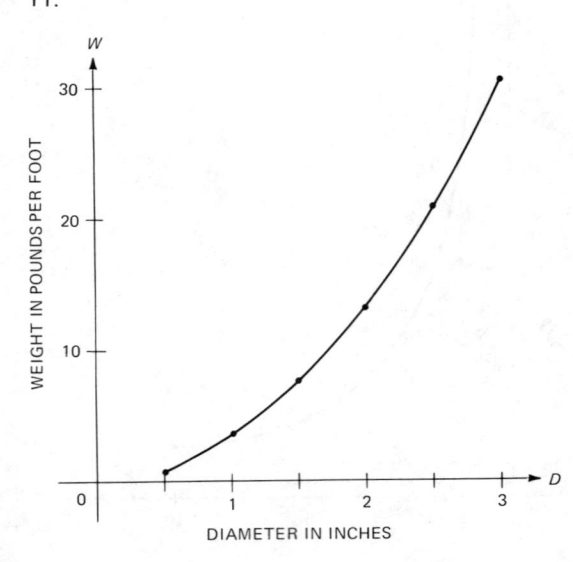

13.

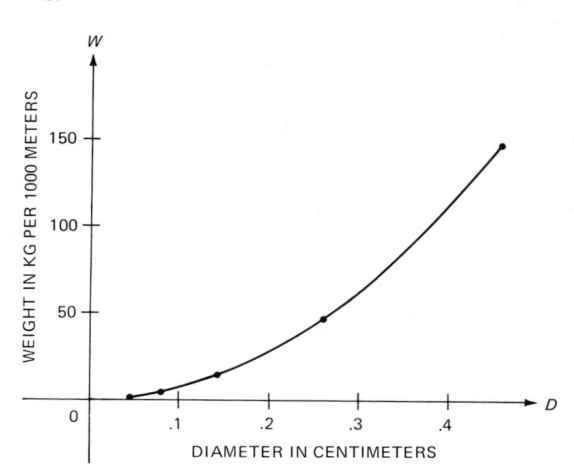

(Graph: W = weight in kg per 1000 meters vs D = diameter in centimeters; curve passing through points near (.05, ~2), (.1, ~8), (.15, ~15), (.25, ~45), (.45, ~145))

15.

$a = 70x - x^2$, with maximum $M(35, 1225)$

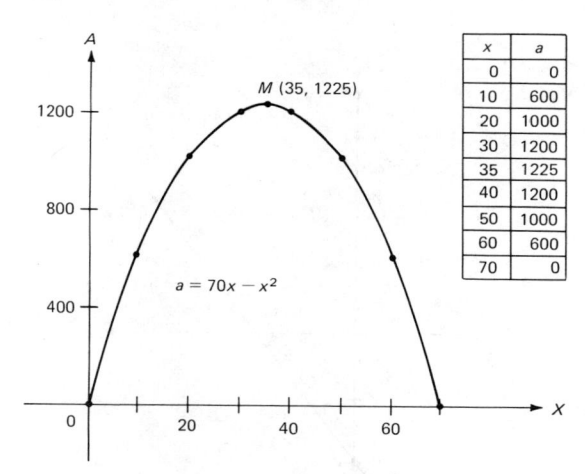

x	a
0	0
10	600
20	1000
30	1200
35	1225
40	1200
50	1000
60	600
70	0

Pages 283–284

1.

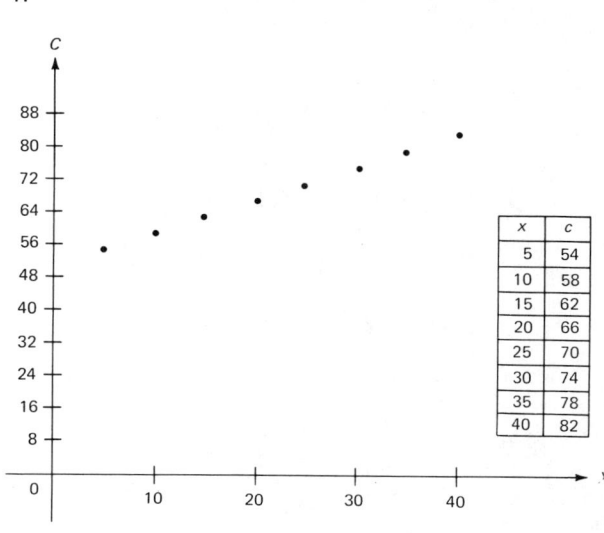

x	c
5	54
10	58
15	62
20	66
25	70
30	74
35	78
40	82

3. a.

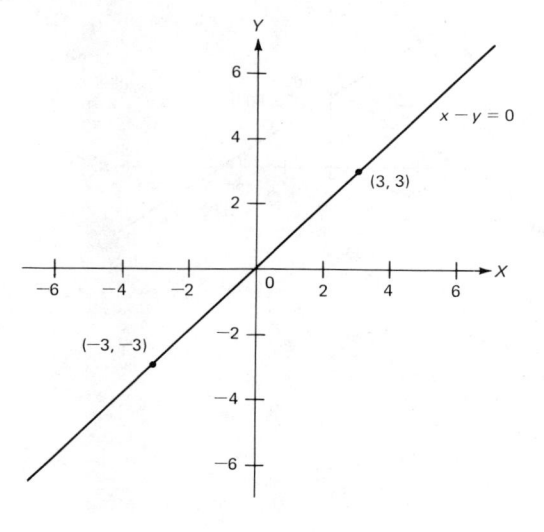

Line $x - y = 0$ through $(-3, -3)$ and $(3, 3)$.

Pages 283–284 (cont.)

3. b.

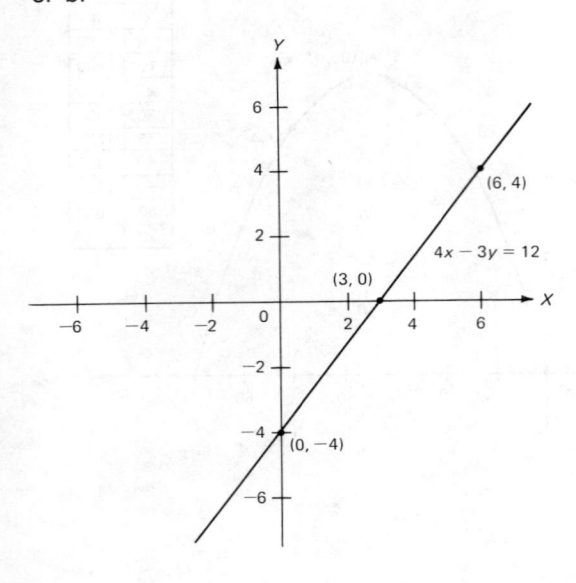

3. c.

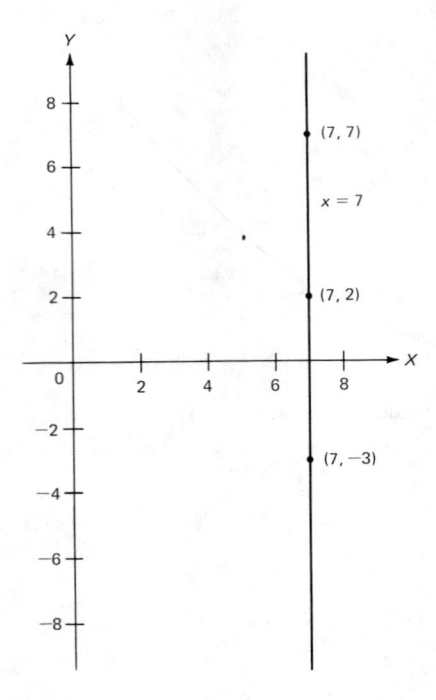

3. d.

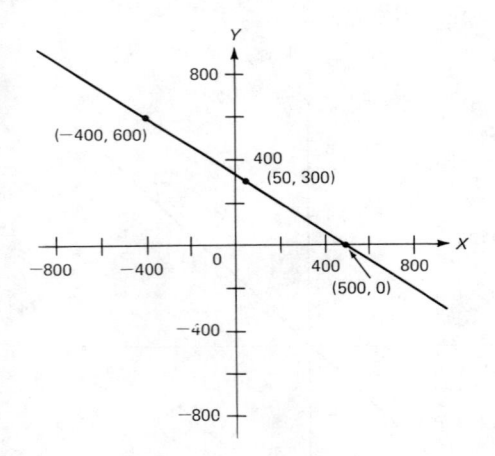

5. a. $x = 8, y = 12$ b. $x = .6, y = 5.8$ c. $x = 140, y = 180$

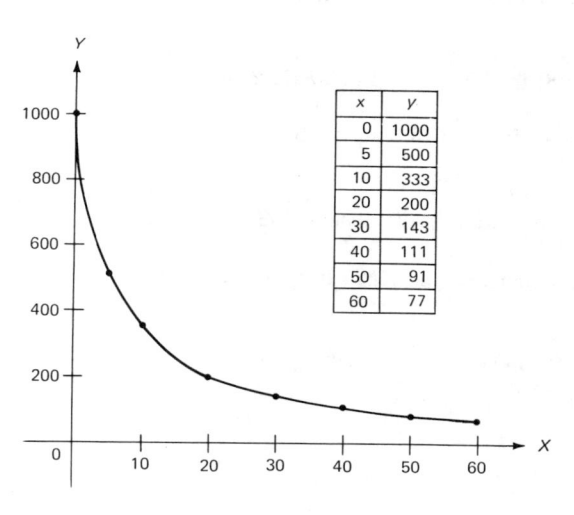

7.

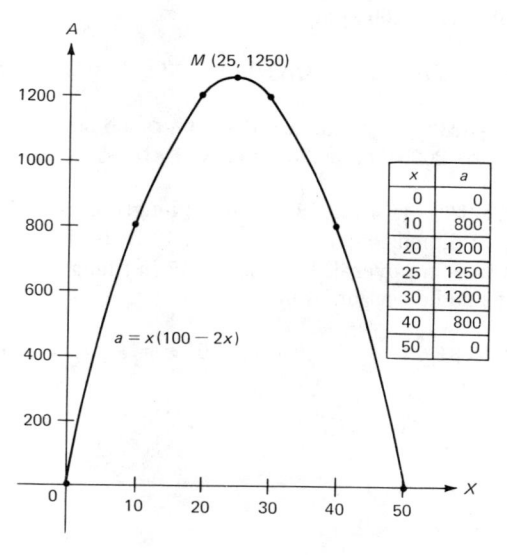

x	y
0	1000
5	500
10	333
20	200
30	143
40	111
50	91
60	77

9. M (25, 1250) $a = x(100 - 2x)$

x	a
0	0
10	800
20	1200
25	1250
30	1200
40	800
50	0

CHAPTER 10 GEOMETRY

Pages 286–287 1. a. 55° b. 115° c. 125° d. 65° e. 60° 3. 45° and 45°

Page 288 1. The central angle is 90°. 3. The central angle is 90°.

Pages 290–291 1. 90° 3. 45° 5. 98°

Page 297 3. a square

Page 300 5. They meet at a point.

Page 302 3. The sum of the measures is 180°.

Page 304 5. XZ is 12 in., YZ is 6 in.

Page 307 1. 5.8 cm 3. 6 in. 5. a. 7.1 cm b. 2.8 in. c. 4.2 ft 7. 5.7 cm

Page 308 3. 96° 11. a is 6 in.

CHAPTER 11 RATIO, PROPORTION, SCALE

Page 311 1. $\frac{3}{4}$ 3. $\frac{1}{4}$ 5. $\frac{2}{9}$ 7. $\frac{66}{47}$ 9. $\frac{12}{11}$ 11. $\frac{47}{72}$ 13. a. $\frac{2}{5}$ b. $\frac{1}{4}$ c. $\frac{1}{3}$ d. $\frac{2}{9}$ e. $\frac{2}{9}$ 15. $\frac{1}{2}$ 17. $\frac{15}{1}$ 19. $\frac{2}{1}$ 21. e:c = $\frac{46}{25}$ 23. a:b = $\frac{21}{13}$ 25. e:a = $\frac{46}{35}$

Page 313 1. a. .0500 in. b. .0162 in. 3. .03 in.

Page 316 1. $x = 1\frac{1}{2}$ 3. $\ell = 21$ 5. $x = 1\frac{1}{5}$ 7. $a = \frac{1}{8}$ 9. A: $\frac{7}{16}''$; B: $\frac{9}{16}''$; C: $\frac{7}{16}''$

Pages 318–319 1. 192 rpm 3. 7 5. 480 rpm 7. 3″ 9. 525 rpm 11. 720 rpm 13. 1166$\frac{2}{3}$ rpm
15. 240 lb 17. 2160 rpm

Page 320 1. *MO*: 12 in.; *ON*: 14 in. 3. *MO*: 11$\frac{1}{4}$ in.; *ON*: 8$\frac{1}{4}$ in. 5. *PQ*: 24 cm; *PR*: 21 cm

Pages 322–323 1. $\frac{16}{9}$ 3. a. 16 sq ft b. 36 sq ft c. 784 sq ft d. 373.8 sq ft 5. 4 7. 4
9. a. 16 b. 2025 c. 30.25 11. a. $\frac{1}{4}$ b. $\frac{9}{25}$ c. $\frac{1}{9}$ 13. 3

Pages 329–330 1. 4 in. 3. ratio: $\frac{1}{12}$; length of line: 7$\frac{1}{2}$ in. 5. ratio: $\frac{1}{48}$; dimension: 198 in. 7. 12$\frac{1}{8}$ in.
9. ratio: $\frac{1}{64}$; dimension: 78$\frac{2}{3}$ ft 11. *A*: 10 ft; *B*: 17 in.; *C*: 3$\frac{2}{3}$ ft
13. width: .5 in.; overall length: 3.9375 in.; length of rectangular section: .5625 in.; length of foot: .375 in.; length of middle section: 3 in.

Pages 330–331 1. $\frac{8}{1}$ 3. .036 in. 5. $x = \frac{1}{3}$ 7. $x = 12\frac{1}{2}$ 9. 4500 rpm 11. 640 sq cm 13. 19′8″

CHAPTER 12 TRIGONOMETRY

Page 339 1. $b = 23.56$ in. 3. $b = 30.08$ ft 5. $a = 47.02$ yd 7. $\angle A = 45°$; $b = 59$ in. 9. 18.5° (approx.)
11. a. 68° b. 22° c. 21.54 ft 13. $b = 58.90$ cm 15. $b = 9.23$m 17. $a = 43.01$ m
19. $\angle A = 45°$; $b = 150$ cm 21. 18° 23. a. 68° b. 22° c. 6.5 m

Pages 345–347 1. $\angle A = 37°$; $\angle B = 53°$ 3. $\angle B = 60°$; $a = 1.625$ in.; $b = 2.81$ in.
5. $\angle A = 32°$; $\angle B = 58°$; $b = 44.9$ ft 7. $\angle CDE = 23°$; $\angle FED = 67°$; $\angle ECD = 67°$ 9. .559 in.
11. $\angle A = 37°$; $\angle B = 53°$ 13. $\angle B = 60°$; $a = 4.125$ cm; $b = 7.14$ cm 15. $\angle A = 32°$; $\angle B = 58°$; $b = 13.7$ m
17. $\angle CDE = 21°$; $\angle FED = 69°$; $\angle ECD = 69°$ 19. 1.25 cm

Pages 352–353 1. $BD = 8.66$ in.; $AD = 5$ in. 3. $AB = 16.2$ in.; $AD = 8.1$ in. 5. $AB = 13$ in.; $BD = 11.3$ in.
7. $\angle A = 144°$; $BC = 2.85$ ft; $AD = .46$ ft 9. $\angle A = 40°$; $d = 17$ in.; $BC = 11.6$ in. 11. 5.20 in.
13. $BD = 21.65$ cm; $AD = 12.5$ cm 15. $AB = 41.1$ cm; $AD = 20.55$ cm 17. $AB = 33$ cm; $BD = 28.58$ cm
19. $\angle A = 144°$; $BC = 85.6$ cm; $AD = 13.9$ cm 21. $\angle A = 40°$; $BC = 29.8$ cm; $d = 43.6$ cm 23. 12.99 cm

Page 359 1. $\sin 105° = .9659$; $\cos 105° = -.2588$; $\tan 105° = -3.7321$
3. $\sin 98° = .9903$; $\cos 98° = -.1392$; $\tan 98° = -7.1154$ 5. $\angle C = 33°$; $a = 5.50$ in.; $c = 3.39$ in.
7. $\angle C = 120°$; $b = 150$ ft; $c = 259.8$ ft 9. $b = 23.86$ cm; $c = 12.69$ cm
11. $\angle C = 101°$; $a = 30.08$ cm; $b = 28.22$ cm 13. $BC = 11.76$ cm

Page 364 1. 15.7 in. 3. 24.33 ft 5. 414 ft 7. 358.8 ft 9. 29.7 cm 11. 7.81 m 13. 142 m 15. 109.4 m

Pages 365–366 1. .5385 3. .9615 5. $b = 12.61$ in.; $c = 18.84$ in. 7. $BC = 8.75$ in.; $AC = 15.16$ in.
9. $a = 10.76$ in. 11. .2143 13. .4632 15. $b = 27.9$ cm; $c = 41.7$ cm 17. $BC = 22.25$ cm; $AC = 38.54$ cm
19. $a = 15.15$ mm

Pages 367–368 1. $7x$ 3. 49 5. 25 7. $x = 2.8$ 9. $s = .76$ 11. $d = \frac{9}{32}$ 13. 1200 ft 15. 1962.5 sq m
17. 63 cu cm 19. 25.4 21. 55 23. 232.8°C 25. $2x^3y^5$ 27. $(5x + 6y)(5x - 6y)$ 29. $x = 10, x = -8$

31.

33.

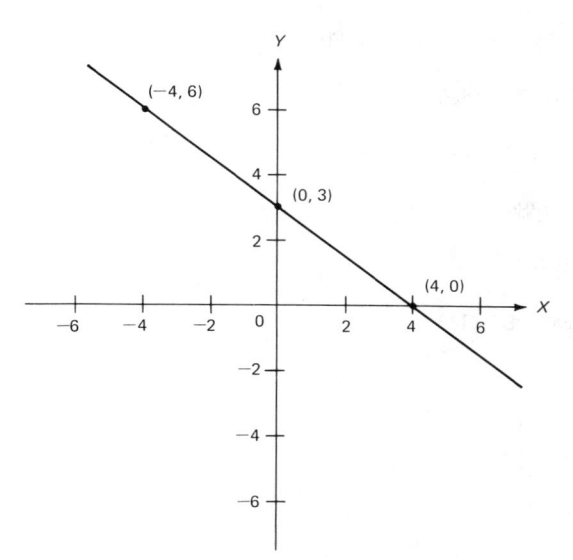

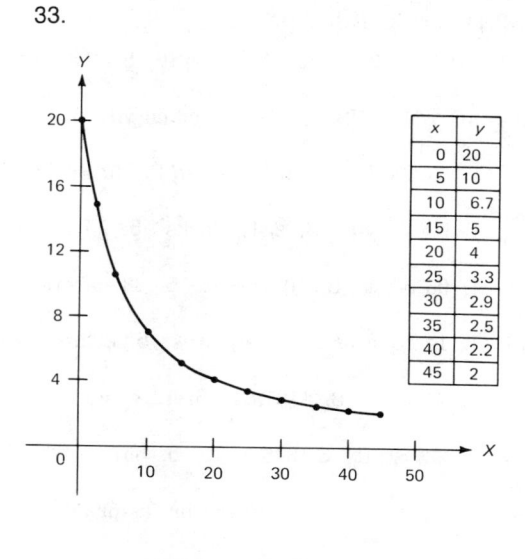

x	y
0	20
5	10
10	6.7
15	5
20	4
25	3.3
30	2.9
35	2.5
40	2.2
45	2

37. $\frac{25}{32}$ 39. $x = 1.6$ 41. 400 rpm 43. 21 in. 45. .7002 47. 7 in. 49. 9.68 cm

CHAPTER 13 POWER AND ENERGY

Pages 374–375 1. 900 ft-lb 3. 120 ft-lb per sec 5. 420 ft-lb per sec 7. 150 ft-lb 9. 100 ft 11. 300 newton-meters 13. 42 watts 15. 327 watts 17. 27 meters

Pages 377–378 1. 514 hp 3. 91 hp 5. 327 hp 7. 209 hp

Page 381 1. 170 cu in. 3. 254 cu in. 5. 4.4 in. 7. a. 3563 cu cm b. 3.563 l 9. a. 2778 cu cm b. 2.778 l 11. 8.0 cm

Pages 383–384 1. 8:1 3. 5.6 cu in. 5. 10 cu in. 7. 192 cu in. 9. 41 cu cm 11. 9½:1 13. 400 cu cm

Pages 387–388 1. 3:1 3. 3:1 5. 27 teeth 7. 2400 rpm 9. 12:1

Page 389 1. 4:13 3. 2:7 5. 8″ 7. 1943 rpm

Page 392 1. 10 psi 3. 450 lb 5. 38 psi 7. 24 sq in. 9. 7.2″

Page 395 1. 60 psi 3. 150 psi 5. 750 lb 7. 1296 psi

Page 397 1. 120 volts 3. 40 amperes 5. 96 watts 7. 12 kilowatts 9. 1100 watts 11. 6.2 amperes

Pages 398–399 1. 1500 ft-lb 3. 213 hp 5. 352 cu in. 7. 3435 cu cm 9. 1:3 11. 500 rpm 13. 160 lb 15. 22 ohms

CHAPTER 14 CONSTRUCTION

Pages 401–402   1. 198 sq ft   3. 40 sq ft   5. 108 sq ft   7. $38\frac{2}{3}$ sq yd   9. $115.20

Pages 402–403   1. $2773\frac{1}{3}$ cu yd   3. 160 cu yd   5. 436 cu yd   7. $1620

Page 405   1. 63,000 lb   3. about 28 sq ft   5. $6.98   7. about 12,467 lb

Page 408   1. 9.3 cu yd.   3. $281\frac{1}{4}$ blocks   5. 20 cu yd   7. $1680

Page 411   1. 96 bd ft   3. $108\frac{3}{4}$ bd ft   5. 33 rafters   7. $53\frac{1}{3}$ bd ft   9. about 14°

Page 413   1. 14 squares   3. 3 squares   5. $82.50   7. 1 roll   9. $111.35

Page 415   1. $495.88   3. $191.84   5. 12 windows

Page 416   1. 220 sq ft   3. 345 sq ft   5. 557 sq ft

Page 418   1. 1310 bricks   3. 10,784 bricks or 36 flats

Page 419   1. 1800 watts   3. 195.5 watts

Page 420   1. $1.13   3. $249.75   5. $49.85

Page 423   1. 7200 sq ft   3. 6 rolls   5. about 6 ($5\frac{2}{3}$) gal   7. 1875 pieces   9. $67.80

Page 425   1. $280.00   3. about 56 cu yd   5. $243.75   7. $450.00   9. $148.44

Pages 426–427   1. $11,200   3. $16\frac{7}{8}$ sq ft   5. 9 squares   7. 685 sq ft   9. 3600 watts   11. $270.00

CHAPTER 15 MANUFACTURING

Page 429   1. a. 444.5 lb   b. 500 castings   3. 10 min   5. 200 lb   7. 180.4 lb

Page 431   1. 38 tons   3. 100 ft   5. 125 tons   7. 5 times

Page 432   1. 1650°   3. 5 sec

Page 433   1. 10,839 fpm   3. 4241 fpm   5. 65 fpm   7. 153 rpm   9. 576 fpm

Page 434   1. 1080 in.   3. 60 milligrams   5. 40 times

Page 435   1. 3.75 gal   3. $7\frac{1}{2}$ hr   5. 15 hr   7. 2300 watts

Pages 435–436   1. about $\frac{1}{2}$ ($\frac{12}{25}$) qt   3. .6 lb   5. 5 rods

Page 437   1. $12.05   3. about 62¢   5. 36 bolts   7. 63 threads

Page 438   1. 18,000 bottles   3. 48,000 mi   5. 15 parts   7. .2555″ max; .2445″ min

480   Answers to selected exercises

Page 439 1. 240 cu in. 3. 20″ 5. 7″ × 5″ 7. 320 cu cm 9. 12 cm 11. 32 sq cm

Page 440 1. $3.50 3. $2.60 5. about 10%

Page 442 1. 3,494,322 lb 3. 240 tons 5. 80 hr 7. 525,000 staples 9. 8 items 11. $6.23

CHAPTER 16 GRAPHIC ARTS

Page 445 1. 3 in. 3. 7 in. 5. 6 in. 7. $3\frac{1}{3}''$ 9. 27 picas × 39 picas 11. 6 picas; 1 inch

Page 446 1. 31 ems 3. 108 points 5. 91 lines 7. $2993\frac{1}{2}$ ems 9. 1944 ems

Page 447 1. 3 column inches 3. 6 column inches 5. 164 column inches 7. $33.60 9. $90.00 11. $779.00

Page 448 1. $15.45 3. 27.5 hr

Page 449 1. $72.00 3. $5.10 5. 6 min

Page 450 1. 120,000 copies 3. $65.00 5. $176.00

Page 451 1. 40 min 3. $7.20 5. 6 min 55 sec 7. $.70 per page (yields $3.00 more) 9. $31.00

Page 453 1. 32 hr 3. 2400 ems 5. 1728 ems per hour

Page 454–455 1. $8450 3. $3\frac{9}{32}$ in. 5. 50% 7. 64% 9. 106.25 mm 11. 25%

Page 457 1. 80 lb 3. $135\frac{11}{19}$ lb 5. 28 7. 4000 letterheads 9. $50.81 11. 25 sq cm; 12.5 sq cm; 6.25 sq cm

Pages 458–459 1. twice 3. 2700 folds 5. $116.67 7. 10 hr 9. 380,900 kg

Page 460 1. 27 picas 3. $144.00 5. $20.00 7. $150.00 9. 150 mm 11. 1250 books

Pages 460–462 1. 700 ft-lb 3. 8:1 5. 5400 lb 7. 8.4 hr 9. $2.00 11. 2300 watts 13. $480.00 15. 18,000 tons 17. 3875 fpm 19. 2.5 parts 21. $3.43 23. 6 column inches 25. $682.67 27. 20,000 lb

Glossary

Absolute value The number after the sign of an integer has been removed. (**p. 42**) For example, 6 is the absolute value of +6 and 6 is the absolute value of −6.

Adjacent angles Angles that have the same vertex and a side in common. (**p. 288**)

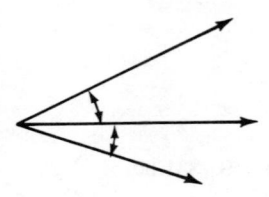

Aligned dimensioning Dimensioning in a technical drawing in which dimensions are readable from either the bottom or the right side of the drawing. (**p. 29**)

Ampere The basic unit of measure of electric current. (**p. 395**)

Arc A part of a circle. (**p. 287**)

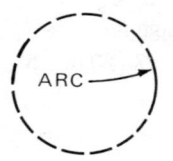

Area The measure of the inside of a flat closed figure. (**p. 191 and p. 194**)

Bar graph A presentation of data in which each quantity to be displayed is represented by the length of a bar. (**p. 9**)

Bearing capacity The amount of weight per unit area that the soil of a given building site will support. (**p. 403**)

Binomial A polynomial which contains two terms. (**p. 233**) For example, $3x + 2y$.

Bisector of an angle A line that divides an angle into two equal angles. (**p. 298**)

Board foot A measure of lumber equivalent to a piece one foot by one foot by one inch. (**p. 408**)

Brake horsepower The amount of work per unit time an engine can do against a braking force. (**p. 375**)

Break lines Jagged lines in a technical drawing used to show a break in an object. (**p. 27**)

Broken-line graph A presentation of data well adapted to showing changes in trends. If lines are drawn joining the tops of bars in a bar graph, a broken line graph is obtained. (**p. 12**)

Cancellation A short cut used in multiplication of fractions. (**p. 80**) For example,

$$6\frac{1}{5} \times 23\frac{3}{4} = \frac{31}{5} \times \frac{95}{4}$$

$$= \frac{31 \times 19 \times \cancel{5}}{\cancel{5} \times 4} = \frac{589}{4} = 147\frac{1}{4}$$

Celsius scale A temperature scale based on the temperature range from the freezing point of water (0°C) to the boiling point of water (100°C). (**p. 222**)

Center lines Alternate short and long dashes in a technical drawing that show the center of a (usually) rounded object. (**p. 27**)

Centi- A prefix meaning $\frac{1}{100}$. (**p. 181**)

Central angle An angle whose vertex is at the center of a circle. (**p. 287**)

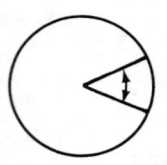

Chord of a circle A line segment that has its endpoints on the circle. (**p. 296**)

Circumference The distance around a circle. (**p. 189**)

Column inch A standard unit of area measure in newspaper publishing equivalent to an area one column wide by one inch deep. (**p. 446**)

Complementary angles Two angles, the sum of whose measures is 90°. (**p. 288**)

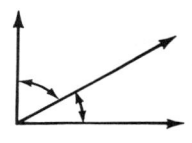

Compression ratio In the cylinder of an engine, the ratio of the maximum space to the minimum space. (**p. 382**)

Computer A calculator that can perform long and complex calculations much faster than ordinary calculators. (**p. 141**)

Cone A solid figure with one circular base and a curved region that comes to a point. (**p. 206**)

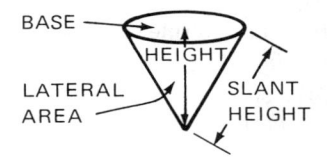

Congruent figures Figures that have the same size and shape. (**p. 303**)

Coordinates of a point An ordered pair of real numbers that indicates the position of a point graphed on a coordinate system. (**p. 260**)

Cosine of an angle For right triangle ABC, the cosine of $\angle A$ is the ratio $\frac{AC}{AB}$. (**pp. 340-341**)

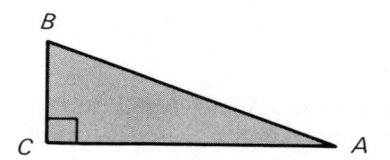

Course (of blocks) A building block together with its mortar joint. (**p. 406**)

Curved-line graph A presentation of data in which quantities are plotted as a series of dots that are then connected by a smooth curve. (**p. 14**)

Cutting speed The rate at which a cutting edge moves across a surface, usually expressed in feet per minute (fpm). (**p. 432**)

Cylinder A solid figure with two circular bases and a curved surface joining them. (**p. 204**)

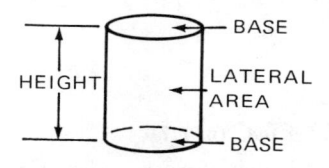

Datum line A fixed line in a technical drawing from which a set of dimensions is referenced. (**p. 28**)

Deci- A prefix meaning $\frac{1}{10}$. (**p. 181**)

Decimal A number that uses place value and a decimal point. For example: 45.32. (**p. 96**)

Degree A unit of measure of angles. The angular measure of a straight line is 180°. (**p. 30 and p. 286**)

Denominator The lower number, or divisor, of a fraction. In $\frac{3}{4}$, 4 is the denominator. (**p. 16**)

Desk calculator A calculator featuring larger keys and a larger display. (**p. 141**)

Diameter A line segment connecting two points on a circle, through the center. (**p. 189**)

Glossary 483

Difference The answer to a subtraction problem. (**p. 36**) For example,

$$\begin{array}{r} 843 \\ -271 \\ \hline 572 \end{array} \text{ difference}$$

Difference of two squares An algebraic expression that can be written in the form $a^2 - b^2$. (**p. 244**) For example,

$$9x^2 - 4y^2 = (3x)^2 - (2y)^2$$

Dimension limits Measurements indicating the acceptable range of lengths or sizes. (**p. 119**)

Dimension lines Lines in a technical drawing showing the specified length of a part of the object. (**p. 28**)

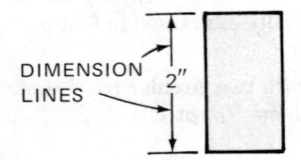

DIMENSION LINES — 2″

Discount price See *Net price*. (**p. 117**)

Dyne-centimeter A metric unit of work in the CGS system. (**p. 373**)

Efficiency The ratio of useful energy output to total energy output of a system; especially, the ratio of the energy delivered by a machine to the energy supplied for its operation. (**p. 123**)

Em A unit of measure of area in the printing trade equal to the square space occupied by the letter M in any given type face. (**p. 445**)

Energy The ability to do work. (**p. 370**)

Engine displacement The amount of space through which one piston in the engine travels in one stroke multiplied by the number of pistons in the engine. (**p. 378**)

Equation A mathematical sentence stating that two expressions are equal. (**p. 148**)

Equivalent fractions Fractions that have the same value. (**p. 18**)

Erg A dyne-centimeter. (**p. 373**)

Estimate A reasonable guess. (**p. 85**)

Exponent A small number written above and to the right of another number. The exponent tells how many times the second number is to be used as a factor. (**p. 153**)

Expression One or more terms connected by addition or subtraction signs. (**p. 148**)

Extension lines In a technical drawing, lines extending beyond the edges of an object. The dimension lines are drawn between the extension lines. (**p. 28**)

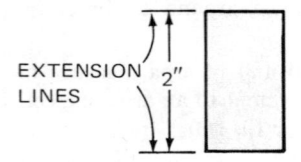

EXTENSION LINES — 2″

Extremes In a proportion, the numerator of the first fraction and the denominator of the second fraction. (**p. 314**) For example,

$$\frac{\mathbf{2}}{5} = \frac{4}{\mathbf{10}}$$

Factoring an algebraic expression Writing the expression as the product of two or more algebraic expressions. (**p. 237**)

Flat (of bricks) 300 bricks. (**p. 417**)

Floor plan The view of a horizontal cut through a building, showing the layout of the walls and the room arrangement. (**p. 400**)

Foot-pound A unit of work equivalent to that done by a one-pound force moving through a distance of one foot. (**p. 371**)

Formula An equation that states a rule or relationship of physical quantities. (**p. 172**)

Gear ratio The ratio of the speeds of two gears. It is the reciprocal of the ratio of the number of teeth of each gear. (**p. 385**)

Gram A unit of weight in the CGS system (1 g = 0.03502 oz). (**p. 218**)

Hand-held calculator A device used to make rapid mathematical calculations. (**p. 128**)

Hidden lines Dashed lines in a technical drawing that indicate a part of the figure that cannot be seen in a particular view. (**p. 27**)

Horsepower A unit of power equivalent to 550 foot-pounds per second, or 33,000 foot-pounds per minute. (**pp. 371-372**)

Hydraulic power Power transmitted by a liquid. (**p. 389**)

Hypotenuse The side of a right triangle opposite the right angle. (**p. 304**)

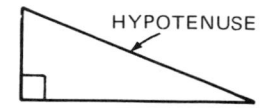

Indicated horsepower A measure of horsepower based on the power input through the pistons to an engine. (**p. 376**)

Integers All the whole numbers and their opposites. For example, 0, $^+1$, $^-1$, $^+2$, $^-2$, $^+3$, $^-3$, ... (**p. 39**)

Inverse proportion A proportion in which the ratios are based on reciprocal relationships. (**p. 316**)

Joule A newton-meter. (**p. 373**)

Kilo- A prefix meaning 1000. (**p. 181**)

Kilowatt-hours (kWh) The basic unit of measure of electrical power consumption. (**p. 2**)

Law of cosines For any triangle ABC,

1. $c^2 = a^2 + b^2 - 2ab \cos C$
2. $b^2 = a^2 + c^2 - 2ac \cos B$
3. $a^2 = b^2 + c^2 - 2bc \cos A$,

where a, b, and c are the lengths of the sides of the triangle opposite $\angle A$, $\angle B$, and $\angle C$, respectively. (**p. 360**)

Law of sines For any triangle ABC,

$$\frac{\sin A}{a} = \frac{\sin B}{b} = \frac{\sin C}{c}$$

where a, b, and c are the lengths of the sides opposite $\angle A$, $\angle B$, and $\angle C$, respectively. (**p. 354**)

Leader In a technical drawing, a line, usually curved, used to direct information and symbols to a place in the drawing. (**p. 28**)

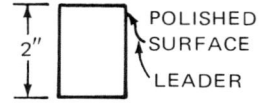

Least common denominator (LCD) The smallest number that can be divided evenly by the denominators of two or more fractions. (**p. 57**)

Linear equation An equation that can be written in the form $ax + by + c = 0$, where x and y are variables; a, b, and c are constants, and a and b are not both zero. (**p. 264**)

Liter The basic unit of volume in the metric system (1 l = 1000 cu cm). (**p. 181**)

Lowest terms A fraction is in lowest terms when its numerator and denominator cannot be divided evenly by the same number. (**pp. 20-21**)

Mass The amount of material of which an object is composed. (**p. 217**)

Means In a proportion, the denominator of the first fraction and the numerator of the second fraction. (**p. 314**) For example,

$$\frac{2}{\mathbf{3}} = \frac{\mathbf{8}}{12}$$

Meter The basic unit of length in the metric system (1 m = 39.37 in.). (**p. 181**)

Micro- A prefix meaning $\frac{1}{1,000,000}$. (**p. 184**)

Micrometer An instrument used in making precise measurements. (**p. 107**)

Milli- A prefix meaning $\frac{1}{1000}$. (**p. 181**)

Monomial An algebraic term that consists of the product of a constant and powers of variables. (**p. 227**) For example, $4x^2yz^3$.

Net price Price after deductions, such as dealer discounts, have been made; discount price. (**p. 117**)

Newton-meter A metric unit of work in the MKS system. (**p. 372**)

Numerator The upper number, or dividend, in a fraction. For example, in $\frac{3}{4}$, 3 is the numerator. (**p. 16**)

Ohm The basic unit of measure of electrical resistance. (**p. 121 and p. 395**)

Parallel lines Lines on a flat surface that do not meet no matter how far they are extended. (**p. 301**)

Parallelogram A four-sided, closed figure with opposite pairs of sides parallel. (**p. 194**)

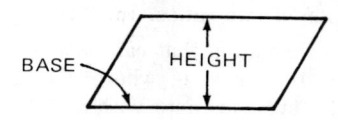

Percent Per hundred. (**p. 114**) For example,

$$35\% = \frac{35}{100} = .35$$

Perimeter The distance around a figure (**p. 63 and p. 186**)

Perpendicular lines Lines that meet at right angles. (**p. 291**)

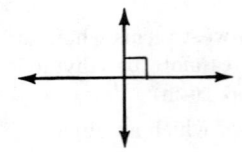

Phantom line A line composed of a long dash and two short dashes that shows the alternate position of a moving part in a technical drawing. (**p. 27**)

Pica A unit used to measure distances in printed matter, equal to about $\frac{1}{6}$ inch. (**p. 443**)

Pitch The ratio of the rise of a roof to its span. (**p. 311**)

Pneumatic power Power transmitted through a gas. (**p. 389**)

Point A unit used to measure the size of printing type, equal to .01384 inches or $\frac{1}{12}$ pica. (**p. 443**)

Polynomial An algebraic expression that consists of an indicated sum or difference of more than one monomial. (**p. 227**) For example, $3x^2 - 6x + 7$.

Pound The basic unit of weight or force in the U.S. system. (**p. 218**)

Power The rate of doing work. (**p. 371**)

Pressure The force per unit area exerted by a fluid or gas. (**p. 389**)

Product The answer to a multiplication problem. (**p. 38**) For example,

```
   32
  ×11
   32
   32
  352  product
```

Proportion An equation of two ratios. (**p. 314**)

Quadratic equation An equation that can be written in the form $ax^2 + bx + c = 0$, where x is a variable and a, b, and c are constants with $a \neq 0$. (**p. 247**)

Quadratic formula The formula for finding the solutions to the quadratic equation $ax^2 + bx + c = 0$. (**p. 251**)

$$x_1 = \frac{-b + \sqrt{b^2 - 4ac}}{2a} \qquad x_2 = \frac{-b - \sqrt{b^2 - 4ac}}{2a}$$

Quotient The answer to a division problem. (**p. 38**) For example,

```
      252   quotient
  3)756
    6
    15
    15
     6
     6
     0
```

Radius A line segment from any point on a circle to its center. (**p. 202**)

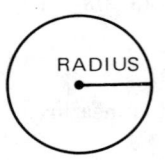

Ratio A comparison of numbers by division. (**p. 309**)

Ream A standard quantity of paper equal to 500 sheets of a basic size. (**p. 455**)

Rear axle ratio The ratio of the drive shaft speed to the rear axle speed. (**p. 386**)

Reciprocals Two numbers whose product is 1. (**p. 81**) $\frac{5}{4}$ and $\frac{4}{5}$ are reciprocals, since

$$\frac{5}{4} \times \frac{4}{5} = 1$$

Rectangle A parallelogram with four right angles. (**p. 194**)

Rectangular pyramid A solid figure with a rectangular base and four triangular faces. (**p. 215**)

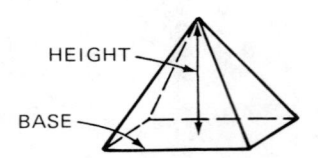

Relative error The tolerance of a dimension of a manufactured article compared to the magnitude of the dimension itself. **(p. 119)**

$$\text{Relative error} = \frac{1}{2} \times \frac{\text{Tolerance}}{\text{Basic size}}$$

Right angle An angle whose measure is 90°. **(p. 286)**

Rule of Pythagoras "In every right triangle, the square of the hypotenuse equals the sum of the squares of the other two sides." **(p. 304)**

Scale The ratio of the length of a line in a technical drawing to the corresponding length of the actual object. **(p. 323)**

Similar figures Figures that have the same shape. **(p. 303)**

Similar terms Terms in the same unknown or terms containing no unknown. **(p. 149)**

Sine of an angle For right triangle ABC, the sine of $\angle A$ is the ratio $\frac{BC}{AB}$. **(p. 340)**

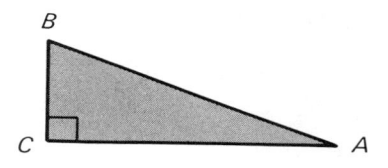

Square A rectangle that has four sides of the same length. **(p. 194)**

Square (of shingles) The amount of shingles that will cover an area of 100 square feet. **(p. 412)**

Straight angle An angle whose measure is 180°. **(p. 286)**

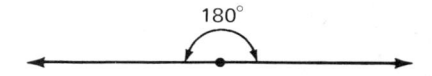

Sum The answer to an addition problem. **(p. 36)** For example,

```
  123
 +355
  478   sum
```

Supplementary angles Two angles the sum of whose measures is 180°. **(p. 288)**

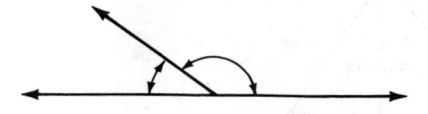

Tangent A line that touches a circle at only one point. **(p. 295)**

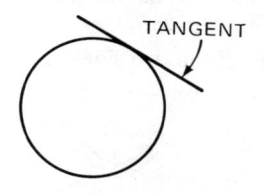

Tangent of an angle For right triangle ABC, the tangent of $\angle A$ is the ratio $\frac{BC}{AC}$. **(p. 334)**

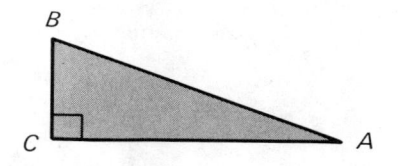

Term A known or unknown quantity in an equation. Terms are connected by addition or subtraction signs. **(p. 148)**

Tolerance The difference between dimension limits. **(p. 119)**

Transmission ratio The ratio of the engine speed to the drive shaft speed. **(p. 385)**

Trapezoid A four-sided figure with exactly one pair of parallel sides, called the *bases*. **(p. 196)**

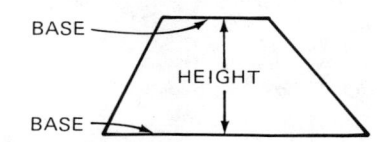

Triangle A three-sided closed figure. **(p. 196)**

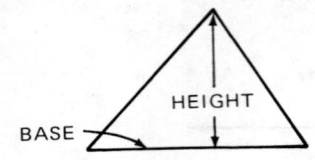

Trinomial A polynomial containing three terms. **(p. 238)** For example, $x^2 + 3x - 6$.

Unidirectional dimensioning Dimensioning in a technical drawing in which all dimensions can be read from the bottom of the drawing. **(p. 28)**

Vertex The common endpoint of the two sides of an angle. **(p. 288)**

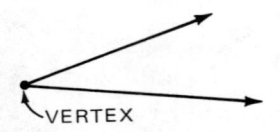

Volt The basic unit of measure of electrical potential. Voltage is analogous to pressure in that it causes electric current to flow. **(pp. 395-396)**

Volume The measure of space inside a solid closed figure. **(p. 210)**

Watt The basic unit of electrical power; the amount of power produced when one volt produces a current of one ampere. It is also a newton-meter. **(p. 373 and p. 396)**

Weight The amount of force that gravity exerts on an object. **(p. 217)**

Whole numbers The numbers that include 0, 1, 2, 3, 4, 5, ... **(p. 5)**

Work The transfer of energy that occurs when a force moves an object through some distance. **(p. 370)**

Index

Absolute value, 42
Addend, 36
Adding and subtracting
 decimals, 101–102
 fractions, 52–56
  with different denominators, 52–56
  with the same denominators, 52–54
 mixed numbers, 60–67
 polynomials, 228
 solving equations by, 158–161
Adhesive bonding, 435
Adjacent angles, 288
Advertisements, 446
Aligned dimensioning, 29
Allowances, 71, 72
Amperage, 418
Ampere, 395
Angle measure, see Degree
Angles, 285–291
 adjacent, 288
 central, 287
 complementary, 288
 construction of, 298–300
 cosine of, 340–341
 dimensioning, 30
 measuring, 30, 285
 sum in triangle, 289
 supplementary, 288
 units of measure of, 30
Architect's scale, 323
Arcs
 and central angles, 287
 dimensioning, 30

Area
 of a circle, 202
 of a cone, 206
 conversions, 191
 of a cylinder, 204
 equivalent units of, 191
 of a parallelogram, 194
 of a rectangle, 194
 of special regions, 198–200
 of a sphere, 205
 of a square, 194
 of a trapezoid, 196
 of a triangle, 196
 units of measure for, 191–207

Bar graph, 9–12
Bearing capacity, 403
Binding, 457
Board feet, 408, 409
 one thousand (M), 408
Bond, 455
Bonding, 435
 adhesive, 435
 fusion, 435
Bore, 378
Brake horsepower, 375–377
Brake, prony, 375
Branch circuit, 419
Break lines, 27
Brick
 common, 417
 face, 417
 standard size, 417
Brickwork, 417
Broken-line graph, 12, 13

Calculator
 desk, 141
 hand-held, 129
Camera copy, 453
Cancellation, 79
Capacity
 conversion, 220
 measurement, 220
 units of, 220
Casting, 428
cc, see Cubic centimeter
CCF, see Cubic feet, one hundred
Celsius temperature scale, 222, 223
Center lines, 27
Centi-, 181
Centimeter, 24, 182, 191, 207
Centimeter-gram-second (CGS), 373
Central angle, 287
 and arcs, 287
CGS, see Centimeter-gram-second
Chemical conditioning, 431
Chemical separating, 433
Chip removal, 432
Chord, 296
Circle(s)
 area, 202
 dimensioning, 30
 circumference, 189
 locating the center of, 297
 tangents to, 295
Circuit breaker, 418

489

Circumference, 189
Color code for resistors, 121
Column inch, 446
Combination of materials, 434–437
Common brick, see Brick
Complementary angles, 288
Complex numbers, 254
Composition, 452
Compression ratio, 381–383
Computer, 141
Computer phototypesetting, 452
Conditioning, 431
  thermal, 431
Cone
  lateral area of, 206
  surface area of, 206
  volume of, 215
Congruent triangles, 303
Conservation of materials, 71–72
Construction, 400–427
  estimated cost of, 401
  see also Geometric constructions
Conversion(s)
  area, 191
  capacity, 220
  factor(s), 182, 184, 208
  feet to feet and inches, 68–69
  linear, 182
  temperature, 222
  volume, 208
  weight, 218
Coordinate system, 259
Cosine(s)
  of an angle, 340–345
  law of, 360
  ratio, trigonometric, 340–345
Course (of blocks), 406
Cubic centimeter (cc), 207–208
Cubic feet, one hundred (CCF), 4
Cubic inch, 207
Current, 395
Curved-line graph, 14, 15
Curves, dimensioning of, 30
Cutting speed, 432
Cylinder
  lateral area of, 204, 205
  surface area of, 205
  volume of, 213
Cylinder press, 448

Datum line, 28
Deci-, 181

Decimal(s), 95–114
  adding and subtracting, 101
  division with, 111
  equivalents (to fractions), 97
  multiplication, 104
  and percent, 114
  point, 96
  reading and writing, 95
Decimeter, 182
Degree
  as an angle measure, 30, 285
  of a monomial, 228
  of a polynomial, 228
Denominator, 16
Destructive testing, 437
Diameter, 30, 189
Difference, 36
Difference of two squares, 244
  factoring, 244
Dimension
  in feet and inches, 68–70
  limits, 119
  lines, 28
Dimensioning, 28–30
  aligned, 29
  angles, 30
  arcs, 30
  curves, 30
  unidirectional, 28
Discount price, see Net price
Displacement, 378–380
Distribution panel, 418
Dividend, 38
Division
  with decimals, 111–114
  by a fraction, 84–85
  of a fraction by a whole number, 81–82
  solving equations by, 163, 164
Divisor, 38
Drill gauge, 23
Dynamometer, 375
Dyne-centimeter, 373

Efficiency, 123
Electrical power, 395–397
Electrical tolerance, 121, 122
Electrical wiring, 418
Electric meters, 2, 3
Electrochemical separating, 433
Electroplating, 434
Em, 445
Energy, 370
English engineering systems, 373

Equation(s)
  definition, 148, 149
  linear, 264
  maximum value of, 280
  minimum value of, 284
  nonlinear, 275–282
  pairs of, 267, 270–273
  quadratic, 247
  solving, 158–171, 247
Equilateral triangles, 348
Equivalent decimals and fractions, 97–99
Equivalent fractions, 18–21
Equivalent units
  of area, 191
  of capacity, 220
  linear, 182, 184
  of volume, 208
  of weight, 218
Erg, 373
Erosion, thermal, 433
Estimated cost (of construction), 401
Estimation, 85–87
Excavation, 402
Exponent, 153
Expression, 148
Extension lines, 28
Extremes (of a proportion), 314
Extrusion, 52

Face brick, 417
Factor, 38
Factoring
  difference of two squares, 244
  polynomials, 237
  quadratic equations, 247
  trinomials, 239
Fahrenheit temperature scale, 222, 223
Fasteners
  mechanical, 436
  threaded, 436
Finishing, 457, 458
Flats, 417
Floor plan, 400–401
Footing, 403
Foot-pound, 371
Forging, 430
Forming, 428–430
Formula(s), 172
  for perimeter, 186, 187
  solving with, 174, 175
  for temperature conversion, 222

490 Index

Formula (*cont.*)
  for volume, 213–216
  writing, 171–173
Foundations, 405–408
Fractions
  adding and subtracting, 52–59
  and decimals, 95–100
  division of, 81–85
  equivalent, 18–21
  in lowest terms, 20–21
  multiplication with, 73–80
  and percents, 115–116
  solving problems with, 88–89
  see also Mixed numbers
Framing, 408–411
Fuse, 418
Fusion bonding, 435

Gable roof, 412
Gas meters, 4–5
Gear, 384–387
  pinion, 386
  ring, 386
Gear ratio, 385
Geometric construction
  of an angle bisector, 298
  of a given angle, 299
  of a parallel line, 301
  of a perpendicular, 292
  of a perpendicular bisector, 294
  of a tangent, 296
Gram, 181, 218
Graphic arts, 443–459
  metric standards in, 455–456
Graphs
  bar, 9–12
  broken-line, 12–13
  curved-line, 14–15
  drawing, 258–283
  of linear equations, 267
  of nonlinear equations, 277
  reading, 9–15
Gravure printing, 449
Greatest common factor, 238

Halftone, 453
Hidden lines, 27
Horsepower, 371, 372
  brake, 375, 376
  indicated, 376, 377
Hydraulic power, 389–392
Hypotenuse, 304

Imposition, 457
Impression cylinder, 448–449
Indicated horsepower, 376
Induced fracture separating, 443
Integers
  negative, 39
  positive, 39
International Standards Organization, 456
Inverse proportion, 316–318
Isosceles triangles, 349–351

Joist, 409
Joule, 373

Kilo-, 181
Kilogram, 218
Kiloliter, 220
Kilometer, 182
Kilowatt, 396
Kilowatt-hour, 2

Landscaping, 423
Lateral area
  of a cone, 206
  of a cylinder, 204
Law of cosines, 360
Law of sines, 354
Leader lines, 28
Least common denominator, 57, 58
Length
  comparing, 181–183
  and rules, 15–16
  units of, 24, 180–190
Letterpress, 447
Linear conversions, 184
Linear equations, 264
  pairs of, 267–274
Line plate, 453
Lines
  break, 27
  center, 27
  datum, 28
  dimension, 28
  extension, 28
  hidden, 27
  leader, 28
  parallel, 301
  perpendicular, 291–294
  phantom, 27
  in technical drawings, 26–28
Linofilm, 452

Linotype, 452
Liter, 181, 220
Lithography, 448
Lowest terms, 20

M, *see* Board feet, one thousand
Machinist's rule, 16, 99
Manufacturing, 428–442
Marketing, 439–440
Mass, 217
Materials
  combination of, 434–437
  conservation of, 71
  separating, 432–433
Maximum value of an equation, 280
Mean (of a proportion), 314
Measurement, United States and metric
  area, 191–207
  capacity, 220–222
  linear, 181–191
  temperature, 222
  volume, 207–217
  weight, 218
Mechanical fasteners, 436
Mega-, 184
Meter, 181, 182
Meter-kilogram-second (MKS), 373
Meters
  electric, 2–3
  gas, 4–5
  reading, 2–7
  tachometer, 8–9
Metric system, 23–24, 180–226
  prefixes in, 181
  units of measure, *see* Units
Micro-, 184
Micrometer (instrument), 107–110
Micrometer (unit of measure), 184
Milli-, 181
Milligram, 218
Millimeter, 24, 182
Minimum value of an equation, 284
Minuend, 36
Minute, 30
Mixed numbers
  adding, 63–66
  multiplying, 74–78
  subtracting, 66–68
  whole numbers and, 60–63

Index 491

MKS, see Meter-kilogram-
    second
Monomials
    degree of, 228
    multiplying, 321
Multiplication
    with cancellation, 79–80
    of decimals, 104–105
    of a fraction and a whole
        number, 73–74
    of fractions, 77–78
    of a mixed number and a
        whole number, 74–75
    of monomials, 231
    of polynomials, 231
    solving equations by, 165, 166

Net price, 117
Newton-meter, 372–373
Nonlinear equations, 275
Numerator, 16

o.c., see On center
Offset duplicator, 449
Offset printing, 448
Ohm, 121, 395
On center (o.c.), 410
Order of operations, 156, 157
Ounce, 220

Packaging, 438
Paper
    classifications, 455
    folding, 458
    sizes, 455, 456
    weights, 455, 456
Parallel lines, 301
Parallelogram, 194
    area of, 194
Percent, 114
    and fractions, 115
Perimeter, 63, 186
    formula for, 186
    of a rectangle, 187
    of a square, 187
    of a triangle, 187
Perpendicular lines, 291–295
Phantom lines, 27
Photocomposition, 452
Photoengraving, 453
Photo-offset printing, 449
Physical conditioning, 431
Pica, 443

Pinion gear, 386
Place value, 6
Plate-making, 453
Platen press, 448
Plumbing, 419
Pneumatic power, 389, 393–394
Point, 443
Point system, 443
Polynomials
    adding, 228
    factoring, 237
    multiplying, 231
    subtracting, 229
Pound, 218
Power, 370–378, 389–397
    electrical, 395–397
    hydraulic, 389–392
    pneumatic, 389, 393–394
Prefixes (in the metric system),
    181
Pressure, 389
Printing
    processes, 447–458
    steps in, 451–458
Product, 38
Prony brake, 375
Proportion, 313–318, 454
    mean of, 314
    terms of, 314
Protractor, 286
Psi, 390
Pulley, 388
Pulley-speed ratio, 388
Pyramid, rectangular, 215
Pythagoras, rule of, 304

Quadratic equations
    solving by factoring, 247
    solving by the quadratic for-
        mula, 251
Quadratic formula, 251
Quality control, 437
Quotient, 38

Radius, 30
Rafter, 409
Ratio
    cosine, 340–341
    gear, 385
    meaning of, 309
    pulley-speed, 388
    rear-axle, 386
    sine, 340–341
    tangent, 332

    transmission, 385
    trigonometric, 332–347
Ream, 455
Rear-axle ratio, 386
Reciprocal, 81
Rectangle, 194
    area of, 194
    perimeter of, 187
Rectangular pyramid, 215
Rectangular solid, 210–211
Relative error, 119
Remainder, 38
Resistance, 395
Resistor, color code for, 121
Right angle, 286
Right triangles, 304, 332–338
Ring gear, 386
Rolling, 430
Roofing, 412
Rotary press, 448
Rotogravure, 450
Rounding numbers, 7
Rules
    and formulas, 171
    and lengths, 15–16
    machinist's, 16
    metric, 23–25
    reading, 15–25
Running bond, 417

Scale, 15
    architect's, 323
    drawing, 326–329
    metric, 23–25
Screw thread depth, 312
Second, 30
Selling price, 439
Separating materials, 432–433
Shearing, 432
Shed roof, 412
Siding, 415
Signature, 457
Similar figures, 303, 320–322
Similar triangles, 303, 321
Sine(s)
    of an angle, 340–345
    law of, 354
    ratio, trigonometric, 340–345
Sod, 423
Solder, 435
Solid, rectangular, 210–211
Solving equations, 158–171
    by addition, 161
    by division, 163
    by multiplication, 165

492   Index

Solving equations (cont.)
  pairs of linear, 267–274, 270–273
  by the quadratic formula, 251
  by subtraction, 158
Solving problems
  with formulas, 174–176
  with fractions, 88–89
Sphere, 216
  surface area of, 205
  volume of, 216
Square
  area of, 194
  perimeter of, 186–187
Square centimeter, 191
Square inch, 191
Square of shingles, 412
Square root table, 306
Stamping, 430
Standard size
  of brick, 417
Straight angle, 286
Stroke, 378
Stud, 409
Substance, 455
Subtraction, *see* Adding and subtracting
Subtrahend, 36
Sum, 36
Supplementary angles, 288
Surface area
  of a cone, 206
  of a cylinder, 205
  of a sphere, 205

Tachometer, 8
Tangent, 295
  of an angle, 334
  ratio, trigonometric, 332–338
Technical drawings
  dimensioning, 28–30

  lines, 26–27
  reading, 26–31
Temperature
  Celsius, 222
  conversions, 222
    formula for, 222
  Fahrenheit, 222
  measurement, 222
  scales, 222
  units of, 222
Template, 23
Term, 148, 314
Thermal conditioning, 431
Thermal erosion, 433
Threaded fasteners, 436
Tolerance
  electrical, 121
  mechanical, 119
Ton, 218
  metric, 218
Topsoil, 423
Total surface area
  of a cone, 206
  of a cylinder, 205
  of a sphere, 205
Transmission ratio, 385
Trapezoid, 196
Triangle(s)
  area of, 196
  congruent, 303
  equilateral, 348
  45°–45°–90°, 351
  isosceles, 349–351
  perimeter of, 187
  right, 304
  similar, 303
  30°–60°–90°, 348
Trigonometric ratios
  cosine, 340–345
  sine, 340–345
  tangent, 332–338
Trinomials
  factoring, 239

Type, 443–447
Typeface, 444

Unidirectional dimensioning, 28
Units
  of angular measure, 30
  of area, 191, 445, 446
  of capacity, 181, 220
  of length, 24, 180–190, 443
  of temperature, 222
  of volume, 207–216
  of weight, 181, 217–219

Value, *see* Absolute value
Vertex, 288
Voltage, 395
Volume
  of a cone, 215
  conversions, 208
  of a cylinder, 213
  formula for, 224–225
  measurement, 207–217
  of a rectangular pyramid, 215
  of a rectangular solid, 213
  of a sphere, 216
  units of, 217–209

Watt, 396, 418
Watt, James, 371
Web-fed press, 448
Weight
  conversions, 218
  equivalent units, 218
  measurement, 218
  units of, 218
Whole numbers, 5–7, 35–39
  and mixed numbers, 60–63
  place value, 6
  rounding, 7
Work, 370–374
Working drawing, 400

Credits

Chapter 1

- 8 Photo Courtesy of Stewart-Warner Corporation
- 15 Courtesy The L. S. Starrett Company
- 16 Courtesy The L. S. Starrett Company
- 16 Courtesy The L. S. Starrett Company
- 17 Courtesy The L. S. Starrett Company
- 17 Courtesy The L. S. Starrett Company
- 18 Courtesy The L. S. Starrett Company
- 19 Courtesy The L. S. Starrett Company
- 22 Courtesy The L. S. Starrett Company
- 23 Courtesy The L. S. Starrett Company
- 24 Courtesy The L. S. Starrett Company
- 24 Courtesy The L. S. Starrett Company
- 24 Courtesy The L. S. Starrett Company
- 25 The Lamson & Sessions Co., Cleveland, Ohio
- 25 New Britain Tool Co.
- 25 Courtesy The L. S. Starrett Company

Chapter 3

- 55 The Lamson & Sessions Co., Cleveland, Ohio
- 60 National Machinery Company
- 60 Gen. Motors Corp.
- 61 Radio Shack, a Tandy Corporation Company
- 66 The C-Thru Ruler Company, Bloomfield, Conn. 06002
- 74 The Lamson & Sessions Co., Cleveland, Ohio
- 76 Radio Shack, a Tandy Corporation Company
- 77 Courtesy The L. S. Starrett Company

Chapter 4

- 95 Courtesy The L. S. Starrett Company
- 97 The Lamson & Sessions Co., Cleveland, Ohio
- 101 Courtesy Stanley Tools
- 104 ROCKWELL INTERNATIONAL, Power Tool Division
- 107 Courtesy The L. S. Starrett Company
- 108 Courtesy The L. S. Starrett Company
- 108 Photo by Jim Ritscher
- 108 Photo by Jim Ritscher
- 108 Courtesy The L. S. Starrett Company
- 109 Photo by Jim Ritscher
- 109 Photo by Jim Ritscher
- 109 Photo by Jim Ritscher
- 109 Photo by Jim Ritscher
- 109 Photo by Jim Ritscher
- 109 Photo by Jim Ritscher
- 110 Photo by Jim Ritscher
- 110 Photo by Jim Ritscher
- 110 Photo by Jim Ritscher
- 110 Photo by Jim Ritscher
- 110 Photo by Jim Ritscher
- 110 Photo by Jim Ritscher
- 110 Photo by Jim Ritscher
- 110 Photo by Jim Ritscher
- 117 Courtesy The L. S. Starrett Company
- 126 Photo by Jim Ritscher
- 126 Photo by Jim Ritscher

Chapter 5

- 129 Texas Instruments Incorporated
- 129 Texas Instruments Incorporated
- 129 Texas Instruments Incorporated
- 129 Texas Instruments Incorporated
- 141 Texas Instruments Incorporated
- 141 Texas Instruments Incorporated
- 145 Armstrong Bros. Tool Co., Chicago, U.S.A.

Chapter 6

- 152 Courtesy The L. S. Starrett Company
- 158 Gen. Motors Corp.
- 165 ROCKWELL INTERNATIONAL, Power Tool Division
- 173 The Lamson & Sessions Co., Cleveland, Ohio

Chapter 7

182 Courtesy The L. S. Starrett Company
183 Courtesy The L. S. Starrett Company
186 Bulova Watch Co., Inc.
204 Photo courtesy L. B. Foster Company
205 Courtesy of Exxon Corporation
206 Ohio Blow Pipe Co., Cleveland, Ohio
213 Courtesy of American Can Company

Chapter 10

286 The C-Thru Ruler Company, Bloomfield, Conn. 06002
286 Courtesy The L. S. Starrett Company
287 The C-Thru Ruler Company, Bloomfield, Conn. 06002
287 Bulova Watch Co., Inc.

Chapter 11

309 Union Gear and Sprocket Corporation
310 Courtesy Portland Cement Association

316 Bulova Watch Co., Inc.
323 Photo by Jim Ritscher
324 Photo by Jim Ritscher
324 Photo by Jim Ritscher
325 Photo by Jim Ritscher
325 Photo by Jim Ritscher

Chapter 13

386 Chrysler Corporation

Chapter 15

430 Bethlehem Steel Corporation
436 The Lamson & Sessions Co., Cleveland, Ohio